W9-BVK-341

ANNUAL REVIEW OF FLUID MECHANICS

VOLUME 16, 1984

MILTON VAN DYKE, *Co-Editor*
Stanford University

J. V. WEHAUSEN, *Co-Editor*
University of California, Berkeley

JOHN L. LUMLEY, *Associate Editor*
Cornell University

ANNUAL REVIEWS INC. 4139 EL CAMINO WAY PALO ALTO, CALIFORNIA 94306 USA

ANNUAL REVIEWS INC.
Palo Alto, California, USA

International Standard Serial Number: 0066-4189
International Standard Book Number: 0-8243-0716-X
Library of Congress Catalog Card Number: 74-80866

TYPESET BY A.U.P. TYPESETTERS (GLASGOW) LTD., SCOTLAND
PRINTED AND BOUND IN THE UNITED STATES OF AMERICA

PREFACE

The editors of the *Annual Review of Fluid Mechanics* hope, and would like to believe, that each article is an authoritative, clear, and well-documented exposition of the subject treated, and that each is read by many fluid dynamicists interested in broadening their overview of the field. Although authors are cautioned not to write just for other experts in the subject of a prospective article, some articles do turn out to provide new insights or to be remarkably clear formulations of the fundamentals and problems of a given field. One measure of the occurrence of such articles and of their impact is the frequency with which they are cited. In the preface to Volume 9 a count was made of the citations of the articles in Volume 1 for the years 1970–1975. For this the *Science Citation Index* was used. A companion to this index, the *Journal Citation Index*, allows one to obtain with little effort the same information for all volumes for the years starting with 1974. As in the earlier preface, the impact of individual articles in a particular volume is not shown, only the total effect of all articles in the volume. The available information is displayed in Table 1, which shows the number of citations in year m to articles in volume n. In the boxes is also shown the average number of citations per article. The *Journal Citation Index* gives citation counts for individual years only for the ten years preceding the year of the index and lumps all earlier years. These lumped numbers have been distributed approximately equally over the relevant years, and the numbers put in parentheses.

From examination of the table it is evident that certain volumes show a consistently higher number of citations each year than other volumes, and that almost all show a relatively constant number after the first year. This latter fact seems to indicate that individual volumes are all maintaining their value over the years, at least according to this measure of value.

THE EDITORS

Table 1 Number of citations in year m to volume n

m \ n	1	2	3	4	5	6	7	8	9	10	11	12
	30	22	26	31	13	5						
1974	1.76	1.47	1.63	1.82	0.72	0.38						
	28	26	29	39	27	32	5					
1975	1.65	1.73	1.81	2.29	1.50	2.46	0.33					
	40	40	30	38	38	55	32	12				
1976	2.35	2.66	1.87	2.23	2.11	4.65	2.13	0.80				
	44	25	21	56	22	36	33	38	8			
1977	2.59	1.67	3.29	3.29	1.22	2.77	2.20	2.53	0.44			
	21	21	23	37	28	39	36	39	21	10		
1978	1.24	1.40	1.44	2.18	1.56	3.00	2.40	2.50	1.17	0.50		
	29	21	25	49	23	45	48	49	28	30	11	
1979	1.71	1.40	1.56	2.88	1.28	3.46	3.20	3.27	1.56	1.50	0.61	
	(23)	(24)	30	34	20	41	41	40	25	31	53	11
1980	1.35	1.60	1.88	2.00	1.11	3.15	2.73	2.27	1.39	1.55	2.94	0.69
	(21)	(22)	(22)	31	20	34	33	55	42	42	71	32
1981	1.24	1.47	1.48	1.82	1.11	2.62	2.20	3.67	2.33	2.10	3.94	2.00

Annual Review of Fluid Mechanics
Volume 16, 1984

CONTENTS

ERRATUM

Volume 15 (*1983*)

In "Autorotation" by Hans J. Lugt, Figure 5 (p. 128) should be as follows:

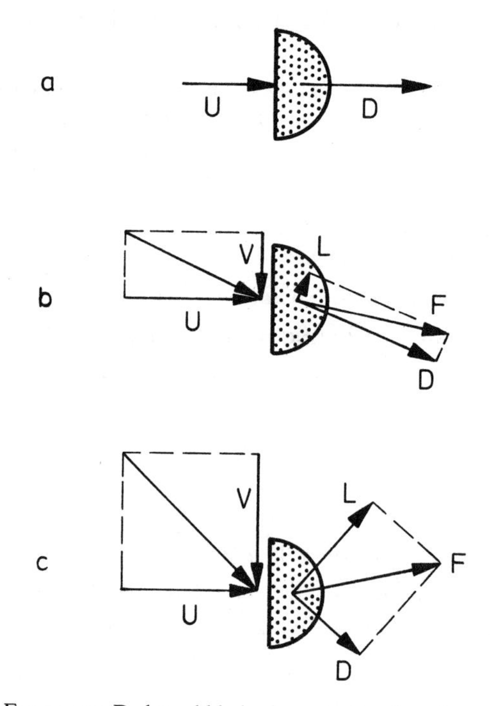

Figure 5 Forces on a D-shaped blade element of the Lanchester propeller.

SOME RELATED ARTICLES IN OTHER *ANNUAL REVIEWS*

From the *Annual Review of Earth and Planetary Sciences*, Volume 11 (1983)

Creep Deformation of Ice, Johannes Weertman

Recent Developments in the Dynamo Theory of Planetary Magnetism, F. H. Busse

The Atmospheres of the Outer Planets, Garry E. Hunt

From the *Annual Review of Physical Chemistry*, Volume 34 (1983)

My Adventures in Theoretical Chemistry, Joseph O. Hirschfelder

ANNUAL REVIEWS INC. is a nonprofit scientific publisher established to promote the advancement of the sciences. Beginning in 1932 with the *Annual Review of Biochemistry*, the Company has pursued as its principal function the publication of high quality, reasonably priced *Annual Review* volumes. The volumes are organized by Editors and Editorial Committees who invite qualified authors to contribute critical articles reviewing significant developments within each major discipline. The Editor-in-Chief invites those interested in serving as future Editorial Committee members to communicate directly with him. Annual Reviews Inc. is administered by a Board of Directors, whose members serve without compensation.

1979

Karl Pohlhausen

Ann. Rev. Fluid Mech. 1984. 16: 1–10

KARL POHLHAUSEN, AS I REMEMBER HIM

Knox Millsaps

Department of Engineering Sciences, University of Florida, Gainesville, Florida 32611

I have had the rare privilege during my lifetime to have known intimately some of the people who have made significant contributions to modern aeronautics: Harry Bateman, Gottfried Guderley, Theodore von Kármán, Harold Kaufman, Hans von Ohain, Karl Pohlhausen, G. I. Taylor, Orville Wright, and Fritz Zwicky, to name a few. Bateman and Zwicky were two of my teachers at Caltech. Kaufman was one of my Colorado State doctoral students of whom I am justifiably proud. Guderley, von Ohain, and Pohlhausen were Wright Field colleagues. Wright quite inexplicably took a young lieutenant to heart when I was assigned to Wright Field in his hometown, Dayton, Ohio; and Taylor and von Kármán were simply close personal friends.

The Herculean feats of Wright and von Ohain will be immortalized by articles, stories, novels, and films, done and redone, over and over again. Legions of pupils and associates and their intellectual heirs will perpetuate the distinct personalities and overwhelming accomplishments of Taylor and von Kármán; I even have a few original anecdotes involving G. I. and Todor that I relish in telling with eloquent gestures and approximately correct accents. Proper biographies of Bateman and Zwicky would require the talents of a Dickens, although Fritz would object with unbelievable vulgarity that the author was incompetent because of his lack of Swiss citizenship. Gottfried and Harold are reserved for another time, another place. But if the story of Karl Pohlhausen is to be told, I must tell it, for during the last three and a half decades of his life, I was his only close friend who had comparable technical training—one could add that there were precious few other friends with comparable technical backgrounds during the first five and a half decades of his life. Dr. P. was an extreme introvert and an incredibly private person; hence, although we spent, whenever we

0066-4189/84/0115-0001$02.00

were living in the same town, from one to four hours every day in each other's company, 365 days a year for 34 years (the last 13 years of his life he lived in an apartment only a few blocks from my house), I am not sure that I really knew him or that he really knew me. On some days we said nothing except a greeting to each other; we simply drank coffee or tea, went to lunch together, and passed papers, articles, and books back and forth while we nodded our heads in approval or expressed marked disapproval by oscillating our heads with peak angular velocities directly proportional to the reputation of the offenders. On other days we talked at the same time in loud voices while both of us wrote furiously and simultaneously on the blackboard without respecting the other's territory, both of us correcting the errors and alleged errors of the other; and, often, we just reminisced. We shared a common interest in and similar reactions to nearly every area of human activity from philosophy and music to horticulture and diabetes mellitus with one overriding exception, economics. I have the inborn contempt of a proper Southern gentleman for all fiscal matters, while Dr. P. thought and frequently said, "The only problem in applied mathematics worth anyone's salt is the prediction of the course of the stock market," and he lectured me without beginning or end on the sins of my prodigal behavior while extolling the virtues of miserly ways. Yet, in spite of our similarities, there were great differences: his language was Saxonian German, while I am locked on a single-track linguistic facility in English; our ages differed by three decades; Kármán, Ollendorff, Rüdenberg, and Trefftz were his contemporaries and are listed among my idols; Hilbert, Prandtl, and Runge were his teachers and friends, while they are merely gigantic historical figures to me; and, most importantly, I read about and studied in finished form lifting-line theory, the analyses of the transient performance of electrical networks, and boundary-layer theory, while Karl Pohlhausen helped create them as the solutions to what he called, "good, hard and important problems." On one hand, it is quite clear that applied mathematicians and electrical and aeronautical engineers will not forget the monumental work of Karl Pohlhausen for many, many years to come; on the other hand, I shall remember him during my lifetime without thinking about his major contributions to modern technology as a close and dear friend and collaborator with whom I shared thousands of hours of pure intellectual joy.

HIS LIFE

Karl Pohlhausen was born on 17 May 1892 in Mittweida, a little village that is located a few kilometers southwest of Dresden. His mother, who died shortly after the birth of her second child, had the maiden name of Hedwiga

Eugen, and his father, August Pohlhausen, was a licensed engineer and a teacher in the Technikum. His father's family came from East Prussia. He had a younger brother, Ernst, who also became a pupil of Prandtl, and who became a celebrated lecturer on applied mechanics and an ordinary professor and rector at the University of Danzig. Sibling rivalry, a synonym for pure hatred, characterized their relationship; in later years Karl would not acknowledge his brother's existence (needless to say, this drove the security people up the wall—I know all too well), although he made frequent and generous gifts to help his brother's first wife until her death.

When Dr. P. was a very young boy, his father, after his mother's death, transferred to a Technikum in Dresden where Dr. P. went to the famous Dreikönigschule. He never tired of relating six things from his boyhood: stories about his French governess, the long walks to school in winter after washing in cold water, the debut of "Der Rosenkavalier," the camellias of Dresden (Dresden single red and Frau Minna Seidel, which we know as Pink Perfection), a proper recipe for potato pancakes, and Christmas stollen.

In 1911 after completing the gymnasium in Dresden, he went to the University of Jena, where he intended to study optics with the lesser Wien, who proved to be far less than Dr. P. could stomach. In 1913 he moved to Göttingen to study under Carl Runge, who had just returned from Columbia in New York City; however, once in Göttingen, Dr. P. was attracted to Ludwig Prandtl, who quickly became and remained his hero and idol. As a result of innumerable conversations with him I arrived at the conclusion that Dr. P. classified scientists and engineers into three categories: virtuosos, first-chair men, and good section men. Dr. P. had Toscanini's standards in that his first-chair men were von Kármán, von Mises, Runge, Sommerfeld, and even Hilbert, although he thought that they all might occasionally give a virtuoso performance, but the only certain virtuoso was Ludwig Prandtl.

Among Pohlhausen's talents was his ability as a draftsman and cartoonist; some old gold-process photographs of his cartoons from the pre-WWI Göttingen days survive. I have had two of these enhanced by computer techniques in the Applied Optics Laboratory of the Department of Engineering Sciences of the University of Florida (Figure 1). One of these is an academic procession showing Klein, Prandtl, Hilbert, and others; the other is of Hilbert on a bicycle and Kuhn collecting biological specimens. Dr. P. denied that he was the artist, but Kármán and Paula Kyropoulos were certain of the artist's identity.

Many interesting stories have been recalled by people who lived in Göttingen in the Early Teens, and I shall retell two of these. It was alleged by the local citizenry of Göttingen that the Pohlhausen brothers knew

Figure 1 Some examples of Pohlhausen's ability as a cartoonist. See text for a description of the frames.

everything, and upon an occasion when Karl couldn't answer a question, he retorted, "Ask my brother." The other story is more serious. A young privatdocent named Todor von Kármán was given the elementary course on applied mechanics to teach, a very thoughtful thing for Prandtl to have done since large numbers of students took the course at so much per student, all of which went to the young, needy Hungarian. A slight difficulty arose in that all of the students signed a petition to Prandtl asking for Kármán's removal on the grounds that (*a*) Todor spoke German with such an intolerable Hungarian accent that effective communication was impossible; and (*b*), in any case Kármán did not understand and/or know elementary mechanics. Prandtl reacted with his characteristic intellectual honesty by saying, "(*a*) is probably true, but (*b*) is false"; he assigned young Karl to write Todor's lectures "in good German." Pohlhausen liked this story; Kármán did not. Incidentally, by 1925 all of the signers of the petition had vanished from the face of the Earth. Nevertheless, the story is true, for Kármán's papers contain Pohlhausen's notes on his 1913 lectures on elementary mechanics.

In August of 1914, young reserve Lt. Karl Pohlhausen was sent with his reserve regiment to serve Fatherland and Kaiser in their attack on France. He later said, "My patriotism was eroded by every kilometer we walked into Belgium and France, and it evaporated at the Marne." Although he was awarded two Iron Crosses (two of the second class or one each of the first and second class—the records are not clear on this matter), he never forgot the carnage of trench warfare in WWI; he still had terrifying nightmares from his military experiences during his final illness nearly six decades later.

By early 1916 the airplane had assumed considerable importance as an observation device, and Prandtl was assigned the role of principal German investigator for this American invention. Prandtl was extended the courtesy of selecting a military officer to be the uniformed head of his laboratory, and to everyone's surprise he selected Lt. Karl Pohlhausen, who returned from the Western front to assume his duty. It is no wonder that Pohlhausen loved Prandtl so much, for Dr. P. was the only officer of his regiment to survive World War I, a fact that he freely acknowledged with the observation, "Prandtl saved my life." The group that Lt. Pohlhausen bossed included Prandtl, Betz, Grammel, Munk, Wieselsberger, Walter Baade (the astronomer), and Peter Debye (the physical chemist)—to name a few. The outstanding accomplishments of their efforts were the formulation of lifting-line theory and the development of the Göttingen profiles. Although history has disagreed with him, Pohlhausen always thought that lifting-line theory, not boundary-layer theory, was Prandtl's masterpiece. Max Munk, an old and dear friend of mine, who will surely forgive me for recording the facts, was Pohlhausen's administrative cross to carry. Once, Major R. von Mises and Capt. L. Hopf, whom Lt. P. reported to in Berlin, complained that some of Munk's data were inconsistent; when Lt. P. relayed the complaint to Munk, he retorted, "I know it, and never again shall I put enough data in a report to permit a cross-check." Years later both Munk and Pohlhausen worked for me at Wright Field, and by then they had forgiven one another for their wrangling during World War I. At least, when a group of Prandtl's former students celebrated in Gainesville on 4 February 1975 the one-hundredth anniversary of the Master's birth, they drank champagne toasts together.

At the end of World War I, Pohlhausen returned to Göttingen and received his doctorate in 1919 from Prandtl. His dissertation on experimental verification of boundary-layer theory contained as little as possible, since he could not afford the printing costs for a voluminous document.

When Kármán went to Aachen in 1920 as the ordinary professor of applied mechanics because of the very active participation of Felix Klein, he took with him his old amanuensis, Karl Pohlhausen, a student of Prandtl; two of Sommerfeld's brilliant students, C. Wieselsberger and L. Hopf; and von Mises' entry, E. Trefftz. While in Aachen, Kármán and Pohlhausen worked closely together and produced their two famous papers on boundary-layer theory that complement each other so beautifully; they were permitted to work, under the terms of the Versailles Treaty, on boundary-layer theory by the French occupation troops, since everyone knew that boundary-layer theory had nothing to do with military aeronautics. Trefftz and Pohlhausen became close personal friends, and Trefftz's early death was felt very deeply by Dr. P.

Although Kármán tried to dissuade him, Dr. P. felt that academic aeronautics had no future, and, although he seriously considered an offer from Count Zeppelin to become his chief engineer, Reinhold Rüdenberg persuaded Pohlhausen to join Siemens-Schukert in Berlin on 18 May 1922 at a salary of 500 marks per month as a replacement for Fritz Nöther, although their employment by Siemens overlapped. He knew Nöther's sister, Emma, and he rephrased Weyl's observation that the Graces did not preside at her cradle with his own, "Genius she was, but the X and Y chromosomes were mixed $X^{n+1}Y^n$, where n is an arbitrarily large number." After Rüdenberg went to England in 1935 because of Nazi pressure, Dr. P. became the Chief Electrician at a salary of 55,000 RM per year; he remained with Siemens until he left Germany in 1946. He was a complete recluse because of his dislike of the Nazis, and this was the only way to live in Germany if you didn't agree with Der Fuehrer. Some idea of the quality of the research staff at Siemens may be inferred by recalling that some Nobel Laureates—G. Hertz, D. Gabor, and G. von Békésy—were scientific members and that Rüdenberg, Ollendorff, and Duffing were engineering members. Siemens hosted Pohlhausen's Jubilarium on 18 May 1947 in Dayton, Ohio, and tried to persuade him to return to Germany; he refused with the simple statement, "You can't go home again; it's never the same."

In 1946 Karl Pohlhausen came to the United States as a participant in Operation Paperclip in order "to start a new life in a free country"; he became a citizen in 1952. He worked in the Armament Laboratory, the Office of Air Research, the Flight Research Laboratory, and the Aeronautical Research Laboratory at Wright Field from 1946 until 1965 (staying until 1967 under a contract with Ohio State University); he was one of the PL-313 scientists of the US Air Force. In early 1948 Dr. P. and I began an almost exclusive collaboration that lasted until his death, although he wrote two reports, one with T. Butler and the other with L. Blichter, that had their bases in our joint work. Most of our joint work dealt with heat transfer to fluids, although we did a few extraneous things. On 17 Nov. 1967 he retired to Gainesville, Florida, at my suggestion, and thereafter he rarely missed a daily coffee, lunch, tea, or dinner at my home—sometimes taking in all four. He was a familiar figure on the streets of Gainesville, where he took daily five- to ten-mile hikes as long as he was able. In 1978 he discovered he had cancer, which led to his death on 18 Nov. 1980. No funeral service was held, in accordance with his will, but a memorial seminar was convened at the University of Florida on 25 Feb. 1981 with Raymond L. Bisplinghoff as Moderator and with two talks (Ray Dandl on "Uncle Elmo's Bumpy Torus" and M. S. Longuet-Higgins on "Vortex Ripples in Sand") in honor of Pohlhausen's life work in electrodynamics and fluid mechanics.

Physically, Dr. Pohlhausen was an average man, 68 inches tall and 150 pounds in weight. His eyes and hair were brown, and he had a birthmark on his right cheekbone and temple that was removed by surgery in the early sixties. He was quite active and strong, and enjoyed mountain climbing for a hobby. He was left-handed, although he wrote right-handed. (His draftsmanship was superb and unbelievably precise.) He admired and collected Persian rugs, although most of his collection was destroyed in the Berlin air raids, and he also liked German nineteenth-century landscapes, of which three fine examples were part of his estate.

Personally, one quickly recalls two ambivalent attributes: his acerbity and associated sentimentality, and his miserly acts and associated generous charity. For example, he once asked a colleague at Wright Field what aircraft companies he had previously worked for, adding, "I want to buy some of their stock, for it is bound to go up now that you are no longer employed by those companies." He also once told a colleague, "Yes, it's good, and Legendre thought so too when he first discovered it." I once asked him what the doctor had told him during a physical examination and without hestiation he replied, "He told me that he's gone up on his prices." He once noticed that my family tipped a waitress too much, and he reduced the tip by putting some change in his pocket. "Separate checks" was a well-known instruction. The other side of the coin was that he refused to go back to Göttingen after his wife's burial there after the Berlin flu epidemic of 1929 and that he gave more to charity drives than the rest of the office put together. He made contributions in the hundreds of thousands of dollars to worthy causes; yet, he gave up a favorite soup because of a two-cent increase in price. Finally, he left two thirds of his estate to a religious charity while insisting that he was a freethinker.

As for his scientific ability, I can only echo Todor von Kármán, who once told his sister Pippa quite emphatically in my presence, "Karl Pohlhausen is the most talented German that I have ever known"; it even stopped Pippa.

HIS WORK

Pohlhausen's list of publications in the open literature is surprisingly small for such a well-known man, although one must bear in mind that he stayed with Siemens for a quarter of a century, where a major portion of his duties was the direction of the work of others.

He wrote in its entirety one book on the methods of applied mathematics (Pohlhausen 1960), based on a series of lectures given in Cloudcroft, New Mexico. He also translated into German the standard book on transforms by Carson (1929), and he contributed sections to two other books: a chapter on long transmission lines in a work on high-tension wires edited by Rüdenberg (Pohlhausen 1932), and, most importantly, a chapter on aerodynamical applications in the classic work by Rothe, Ollendorff & Pohlhausen on the uses of complex variables in engineering (Rothe et al. 1931), still, far and away, the best book on the subject ever published.

The most frequently quoted work by Pohlhausen in the literature of electrical engineering is his article on ground effects on transmission lines (Pohlhausen 1927).

It is rather curious that two of his research papers on aerodynamics are only known throughout the world because of secondary references to them by Prandtl (1918, 1919), and in these days of extensive publication of pure trash it is even more curious that these papers have not been republished. When he was the uniformed head of Prandtl's laboratory during World War I, Pohlhausen used his facility with elliptic functions to show that an elliptical loading was the optimal one in terms of form drag for wings with a finite span and elliptic planform. The result was generalized by Max Munk in his dissertation, and Munk's simple proof of the general result may explain the obscurity of Pohlhausen's suggestive and first limited result. Pohlhausen was also the first to extend Prandtl's lifting-line theory to biplanes and triplanes, since the Fokker triplane was in field use by the German army.

When Pohlhausen first came to the United States as one of the Paperclip scientists, he worked on inertial guidance and wrote with F. Wazelt and W. Kerris an important report on error accumulation in inertial guidance systems. Somewhat later, in 1948, Pohlhausen and I began a close collaboration that lasted until his death, and, although we wandered from time to time into other fields [e.g. the metabolism of glucose (Pohlhausen & Millsaps 1975), the rail gun (Pohlhausen & Millsaps 1960), and turbulence, a phenomenon that we never tired of discussing and cussing], our principal area of investigation was forced and free heat convection to fluids. In the course of our collaboration in this area we produced articles on heat transfer to rotating plates, in converging and diverging channels, in circular pipes, and from vertical cylinders (Pohlhausen & Millsaps 1952, 1953, 1956, 1958).

Nevertheless, the aforementioned work amounts to little when it is compared with one article (Pohlhausen 1921) published in the first volume of *Zeitschrift für angewandte Mathematik und Mechanik*, with von Kármán, von Mises, and Prandtl as editors. Pohlhausen's reputation as a legendary figure in fluid mechanics is firmly based on this article, for this one paper contains so many important results that it is virtually impossible to write a book on fluid mechanics, either at the most elementary level or at the height of sophistication, without the use of its results. Let me list some of the topics contained in this paper:

1. The first clear mathematical proof that the Prandtl boundary-layer equations are asymptotic forms of the Navier-Stokes equation for large Reynolds numbers.
2. The demonstration that the Kármán momentum equation can be derived from the boundary-layer equation by a simple integration.
3. The usefulness of the Kármán momentum theory, shown by approximating the velocity profiles by a quartic and by deriving simple numerical results for the pertinent physical quantities. The current reader must remember that until Pohlhausen's paper boundary-layer theory treated only two cases, the flat plate (very well) and the circular cylinder (very poorly).
4. The smoothing conditions at the outer edge of the boundary layer. Kármán thought this to be particularly important.
5. The concept of weighing various sublayers within the boundary layer, and the concept of inner and outer boundary layers.
6. The possible simplifications in boundary-layer theory when the external velocity profiles vary as a power of x. The flow in a converging channel ($U = a/x$) is solved in closed form as an example.

Although the first two topics are now standard parts of all books on fluids, widespread use of the third item as the Kármán-Pohlhausen method has overwhelmed the recognition of the importance of the other items; however, the fourth and fifth items will yield even more results in the future as their implications are explored. The sixth topic led to the Falkner-Skan profiles and the enormous literature associated with them.

Now, it is a widely held belief among aerodynamicists and other good fellows that St. Peter gives regularly scheduled lectures on turbulence. Incidentally, in addition to his role as keeper of the keys, St. Peter, a simple fisherman, is the ideal lecturer on turbulence since the idea behind turbulence is undoubtedly very elementary, although it is still elusive to us, and since a commercial fisherman would have to be interested in turbulent flows. I only hope that the good saint also includes plenty of "good, hard and important problems" for his angelic audience. But, fantasies aside, Pohlhausen's earthly memorial is his paper on the approximate integration of the boundary-layer equations; it is an impressive and enduring monument, one of the true milestones placed by a legendary figure on the ascent to the summit of human intellectual activity, fluid mechanics.

Literature Cited

Carson, J. R. 1929. *Elektrische Ausgleichsvorgänge und Operatorenrechnung.* Berlin: Springer. Trans. into German by K. Pohlhausen and F. Ollendorff.

Pohlhausen, K. 1921. Zur näherungsweisen Integration der Differentialgleichung der laminaren Grenzschicht. *Z. Angew. Math. Mech.* 1:252–68. Trans. R. C. Anderson, 1965, Dept. of Eng. Sci., Univ. Fla., Gainesville

Pohlhausen, K. 1927. Grundlagen der Bemessung von Starkstromerdern. *VDE-Fachber.*

Pohlhausen, K. 1932. Theorie der langen Leitungen. In *Hochleistungsübertragung,* ed. R. Rüdenberg, Chap. 2. Berlin: Springer

Pohlhausen, K. 1960. *Mathematica Memorabilia.* Holloman AFB, N.M: Air Force Missile Dev. Cent.

Pohlhausen, K., Millsaps, K. 1952. Heat transfer by laminar flow from a rotating plate. *J. Aeronaut. Sci.* 19:120–26

Pohlhausen, K., Millsaps, K. 1953. Thermal distributions in Jeffery-Hamel flows between nonparallel plane walls. *J. Aeronaut. Sci.* 20:187–96

Pohlhausen, K., Millsaps, K. 1956. Heat transfer to Hagen-Poiseuille flows. *Proc. Conf. Differ. Equations, College Park, Md.,* Chap. 19

Pohlhausen, K., Millsaps, K. 1958. The laminar free-convective heat transfer from the outer surface of a vertical cylinder. *J. Aeronaut. Sci.* 25:357–60

Pohlhausen, K., Millsaps, K. 1960. The linear acceleration of large masses by electrical means. *Tech. Rep. TR-60-11,* Air Force Missile Dev. Cent., Holloman AFB, N.M. Originally published as *OAR-50-13,* Off. Air Res., Wright Field, Dayton, Ohio

Pohlhausen, K., Millsaps, K. 1975. A mathematical model for glucose-insulin interactions. *Math. Biosci.* 23:237–51

Prandtl, L. 1918. Tragflügeltheorie. I. *Nachr. Ges. Wiss. Göttingen* 1918:151–77

Prandtl, L. 1919. Tragflügeltheorie. II. *Nachr. Ges. Wiss. Göttingen* 1919:107–37

Rothe, R., Ollendorff, F., Pohlhausen, K. 1931. *Funktionentheorie und ihre Anwendung in der Technik.* Berlin: Springer. Trans., 1933, as *Theory of Functions as Applied to Engineering Problems.* Cambridge, Mass: MIT Press

Ann. Rev. Fluid Mech. 1984. 16: 11–44

WAVE ACTION AND WAVE-MEAN FLOW INTERACTION, WITH APPLICATION TO STRATIFIED SHEAR FLOWS

R. Grimshaw

Department of Mathematics, University of Melbourne, Parkville, Victoria 3052, Australia

1. GENERAL THEORY

Over the last two decades wave action principles and the associated wave-mean flow interaction theorems have become increasingly important for the study of the various kinds of waves that occur in fluid mechanics. Action density is a well-established entity in classical mechanics and plays a central role in the Lagrangian and Hamiltonian development of that subject. However, only relatively recently have the corresponding entities in fluid mechanics been identified and exploited. This is partly because fluid mechanics has traditionally been developed in an Eulerian framework and wave action principles are more obvious in a Lagrangian framework, and partly because the classes of waves for which wave action principles are particularly useful (e.g. internal gravity waves in stratified shear flows) have only recently received much attention.

The current interest in wave action began with the pioneering work of Whitham (1965, 1970), who introduced the wave action equation through the averaged variational principle. Although the initial motivation was the study of finite-amplitude waves, it was soon recognized that the wave action equation was also useful for the study of linearized waves on a mean flow (Bretherton & Garrett 1968). However, because these early theories were Lagrangian in concept, it became necessary to develop Lagrangian equations of motion in contrast to the more familiar Eulerian equations of motion. The key concept here is a correct definition of the Lagrangian-mean flow with respect to which particle displacements can be defined;

0066-4189/84/0115-0011$02.00

these particle displacements then serve as the appropriate field variables in a Lagrangian formulation. The preliminary ideas were developed by Dewar (1970) and Bretherton (1971) and culminated in the generalized Lagrangian-mean formulation of Andrews & McIntyre (1978a,b).

In this review, we describe wave action principles and wave-mean flow interaction theorems in three stages. In this section, we present a general theory based on a Lagrangian formulation of the equations of motion. Our treatment is in the spirit of Whitham's approach (Whitham 1965, 1970), but follows the development by Hayes (1970) more closely. Because wave action and wave-mean flow interaction are intertwined concepts, we complement the discussion on wave action by introducing the radiation stress tensor and describing its role in the wave energy equation and the mean flow equation. Our treatment is based on the ideas of Dewar (1970) and Bretherton (1971), but goes beyond their results in that there is no restriction to slowly varying linearized waves.

In order to apply this general theory to fluids, we turn in Section 2 to a description of the generalized Lagrangian-mean formulation of Andrews & McIntyre. Although the results of this section are complete, they are obtained in a form where their application to specific problems generally requires more discussion. Rather than give a catalog of all the contexts in fluid mechanics where wave action principles have been invoked, we instead give in Section 3 a brief account of internal gravity waves in a stratified shear flow. Our purpose here is didactic; that is, our concern is not so much to present some specific results as to illustrate how the general theory is adapted in a specific case.

Formulation

We begin by supposing that the physical system is specified by the vector-valued field $\phi\ (x_i)$, where x_i $(i = 0, 1, 2, 3)$ are the independent variables. Subsequently it will be useful to distinguish between $x_0 = t$, a timelike coordinate, and x_α $(\alpha = 1, 2, 3)$, which are spacelike coordinates. Throughout we employ the dual summation convention that Latin indices are summed over the range 0 to 3, but Greek indices are summed over the range 1 to 3. In the absence of dissipation, we suppose that the physical system obeys a variational principle with a Lagrangian density $L\,(\phi_i, \phi\,; x_i)$, where ϕ_i denotes the partial derivative $\partial\phi/\partial x_i$. Then the equations of motion are

$$\frac{\partial}{\partial x_i}\left(\frac{\partial L}{\partial \phi_i}\right) - \frac{\partial L}{\partial \phi} = Q. \tag{1.1}$$

Here the generalized force Q represents the effects of dissipation. A useful consequence of this formulation (see Hayes 1970) is that if ψ is any field with the same dimension as ϕ, then

$$\frac{\partial}{\partial x_i}\left(\psi\frac{\partial L}{\partial \phi_i}\right) = \psi_i\frac{\partial L}{\partial \phi_i} + \psi\frac{\partial L}{\partial \phi} + \psi Q. \tag{1.2}$$

For instance, putting $\psi = \phi_j$ in (1.2), it follows that

$$\frac{\partial T_{ji}}{\partial x_i} = -\frac{\partial L}{\partial x_j} + \phi_j Q, \tag{1.3a}$$

where

$$T_{ji} = \phi_j\frac{\partial L}{\partial \phi_i} - L\,\delta_{ji}. \tag{1.3b}$$

Here $\partial L/\partial x_j$ on the right-hand side of (1.3a) is the explicit derivative of L with respect to x_j, and T_{ji} can be identified as the energy-momentum tensor of classical theoretical physics (see, for example, Landau & Lifshitz 1962). Although the precise physical interpretation of the components of T_{ji} will depend *inter alia* on the choice of Lagrangian density, we shall find it useful to identify T_{00} as the energy density and $T_{0\alpha}$ as its flux, and $T_{\alpha 0}$ as the momentum density and $T_{\alpha\beta}$ as the corresponding fluxes.

WAVE ACTION In order to define wave action density and its flux, we must first introduce the notion of an ensemble average $\langle\ \rangle$. We base our discussion on the ideas of Sturrock (1962) and Hayes (1970), which were further developed by Andrews & McIntyre (1978b). The relationship with the more specialized notions of Whitham (1965, 1970), Dewar (1970), Dougherty (1970), and Bretherton (1971) are developed below. We suppose that $\phi(x_i, \theta)$ depends smoothly on the ensemble parameter θ, such that

$$\phi(x_i, \theta+2\pi) = \phi(x_i, \theta). \tag{1.4}$$

We then define the averaging operator

$$\langle\ \rangle = \frac{1}{2\pi}\int_0^{2\pi}(\qquad)\,d\theta. \tag{1.5}$$

For simplicity, we sometimes denote the mean field $\langle\phi\rangle$ by $\bar{\phi}$, and we note the important observation that all mean quantities are independent of θ. The averaging operator commutes with $\partial/\partial x_i$, and has other simple and obvious properties [see Andrews & McIntyre (1978a,b) for an extensive discussion].

We next define the wave perturbation or disturbance field $\hat{\phi}(x_i, \theta)$ of ϕ by

$$\phi = \bar{\phi} + \hat{\phi}. \tag{1.6}$$

Clearly $\hat{\phi}$ has a zero mean ($\langle\hat{\phi}\rangle = 0$). Although we have called $\hat{\phi}$ the wave perturbation of ϕ, there is at present no restriction on the magnitude of $\hat{\phi}$ vis-à-vis that of $\bar{\phi}$, nor on their relative scales. Next put $\psi = \hat{\phi}_\theta$ in (1.2), where $\hat{\phi}_\theta = \partial\hat{\phi}/\partial\theta$, and apply the averaging operator $\langle \ \rangle$. It follows that (see Hayes 1970)

$$\frac{\partial}{\partial x_i}\left\langle \hat{\phi}_\theta \frac{\partial L}{\partial \phi_i} \right\rangle = \langle \hat{\phi}_\theta Q \rangle. \tag{1.7}$$

This is the wave action equation. In the absence of dissipation ($Q = 0$), it gives a local conservation law. It is a consequence of the invariance of the mean Lagrangian $\langle L \rangle$ to changes in θ. We shall find it useful to identify

$$\mathbf{A} = \left\langle \hat{\phi}_\theta \frac{\partial L}{\partial \phi_t} \right\rangle, \qquad \mathbf{B}_\alpha = \left\langle \hat{\phi}_\theta \frac{\partial L}{\partial \phi_\alpha} \right\rangle \tag{1.8}$$

as the wave action density and flux, respectively. Here and subsequently, we write ϕ_t in place of ϕ_0, which denotes the partial derivative $\partial\phi/\partial t$. From (1.8) it is clear that both $\mathbf{A}$ and $\mathbf{B}_\alpha$ are $O(a^2)$ in the limit of small wave amplitude a. Here the wave amplitude parameter a is a measure of the magnitude of $\hat{\phi}$ and its derivatives. Consequently, $\mathbf{A}$ and $\mathbf{B}_\alpha$ are wave properties and are the appropriate general measures of wave activity. The analogue of (1.7) in classical mechanics is obtained by restricting the independent variables to t alone. The action density $\mathbf{A}$ can then be recognized as $\oint p \, d\phi$, where p is the momentum conjugate to ϕ, and the integral is over one cycle. The action equation (1.7) is then the basis for the study of adiabatic invariants (for a lucid discussion, see Landau & Lifshitz 1960).

It may be shown that both the density and flux are unaffected by smooth, monotonic transformations of the ensemble parameter θ, and by the addition of flux terms to the Lagrangian density L that leave the equations of motion (1.1) unchanged. Further, under a Galilean transformation $x'_\alpha = x_\alpha - U_\alpha t, t' = t$, where U_α is a constant velocity, $\mathbf{A}$ is invariant, and $\mathbf{B}_\alpha$ transforms to $\mathbf{B}_\alpha - U_\alpha \mathbf{A}$; thus (1.7) is left invariant. Under a Lorentz transformation, the four-vector $(\mathbf{A}, \mathbf{B}_\alpha)$ transforms according to the usual laws for a relativistic four-vector. These properties are in strong contrast to the corresponding properties of the energy-momentum tensor and Equation (1.3a).

The application of (1.7) depends upon the delineation of the family $\phi(x_i, \theta)$. One possibility that has received much attention is to identify θ as a phase shift in the phase of the wave; the averaging operator (1.5) is then an average over this phase. Thus we put

$$\hat{\phi} = \hat{\phi}(x_i, s(x_i) - \theta), \tag{1.9a}$$

and

$$\kappa_i = \frac{\partial s}{\partial x_i}. \tag{1.9b}$$

Here $s(x_i)$ is the phase, and ϕ is periodic in s. Let κ_α be the wavenumber components, and $\omega = -\kappa_0$ the frequency. It follows that

$$\mathbf{A} = \frac{\partial}{\partial \omega}\langle L \rangle, \qquad \mathbf{B}_\alpha = -\frac{\partial}{\partial \kappa_\alpha}\langle L \rangle, \tag{1.10}$$

and the wave action equation (1.7) takes the form obtained by Whitham (1970). In this form, the wave action equation can be recognized as the Euler equation that is obtained when the mean Lagrangian $\langle L \rangle$ is subjected to variations in the phase $s(x_i)$, and is an application of Whitham's averaged variational principle (Whitham 1965, 1970), here extended to include dissipative effects (Ostrovsky & Pelinovsky 1972).

It is a remarkable fact that the wave action equation (1.7) is formally exact. It is valid without any restriction in wave amplitude, or without any assumption that the mean field is slowly varying with respect to the waves. However, the utility of this is reduced in practice by the presence of two kinds of error. The first of these has been called by Hayes (1970) the *identification* error, and occurs whenever the ensemble parameter θ is interpreted as a phase shift. It arises as a result of the identification of the family (1.9a) with a particular solution of interest, and can in principle be made arbitrarily small with respect to a small parameter characterizing the difference in scale between the rapidly varying phase of the waves and other variations, such as those in the mean field. The second kind of error is due to the implicit hypothesis that only a single wave is present, and is particularly severe in strongly nonlinear systems. To some extent it can be removed by allowing θ to be vector-valued (see Hayes 1970), so that the family $\phi(x_i, \theta)$ describes a multiple wave system. This aspect has been largely neglected in the wave action literature, although this is compensated by the extensive literature on wave interactions. In the extreme case of strongly nonlinear random wave interactions, the action density obeys a diffusion equation in phase space (see, for instance, Abarbanel 1981).

PSEUDOMOMENTUM If the averaging operator (1.5) is applied to (1.3a) we obtain equations for the total energy $\langle T_{00} \rangle$ and total momentum $\langle T_{\alpha 0} \rangle$. However, these equations are not generally as useful as the wave action equation in determining wave properties, as they contain both wave and mean flow expressions; in particular, the mean flow expressions will contain $O(a^2)$ contributions due to the waves. To obtain a mathematical analogue of $\langle T_{ij} \rangle$ for the waves, we follow the procedure of Andrews & McIntyre (1978b). First define the "undisturbed" Lagrangian by

$$L_0 = L(\bar{\phi}_i, \bar{\phi}; x_i) \tag{1.11}$$

and then put

$$L_1(\hat{\phi}_i, \hat{\phi}; x_i) = L - L_0. \tag{1.12}$$

Note that L_1 is an $O(a^2)$ wave property, and that the explicit dependence of L_1 on x_i includes the dependence of L_1 on the mean field $\bar{\phi}$ and its derivatives $\bar{\phi}_i$. Then put $\psi = \hat{\phi}_j$ in (1.2) and apply the averaging operator (1.5). The result is

$$\frac{\partial}{\partial x_i} \mathbf{T}_{ji} = -\left\langle \frac{\partial L_1}{\partial x_j} \right\rangle + \langle \hat{\phi}_j Q \rangle, \tag{1.13a}$$

where

$$\mathbf{T}_{ji} = \left\langle \hat{\phi}_j \frac{\partial L_1}{\partial \phi_i} - L_1 \, \delta_{ji} \right\rangle. \tag{1.13b}$$

$\mathbf{T}_{ji}$ is an $O(a^2)$ wave property. We shall call $\mathbf{T}_{00}$ the pseudoenergy, $\mathbf{T}_{0\alpha}$ its flux, $-\mathbf{T}_{\alpha 0}$ the pseudomomentum, and $-\mathbf{T}_{\alpha\beta}$ the corresponding flux. The sign conventions have been chosen to agree with historical convention (see Andrews & McIntyre 1978b). Note that $\mathbf{T}_{ji}$ can be identified as the averaged energy-momentum tensor for the disturbance Lagrangian L_1. Unlike the wave action density, $\mathbf{T}_{00}$ is not Galilean invariant and transforms to $\mathbf{T}_{00} + U_\alpha \mathbf{T}_{\alpha 0}$; however, the pseudomomentum $-\mathbf{T}_{\alpha 0}$ is Galilean invariant. Equation (1.13a) is a conservation equation only in the absence of dissipation ($Q = 0$) and when L_1 is explicitly independent of x_j; in particular, this latter condition requires the mean field $\bar{\phi}$ to be independent of x_j.

The relationship between (1.13a) and the wave action equation is obtained by observing that when the mean field $\bar{\phi}$ is independent of a particular coordinate x_j, then, by invoking a suitable ergodic principle, we may replace the averaging operator (1.5) with averaging over that coordinate and so identify θ with $-x_j$. The wave action equation then reduces to (1.13a), the wave action density to $-\mathbf{T}_{j0}$ for $j \neq 0$, and the flux $\mathbf{B}_\alpha$ to $-\mathbf{T}_{j\alpha}$ for $\alpha \neq j$; note that the diagonal term $\mathbf{T}_{jj}$ is now absent from (1.13a), being independent of x_j by definition. This application of the wave action equation is quite common, although in the literature it has not always been recognized as such. It is important to note the distinction between energy and momentum on the one hand, and pseudoenergy and pseudomomentum on the other. The former quantities are conserved, in the absence of dissipation, whenever the total system, represented by the full

Lagrangian L, is independent of t or x_α respectively. The latter quantities are conserved, again in the absence of dissipation, whenever the mean field, represented by $\langle L_1 \rangle$, is independent of t or x_α, respectively.

WAVE ENERGY Wave energy is a constantly recurring theme in the literature, although as we show below it is generally not as useful a concept as wave action. This arises, in part, from ambiguities in its definition. Bretherton & Garrett (1968) have given a comprehensive discussion of an appropriate definition for wave energy in the context of linearized waves. Adapting their conclusions, we define the wave energy density **E** as the pseudoenergy in a frame with respect to which the mean state is locally at rest. Specifically, let us now postulate that the mean field $\bar{\phi}$ consists of a mean velocity $\bar{u}_\alpha$ and a vector-valued mean field $\bar{\lambda}$. In applications, $\bar{\lambda}$ will incorporate the mean density, mean magnetic field, mean fluid depth, etc. Thus we define

$$\mathbf{E} = \mathbf{T}_{00} + \bar{u}_\alpha \mathbf{T}_{\alpha 0}, \tag{1.14a}$$

or

$$\mathbf{E} = \left\langle \frac{d\hat{\phi}}{dt}\frac{\partial L_1}{\partial \phi_t} - L_1 \right\rangle, \tag{1.14b}$$

where

$$\frac{d}{dt} = \frac{\partial}{\partial t} + \bar{u}_\alpha \frac{\partial}{\partial x_\alpha}. \tag{1.14c}$$

Note that d/dt is the time derivative following the mean motion. The corresponding definition of wave energy flux is

$$\mathbf{F}_\alpha = \left\langle \frac{d\hat{\phi}}{dt}\frac{\partial L_1}{\partial \phi_\alpha} - \bar{u}_\alpha(L_1 + \mathbf{E}) \right\rangle. \tag{1.15}$$

Then from (1.13a), or more directly by putting $\psi = d\hat{\phi}/dt$ in (1.2), it follows that the wave energy equation is

$$\frac{\partial \mathbf{E}}{\partial t} + \frac{\partial}{\partial x_\alpha}(\bar{u}_\alpha \mathbf{E} + \mathbf{F}_\alpha) = -\left\langle \frac{\partial L_1}{\partial t} + \bar{u}_\alpha \frac{\partial L_1}{\partial x_\alpha} \right\rangle + \frac{\partial \bar{u}_\alpha}{\partial x_i}\mathbf{T}_{\alpha i} + \left\langle \frac{d\hat{\phi}}{dt} Q \right\rangle. \tag{1.16}$$

Note that here, as in (1.13a), $\partial L_1/\partial x_i$ denotes the explicit derivative of L_1 with respect to x_i, including the dependence of L_1 on x_i through $\bar{u}_\alpha$ and $\bar{\lambda}$. Equation (1.16) demonstrates that whenever the mean state is varying, wave energy is not conserved, and is generally exchanged with the mean flow. This contrasts unfavorably with the wave action equation, which states that wave action is conserved in the absence of dissipation, regardless of the variability of the mean state.

RADIATION STRESS At this point is is useful to make some further hypotheses concerning the Lagrangian L. These are motivated by the generalized Lagrangian-mean description of fluid flow introduced by Andrews & McIntyre (1978a) (see also Dewar 1970, Bretherton 1971), which we describe in more detail in Section 2. We suppose that L_1 depends on $\hat{\phi}_t$ only through its dependence on $d\hat{\phi}/dt$; further, apart from the dependence of L_1 on $\bar{u}_\alpha$ through $d\hat{\phi}/dt$, any other explicit dependence of L_1 on $\bar{u}_\alpha$ is bilinear in $\bar{u}_\alpha$ and the disturbance variables $\hat{\phi}$ and $\hat{\phi}_i$. This last property arises from the fact that the full Lagrangian is usually at most quadratic in the velocity field. It then follows that

$$\left\langle \frac{\partial L_1}{\partial \bar{u}_\alpha} \right\rangle = \mathbf{T}_{\alpha 0}. \tag{1.17}$$

Next, following Garrett (1968) and Dewar (1970), we observe that in many cases of interest λ will satisfy an equation of the form

$$\frac{d\lambda}{dt} + \Lambda_{\alpha\beta}\lambda \frac{\partial \bar{u}_\alpha}{\partial x_\beta} = \sigma. \tag{1.18}$$

Here σ represents dissipative effects. When σ is zero, λ is a mean quantity ($\lambda = \bar{\lambda}$); however, it is useful for the applications to be discussed later to allow λ to have a fluctuating component $\hat{\lambda}$ when σ is nonzero. These dissipative components are not included in the disturbance components $\hat{\phi}$, but are included in the radiation stress tensor defined below. The wave energy equation then becomes

$$\frac{\partial \mathbf{E}}{\partial t} + \frac{\partial}{\partial x_\alpha}(\bar{u}_\alpha \mathbf{E} + \mathbf{F}_\alpha) = -R_{\alpha\beta}\frac{\partial \bar{u}_\alpha}{\partial x_\beta} - \left\langle \frac{dL_1}{dt} \right\rangle_{\mathrm{e}} + \left\langle \frac{d\hat{\phi}}{dt} Q \right\rangle - \left\langle \frac{\partial L_1}{\partial \lambda}\sigma \right\rangle, \tag{1.19a}$$

where

$$R_{\alpha\beta} = -\mathbf{T}_{\alpha\beta} + \bar{u}_\beta \mathbf{T}_{\alpha 0} - \left\langle \Lambda_{\alpha\beta}\lambda \frac{\partial L_1}{\partial \lambda} \right\rangle. \tag{1.19b}$$

Although the nomenclature is not universal, we shall call $R_{\alpha\beta}$ the radiation stress tensor. Here $\langle \partial L_1/\partial x_i \rangle_{\mathrm{e}}$ denotes the explicit derivative of L_1 with respect to x_i when the disturbance variables $\hat{\phi}$, $\hat{\phi}_i$ and the mean variables $\bar{u}_\alpha$ and λ are all held constant. Also note the useful results in this context that

$$\mathbf{B}_\alpha = \bar{u}_\alpha \mathbf{A} + \left\langle \hat{\phi}_\theta \left(\frac{\partial L_1}{\partial \hat{\phi}_\alpha} \right)_{\mathrm{d}} \right\rangle, \tag{1.20a}$$

$$\mathbf{F}_\alpha = \left\langle \frac{d\hat{\phi}}{dt} \left(\frac{\partial L_1}{\partial \hat{\phi}_\alpha} \right)_{\mathrm{d}} \right\rangle, \tag{1.20b}$$

$$\mathbf{T}_{\alpha\beta} = \bar{u}_\beta \mathbf{T}_{\alpha 0} + \left\langle \hat{\phi}_\alpha \left(\frac{\partial L_1}{\partial \hat{\phi}_\beta} \right)_{\mathrm{d}} - L_1 \, \delta_{\alpha\beta} \right\rangle. \tag{1.20c}$$

Here $(\partial L_1/\partial \hat{\phi}_\alpha)_{\mathrm{d}}$ denotes the derivative with respect to $\hat{\phi}_\alpha$ when $d\hat{\phi}/dt$ is held constant. An important consequence of these equations is that when the averaging operator (1.5) can be interpreted as an average over the coordinate x_γ (i.e. θ can be identified with $-x_\gamma$), then the off-diagonal components of $R_{\gamma\beta}$ differ from $\mathbf{B}_\beta - \bar{u}_\beta \mathbf{A}$ only by the terms involving $\Lambda_{\gamma\beta}$; in particular, if $\Lambda_{\gamma\beta}$ is itself diagonal, then $R_{\gamma\beta}$ for $\gamma \neq \beta$ is exactly equal to $\mathbf{B}_\beta - \bar{u}_\beta \mathbf{A}$.

MEAN FLOW To complete the description of the interaction between the waves and the mean flow, equations describing the forcing of the mean flow by the waves are needed. These can be obtained by applying the averaging operator (1.5) to (1.1) or, alternatively, to (1.3). In the present context, when the mean field consists only of the mean velocity $\bar{u}_\alpha$ and the vector-valued mean field $\bar{\lambda}$ that satisfies (1.18), the most convenient result is

$$\frac{\partial}{\partial t}\left(\frac{\partial L_0}{\partial \bar{u}_\alpha} \right) + \frac{\partial}{\partial x_\beta}\left(\bar{u}_\beta \frac{\partial L_0}{\partial \bar{u}_\alpha} - \Lambda_{\alpha\beta} \bar{\lambda} \frac{\partial L_0}{\partial \bar{\lambda}} + L_0 \, \delta_{\beta\alpha} \right) - \left\langle \frac{\partial L}{\partial x_\alpha} \right\rangle_{\mathrm{e}} = -\frac{\partial R_{\alpha\beta}}{\partial x_\beta} + \langle Q_\alpha \rangle. \tag{1.21}$$

Since L_0 is quadratic in $\bar{u}_\alpha$, this is just the mean momentum equation and can be obtained by averaging the full-momentum equation [i.e. those equations in (1.1) corresponding to the components of ϕ for which $d\phi/dt$ is the total velocity u_α].

However, a simpler and more revealing derivation of (1.21) is to apply Whitham's averaged variational principle (Whitham 1965, 1970, Ostrovsky & Pelinovsky 1972)

$$\delta \int \langle L \rangle \, dx_\alpha \, dt = -\int \langle Q \delta \phi \rangle \, dx_\alpha \, dt \tag{1.22}$$

with respect to variations in $\bar{u}_\alpha$ and λ. This is the procedure used by Dewar (1970) and Bretherton (1971) in the context of small-amplitude waves. The variations in $\bar{u}_\alpha$ and λ are obtained by considering variations Δx_α in x_α, where Δx_α is a Lagrangian variation, or a variation incurred on a given fluid particle moving with the mean velocity. The corresponding Lagrangian variations $\Delta \bar{\phi}$ in any mean quantity $\bar{\phi}$ must be distinguished from the Eulerian variation $\delta \bar{\phi}$, the variation incurred at a given point x_α [see

Bretherton (1970) for a lucid discussion of this point]. They are related by the expression

$$\delta\bar{\phi} = \Delta\bar{\phi} - \Delta x_\alpha \frac{\partial \phi}{\partial x_\alpha}. \tag{1.23}$$

Thus the variations in $\bar{u}_\alpha$ and λ are given by

$$\delta\bar{u}_\alpha = \frac{d}{dt}(\Delta x_\alpha) - \Delta x_\beta \frac{\partial \bar{u}_\alpha}{\partial x_\beta}, \tag{1.24a}$$

$$\delta\lambda = -\Lambda_{\alpha\beta}\lambda \frac{\partial}{\partial x_\beta}(\Delta x_\alpha) - \Delta x_\beta \frac{\partial \lambda}{\partial x_\beta}. \tag{1.24b}$$

The expression (1.24b) is valid only for a restricted class of tensors $\Lambda_{\alpha\beta}$ (see Dewar 1970), which, however, includes all cases so far encountered in the literature. Applying the principle (1.22), it now follows that

$$\frac{\partial}{\partial t}\left(\frac{\partial}{\partial \bar{u}_\alpha}\langle L\rangle\right) + \frac{\partial}{\partial x_\beta}\left(\bar{u}_\beta \frac{\partial}{\partial \bar{u}_\alpha}\langle L\rangle - \left\langle \Lambda_{\alpha\beta}\lambda \frac{\partial L}{\partial \lambda}\right\rangle\right) + \frac{\partial \bar{u}_\beta}{\partial x_\alpha}\frac{\partial}{\partial \bar{u}_\beta}\langle L\rangle + \left\langle \frac{\partial \lambda}{\partial x_\beta}\frac{\partial L}{\partial \lambda}\right\rangle = \langle Q_\alpha\rangle + \langle \hat{\phi}_\alpha Q\rangle. \tag{1.25}$$

Decomposing L into L_1 and L_0 (1.12), we can derive (1.21) by using (1.13a), (1.17), and (1.19b). An alternative to (1.21) that does not involve the radiation stress tensor $R_{\alpha\beta}$ but instead contains the pseudomomentum $-\mathbf{T}_{\alpha 0}$ can be derived from (1.25) by decomposing L into L_1 and L_0 (1.12) and then using only (1.13a). This latter form is the one preferred by Andrews & McIntyre (1978a) and is often more useful, particularly for irrotational flow and for situations where one component of the divergence of the radiation stress tensor is larger, by an order of magnitude in some small parameter, than all the other terms in (1.21). For examples of this, see Andrews & McIntyre (1978a) or Grimshaw (1979). An important application of this alternative procedure arises when the averaging operator (1.5) can be interpreted as an average over the coordinate x_γ (i.e. θ can be identified with $-x_\gamma$). Then if $\Lambda_{\alpha\beta}$ is diagonal, the off-diagonal components of $R_{\gamma\beta}$ are just $\mathbf{B}_\beta - \bar{u}_\beta \mathbf{A}$, and so in the γ-component of (1.21) the divergence of the radiation stress tensor is given by

$$-\frac{\partial R_{\gamma\beta}}{\partial x_\beta} = \frac{\partial \mathbf{A}}{\partial t} + \frac{\partial}{\partial x_\beta}(\bar{u}_\beta \mathbf{A}) - \langle \hat{\phi}_\gamma Q\rangle. \tag{1.26}$$

If (1.21) is multiplied by $\bar{u}_\alpha$ and the result is added to (1.19a), we obtain the mean-energy equation

$$\frac{\partial \mathbf{E}^{\mathrm{T}}}{\partial t}+\frac{\partial}{\partial x_\beta}\left\{\bar{u}_\beta \mathbf{E}^{\mathrm{T}}+\bar{u}_\alpha\left(-\Lambda_{\alpha\beta}\lambda\frac{\partial L_0}{\partial \lambda}+L_0\,\delta_{\beta\alpha}\right)+\mathbf{F}_\beta+\bar{u}_\alpha R_{\alpha\beta}\right\}$$
$$+\left\langle\frac{\partial L}{\partial t}\right\rangle_{\mathrm{e}}=\left\langle \bar{u}_\alpha Q_\alpha+\frac{d\hat{\phi}}{dt}Q-\frac{\partial L}{\partial \lambda}\sigma\right\rangle, \quad (1.27\mathrm{a})$$

where

$$\mathbf{E}^{\mathrm{T}}=\bar{u}_\alpha\frac{\partial \bar{L}_0}{\partial \bar{u}_\alpha}-L_0+\mathbf{E}. \quad (1.27\mathrm{b})$$

Here $\mathbf{E}^{\mathrm{T}}$ is the total mean energy. In some applications this equation is more useful than the wave energy equation (1.19a), as it is in conservation form in the absence of dissipation and explicit time dependence.

The mean field equations are thus (1.18) and (1.21), and apart from the dissipative terms and the term representing the effects associated with external forces $(\langle \partial L/\partial x_\alpha \rangle)_{\mathrm{e}}$, the radiation stress tensor $R_{\alpha\beta}$ represents the sole effect due to the waves. Note that $R_{\alpha\beta}$ is generally asymmetric and has no simple relationship to the Reynolds stresses and buoyancy fluxes encountered in Eulerian formulations of the mean flow equations. These mean-field equations are complemented by the wave action equation (1.7), which is generally the most useful way of describing the effect of the mean field on the waves. For finite-amplitude waves the two sets of equations are coupled. However, for linearized wave motion the action density and flux can be evaluated correct to $O(a^2)$, with the mean field fixed at the basic state values. The mean field changes due to the waves can then be calculated from (1.21), where $R_{\alpha\beta}$ can be evaluated correct to $O(a^2)$ independently of the mean field changes. Much of the literature on wave action and wave-mean flow interaction has considered only this special case of linearized wave motion.

SLOWLY VARYING WAVES Slowly varying, almost-plane waves have the representation (1.9a) where the dependence on the phase $s(x_i)$ is rapidly varying relative to the explicit dependence on x_i (although Rossby waves on a β-plane are an important exception). From (1.9a) it follows that

$$\hat{\phi}_i=-\kappa_i\hat{\phi}_\theta+\frac{\partial\hat{\phi}}{\partial x_i}, \quad (1.28)$$

where the explicit derivative $\partial\hat{\phi}/\partial x_i$ can be neglected compared to $\kappa_i\hat{\phi}_\theta$. The following approximate expressions can then be derived from (1.8), (1.13b),

(1.14a), (1.15), and (1.19b):

$$\mathbf{T}_{00} \approx \omega\mathbf{A} - \bar{L}_1, \qquad \mathbf{T}_{0\alpha} \approx \omega\mathbf{B}_\alpha, \tag{1.29a}$$

$$\mathbf{T}_{\alpha 0} \approx -\kappa_\alpha\mathbf{A}, \qquad \mathbf{T}_{\alpha\beta} \approx -\kappa_\alpha\mathbf{B}_\beta - \bar{L}_1\,\delta_{\alpha\beta}, \tag{1.29b}$$

$$\mathbf{E} \approx \omega^*\mathbf{A} - \bar{L}_1, \qquad \mathbf{F}_\alpha \approx \omega^*(\mathbf{B}_\alpha - \bar{u}_\alpha\mathbf{A}), \tag{1.29c}$$

$$R_{\alpha\beta} \approx \kappa_\alpha(\mathbf{B}_\beta - \bar{u}_\beta\mathbf{A}) + \bar{L}_1\,\delta_{\alpha\beta} - \left\langle \Lambda_{\alpha\beta}\lambda\frac{\partial L_1}{\partial\lambda} \right\rangle, \tag{1.29d}$$

where

$$\omega^* = \omega - \kappa_\alpha\bar{u}_\alpha. \tag{1.29e}$$

Here ω^* is the intrinsic wave frequency.

SLOWLY VARYING LINEARIZED WAVES For small-amplitude waves further simplifications are possible. First note that if we put $\psi = \hat{\phi}$ in (1.2), then it follows that

$$\frac{\partial}{\partial x_i}\left\langle \hat{\phi}\frac{\partial L_1}{\partial \phi_i} \right\rangle = \left\langle \hat{\phi}_i\frac{\partial L_1}{\partial \hat{\phi}_i} + \hat{\phi}\frac{\partial L_1}{\partial \hat{\phi}} \right\rangle + \langle \hat{\phi} Q \rangle. \tag{1.30}$$

This can be regarded as a virial theorem (Hayes 1974, Andrews & McIntyre 1978b). For linearized wave motion, L_1 is at most quadratic in the disturbance quantities $\hat{\phi}$ and $\hat{\phi}_i$. Hence the first term on the right-hand side is just $2\bar{L}_1$. For slowly varying waves, the left-hand side can be neglected and, assuming that the dissipative term can likewise be neglected, it follows that $\bar{L}_1 \approx 0$. This in turn implies equipartition of energy in nonrotating systems. Thus for slowly varying linearized waves, (1.29c) shows that the action density $\mathbf{A}$ is given by the classical result $\mathbf{E}/\omega^*$. In the context of fluid mechanics, this result was first derived by Bretherton & Garrett (1968) using the averaged variational principle (Whitham 1965), although the identification of action density in terms of an energy density divided by a local frequency has antecedents in the classical theory of adiabatic invariants (Landau & Lifshitz 1960). Analogous results in the context of plasma physics were developed by Dewar (1970) and Dougherty (1970).

Next, for linearized waves, $\bar{L}_1$ will be quadratic in the wave amplitude, and hence given by an expression of the form

$$\bar{L}_1 \approx D(\omega^*, \kappa_\alpha; \lambda)a^2, \tag{1.31}$$

where the explicit dependence on ω^*, rather than just ω, is a consequence of the hypothesis that the mean velocity $\bar{u}_\alpha$ is slowly varying. Hence, assuming Galilean invariance, the averaged Lagrangian $\bar{L}_1$ can be evaluated approximately in a frame with respect to which the mean state is locally at rest. But $\bar{L}_1 \approx 0$ and so $D \approx 0$; this must be equivalent to the local dispersion relation

$$\omega^* = W^*(\kappa_\alpha; \lambda). \tag{1.32}$$

But from (1.10) it now follows that

$$\mathbf{B}_\alpha \approx c_\alpha \mathbf{A}, \tag{1.33a}$$

where

$$c_\alpha = \bar{u}_\alpha + c_\alpha^*, \tag{1.33b}$$

and

$$c_\alpha^* = \frac{\partial W^*}{\partial \kappa_\alpha}. \tag{1.33c}$$

Here c_α^* is the intrinsic group velocity. The wave action equation (1.7) now reduces to the form proposed by Bretherton & Garrett (1968). In the absence of dissipation this is

$$\frac{\partial}{\partial t}\left(\frac{\mathbf{E}}{\omega^*}\right) + \frac{\partial}{\partial x_\alpha}\left([\bar{u}_\alpha + c_\alpha^*]\frac{\mathbf{E}}{\omega^*}\right) \approx 0. \tag{1.34}$$

With the same approximations, the energy flux $\mathbf{F}_\alpha \approx c_\alpha^* \mathbf{E}$, and the wave energy equation (1.19a) becomes

$$\frac{\partial \mathbf{E}}{\partial t} + \frac{\partial}{\partial x_\alpha}([\bar{u}_\alpha + c_\alpha^*]\mathbf{E}) \approx -R_{\alpha\beta}\frac{\partial \bar{u}_\alpha}{\partial x_\beta}, \tag{1.35a}$$

where

$$R_{\alpha\beta} \approx \frac{\mathbf{E}}{\omega^*}\left\{\kappa_\alpha c_\beta^* + \Lambda_{\alpha\beta}\lambda\frac{\partial W^*}{\partial \lambda}\right\}. \tag{1.35b}$$

It is readily verified (see Garrett 1968) that (1.34) and (1.35a) are equivalent. Finally, we note that the pseudomomentum $-\mathbf{T}_{\alpha 0}$ is approximately given by $\kappa_\alpha \mathbf{E}/\omega^*$.

MODAL WAVES In many applications the waves are confined to a waveguide by the presence of boundaries. Consequently, the waves possess a propagating character only with respect to coordinates that vary along the waveguide, and have a modal character across the waveguide. Following the notions of Hayes (1970) and Andrews & McIntyre (1978b), we suppose that a boundary Σ to the waveguide is undisturbed and impermeable to the fluid. The appropriate boundary condition on Σ is then

$$\text{either} \quad \hat{\phi} = 0 \quad \text{on} \quad \Sigma, \tag{1.36a}$$

$$\text{or}\quad n_i \frac{\partial L}{\partial \hat{\phi}_i} = 0 \quad \text{on} \quad \Sigma. \tag{1.36b}$$

Here n_i are the components of the normal to Σ; for instance, if Σ is given by $F(x_i) = 0$, then $n_i \propto \partial F/\partial x_i$. We also allow for the possibility that (1.36a) holds for some components of $\hat{\phi}$, and (1.36b) for the remaining components. It follows that

$$\hat{\phi}_\theta n_i \frac{\partial L}{\partial \hat{\phi}_i} = 0 \quad \text{on} \quad \Sigma. \tag{1.37}$$

Thus, the wave action flux normal to Σ vanishes on Σ. For simplicity, we are considering only nondissipative boundary conditions on Σ; for cases where dissipative boundary conditions are discussed, see Grimshaw (1981, 1982).

Let us now suppose, for simplicity, that the x_3-coordinate varies across the waveguide, which is bounded above and below by the surfaces $x_3 = F_\pm(t, x_1, x_2)$, respectively. The coordinates t, x_1, and x_2 thus characterize the propagation space. The wave action equation (1.7) continues to hold locally. However, the x_3-derivative in this equation will generally be the dominant term, and it is useful to remove it by integrating across the waveguide. Using the boundary condition (1.37), it follows that

$$\frac{\partial \mathscr{A}}{\partial t} + \frac{\partial \mathscr{B}_1}{\partial x_1} + \frac{\partial \mathscr{B}_2}{\partial x_2} = \int_{F-}^{F+} \langle \hat{\phi}_\theta Q \rangle \, dx_3. \tag{1.38a}$$

where

$$\mathscr{A} = \int_{F-}^{F+} \mathbf{A} \, dx_3, \tag{1.38b}$$

$$\mathscr{B}_\alpha = \int_{F-}^{F+} \mathbf{B}_\alpha \, dx_3. \tag{1.38c}$$

Equation (1.38a) is a global form of the wave action equation appropriate for modal waves; $\mathscr{A}$ and $\mathscr{B}_\alpha$ are the global wave action and flux, respectively. The analogue of (1.9a) for modal waves is obtained by restricting the phase s to be a function of only the variables t, x_1, and x_2. Since integration across the waveguide commutes with the averaging operator (1.5), it follows from (1.10) that

$$\mathscr{A} = \frac{\partial \mathscr{L}}{\partial \omega}, \qquad \mathscr{B}_\alpha = -\frac{\partial \mathscr{L}}{\partial \kappa_\alpha}, \tag{1.39a}$$

where

$$\mathscr{L} = \int_{F-}^{F+} \langle L \rangle \, dx_3. \tag{1.39b}$$

Thus, the wave action equation (1.38a) can be obtained from Whitham's averaged variational principle applied directly to $\mathscr{L}$.

The analogous global results for other quantities, such as the wave energy, the pseudomomentum, and the mean field, can also be obtained by integration across the waveguide. However, simple results analogous to (1.38a) are not generally obtained. For slowly varying waves, $\partial\hat{\phi}/\partial x_i$ can be neglected compared with $\kappa_i\hat{\phi}_\theta$ for $i = 0, 1, 2$, but $\partial\hat{\phi}/\partial x_3$ cannot be neglected. Nevertheless, the relations (1.29a–d) will continue to hold, provided the indices α, β are restricted to the values 1 and 2. Quantities such as the global wave energy, etc., can then be obtained by integrating (1.29a–d) across the waveguide. In this context, it is useful to note that the virial theorem (1.30) holds locally. By considering linearized waves, integrating across the waveguide, and using the boundary conditions (1.36a, b), it may be shown that the integral across the waveguide of $\bar{L}_1$ (i.e. $\mathscr{L}_1$) is approximately equal to zero. Thus for slowly varying, linearized modal waves whose dispersion relation is $\omega = W(\kappa_\alpha; t, x_1, x_2)$, we have $\mathscr{B}_\alpha = c_\alpha \mathscr{A}$, where c_α is the total group velocity $\partial W/\partial\kappa_\alpha$. This is the counterpart of (1.33a) for modal waves. Also, if we define

$$\mathscr{E} = (\omega - \kappa_\alpha v_\alpha)\mathscr{A} = \int_{F-}^{F+} \omega^* \mathbf{A}\, dx_3, \tag{1.39c}$$

where these equations also act as the definition of the mean velocity v_α, then we obtain the counterparts of (1.34) and (1.35a) for modal waves, i.e. replace E with $\mathscr{E}$ and ω^* with $\omega - \kappa_\alpha v_\alpha$, etc. The terms involving the mean field $\bar{\lambda}$ must of course be interpreted to apply to a different quantity that obeys a relation analogous to (1.18), with $\bar{u}_\alpha$ replaced with v_α and α, β restricted to the values 1 and 2. Also, the dispersion relation is assumed to take the form (1.32), with ω^* replaced with $\omega - \kappa_\alpha v_\alpha$.

2. FLUIDS AND THE GENERALIZED LAGRANGIAN-MEAN FORMULATION

An important feature of the general theory of Section 1 is that a Lagrangian formulation of the problem is an essential preliminary step to the efficient derivation of the wave action equation and the mean flow equations. Thus, in order to apply the results of Section 1 to specific cases involving fluids, the following points should be noted:

1. The problem should be formulated in terms of particle displacements ξ_α from a mean position that moves with the mean velocity $\bar{u}_\alpha$.
2. The wave action equation (1.7) is obtained by scalar multiplication of the momentum equation with $\partial\xi_\alpha/\partial\theta$, and averaging.

3. The wave energy equation (1.19a) is obtained by scalar multiplication of the momentum equation with $d\xi_\alpha/dt$, and averaging.
4. The mean flow equations should take the form (1.18) and (1.21), where the latter is obtained from averaging the momentum equation.

If these procedures are followed, it is often not necessary to identify the Lagrangian specifically, although its existence underlies the general theory. In particular, the radiation stress tensor is often most conveniently obtained by deriving the wave energy equation (1.19a) and the mean-flow equation (1.21) and consequently identifying $R_{\alpha\beta}$.

Lagrangian-Mean Formulation

The equations of motion for a conducting, compressible fluid in the nonrelativistic case are

$$\rho\frac{du_\alpha}{dt} + 2\rho\varepsilon_{\alpha\beta\gamma}\Omega_\beta u_\gamma + \rho\frac{\partial\Phi}{\partial x'_\alpha} + \frac{\partial q}{\partial x'_\alpha} - \frac{B_\beta}{\mu}\frac{\partial B_\alpha}{\partial x'_\beta} = \rho X_\alpha, \tag{2.1a}$$

$$\frac{d\rho}{dt} + \rho\frac{\partial u_\alpha}{\partial x'_\alpha} = 0, \tag{2.1b}$$

$$\frac{dS}{dt} = h, \tag{2.1c}$$

$$\frac{dB_\alpha}{dt} - B_\beta\frac{\partial u_\alpha}{\partial x'_\beta} + B_\alpha\frac{\partial u_\beta}{\partial x'_\beta} = j_\alpha, \tag{2.1d}$$

where

$$q = p + \frac{1}{2\mu}B_\alpha B_\alpha \tag{2.1e}$$

and

$$\frac{d}{dt} = \frac{\partial}{\partial t} + u_\alpha\frac{\partial}{\partial x'_\alpha}. \tag{2.1f}$$

Here x'_α is the Eulerian coordinate such that a fluid particle at x'_α has velocity u_α. The notation is standard; in particular, Ω_α is the constant angular velocity of the frame of reference, $\Phi(x'_\alpha)$ is the potential for both the gravitational and centrifugal forces, $p(\rho, S)$ is the thermodynamic pressure, S is the entropy, and B_α is the magnetic field. The terms X_α, h, and j_α represent, respectively, the effects of nonconservative and dissipative forces, nonadiabatic motion, and finite magnetic conductivity. In particular, note that $\partial j_\alpha/\partial x'_\alpha = 0$, so that $\partial B_\alpha/\partial x'_\alpha = 0$ is a consequence of (2.1d).

The appropriate Lagrangian formulation of these equations is the generalized Lagrangian-mean formulation of Andrews & McIntyre (1978a) (see also Dewar 1970, Bretherton 1971). For a comprehensive account and justification of this theory in the absence of a magnetic field, the reader is referred to Andrews & McIntyre (1978a) [see also McIntyre (1977, 1980), Grimshaw (1979), or Dunkerton (1980) for simplified versions], as here we give only a brief outline. Let x_α be generalized Lagrangian coordinates and let $\xi_\alpha(t, x_\beta)$ be the particle displacements, defined so that

$$x'_\alpha = x_\alpha + \xi_\alpha. \tag{2.2}$$

Then, for any given u_α there is a unique "reference" velocity $\bar{u}_\alpha(t, x_\beta)$, such that when the point x_α moves with velocity $\bar{u}_\alpha$ the point x'_α moves with velocity u_α. It follows that the material time derivative (2.1f) is also given by

$$\frac{d}{dt} = \frac{\partial}{\partial t} + \bar{u}_\alpha \frac{\partial}{\partial x_\alpha}. \tag{2.3}$$

The generalized Lagrangian-mean formulation is now obtained by letting $\bar{u}_\alpha$ be the mean velocity, precisely that introduced in Section 1, and requiring that

$$\langle \xi_\alpha(t, x_\beta) \rangle = 0. \tag{2.4}$$

Note, in particular, that (2.3) agrees with our previous definition (1.14c). The reader should also note that our notation differs in one important respect from that of Andrews & McIntyre (1978a,b); here $\bar{u}_\alpha$ denotes the Lagrangian-mean velocity, rather than $\bar{u}_\alpha^L$ used in Andrews & McIntyre (1978a,b), who use the single overbar to denote Eulerian means. No confusion should arise, as Eulerian means are not discussed in this article.

Next we define a mean density $\tilde{\rho}$, so that

$$\frac{d\tilde{\rho}}{dt} + \tilde{\rho} \frac{\partial \bar{u}_\alpha}{\partial x_\alpha} = 0. \tag{2.5}$$

It is an immediate consequence of (2.1b) and (2.5) that

$$\rho J = \tilde{\rho}, \tag{2.6a}$$

where

$$J = \det\, [\partial x'_\alpha / \partial x_\beta]. \tag{2.6b}$$

For the magnetic field, we define a new variable H_α by

$$J B_\alpha = H_\beta \frac{\partial x'_\alpha}{\partial x_\beta}, \quad \text{or} \quad H_\alpha = B_\beta K_{\beta\alpha}. \tag{2.7}$$

Here $K_{\alpha\beta}$ is the α, β-cofactor of J, and so

$$K_{\alpha\beta}\frac{\partial x'_\alpha}{\partial x_\gamma} = \delta_{\beta\gamma}J = K_{\beta\alpha}\frac{\partial x'_\gamma}{\partial x_\alpha}. \tag{2.8}$$

It is useful to note that $K_{\alpha\beta}$ is the derivative of J with respect to $\partial x'_\alpha/\partial x_\beta$ and that $\partial K_{\alpha\beta}/\partial x_\beta = 0$. With the definitions (2.7), it can now be shown that (2.1d) becomes

$$\frac{dH_\alpha}{dt} - H_\beta\frac{\partial \bar{u}_\alpha}{\partial x_\beta} + H_\alpha\frac{\partial \bar{u}_\beta}{\partial x_\beta} = k_\alpha = j_\beta K_{\beta\alpha}. \tag{2.9}$$

Also, $\partial k_\alpha/\partial x_\alpha = 0$, so that $\partial H_\alpha/\partial x_\alpha = 0$ is a consequence of (2.9). The entropy equation (2.1c) is left unchanged, and the final step is the converting of the momentum equation (2.1a) to its Lagrangian form. The result is

$$\tilde{\rho}\frac{du_\alpha}{dt} + 2\tilde{\rho}\varepsilon_{\alpha\beta\gamma}\Omega_\beta u_\gamma + \tilde{\rho}\frac{\partial \Phi}{\partial x'_\alpha} + \frac{\partial}{\partial x_\beta}\left(qK_{\alpha\beta} - \frac{B_\alpha H_\beta}{\mu}\right) = \tilde{\rho}X_\alpha. \tag{2.10}$$

Here the velocity u_α is given by

$$u_\alpha = \bar{u}_\alpha + \frac{d\xi_\alpha}{dt}. \tag{2.11}$$

In summary, the generalized Lagrangian-mean equations are (2.10), the entropy equation (2.1c), the magnetic equation (2.9), and the mean density equation (2.5). They can be identified as the Euler equations (1.1) for the Lagrangian

$$L(u_\alpha, x'_\alpha, \partial x'_\alpha/\partial x_\beta, \tilde{\rho}, S, H_\alpha)$$
$$= \tilde{\rho}\{\tfrac{1}{2}u_\alpha u_\alpha + \varepsilon_{\alpha\beta\gamma}\Omega_\alpha x'_\beta u_\gamma - \Phi(x'_\alpha) - E(\rho, S)\} - \frac{J}{2\mu}B_\alpha B_\alpha. \tag{2.12}$$

Here we recall that ρ and B_α are defined in terms of $\tilde{\rho}$ and H_α by (2.6a) and (2.7), respectively. Also, $E(\rho, S)$ is the internal energy per unit mass, and

$$\frac{\partial E}{\partial \rho} = \frac{p}{\rho^2}, \qquad \frac{\partial E}{\partial S} = T, \tag{2.13}$$

where T is the temperature. Variations in x'_α (or equivalently ξ_α) then give (2.10), with $Q_\alpha = \tilde{\rho}X_\alpha$. Equation (2.5) for $\tilde{\rho}$ involves only mean quantities, and so acts as a constraint on the Lagrangian variations Δx_α, which determine the mean flow equation (1.21). Since S and H_α are mean quantities when the dissipative terms h and k_α vanish, Equations (2.1c) and (2.9) are in the same category. However, in order to keep the correspondence with the general theory of Section 1 as close as possible, we define the generalized forces Q_S and Q_{H_α} so that the corresponding Euler equation is an identity:

$$Q_S = \tilde{\rho}T, \qquad Q_{H_\alpha} = \frac{1}{\mu} B_\beta \frac{\partial x'_\beta}{\partial x_\alpha}. \tag{2.14}$$

Finally, we identify λ as the 5-vector whose components are $\tilde{\rho}$, S, and H_α. Then it may be verified that each of the equations (2.1c), (2.5), and (2.9) leads to an equation of the form (1.18) for λ (Dewar 1970), and that the variations in λ then satisfy (1.24b) as required.

WAVE ACTION This can now be obtained directly from (1.7), or by following the procedure of Andrews & McIntyre (1978b) and multiplying (2.10) by $\partial\xi_\alpha/\partial\theta$ and averaging. The result is

$$\frac{\partial \mathbf{A}}{\partial t} + \frac{\partial \mathbf{B}_\alpha}{\partial x_\alpha} = \mathbf{D}, \tag{2.15a}$$

where

$$\mathbf{A} = \left\langle \frac{\partial \xi_\alpha}{\partial \theta}\left(\tilde{\rho}\frac{d\xi_\alpha}{dt} + \tilde{\rho}\varepsilon_{\alpha\beta\gamma}\Omega_\beta\xi_\gamma\right)\right\rangle, \tag{2.15b}$$

$$\mathbf{B}_\alpha = \bar{u}_\alpha \mathbf{A} + \left\langle \frac{\partial \xi_\beta}{\partial \theta}\left(qK_{\beta\alpha} - \frac{B_\beta H_\alpha}{\mu}\right)\right\rangle, \tag{2.15c}$$

$$\mathbf{D} = \left\langle \frac{\partial \xi_\alpha}{\partial \theta}\tilde{\rho}X_\alpha + \frac{\partial S}{\partial \theta}\tilde{\rho}T + \frac{\partial H_\alpha}{\partial \theta}\frac{B_\beta}{\mu}\frac{\partial x'_\beta}{\partial x_\alpha}\right\rangle. \tag{2.15d}$$

That $\mathbf{D}$ represents the effects of dissipation follows from the identification of X_α as representing nonconservative and dissipative forces, and from the fact that S and H_α are disturbance quantities only when the dissipative terms h and k_α are nonzero. The expressions (2.15b–d) agree with those obtained by Andrews & McIntyre (1978b) in the absence of a magnetic field, although the dissipative term has been written in a different form here.

For linearized wave motion, it is useful to introduce the Eulerian pressure perturbation

$$q' = \hat{q} - \xi_\alpha \frac{\partial \bar{q}}{\partial x_\alpha} + O(a^2). \tag{2.16}$$

Then it may be shown that [see Andrews & McIntyre (1978b) or Grimshaw (1980)]

$$\mathbf{B}_\alpha - \bar{u}_\alpha \mathbf{A} = \left\langle q'\frac{\partial \xi_\alpha}{\partial \theta}\right\rangle + \frac{\partial}{\partial x_\beta}\left\langle \bar{q}\xi_\beta \frac{\partial \xi_\alpha}{\partial \theta}\right\rangle - \frac{1}{\mu}\left\langle H_\alpha H_\beta \frac{\partial \xi_\gamma}{\partial \theta}\left\{\frac{\partial \xi_\gamma}{\partial x_\beta} + \delta_{\gamma\beta}\left(1 - \frac{\partial \xi_\sigma}{\partial x_\sigma}\right)\right\}\right\rangle + O(a^3). \tag{2.17}$$

Note that the second term here is identically nondivergent and can be omitted from (2.15a). In many applications only the first term of (2.17) is significant; for instance, if all mean quantities depend only on a single coordinate, say x_3, and H_α is normal to this direction ($H_3 = 0$), then the only relevant component of $\mathbf{B}_\alpha$ is $\mathbf{B}_3$ and this is just $\langle q' \partial\xi_3/\partial\theta \rangle$. Equation (2.15a) can be rederived correct to $O(a^2)$ without invoking the generalized Lagrangian-mean formulation; the linearized momentum equation is multiplied by $\partial\xi_\alpha/\partial\theta$ and then averaged. If one invokes the basic flow equations (i.e. the mean flow equations to zeroth order in a), Equation (2.15a) follows with $\mathbf{A}$ given by (2.15b) and $\mathbf{B}_\alpha$ by (2.17) (McIntyre 1977, 1980, or Grimshaw 1980). This derivation also applies the useful result that the Eulerian velocity perturbation is given by

$$u'_\alpha = \frac{d\xi_\alpha}{dt} - \xi_\beta \frac{\partial \bar{u}_\alpha}{\partial x_\beta} + O(a^2). \tag{2.18}$$

LAGRANGIAN-MEAN FLOW To conform with the definitions of Section 1, we define

$$L_0 = L(\bar{u}_\alpha, x_\alpha, \delta_{\alpha\beta}, \tilde{\rho}, \bar{S}, \bar{H}_\alpha), \tag{2.19}$$

where L is given by (2.12). The mean flow equation is then obtained from (1.21), or more directly by averaging (2.10) (Andrews & McIntyre 1978a). The result is

$$\tilde{\rho}\frac{d\bar{u}_\alpha}{dt} + 2\tilde{\rho}\varepsilon_{\alpha\beta\gamma}\Omega_\beta\bar{u}_\gamma + \tilde{\rho}\left\langle \frac{\partial\Phi}{\partial x'_\alpha} \right\rangle + \frac{\partial\tilde{q}}{\partial x_\alpha} - \frac{1}{\mu}\frac{\partial}{\partial x_\beta}(\bar{H}_\alpha\bar{H}_\beta) = -\frac{\partial R_{\alpha\beta}}{\partial x_\beta} + \langle \tilde{\rho} X_\alpha \rangle, \tag{2.20a}$$

where

$$\tilde{q} = p(\tilde{\rho}, \bar{S}) + \frac{1}{2\mu}\bar{H}_\alpha\bar{H}_\alpha, \tag{2.20b}$$

and

$$R_{\alpha\beta} = \delta_{\alpha\beta}\langle qJ - \tilde{q}\rangle - \left\langle q\frac{\partial\xi_\gamma}{\partial x_\alpha}K_{\gamma\beta}\right\rangle - \frac{1}{\mu}\langle B_\alpha H_\beta - \bar{H}_\alpha\bar{H}_\beta\rangle. \tag{2.20c}$$

It can be verified that $R_{\alpha\beta}$ is the radiation stress tensor defined by (1.19b). An alternative form of (2.20a), involving the pseudomomentum $-\mathbf{T}_{\alpha 0}$, can be obtained by first multiplying (2.10) by $\partial x'_\alpha/\partial x_\beta$ and then averaging (see Andrews & McIntyre 1978a). Here, from (1.13b),

$$\mathbf{T}_{\alpha 0} = \left\langle \frac{\partial\xi_\gamma}{\partial x_\alpha}\left(\tilde{\rho}\frac{d\xi_\gamma}{dt} + \tilde{\rho}\varepsilon_{\gamma\beta\delta}\mathbf{\Omega}_\beta\xi_\beta\right)\right\rangle, \tag{2.21a}$$

and

$$\mathbf{T}_{\alpha\beta} = \bar{u}_\beta \mathbf{T}_{\alpha 0} + \left\langle \frac{\partial \xi_\gamma}{\partial x_\alpha} \left(qK_{\gamma\beta} - \frac{B_\gamma H_\beta}{\mu} \right) \right\rangle - \langle L_1 \rangle \, \delta_{\alpha\beta}, \tag{2.21b}$$

where we recall that L_1 is $L - L_0$ [see (2.12) and (2.19)]. The wave energy density $\mathbf{E}$ (1.14b) and flux $\mathbf{F}_\alpha$ (1.15) are given by

$$\mathbf{E} = \left\langle \tilde{\rho} \left\{ \frac{1}{2} \frac{d\xi_\alpha}{dt} \frac{d\xi_\alpha}{dt} + \Phi(x_\alpha + \xi_\alpha) - \Phi(x_\alpha) + E(\tilde{\rho} J^{-1}, S) - E(\tilde{\rho}, \bar{S}) \right\} + \frac{J}{2\mu} B_\alpha B_\alpha - \frac{1}{2\mu} \bar{H}_\alpha \bar{H}_\alpha \right\rangle, \tag{2.22a}$$

$$\mathbf{F}_\alpha = \left\langle \frac{d\xi_\beta}{dt} \left(qK_{\beta\alpha} - \frac{B_\beta H_\alpha}{\mu} \right) \right\rangle. \tag{2.22b}$$

The wave energy equation can now be obtained from (1.19a) or by multiplying (2.10) by $d\xi_\alpha/dt$ and averaging. The result is

$$\frac{\partial \mathbf{E}}{\partial t} + \frac{\partial}{\partial x_\alpha} (\bar{u}_\alpha \mathbf{E} + \mathbf{F}_\alpha) = -R_{\alpha\beta} \frac{\partial \bar{u}_\alpha}{\partial x_\beta} + \left\langle \tilde{\rho} \bar{u}_\alpha \left(\frac{\partial \Phi}{\partial x_\alpha} (x_\beta + \xi_\beta) - \frac{\partial \Phi}{\partial x_\alpha} (x_\beta) \right) \right\rangle + \mathbf{D}^{\mathrm{E}}, \tag{2.23a}$$

where

$$\mathbf{D}^{\mathrm{E}} = \left\langle \frac{d\xi_\alpha}{dt} \tilde{\rho} X_\alpha + h\tilde{\rho} \{ T(\rho, S) - T(\tilde{\rho}, \bar{S}) \} + \frac{1}{\mu} k_\alpha \left\{ B_\beta \frac{\partial x'_\beta}{\partial x_\alpha} - \bar{H}_\alpha \right\} \right\rangle. \tag{2.23b}$$

Here $\mathbf{D}^{\mathrm{E}}$ represents the effects of dissipation. Finally, the total energy equation is (1.27a), which here becomes

$$\frac{\partial \mathbf{E}^{\mathrm{T}}}{\partial t} + \frac{\partial}{\partial x_\alpha} \left\{ \bar{u}_\alpha (\mathbf{E}^{\mathrm{T}} + \tilde{q}) - \bar{u}_\beta \frac{\bar{H}_\beta \bar{H}_\alpha}{\mu} + F_\alpha + \bar{u}_\beta R_{\beta\alpha} \right\} = \left\langle u_\alpha \hat{\rho} X_\alpha + h\tilde{\rho} T(\rho, S) + \frac{1}{\mu} J B_\alpha j_\alpha \right\rangle, \tag{2.24a}$$

where

$$\mathbf{E}^{\mathrm{T}} = \left\langle \tilde{\rho} \left\{ \frac{1}{2} \bar{u}_\alpha \bar{u}_\alpha + \frac{1}{2} \frac{d\xi_\alpha}{dt} \frac{d\xi_\alpha}{dt} + \bar{\Phi}(x_\alpha + \xi_\alpha) + E(\tilde{\rho} J^{-1}, S) \right\} + \frac{J}{2\mu} B_\alpha B_\alpha \right\rangle. \tag{2.24b}$$

INCOMPRESSIBLE FLOW The corresponding results for an incompressible flow may be obtained by taking a limit in which the local sound speed becomes infinite, although some care should be taken when the Boussinesq approximation is also made due to the presence of a large hydrostatic component in the pressure field [see Grimshaw (1975a) or McIntyre (1977, 1980)]. Alternatively, we may proceed directly from the equations of motion for incompressible flow. These are just (2.1a) and (2.1d), with (2.1b) and (2.1c) replaced with

$$\frac{\partial u_\alpha}{\partial x'_\alpha} = 0, \tag{2.25a}$$

$$\frac{1}{\rho}\frac{d\rho}{dt} = m, \tag{2.25b}$$

where m is the counterpart of h in (2.1c) and represents the effects of nonadiabatic motion. In (2.1e) the pressure p is no longer the thermodynamic pressure, and is instead an independent variable in its own right. In the generalized Lagrangian-mean formulation we again define $\tilde{\rho}$ and J by (2.6a) and (2.6b), respectively. In place of (2.5) we now have

$$\frac{1}{\tilde{\rho}}\frac{d\tilde{\rho}}{dt} + \frac{\partial \bar{u}_\alpha}{\partial x_\alpha} = m. \tag{2.26}$$

Also, J is a mean quanity $\bar{J}$, which satisfies the equation

$$\frac{d\bar{J}}{dt} + \bar{J}\frac{\partial \bar{u}_\alpha}{\partial x_\alpha} = 0. \tag{2.27}$$

Note that because of the dissipative term m in (2.25), $\tilde{\rho}$ will have a fluctuating component. The Lagrangian-mean equation is then (2.10), and we identify λ with the 5-vector $\tilde{\rho}$, $\bar{J}$, and H_α. A suitable Lagrangian is

$$L(u_\alpha, x'_\alpha, \partial x'_\alpha/\partial x_\beta, p, \tilde{\rho}, \bar{J}, H_\alpha)$$
$$= \tilde{\rho}\{\tfrac{1}{2}u_\alpha u_\alpha + \varepsilon_{\alpha\beta\gamma}\Omega_\alpha x'_\beta u_\gamma - \Phi(x'_\alpha)\} + p(J-\bar{J}) - J\frac{B_\alpha B_\alpha}{2\mu}. \tag{2.28}$$

Here the disturbance fields to be varied are x'_α (or equivalently ξ_α) and p. However, Q_α is now given by $\tilde{\rho}X_\alpha + \tilde{\rho}m(u_\alpha + \varepsilon_{\alpha\beta\gamma}\Omega_\beta x'_\gamma)$. We also define $Q_{\tilde{\rho}}$ and Q_{H_α} so that the corresponding Euler equations are identities.

The wave action equation is again (2.15a), with $\mathbf{A}$ and $\mathbf{B}_\alpha$ again given by (2.15b) and (2.15c), respectively. However, the dissipative term $\mathbf{D}$ is now

$$\mathbf{D} = \left\langle \frac{\partial \xi_\alpha}{\partial \theta}\{\tilde{\rho}X_\alpha + \tilde{\rho}m(u_\alpha + \varepsilon_{\alpha\beta\gamma}\Omega_\beta x'_\gamma\}\right\rangle - \left\langle \frac{\partial \tilde{\rho}}{\partial \theta}\{\tfrac{1}{2}u_\alpha u_\alpha + \varepsilon_{\alpha\beta\gamma}\Omega_\beta x'_\gamma - \bar{\Phi}(x'_\alpha)\}\right\rangle$$
$$+ \left\langle \frac{\partial H_\alpha}{\partial \theta}\frac{B_\beta}{\mu}\frac{\partial x'_\beta}{\partial x_\alpha}\right\rangle. \tag{2.29}$$

If we assume for simplicity that m is zero, the mean flow equation is again (2.20a), with the proviso that in the expression (2.20b) for $\tilde{q}$, $p(\tilde{\rho}, \bar{S})$ is replaced with $\bar{p}$. The radiation stress tensor is again given by (2.20c). The pseudomomentum and its flux are still given by (2.21a) and (2.21b), with the proviso that L_0 is now $L(\bar{u}_\alpha, x_\alpha, \delta_{\alpha\beta}, \bar{p}, \tilde{\rho}, \bar{J}, \bar{H}_\alpha)$, where L is given by (2.28). The wave energy density is again given by (2.22a), with the proviso that the terms involving the internal energy E are replaced by $\bar{p}(1-\bar{J})$; the wave energy flux is again given by (2.22b). The wave energy equation is again (2.23a), but the dissipative term now takes a different form from (2.23b), and an extra term $(1-\bar{J})\,d\bar{p}/dt$ must be included on the right-hand side.

CURVILINEAR COORDINATES So far, our results in this section have been expressed in Cartesian coordinates. However, in the general theory of Section 1 we may allow the coordinates x_α to be any set of spacelike coordinates. By way of illustration, let us now consider the case when x_α are the cylindrical polar coordinates r, λ, and z. Analogous results using spherical polar coordinates have been obtained by F. P. Bretherton (personal communication). Thus, we let

$$x_1 = r, \qquad x_2 = \lambda, \qquad x_3 = z, \tag{2.30}$$

be generalized Lagrangian coordinates, whose Eulerian counterparts are r', λ', and z'. The particle displacements are then defined by [see (2.2)]

$$\xi_1 = r' - r, \qquad \xi_2 = \lambda' - \lambda, \qquad \xi_3 = z' - z. \tag{2.31}$$

The velocity components in the Eulerian coordinate directions are

$$u_1 = \frac{dr'}{dt}, \qquad u_2 = r'\frac{d\lambda'}{dt}, \qquad u_3 = \frac{dz'}{dt}, \tag{2.32a}$$

where

$$\frac{d}{dt} = \frac{\partial}{\partial t} + \bar{u}_1\frac{\partial}{\partial r} + \frac{\bar{u}_2}{r}\frac{\partial}{\partial \lambda} + \bar{u}_3\frac{\partial}{\partial z}. \tag{2.32b}$$

Here $\bar{u}_\alpha$ are the mean velocity components in the Lagrangian coordinate directions, which must be carefully distinguished from the Eulerian coordinate directions. The Lagrangian is again given by (2.12); for simplicity, we suppose that the axis of rotation is in the z-direction, that the potential Φ is axisymmetric, and that there is no magnetic field ($B_\alpha = 0$). Thus the Lagrangian is

$$L\left(u_\alpha, x'_\alpha, \frac{\partial x'_\alpha}{\partial x_\beta}, r\tilde{\rho}, S\right) = r\tilde{\rho}\{\tfrac{1}{2}u_\alpha u_\alpha + \Omega r' u_2 - \Phi(r', z') - E(\rho, S)\}, \tag{2.33a}$$

where

$$r'\rho J = r\tilde{\rho}, \tag{2.33b}$$

and J is again defined by (2.6b), but now x'_α and x_α are the cylindrical polar coordinates. Variations in x'_α (or equivalently ξ_α) then give the equations of motion:

$$\tilde{\rho}\left(\frac{du_1}{dt} - \frac{u_2^2}{r'} - 2\Omega u_2 + \frac{\partial \Phi}{\partial r'}\right) + \frac{r'}{r}\frac{\partial}{\partial x_\beta}(pK_{r\beta}) = \tilde{\rho}X_1, \tag{2.34a}$$

$$\tilde{\rho}\left(\frac{du_2}{dt} + \frac{u_1 u_2}{r'} + 2\Omega u_1\right) + \frac{1}{r}\frac{\partial}{\partial x_\beta}(pK_{\lambda\beta}) = \tilde{\rho}X_2, \tag{2.34b}$$

$$\tilde{\rho}\left(\frac{du_3}{dt} + \frac{\partial \Phi}{\partial z'}\right) + \frac{r'}{r}\frac{\partial}{\partial x_\beta}(pK_{z\beta}) = \tilde{\rho}X_3. \tag{2.34c}$$

Here the generalized forces are given by $Q_1 = \tilde{\rho}rX_1$, $Q_2 = \tilde{\rho}rr'X_2$, and $Q_3 = \tilde{\rho}rX_3$. The counterpart of (2.5) is

$$\frac{d}{dt}(r\tilde{\rho}) + r\tilde{\rho}\left\{\frac{\partial \bar{u}_1}{\partial r} + \frac{1}{r}\frac{\partial \bar{u}_2}{\partial \lambda} + \frac{\partial \bar{u}_3}{\partial z}\right\} = 0, \tag{2.35}$$

while S again satisfies (2.1c), where d/dt is now given by (2.32b).

The wave action equation can now be obtained from (1.7), or by multiplying (2.34a), (2.34b), and (2.34c) by $\partial\xi_\alpha/\partial\theta$ and averaging. The result is (2.15a), where now

$$\mathbf{A} = r\tilde{\rho}\left\langle \frac{\partial \xi_\alpha}{\partial \theta} h'_\alpha u_\alpha + \Omega r'^2 \frac{\partial \xi_2}{\partial \theta} \right\rangle, \tag{2.36a}$$

$$\mathbf{B}_\alpha = h_\alpha^{-1}\bar{u}_\alpha \mathbf{A} + \left\langle \frac{\partial \xi_\beta}{\partial \theta} r' p K_{\beta\alpha} \right\rangle, \tag{2.36b}$$

$$\mathbf{D} = r\tilde{\rho}\left\langle \frac{\partial \xi_\alpha}{\partial \theta} h'_\alpha X_\alpha + \frac{\partial S}{\partial \theta} T \right\rangle, \tag{2.36c}$$

where

$$h'_1 = h_1 = 1; \qquad h'_2 = r',\ h_2 = r; \qquad h'_3 = h_3 = 1. \tag{2.36d}$$

For linearized wave motion, we introduce the Eulerian pressure perturbation p' by (2.16), with q replaced by p, and x_α and ξ_α defined by (2.30) and (2.37), respectively. It may then be shown that [compare (2.17)]

$$\mathbf{B}_\alpha - \bar{u}_\alpha \mathbf{A} = \left\langle rp' \frac{\partial \xi_\alpha}{\partial \theta} \right\rangle + \frac{\partial}{\partial x_\beta}\left\langle r\bar{p}\xi_\beta \frac{\partial \xi_\alpha}{\partial \theta} \right\rangle + O(a^3). \tag{2.37}$$

The second term is identically nondivergent and can be omitted from (2.15a). When the basic flow is zonal [i.e. $\bar{u}_1$ and $\bar{u}_3$ are $O(a^2)$] and the averaging operator is interpreted as a zonal average (i.e. θ is identified with $-\lambda$), the wave action equation (2.15a) reduces to the generalized Eliassen-Palm relation derived by Andrews & McIntyre (1978c), and $r^{-1}\mathbf{A}$ is the angular pseudomomentum.

The mean flow equations can now be obtained from (1.21) [after allowing for the presence of geometrical factors involving r in L, (2.33a) and (1.18)], or more directly, by averaging (2.34a–c). For instance, from the azimuthal equation (2.34b), we obtain

$$\tilde{\rho}\frac{dM}{dt}+\frac{\partial\tilde{p}}{\partial\lambda} = -\frac{1}{r}\frac{\partial}{\partial x_\beta}R_{\lambda\beta}+\langle\tilde{\rho}r'X_2\rangle, \tag{2.38a}$$

where

$$R_{\lambda\beta} = \delta_{\lambda\beta}\langle r'pJ-r\tilde{p}\rangle-\left\langle r'p\frac{\partial\xi_\gamma}{\partial\lambda}K_{\gamma\beta}\right\rangle, \tag{2.38b}$$

and

$$M = \langle r'u_2+\Omega r'^2\rangle. \tag{2.38c}$$

Here $\tilde{p}$ is $p(\tilde{\rho}, \bar{S})$, M is the mean specific angular momentum about the z-axis, and $R_{\lambda\beta}$ is the azimuthal component of the radiation stress tensor. In particular, when the averaging operator is interpreted as a zonal average (i.e. θ is identified with $-\lambda$), the off-diagonal components of $R_{\lambda\beta}$ are identical with $\mathbf{B}_\beta-\bar{u}_\beta\mathbf{A}$; note that the diagonal components will not now appear in (2.38a). With the further restriction to linearized waves on a zonal basic flow, (2.38a) reduces to a generalized Charney-Drazin theorem (see the similar results obtained by Andrews & McIntyre 1978c). Since the divergence of the radiation stress tensor in (2.38a) is here given by (1.26a), it follows that the M will change only in response to wave transience or dissipative effects (for an explicit demonstration of this and the relationship between M and the zonal mean flow, see Dunkerton 1980).

These results and their counterparts in spherical polar coordinates are now finding extensive application in stratospheric meteorology. In this context the literature abounds with results on conservation equations for wave activity, derived usually for linearized waves and using various approximations (e.g. quasi-geostrophy, hydrostatic, slowly varying mean flows). These results are now generally called Eliassen-Palm relations after the pioneering work of Eliassen & Palm (1961), and can be recognized as special cases of the wave action equation. The corresponding results for the mean flow, such as (2.38a), are known variously as nonacceleration theorems, or Charney-Drazin theorems after the initial work by Charney &

Drazin (1961). For recent and comprehensive reviews of this now extensive and rapidly growing subject, see Andrews & McIntyre (1978c), McIntyre (1980), Dunkerton (1980), and Uryu (1980). The significant feature of the Eliassen-Palm relations on the one hand and the Charney-Drazin theorems on the other is the equality between the flux terms of the wave action equation and the wave forcing terms in the mean flow equation. The general theory of Section 1 shows that this duality is not a peculiarity of the equations governing stratospheric circulation, but is instead a general property of wave-mean flow interactions.

3. APPLICATIONS TO STRATIFIED SHEAR FLOWS

It is not possible in a single article to cover all instances where the wave action equation has proved a useful tool in elucidating wave-mean flow interaction. Instead, we discuss a specific case that is relatively familiar and sufficiently simple to permit a compact description. In applying the general theory, the reader is reminded that it is preferable to use the principles enunciated at the beginning of Section 2, rather than a slavish use of the subsequent formulae. This is particularly relevant when additional approximations, such as small wave amplitude or slowly varying waves, are being invoked.

Internal Gravity Waves

We consider internal gravity waves propagating on a basic state consisting of a horizontal shear flow $u_0(z)$ in the x-direction and the density profile $\rho_0(z)$. Here z is a coordinate in the vertical direction. We assume that the flow is incompressible and ignore the effects of rotation, magnetic fields, and dissipation. Then the linearized, two-dimensional equations of motion for the particle displacements $\xi(t, x, z)$ and $\zeta(t, x, z)$ in the horizontal and vertical directions, respectively, are

$$\rho_0 \frac{d^2\xi}{dt^2} + \frac{\partial p'}{\partial x} = 0, \tag{3.1a}$$

$$\rho_0 \frac{d^2\zeta}{dt^2} + \frac{\partial p'}{\partial z} + \rho_0 N^2 \zeta = 0, \tag{3.1b}$$

$$\frac{\partial \xi}{\partial x} + \frac{\partial \zeta}{\partial z} = 0, \tag{3.1c}$$

where

$$\frac{d}{dt} = \frac{\partial}{\partial t} + u_0 \frac{\partial}{\partial x}. \tag{3.1d}$$

Here d/dt is the linearized approximation to the material time derivative (2.3), and N^2 is the Brunt-Väisälä frequency $-g\rho_0^{-1}\, d\rho_0/dz$. Note that the equations have been formulated using the Eulerian pressure perturbation p' [see (2.16)], rather than its Lagrangian counterpart $\hat{p}$. For incompressible flow, p' is generally found to be a more convenient entity than $\hat{p}$, which is dominated by a large hydrostatic component [see Grimshaw (1975b) or McIntyre (1977, 1980)]. Also note that the Eulerian velocity perturbations are given by (2.18):

$$u' = \frac{d\xi}{dt} - \zeta \frac{\partial u_0}{\partial z}, \qquad w' = \frac{d\zeta}{dt}. \tag{3.2}$$

The derivation of (3.1a–c) is either from (2.10) (Grimshaw 1979), or from the linearized Eulerian equations after using (3.2).

Next we seek solutions of (3.1a–c) for which

$$\zeta = \psi \exp(ikx - i\omega t - i\theta) + \text{c.c.}, \tag{3.3}$$

with similar expressions for the other variables. At first, suppose that ψ is a function of z alone. Then $\psi(z)$ satisfies the equation

$$\frac{\partial}{\partial z}\left(\rho_0 \omega^{*2} \frac{\partial \psi}{\partial z}\right) + \rho_0 k^2 (N^2 - \omega^{*2})\psi = 0, \tag{3.4a}$$

where

$$\omega^* = \omega - ku_0. \tag{3.4b}$$

Here ω^* is the intrinsic frequency (1.29e). Equation (3.4a) is transformed into the Taylor-Goldstein equation when ψ is replaced by $\phi = \omega^*\psi$. As such, its properties are well known (see, for instance, Booker & Bretherton 1967). Equation (3.4a) has the wave invariant

$$\mathbf{B} = \left\langle p' \frac{\partial \zeta}{\partial \theta} \right\rangle, \tag{3.5a}$$

or

$$k^2 \mathbf{B} = -2\ \mathrm{Im}\left\{\rho_0 \omega^{*2} \frac{\partial \psi}{\partial z} \psi^*\right\} \tag{3.5b}$$

Here $\langle\ \rangle$ is an average over the phase-shift parameter θ [see (1.5)], and from (3.3) is equivalent to an average over a wavelength in the x-direction. From (2.17), $\mathbf{B}$ can be recognized as the vertical component of the wave action flux, correct to $O(a^2)$, where for simplicity we have omitted the subscript 3. It is a constant of the motion except at critical levels where $\omega^* = 0$. Of course, this result is a consequence of (2.15a), where only the z-derivative term survives; it can also be derived directly from (3.4a) or by using (3.5b). Using (3.1a) and (3.2), we can show that

$$k\mathbf{B} = \langle \rho_0 u' w' \rangle, \tag{3.6}$$

which is the vertical flux of horizontal momentum, or the xz-component of Reynolds stress. Historically it was in this form that **B** was first identified, but the general theory of the previous sections shows that (3.5a) is the more fundamental expression.

CRITICAL LEVELS AND OVER-REFLECTION Let us now suppose that $u_0(z) \to U_{1,2}$ and $N \to N_{1,2}$ as $z \to \pm\infty$, respectively. Then

$$\rho_0^{1/2}\psi \sim I \exp(im_2 z) + R \exp(-im_2 z) \quad \text{as} \quad z \to -\infty, \tag{3.7a}$$

$$\rho_0^{1/2}\psi \sim T \exp(im_1 z) \quad \text{as} \quad z \to \infty, \tag{3.7b}$$

where

$$m_{1,2}^2 + \frac{N^4}{4g^2} = \left(\frac{N^2}{(v - U_{1,2})^2} - k^2\right). \tag{3.7c}$$

Here v is the phase speed ωk^{-1} in the x-direction. We suppose that the wave frequency ω and wave number k are such that $m_{1,2}$ are both real, and that the signs of $m_{1,2}$ are chosen so that I, R, and T correspond to incident, reflected, and transmitted waves, respectively. From (3.5b),

$$\mathbf{B} = 2m_2(v - U_2)^2\{|R|^2 - |I|^2\} \quad \text{as} \quad z \to -\infty, \tag{3.8a}$$

and

$$\mathbf{B} = -2m_1(v - U_1)^2|T|^2 \quad \text{as} \quad z \to \infty. \tag{3.8b}$$

Since, from (3.5a), $\mathbf{B} = \omega^*\langle p'w' \rangle$ and $\langle p'w' \rangle$ is the vertical flux of wave energy, it follows that

$$-km_{1,2}(v - U_{1,2}) > 0. \tag{3.9}$$

If there are no critical levels in the flow, then **B** is constant throughout and we can equate (3.8a) with (3.8b). The result is an expression for the conservation of wave action and implies that $|R|^2 < |I|^2$.

However, if there is a critical level, then **B** is constant throughout except at the critical level. Suppose there is a single critical level at $z = 0$, where $\omega^* = 0$. Then, near $z = 0$,

$$\rho_0^{1/2}\psi \approx A(v - u_0)^{-1/2 + i\mu} + B(v - u_0)^{1/2 - i\mu}, \tag{3.10a}$$

where

$$\mu^2 = \left\{N^2\left(\frac{\partial u_0}{\partial z}\right)^{-2} - \frac{1}{4}\right\} \quad \text{at} \quad z = 0. \tag{3.10b}$$

Following Booker & Bretherton (1967), we determine the branch of $v-u_0$ as z passes through zero by assuming that the critical level is viscosity dominated. To be explicit, suppose that $k>0$ and $\partial u_0/\partial z$ is positive at $z=0$; then $v-u_0$ is real and positive for $z<0$, and is given by $|v-u_0|\,e^{i\pi}$ for $z>0$. Suppose first that μ is real and positive (i.e. the local Richardson number is greater than 1/4). Then, from (3.5b),

$$\mathbf{B} = 2\mu\left(\frac{\partial u_0}{\partial z}\right)_0 \{|A|^2-|B|^2\} \quad \text{for} \quad z<0 \tag{3.11a}$$

and

$$\mathbf{B} = -2\mu\left(\frac{\partial u_0}{\partial z}\right)_0 \{|A|^2 \exp(-2\mu\pi)-|B|^2 \exp(2\mu\pi)\} \quad \text{for} \quad z>0. \tag{3.11b}$$

Recalling the sign conventions, it follows that the "A"-wave is upgoing and the "B"-wave is downgoing. In either case a wave passing through the critical level is absorbed (Booker & Bretherton 1967). Further, since **B** is constant throughout $z \gtrless 0$, respectively, we may equate (3.8a) with (3.11a), and (3.8b) with (3.11b). It then follows that $|R|^2<|I|^2$.

However, if $\mu = i\nu$, where $0<\nu<1/2$ (i.e. the local Richardson number is less than 1/4), then

$$\mathbf{B} = 2i\nu\left(\frac{\partial u_0}{\partial z}\right)_0 \{AB^*-A^*B\} \quad \text{for} \quad z<0, \tag{3.12a}$$

$$\mathbf{B} = -2i\nu\left(\frac{\partial u_0}{\partial z}\right)_0 \{AB^* \exp(-2i\nu\pi)-A^*B \exp(2i\nu\pi)\} \quad \text{for} \quad z>0. \tag{3.12b}$$

Again, we may equate (3.8a) with (3.12a), and (3.8b) with (3.12b). As $\nu \to 0$, the critical level looks more like a vortex sheet and **B** is continuous at $z=0$. It follows that $|R|^2>|I|^2$ in this limit, and the incident wave is over-reflected (Acheson 1976). For small but nonzero ν, this argument suggests that there may be over-reflection, since, from (3.12a,b), the jump in **B** across the critical level is $O(\nu)$.

The importance of this discussion in relation to the wave action equation is that it illustrates how a knowledge of local solutions [i.e. (3.7a,b) or (3.10a)], together with a wave invariant (3.5b), enables a number of significant conclusions to be made without necessarily solving the wave equation (3.4a). For a similar account of critical levels when compressibility, rotation, and magnetic effects are included, see Grimshaw (1980), which also contains a review of the extensive literature in linearized wave motion near critical levels.

WAVE ACTION AND ENERGY As a preliminary to discussing the mean flow, we first allow ψ in (3.3) to depend on both z and t. Then the wave action equation (2.15a) is

$$\frac{\partial \mathbf{A}}{\partial t} + \frac{\partial \mathbf{B}}{\partial z} = 0, \tag{3.13a}$$

where

$$\mathbf{A} = \left\langle \rho_0 \left(\frac{\partial \xi}{\partial \theta} \frac{d\xi}{dt} + \frac{\partial \zeta}{\partial \theta} \frac{d\zeta}{dt} \right) \right\rangle, \tag{3.13b}$$

and $\mathbf{B}$ is given by (3.5a). The expressions (3.5b) and (3.6) for $\mathbf{B}$ are no longer valid, although they are first approximations when ψ is a slowly varying function of t. Equation (3.13a) can be derived directly from (3.1a,b) by multiplying with $\partial\xi/\partial\theta$ and $\partial\zeta/\partial\theta$, respectively. Note here that since θ is a phase-shift parameter in the x-direction, we can identify $k\mathbf{A}$ as the x-component of pseudomomentum $-\mathbf{T}_{10}$, and $k\mathbf{B}$ is the corresponding vertical flux.

As a comparison, the wave energy equation (2.23a) is

$$\frac{\partial \mathbf{E}}{\partial t} + \frac{\partial \mathbf{F}}{\partial z} = -R_{13} \frac{\partial u_0}{\partial z}, \tag{3.14a}$$

where

$$\mathbf{E} = \left\langle \tfrac{1}{2}\rho_0 \left\{ \left(\frac{d\xi}{dt} \right)^2 + \left(\frac{d\zeta}{dt} \right)^2 + N^2 \zeta^2 \right\} \right\rangle, \tag{3.14b}$$

$$\mathbf{F} = \langle p'w' \rangle, \tag{3.14c}$$

$$R_{13} = -\langle p'\zeta_x \rangle = k\mathbf{B}. \tag{3.14d}$$

Here R_{13} is the xz-component of the radiation stress tensor (2.20c). Equation (3.14a) is most simply derived from (3.1a,b) by multiplying with $d\xi/dt$ and $d\zeta/dt$, respectively. It can also be obtained from the counterpart of (2.22a,b) for incompressible flow. However, if this approach is followed, (2.22a) yields an expression for $\mathbf{E}$ that differs from (3.14b) by

$$\frac{\partial}{\partial z} \left\{ \langle g\rho_0 \zeta \rangle + \frac{1}{2} \frac{\partial}{\partial z} \langle \rho_0 \zeta^2 \rangle \right\}. \tag{3.15}$$

Note, however, that a corresponding term $(\partial/\partial t)\{\text{——}\}$ occurs in $\mathbf{F}$ (2.22b), and consequently can be omitted in (3.14b,c). This illustrates the fact that expressions such as (2.22a,b) derived from a Lagrangian may not always yield familiar expressions, and emphasizes the desirability of a direct derivation from the particular equation of motion being considered.

When ψ is slowly varying in both z and t, it is readily shown that $\mathbf{E} \approx \omega^* \mathbf{A}$ and $\mathbf{F} \approx \omega^* \mathbf{B}$, in agreement with the general results (1.29c) for slowly varying waves, recalling that for linearized waves $\bar{L}_1 \approx 0$. Further, slowly varying waves have a slowly varying vertical wave number $m(t, z)$, determined from the dispersion relation

$$\omega^{*^2} = N^2 k^2 (k^2 + m^2)^{-1}. \tag{3.16}$$

The vertical group velocity is $c_3 = \partial\omega/\partial m$, and $\mathbf{B} \approx c_3 \mathbf{A}$. In this form the wave action equation (3.13a) holds without restriction on wave amplitude, provided that u_0 is replaced with $\bar{u}$ in ω^* (3.4b), since, for slowly varying waves in incompressible flow, the waves are transverse and expressions such as (3.3) hold without restriction in amplitude (see Grimshaw 1975a).

MEAN FLOW The horizontal component of the Lagrangian-mean flow equation (2.20a) is, correct to $O(a^2)$,

$$\rho_0 \frac{\partial \bar{u}}{\partial t} + \rho_0 \bar{w} \frac{\partial u_0}{\partial z} = -\frac{\partial R_{13}}{\partial z}. \tag{3.17}$$

Also, the vertical component $\bar{w}$ is determined from (2.27), which, correct to $O(a^2)$, is

$$\frac{\partial \bar{J}}{\partial t} + \frac{\partial \bar{w}}{\partial z} = 0, \tag{3.18a}$$

and

$$\bar{J} - 1 = -\frac{\partial^2}{\partial z^2} \langle \tfrac{1}{2} \zeta^2 \rangle. \tag{3.18b}$$

Assuming a state of no disturbance before the arrival of the waves, and using the wave action equation (3.13a), it follows that

$$\rho_0 (\bar{u} - u_0) = k\mathbf{A} - \rho_0 \frac{\partial u_0}{\partial z} \frac{\partial}{\partial z} \langle \tfrac{1}{2} \zeta^2 \rangle. \tag{3.19}$$

Remarkably, this result is exact without any restriction to slowly varying waves. If the slowly varying hypothesis is invoked, the second term on the right-hand side of (3.19) is omitted.

The total energy equation (2.24a) is here most simply obtained by multiplying (3.17) with u_0 and adding the result to (3.14a). We find that

$$\frac{\partial}{\partial t} (\mathbf{E} + \rho_0 u_0 k \mathbf{A}) + \frac{\partial}{\partial z} (\mathbf{F} + u_0 k \mathbf{B}) = 0. \tag{3.20}$$

Here the wave-induced total energy density is $(\mathbf{E} + \rho_0 u_0 k \mathbf{A})$, and (3.20) gives a succinct description of how the wave-induced mean flow term $k\mathbf{A}$

combines with the wave energy **E** to ensure conservation of total energy. In particular, note that for slowly varying waves the total energy density is $\mathbf{E}v(v-u_0)^{-1}$, and the total energy flux is just this quantity multiplied by the vertical group velocity c_3. Acheson (1976) has shown how these expressions provide an energetic explantation of the phenomenon of over-reflection where the wave energy flux is directed away from the critical level in both $z > 0$ and $z < 0$, but the total energy flux is one-signed.

For slowly varying waves, both the wave action equation (3.13a) and the mean flow equation (3.17) are valid without any restriction on wave amplitude (Grimshaw 1975a). Combined with the dispersion relation (3.16), in which ω^* is $\omega - k\bar{u}$, they form a set of three coupled equations for the wave action density **A**, the mean flow $\bar{u}$, and the vertical wave number m. Numerical solutions of this set are described by Grimshaw (1975b), and Dunkerton (1981) has obtained analytic solutions by invoking the hydrostatic approximation in the dispersion relation (3.16). This is one of the rare instances where finite-amplitude wave-mean flow interaction can be analyzed in a simple analytic manner.

Conclusion

This brief account of wave action and wave-mean flow interaction for internal gravity waves is intended as a didactic illustration of the general theory. Although this particular example can also be analyzed using Eulerian means and wave energy arguments, it should be clear that wave action and Lagrangian-mean concepts lead simply and directly to the main conclusions. The advantages that ensue when the wave action equation and Lagrangian means are employed are particularly clear when Coriolis forces are included (Grimshaw 1975a, McIntyre 1980, Andrews 1980).

Dissipative and nonconservative effects are readily incorporated in the above discussion in the manner described in Section 2. However, some caution is needed when the basic state is maintained by nonconservative or diabatic terms that may not appear explicitly in the linearized wave equations [i.e. (3.1a–c) or their counterparts]. In this situation, wave action is not generally conserved. An example of this occurs when the Brunt-Väisälä frequency varies with time, but there is no corresponding basic vertical velocity; the time variation in the basic density profile must then be maintained by diabatic terms and so m in (2.26) is nonzero, and consequently the dissipative term **D** (2.29) in the wave action equation (2.15a) is nonzero (see Rotunno 1977). An analogous situation occurs for Rossby waves on a nonzonal flow (Young & Rhines 1980).

The wave action equation occurs in a variety of other physical systems. The extension of the theory described in this section to include Coriolis forces and its application to stratospheric meteorology has already been referred to at the end of Section 2. Completely contrasting physical systems are sound waves and surface gravity waves, as in both cases the Eulerian flow is irrotational. For a summary of wave action conservation in acoustic waveguides, the reader is referred to Andrews & McIntyre (1978b). The development of wave action concepts in water waves can be found in the pioneering work of Whitham (1965, 1970) and Bretherton & Garrett (1968); applications to finite-amplitude water waves began with Lighthill (1965) and have been extensively developed by Peregrine & Thomas (1979) and Stiassnie & Peregrine (1979). Finally, although it is beyond the scope of this review to delve into applications to plasma physics, the reader may like to consult Dewar (1970, 1972) or Dougherty (1970, 1974) for the development of Lagrangian concepts in that context.

Literature Cited

Abarbanel, H. D. I. 1981. Diffusion of action in nonlinear dynamical systems. *Nonlinear Properties of Internal Waves, AIP Conf. Proc. 76*, ed. B. J. West, pp. 321–38. New York: Am. Inst. Phys.

Acheson, D. J. 1976. On over-reflexion. *J. Fluid Mech.* 77:433–72

Andrews, D. G. 1980. On the mean motion induced by transient inertio-gravity waves. *Pure Appl. Geophys.* 118:177–88

Andrews, D. G., McIntyre, M. E. 1978a. An exact theory of nonlinear waves on a Lagrangian-mean flow. *J. Fluid Mech.* 89:609–46

Andrews, D. G., McIntyre, M. E. 1978b. On wave-action and its relatives. *J. Fluid Mech.* 89:647–64

Andrews, D. G., McIntyre, M. E. 1978c. Generalized Eliassen-Palm and Charney-Drazin theorems for waves on axisymmetric mean flows in compressible atmospheres. *J. Atmos. Sci.* 35:175–85

Booker, J. R., Bretherton, F. P. 1967. The critical layer for internal gravity waves in a shear flow. *J. Fluid Mech.* 27:513–39

Bretherton, F. P. 1970. A note on Hamilton's principle for perfect fluids. *J. Fluid Mech.* 44:19–31

Bretherton, F. P. 1971. The general linearised theory of wave propagation. In *Mathematical Problems in the Geophysical Sciences, Lectures in Applied Mathematics*, 13(1):61–102. Providence: Am. Math. Soc. 383 pp.

Bretherton, F. P., Garrett, C. J. R. 1968. Wavetrains in inhomogeneous moving media. *Proc. R. Soc. London Ser. A* 302:539–54

Charney, J. G., Drazin, P. G. 1961. Propagation of planetary-scale disturbances from the lower into the upper atmosphere. *J. Geophys. Res.* 66:83–109

Dewar, R. L. 1970. Interaction between hydromagnetic waves and a time-dependent, inhomogeneous medium. *Phys. Fluids* 13:2710–20

Dewar, R. L. 1972. A Lagrangian theory for nonlinear wave packets in a collisionless plasma. *J. Plasma Phys.* 7:267–84

Dougherty, J. P. 1970. Lagrangian methods in plasma dynamics. I. General theory of the method of the averaged Lagrangian. *J. Plasma Phys.* 4:761–85

Dougherty, J. P. 1974. Lagrangian methods in plasma dynamics. II. Construction of Lagrangians for plasmas. *J. Plasma Phys.* 11:331–46

Dunkerton, T. 1980. A Lagrangian mean theory of wave, mean-flow interaction with applications to nonacceleration and its breakdown. *Rev. Geophys. Space Phys.* 18:387–400

Dunkerton, T. J. 1981. Wave transience in a compressible atmosphere. Part I: Transient internal wave, mean-flow interaction. *J. Atmos. Sci.* 38:281–97

Eliassen, A., Palm, E. 1961. On the transfer of energy in stationary mountain waves. *Geofys. Publ.* 22(3):1–23

Garrett, C. J. R. 1968. On the interaction between internal gravity waves and a shear flow. *J. Fluid Mech.* 34:711–20

Grimshaw, R. 1975a. Nonlinear internal gravity waves in a rotating fluid. *J. Fluid Mech.* 71:497–512

Grimshaw, R. 1975b. Nonlinear internal gravity waves and their interaction with the mean wind. *J. Atmos. Sci.* 32:1779–93

Grimshaw, R. 1979. Mean flows induced by internal gravity wave packets propagating in a shear flow. *Philos. Trans. R. Soc. London Ser. A* 292:391–417

Grimshaw, R. 1980. A general theory of critical level absorption and valve effects for linear wave propagation. *Geophys. Astrophys. Fluid Dyn.* 14:303–26

Grimshaw, R. 1981. Mean flows generated by a progressing water wave packet. *J. Austral. Math. Soc. Ser. B* 22:318–47

Grimshaw, R. 1982. The effect of dissipative processes on mean flows induced by internal gravity-wave packets. *J. Fluid Mech.* 115:347–77

Hayes, W. D. 1970. Conservation of action and modal wave action. *Proc. R. Soc. London Ser. A* 320:187–208

Hayes, W. D. 1974. Conservation of wave action. In *Nonlinear Waves*, ed. S. Leibovich, A. R. Seebass, Ch. 6, pp. 170–85. Ithaca, N.Y.: Cornell Univ. Press. 331 pp.

Landau, L. D., Lifshitz, E. M. 1960. Adiabatic invariants. In *Mechanics*, Sect. 49, pp. 154–57. Oxford: Pergamon. 165 pp.

Landau, L. D., Lifshitz, E. M. 1962. The energy-momentum tensor. In *The Classical Theory of Fields*, Sect. 32, pp. 87–91. Oxford: Pergamon. 404 pp. Revised 2nd ed.

Lighthill, M. J. 1965. Contributions to the theory of waves in nonlinear dispersive systems. *J. Inst. Math. Its Appl.* 1:269–306

McIntyre, M. E. 1977. Wave transport in stratified, rotating fluids. In *Problems of Stellar Convection, Lecture Notes in Physics*, ed. E. A. Spiegel, J. P. Zahn, 71:290–314. Berlin: Springer. 363 pp.

McIntyre, M. E. 1980. An introduction to the generalized Lagrangian-mean description of wave, mean-flow interaction. *Pure Appl. Geophys.* 118:152–76

Ostrovsky, L. A., Pelinovsky, E. N. 1972. Method of averaging and the generalized variational principle for nonsinusoidal waves. *J. Appl. Math. Mech.* (*Prikl. Math. Mekh.*) 36:63–70

Peregrine, D. H., Thomas, G. P. 1979. Finite-amplitude deep-water waves on currents. *Philos. Trans. R. Soc. London Ser. A* 292:371–90

Rotunno, R. 1977. Internal gravity waves in a time-varying stratification. *J. Fluid Mech.* 82:609–19

Stiassnie, M., Peregrine, D. H. 1979. On averaged equations for finite-amplitude water waves. *J. Fluid Mech.* 94:401–7

Sturrock, P. A. 1962. Energy and momentum in the theory of waves in plasmas. In *Plasma Hydromagnetics, Lockheed Symp., 6th*, ed. D. Bershader, pp. 47–57. Stanford, Calif: Stanford Univ. Press. 146 pp.

Uryu, M. 1980. Acceleration of mean zonal flows by planetary waves. *Pure Appl. Geophys.* 118:661–93

Whitham, G. B. 1965. A general approach to linear and nonlinear dispersive waves using a Lagrangian. *J. Fluid Mech.* 22:273–83

Whitham, G. B. 1970. Two-timing, variational principles and waves. *J. Fluid Mech.* 44:373–95

Young, W. R., Rhines, P. B. 1980. Rossby wave action, enstrophy and energy in forced mean flows. *Geophys. Astrophys. Fluid Dyn.* 15:39–52

Ann. Rev. Fluid Mech. 1984. 16: 45–66

THE DEFORMATION OF SMALL VISCOUS DROPS AND BUBBLES IN SHEAR FLOWS

J. M. Rallison

Department of Applied Mathematics and Theoretical Physics,
University of Cambridge, Cambridge CB3 9EW, England

1. INTRODUCTION

When drops of one fluid are suspended in a second fluid that is caused to shear, the drops will deform, and, if the local shear rate is sufficiently large, will break into two or more fragments. The principal goals of experimental and theoretical work on this topic have been to discover how much distortion a given flow produces, how strong the flow must be to break the drop, and the number and size of the droplets that result from a burst. The chief area of application of such studies is to the formation of emulsions where one fluid phase is to be dispersed throughout a second, and in particular to the determination of the emulsion rheology and to the design of efficient mixing devices. More recently, attention has been focused on problems of oil recovery where oil drops are to be displaced by water in porous rock.

If the drops of the discrete phase are sufficiently small (and in most emulsions they are), the Reynolds number of the fluid motion responsible for deforming the drop is low (even though that appropriate to the motion of the continuous phase as a whole may be high). Considerable theoretical simplifications result when inertia forces are negligible, and (almost all) the work described here has invoked this approximation. Comparable phenomena when inertia forces are important have been reviewed in this series by Wegener & Parlange (1973).

A second simplification can arise from the smallness of the drop if the length scale L of variation of the imposed flow (determined by the overall

0066-4189/84/0115-0045$02.00

size of the mixing device or stirrer) is sufficiently large. If the drop size (say the radius of the spherical drop of equal volume) is a, then the undisturbed flow $\mathbf{U}(x)$ in the neighborhood of the drop can be written

$$\mathbf{U}(x) = \mathbf{U}(0) + \mathbf{x} \cdot \nabla \mathbf{U}(0) + O(a/L)^2, \tag{1}$$

where $\mathbf{x}$ is the distance measured from the drop center (of volume, say) at the origin. When the drop is placed in the ambient flow, Equation (1) represents an outer boundary condition for the Stokes equations, which apply close to the drop [i.e. within distances of $O(a)$]. Now a force-free drop will, in the absence of inertia and whatever its instantaneous shape, move so that its velocity (at leading order in a/L) is $\mathbf{U}(0)$, and so by choosing a frame of reference in which the drop center is always at rest the only external flow term responsible for deforming the drop shape is a linear shear $\mathbf{x} \cdot \nabla \mathbf{U}(0)$.

We note in passing two circumstances that would appear to complicate the picture above. First, if the drop shape is sufficiently unsymmetric at some instant, then the shear will induce a further translational velocity (at order a/L) of the drop center, with the possibility of deformation due to that additional translation. Since, in practice, linear shear fields generate only point-symmetric drop shapes, however, this complication does not arise in experiments and does not appear to have been investigated theoretically. Second, we should consider the case excluded above where a net external force (e.g. gravity) does act on the drop and causes it to move with a nonzero velocity relative to the fluid. This problem has been analyzed in detail for an instantaneously spherical drop of arbitrary viscosity [originally by Hadamard (1911) and Rybczyński (1911); see e.g. Batchelor (1967, Section 4.9)], and the zero-Reynolds-number solution has the remarkable feature that even in the absence of surface tension, there is no tendency for the drop to deform, and so it remains spherical. There is thus little incentive to examine other shapes (unlike the high-Reynolds-number case). The case in which both shear and gravitational forces act on the drop remains open.

In summary, then, small drops in shear flows are deformed and burst solely because of the local velocity gradient they experience, and their dynamics can be analyzed locally on the basis of the creeping-flow equations.

2. DIMENSIONAL ANALYSIS AND GOVERNING EQUATIONS

We denote the viscosities of the drop and the suspending fluid as $\lambda\mu$ and μ, respectively, so that $\lambda = 0$ corresponds to a bubble, and $\lambda \to \infty$ to a highly viscous drop. The corresponding densities are written as $\kappa\rho$ and ρ, and the surface-tension coefficient between the phases is γ. Then if the outer fluid is caused to shear at a rate G, the flow will be viscously dominated inside and near the drop provided the Reynolds numbers $\rho Ga^2/\mu$ and $\kappa\rho Ga^2/\lambda\mu$ are both small. For an air bubble of diameter 0.1 mm in water at a shear rate $1\ \mathrm{s}^{-1}$, both numbers are about 3×10^{-3}. In that case the fluid densities do not enter the problem, and the fluid velocity $\mathbf{u}$ is governed by the Stokes equations (see Figure 1)

$$\nabla \cdot \mathbf{u} = 0 \quad \text{and} \quad \nabla \cdot \boldsymbol{\sigma} = 0 \tag{2}$$

everywhere except on the drop surface S, with the stress $\boldsymbol{\sigma}$ and pressure p given by

$$\boldsymbol{\sigma} = \begin{cases} -p\mathbf{I} + \mu(\nabla\mathbf{u} + \nabla\mathbf{u}^{\mathrm{T}}) & \text{in} \quad \hat{V}, \\ -p\mathbf{I} + \lambda\mu(\nabla\mathbf{u} + \nabla\mathbf{u}^{\mathrm{T}}) & \text{in} \quad V. \end{cases} \tag{3}$$

To complete the specification of the problem we require appropriate boundary conditions. As previously discussed, we have

$$\mathbf{u} \sim \mathbf{x} \cdot \nabla\mathbf{u} = G(\mathbf{e} + \boldsymbol{\omega})\mathbf{x} \quad \text{as} \quad |\mathbf{x}| \to \infty, \tag{4}$$

where $\mathbf{e}$ and $\boldsymbol{\omega}$ are the dimensionless rate-of-strain and vorticity tensors, and at the drop surface S,

$$[\mathbf{u}]_s = 0, \qquad [\boldsymbol{\sigma} \cdot \mathbf{n}]_s = \gamma\mathbf{n}\kappa = \gamma\mathbf{n}\nabla \cdot \mathbf{n}, \tag{5}$$

where $[\ ]_s$ denotes the jump in the bracketed quantity across S, and κ is the surface curvature (equal to the surface divergence of $\mathbf{n}$). We have finally the kinematic condition at S by which the drop shape changes in time t, which may be written symbolically as

$$dS/dt = \mathbf{u} \cdot \mathbf{n} \quad \text{for points of } S. \tag{6}$$

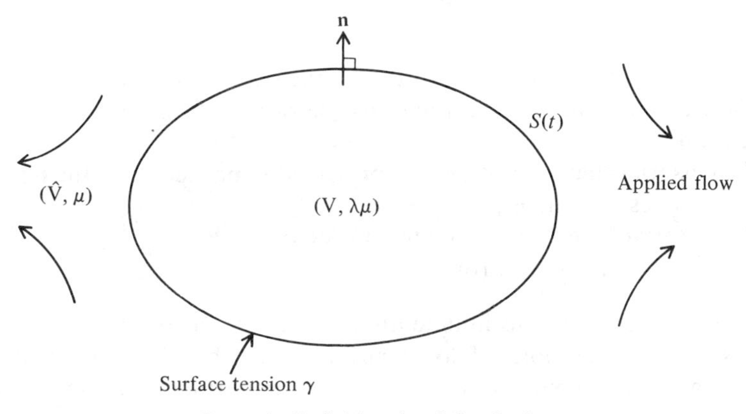

Figure 1 Definition sketch for the drop.

Two useful conclusions stem from the low-Reynolds-number assumption. First, the equations are quasi-static, so that $\mathbf{u}$ is instantaneously and uniquely determined by the drop shape and the imposed flow. Thus the drop shape is uniquely specified at all times by an initial condition for S together with the history of the flow it has experienced. Second, the equations are linear, and hence the instantaneous velocity everywhere in the fluids can be written as a sum of contributions from the two forcings in the problem—the surface tension (γ) and the imposed flow (G). This linear superposition is appropriate only for the time-dependent calculation, of course; the steady deformation (i.e. the shape S for which $\mathbf{u}\cdot\mathbf{n} = 0$) is a highly nonlinear function of G, which may not even be unique (see Section 4).

Now on scaling lengths by a and times by the surface-tension relaxation time $\mu a/\gamma$, so that the fluid velocity is scaled by γ/μ and the pressure by γ/a, the nondimensional form of the boundary conditions (4) and (5) becomes

$$\begin{aligned} &\mathbf{u} \sim \mathrm{Ca}\,(\mathbf{e}+\boldsymbol{\omega})\cdot\mathbf{x} \quad \text{as} \quad |\mathbf{x}| \to \infty, \\ &[\mathbf{u}]_s = 0, \qquad [\boldsymbol{\sigma}\cdot\mathbf{n}]_s = \mathbf{n}\nabla\cdot\mathbf{n}, \end{aligned} \tag{7}$$

where

$$\mathrm{Ca} = \mu G a/\gamma. \tag{8}$$

We note in passing that this choice of time scale is not always the most sensible. It is appropriate when both (a) surface tension rather than flow dominates the dynamics (otherwise G^{-1} might be better) and (b) $\lambda \leq O(1)$. If $\lambda \gg 1$, then since it is the larger of the two viscosities that dominates the relaxation of the shape, $\lambda\mu a/\gamma$ would be a better choice (see Section 4).

In summary, the following dimensionless parameters govern drop deformation and burst:

1. λ, the ratio of drop viscosity to that of the suspending fluid;
2. Ca, a capillary number representing the ratio of flow forces to surface tension;
3. the tensorial character of $\nabla\mathbf{u}$ as represented in particular by the relative magnitudes of $\mathbf{e}$ and $\boldsymbol{\omega}$;
4. the history of the flow as specified by $\nabla U(t)$;
5. the initial shape of the drop.

A full investigation of this infinite-dimensional parameter space is plainly impossible, and the choice of flow type and history has been restricted to a set designed to demonstrate the physical importance of each of the parameters above.

Types of Flow

The principal experimental difficulty as regards choice of flow type is to realize a shear flow that is both spatially homogeneous near the drop and constant in time in a frame of reference moving with the drop. This difficulty has severely restricted the class of flows that have been considered in detail.

The simplest flow to generate, and the one for which the most data have been obtained, is a *simple shear*, usually produced in a Couette device, with (in Cartesian coordinates) $\mathbf{u} = (Gy, 0, 0)$ so that

$$\mathbf{e} = \frac{1}{2}\begin{pmatrix} 0 & 1 & 0 \\ 1 & 0 & 0 \\ 0 & 0 & 0 \end{pmatrix} \quad \text{and} \quad \boldsymbol{\omega} = \frac{1}{2}\begin{pmatrix} 0 & 1 & 0 \\ -1 & 0 & 0 \\ 0 & 0 & 0 \end{pmatrix}. \tag{9}$$

It is thus a flow with "as much" vorticity as rate of strain.

A second flow that has been examined in detail is *plane hyperbolic* (stagnation point) flow $\mathbf{u} = G(x, -y, 0)$ produced in a four-roll mill (Taylor 1934). Here,

$$\mathbf{e} = \operatorname{diag}(1, -1, 0), \qquad \boldsymbol{\omega} = 0, \tag{10}$$

and it is the absence of vorticity that is the significant difference from simple shear.

Finally, one study (Hakimi & Schowalter 1980) has explored the deformation produced by an *orthogonal-rheometer* flow

$$\mathbf{e} = \frac{1}{2}\begin{pmatrix} 0 & 1 & 0 \\ 1 & 0 & 0 \\ 0 & 0 & 0 \end{pmatrix}, \qquad \boldsymbol{\omega} = \frac{1}{2}\begin{pmatrix} 0 & 1 & -2\psi \\ -1 & 0 & 0 \\ 2\psi & 0 & 0 \end{pmatrix}. \tag{11}$$

The appearance of the parameter ψ here permits a change of the relative orientation of vorticity to the principal axes of strain, but unfortunately the fact that the vorticity term is always large implies that large drop deformations are never produced.

The availability of experimental results for these flow fields has motivated theoretical studies for these types of shear. But for analytical purposes, other flows prove more convenient; in particular, a flow which in view of its axisymmetry provides special simplifications is the pure *extension*

$$\mathbf{u} = G(x, -\tfrac{1}{2}y, -\tfrac{1}{2}z),$$

with

$$\mathbf{e} = \operatorname{diag}(1, -\tfrac{1}{2}, -\tfrac{1}{2}) \quad \text{and} \quad \boldsymbol{\omega} = 0. \tag{12}$$

Table 1 Summary of some experimental and theoretical work on drop deformation

Type of flow		Range of λ	Range of Ca	Authors
Simple shear Equation (9)	Experiment	Arbitrary	Arbitrary	Taylor (1934) Bartok & Mason (1959) Rumscheidt & Mason (1961) Torza et al. (1972) Grace (1971)
	Theory	Arbitrary	$\mathrm{Ca} \ll 1 \begin{cases} O(\mathrm{Ca}) \\ O(\mathrm{Ca}^2) \end{cases}$	Taylor (1932) Barthès-Biesel & Acrivos (1973a)
		$\lambda \to \infty$	Arbitrary	Taylor (1934) Cox (1969)
		$\lambda \to 0$	$\mathrm{Ca} \to \infty$	Hinch & Acrivos (1980)
		$\lambda = 1$	Arbitrary	Rallison (1981)
Plane hyperbolic Equation (10)	Experiment	Arbitrary	Arbitrary	Taylor (1934) Rumscheidt & Mason (1961) Grace (1971)
	Theory	Arbitrary	$\mathrm{Ca} \ll 1 \begin{cases} O(\mathrm{Ca}) \\ O(\mathrm{Ca}^2) \end{cases}$	Taylor (1932) Barthès-Biesel & Acrivos (1973a)
		$\lambda \to 0$	$\mathrm{Ca} \to \infty$	Hinch & Acrivos (1979)
		$\lambda = 1$	Arbitrary	Rallison (1981)
Orthogonal rheometer Equation (11)	Experiment	~ 0.1	Arbitrary	Hakimi & Schowalter (1980)
	Theory	$\lambda = 1$	Arbitrary	Rallison (1981)
Axisymmetric pure extension Equation (12)	Theory	Arbitrary	$\mathrm{Ca} \ll 1\ O(\mathrm{Ca}^2)$	Barthès-Biesel & Acrivos (1973a)
		$\lambda \to 0$	$\mathrm{Ca} \to \infty$	Acrivos & Lo (1978)
		$\lambda > 0.3$	Arbitrary	Rallison & Acrivos (1978)
		$\lambda = 0$	Arbitrary	Youngren & Acrivos (1976)
Plane flows Equation (13)	Theory	Arbitrary	$\mathrm{Ca} \ll 1$	Taylor (1932)
		$\lambda = 1$	Arbitrary	Rallison (1981)
		$\lambda \to \infty\ (\chi \neq 1)$	Arbitrary	Taylor (1934) Cox (1969)

In addition, in order to examine the role of the vorticity, the full range of *plane flows* has been studied. This class includes simple shear ($\chi = 0$) and plane hyperbolic ($\chi = 1$) as special cases. Here,

$$\mathbf{e} = \tfrac{1}{2}(1+\chi)\begin{pmatrix} 0 & 1 & 0 \\ 1 & 0 & 0 \\ 0 & 0 & 0 \end{pmatrix}, \qquad \boldsymbol{\omega} = \tfrac{1}{2}(1-\chi)\begin{pmatrix} 0 & 1 & 0 \\ -1 & 0 & 0 \\ 0 & 0 & 0 \end{pmatrix} \tag{13}$$

and $-1 \leq \chi \leq 1$.

For ease of reference, we show in Table 1 a summary of work (both theoretical and experimental) using each of these flow types.

3. OBSERVED DROP DEFORMATIONS IN LINEAR FLOWS

Measurement of Deformation

Deformed drop shapes can be complicated, and it is convenient to have a simple scalar measure of the magnitude of the deformation. Following Taylor (1932), it is usual to define a dimensionless deformation D by

$$D = (\ell - b)/(\ell + b),$$

where ℓ and b are the largest and smallest distances of the drop surface from its center (the "major" and "minor" axes). Thus D vanishes for a sphere, and is asymptotically unity for a long slender drop.

Plane Hyperbolic Flow (References in Table 1)

At small values of Ca, drops whose initial shape is spherical deform into an ellipse whose principal axis is aligned with the x-direction. This is the steady stable shape for all values of λ. At higher rates of flow, the shape remains rounded for moderate and large values of λ (>0.2), but comparatively inviscid drops ($\lambda < 0.2$) develop a characteristic "pointed-end" shape, so that the tips appear conical. In every case the approach to equilibrium is monotonic in time. If the flow rate is increased still further so that Ca exceeds some critical value Ca_c, high-λ drops extend into a long, thin thread that breaks into a large number of small droplets. Low-λ drops are capable of extending to produce long, thin stable shapes for much higher values of Ca, but at sufficiently high flow rates they will break, either by extension into a long, thin thread (if $\mathrm{Ca} \gg \mathrm{Ca}_c$) or, if Ca is only slightly supercritical, by ejecting small droplets at the pointed ends ("tip streaming") so that the drop volume is reduced and a new equilibrium is established (with $\mathrm{Ca} < \mathrm{Ca}_c$ again, in view of the change in drop size a). It is also notable that if the flow is switched off after a drop has significantly extended

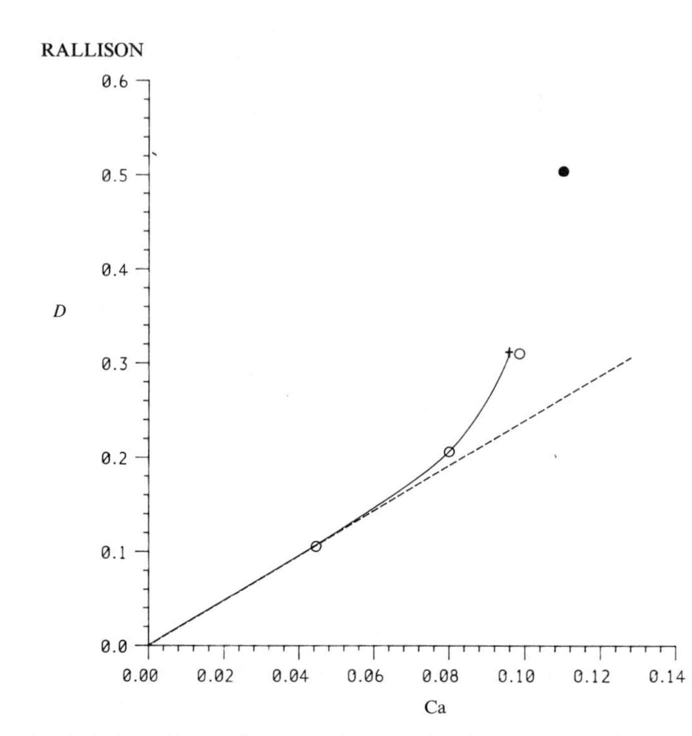

Figure 2 Variation of drop deformation D with Ca for plane hyperbolic flow $\lambda = 6$. ○ Data of Rumscheidt & Mason (1961), ● burst; - - - - $O(\text{Ca})$ theory of Taylor (1932), Equation (15); ——— $O(\text{Ca}^2)$ theory of Barthès-Biesel & Acrivos (1973a), + burst.

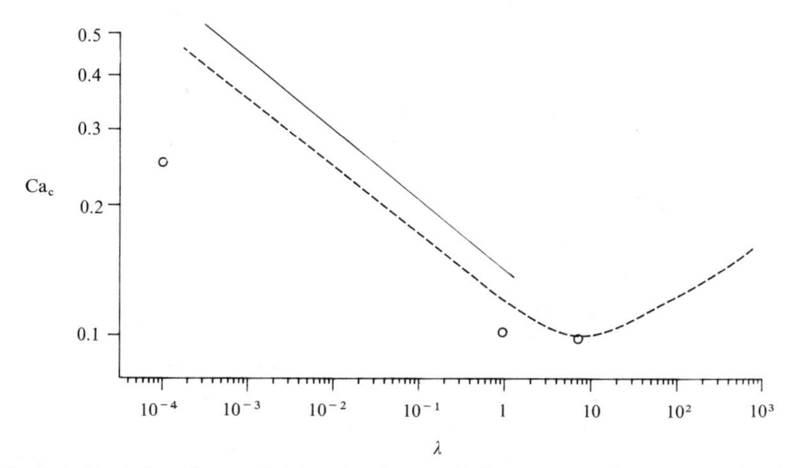

Figure 3 Variation of Ca_c with λ for plane hyperbolic flow. - - - - Experiments taken from Grace (1971); ——— asymptotic theory (for $\lambda \to 0$) of Hinch & Acrivos (1979); ○ $O(\text{Ca}^2)$ theory of Barthès-Biesel & Acrivos (1973a).

(either to an equilibrium if $Ca < Ca_c$, or to an as yet unbroken thread if $Ca > Ca_c$), the drop will break by a Rayleigh-Taylor instability into a chain of drops of uniform size, each pair being separated by three smaller satellite drops. The number of fragments is significantly fewer than when the same drop bursts under a supercritical flow.

In Figure 2 a typical data set is shown for the measured deformation D as Ca increases. Each point is obtained by a gradual (quasi-static) increase of Ca from the previous equilibrium at lower Ca. In Figure 3 is plotted the corresponding variation of Ca_c with λ. The curve has a weak minimum near $\lambda = 1$ and (may have) a plateau as $\lambda \rightarrow \infty$. In the latter case the time scale for breakup increases as λ, and so the applied flow must remain constant for very large times for the results to be useful. The largest stable sustainable distortion also appears to increase slowly with large λ, being about a 5:1 ellipse for $\lambda = 500$. On the other hand, for $\lambda \rightarrow 0$, Ca_c increases rapidly (as $\lambda^{-0.16}$ approximately).

Simple Shear Flow (*References in Table 1*)

The position here is much more complex in that the direction of maximum drop extension varies with both Ca and time. To discuss this directionality, we denote by α the angle between the principal axis of the drop and the positive x-direction. For weak flows ($Ca \ll 1$) the deformation is again elliptical and $\alpha = 45°$ (i.e. the drop is aligned with the direction of principal extension). As the flow rate increases, the equilibrium shape becomes more elongated, and, for each fixed λ, α decreases toward zero as Ca increases.

Furthermore, the transient approach to equilibrium, starting from a sphere, is no longer monotonic. For short times, α is 45° and D increases; then α falls and undershoots the equilibrium (even becoming negative) while simultaneously D overshoots; α then increases beyond the equilibrium again as D falls. This oscillation is gradually damped by surface tension until the equilibrium is attained. A typical time evolution is shown in Figure 4.

For higher flow rates there is a marked difference in behavior between high- and low-viscosity drops. For λ bigger than some critical value (about 4), the drop attains an equilibrium shape—a modestly deformed ellipse with α close to zero—(via the oscillatory wobble), and does not break *however large* Ca. For small λ, on the other hand, a long, thin, pointed-end "S" shape is achieved at moderate Ca, and at sufficiently high flow rates the drop breaks with $Ca_c \approx \lambda^{-0.55}$. In Figure 5 we show the dependence of Ca_c on λ. It may be noted by comparison with Figure 3 that simple shear is less effective in breaking drops than plane hyperbolic flow, in that higher values of Ca are needed.

For low-viscosity drops at high flow rates, the rate at which Ca is

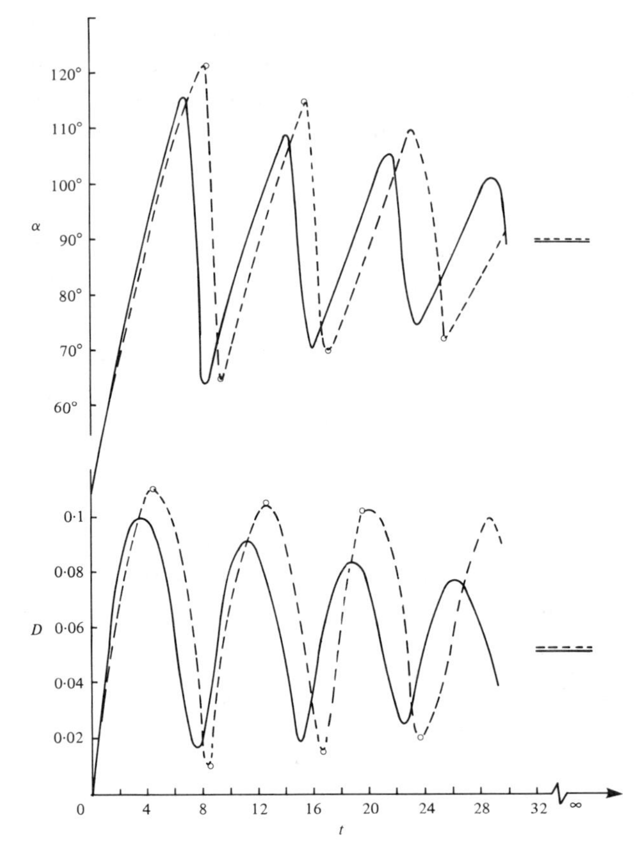

Figure 4 Time evolution of drop shape and orientation for a simple shear flow, $\lambda = 25$, $\mathrm{Ca} = 1.5$. - - - - Experimental results of Torza et al. (1972); ——— $O(\lambda^{-1})$ theory from Rallison (1980).

increased can also be important. In particular, a subcritical flow suddenly applied to the equilibrium drop shape at a lower value of Ca may lead to an equilibrium if the jump is not too large, but may cause the drop to break (by extension into a long thread) if the jump is too big. The results for $\mathrm{Ca_c}$ quoted above and in Figure 5 properly apply, therefore, only where the increase in Ca is made quasi-statically. The breakup mode also depends upon the history of the flow and on the extent by which Ca exceeds $\mathrm{Ca_c}$. If Ca just exceeds $\mathrm{Ca_c}$ and the drop has previously come to equilibrium at, or just below, $\mathrm{Ca_c}$, then a burst is obtained into two large and three small fragments; otherwise, a thread parallel to the x-direction is generated, which breaks up into many small droplets. The number of fragments on burst increases with $\mathrm{Ca/Ca_c}$ and can be as many as 10^4 for $\mathrm{Ca/Ca_c}$ of order

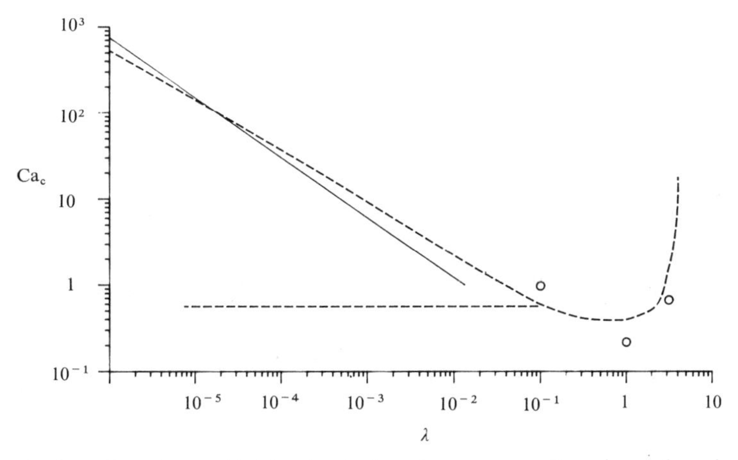

Figure 5 Variation of Ca_c with λ for simple shear flows. – – – – Experimental results from Grace (1971), horizontal-branch burst by tip streaming, main-branch burst by drop fracture; ——— asymptotic theory ($\lambda \to 0$) of Hinch & Acrivos (1980); ○ $O(Ca^2)$ theory of Barthès-Biesel & Acrivos (1973a).

20 (Grace 1971). In addition, the size distribution of the droplets is broadened by increase of Ca. Finally, for low-viscosity drops, a third burst mechanism is possible by tip streaming from the pointed ends (cf. hyperbolic flows).

Orthogonal Rheometer Flows (Hakimi & Schowalter 1980)

The observations in this case are much more limited, with only one drop viscosity (≈ 0.1) having been examined. Even for modest values of Ca (up to about 0.7) very small, steady drop deformations are produced ($D \approx 0.1$), and no critical value for Ca is found.

4. THEORETICAL CALCULATIONS OF DROP DEFORMATION

Three main lines of attack have proved fruitful in studying three-dimensional drop deformation. The first, originating with Taylor (1932, 1934), treats deformations when the distortion of the drop from sphericity is slight. The second, also suggested by Taylor (1964), uses the methods of slender-body theory to examine the case where the drop is pulled into a thin thread. And third, numerical techniques have been devised to bridge the gap. We review these major headings below, but first discuss some more specialized techniques that have proved successful.

1. *Complex variables* have been used by Richardson (1968, 1973) for the

case $\lambda = 0$, and by Buckmaster & Flaherty (1973) for the case $\lambda = 1$. Both analyses permit calculation of the full flow fields for steady drop shapes, but both suffer from the defect that the "drop" and the flow are two-dimensional. Nevertheless, two important conclusions arise from this work and carry over into later work on the three-dimensional problem, namely (*a*) that the case $\lambda = 1$ offers special mathematical simplifying features; and (*b*) there may be more than one steady drop shape for a given rate of shear, though probably only one such shape is stable.

2. Richardson (1968) and Buckmaster (1972, 1973) have studied the flow in the neighborhood of the supposed pointed end of an inviscid drop and come to the conclusion (surprising in view of the observations) that no corner of finite angle represents a possible balance of viscous and surface-tension forces, and that a cusp is therefore the only possibility. The analysis is appropriate within a tiny region [$O(\exp(-\mathrm{Ca}^6))$] near the drop tip as $\mathrm{Ca} \to \infty$, which is too small a region to be distinguishable and thus is still consistent with the appearance of pointed ends. A more recent investigation by Sherwood (1981) based on a spindle-shaped drop suggests that it is the surface tension forces near a supposed conical end that are too strong, and hence that the drop tips are not cusped but rounded. The problem of extending these analyses to explain their probable relevance to the tip-streaming phenomenon remains unsolved.

3. The surface-tension instability of a long, thin thread of fluid under extension has been examined by Tomotika (1936) and by Mikami et al. (1975). The most important qualitative conclusion from this work is that the instability is inhibited by the extension (in agreement with experiment) and the most unstable wavelength reduced by it. It is difficult to translate the result for an infinite thread into a quantitative prediction for the number of fragments produced by the burst of a finite drop.

Small Deformation Theories

At the heart of these analyses is the use of known solutions of the Stokes equations for spherical geometries due to Lamb (1932). Since for a nearly spherical drop shape the boundary conditions can be extrapolated onto a sphere, the flow field inside and outside the drop can be determined as a regular perturbation expansion, and hence the evolution of the distortion can be predicted until such time as it ceases to be small. As noted by Cox (1969), there is no reason why, within the confines of such an analysis, an arbitrary nearly spherical shape should not be chosen for the drop, but since only second harmonics are present in the undisturbed exterior flow, only second-harmonic (i.e. ellipsoidal) distortions are forced in the shape at leading order in the departure from sphericity, and higher harmonics decay under surface tension. Only at higher order in the small-deformation analysis are higher harmonics generated in the shape. Thus, one is led (Taylor 1932) to postulate an ellipsoidal drop shape at $t = 0$ and to verify that this shape is preserved at all times t. The drop surface, therefore, is taken as the near sphere

$$(\mathbf{x}\cdot\mathbf{x})^{1/2} = 1+\varepsilon\mathbf{x}\cdot\mathbf{A}(t)\cdot\mathbf{x}+O(\varepsilon^2) \quad \text{with} \quad \varepsilon \ll 1,$$

and the time evolution of its distortion is found to be

$$\varepsilon\frac{D\mathbf{A}}{Dt} \equiv \varepsilon\frac{\partial\mathbf{A}}{\partial t} - \varepsilon\,\mathrm{Ca}\,\boldsymbol{\omega}\cdot\mathbf{A}+\varepsilon\,\mathrm{Ca}\,\mathbf{A}\cdot\boldsymbol{\omega} = \frac{5\mathrm{Ca}\,\mathbf{e}}{2\lambda+3} - \frac{40(\lambda+1)\varepsilon\mathbf{A}}{(2\lambda+3)(19\lambda+16)} + O(\varepsilon\mathrm{Ca}, \varepsilon^2). \tag{14}$$

This equation, though complicated in appearance, has the simple physical interpretation that the rate of change of the distortion $\mathbf{A}$ as measured in a frame of reference that rotates with the particle (and hence, since the drop is couple-free, with the local fluid angular velocity) is due first to the effect of the ambient strain field $\mathbf{e}$ on a sphere, and second to the restoring effect of surface tension (proportional to $\mathbf{A}$). The neglected terms of order $\varepsilon\mathrm{Ca}$ arise from the straining flow acting on the perturbed shape, and $O(\varepsilon^2)$ terms from harmonics higher than the second.

Equation (14) predicts distortions that remain small for all times in two cases:

1. When the flow is weak, $\mathrm{Ca} \ll 1$, so that ultimately the distortion is limited by the strong surface tension. Thus, ε may be identified with Ca. This gives a linear approximation to the steady deformation

$$\varepsilon\mathbf{A} = \frac{19\lambda+16}{8(\lambda+1)}\mathrm{Ca}\,\mathbf{e} \quad \text{or} \quad D = \frac{19\lambda+16}{16\lambda+16}\mathrm{Ca}\,(e_{\max}-e_{\min}), \tag{15}$$

where $e_{\max}$ and $e_{\min}$ are the largest and smallest principal rates of strain. This result, which predicts *inter alia* a very weak dependence of the steady distortion on λ, is in excellent agreement with experiment for small (and even modest) Ca (see, for example, Figure 2).

2. When both the internal viscosity is high ($\lambda \gg 1$) and the flow is not too strong, $\mathrm{Ca} \leq 1$. There is then possible a balance in (14) between the rotation terms and the weak modified strain $\mathbf{e}/\lambda$, so that ε is to be identified with $1/\lambda$. Such an equilibrium is possible only for flows with moderate vorticity, and its somewhat surprising appearance is due to the fact that as the drop spins (almost as a rigid body for high λ), a fluid particle on its surface "sees" an oscillating strain field with zero mean.

At more moderate λ, on the other hand, the drop can extend further in the extensional quadrants of the flow, and in consequence rotates more slowly. For sufficiently large Ca the deformation can continue to grow and the drop bursts. There is thus a critical value for λ (about 4 for simple shear) beyond which the analysis of this section applies.

In the absence of surface tension a finite oscillation can occur (Cox 1969), but with nonzero surface tension an equilibrium is reached, with the direction of maximum drop extension "lagging" (in the sense of the rotation) the direction of maximum rate of strain. Thus for a simple shear flow, the angle α discussed in Section 3 is 45° for case 1, but close to zero here. For completeness we should note that for the high-λ case, Equation (14), though correct in its physical structure, requires amendment because the $O(\varepsilon\text{Ca})$ terms neglected are as important as the terms retained (Frankel & Acrivos 1970, Barthès-Biesel & Acrivos 1973b, Rallison 1980). Both as regards the steady and the transient behavior, these results are in gratifying qualitative agreement with the simple shear experiments discussed in Section 3. A typical quantitative comparison is shown in Figure 4. The physical explanation given above for the appearance of an equilibrium demonstrates [as may be verified by detailed mathematics (Olbricht et al. 1982)] that the vorticity inhibits the deformation only if its direction is not parallel to the principal extension axis. For in that special case (and sufficiently close to it) the drop will rotate around the direction of extension, but will continue to be extended.

Case 1 above has been pursued to $O(\text{Ca}^2)$ by Barthès-Biesel & Acrivos (1973a), using a computer to do the algebraic manipulation, and they have obtained thereby a quadratic version of (15), which may be written

$$D = \frac{c_0(\lambda)\,\text{Ca}}{1-c_1(\lambda)\,\text{Ca}} + O(\text{Ca}^3),$$

in which $c_1(\lambda)$ depends also on the flow type. Since the $O(\text{Ca}^2)$ version of Equation (14) has been shown by Barthès-Biesel & Acrivos (1973a) to have no solution for $\mathbf{A}$ beyond some critical value of Ca, it is reasonable to regard the expression above as providing an estimate $[c_1(\lambda)]^{-1}$ for Ca_c. Typical results from this theory are shown for hyperbolic and simple shear flows in Figures 3 and 6.

Large Deformation Theories

The analytical technique underlying large deformation analyses of drop deformation is slender-body theory for Stokes flow, by which the influence of the drop is represented as a distribution of singularities along its axis, rather than over its surface. Since highly extended stable drop shapes occur only for low-viscosity drops at high shear rates, these theories are appropriate asymptotically for $\lambda \to 0$ and $\text{Ca} \to \infty$.

In general, the unknowns in the problem are the position of the drop center line and the strengths and types of singularities to be distributed along it. To illustrate the technique, we consider here the simplest case,

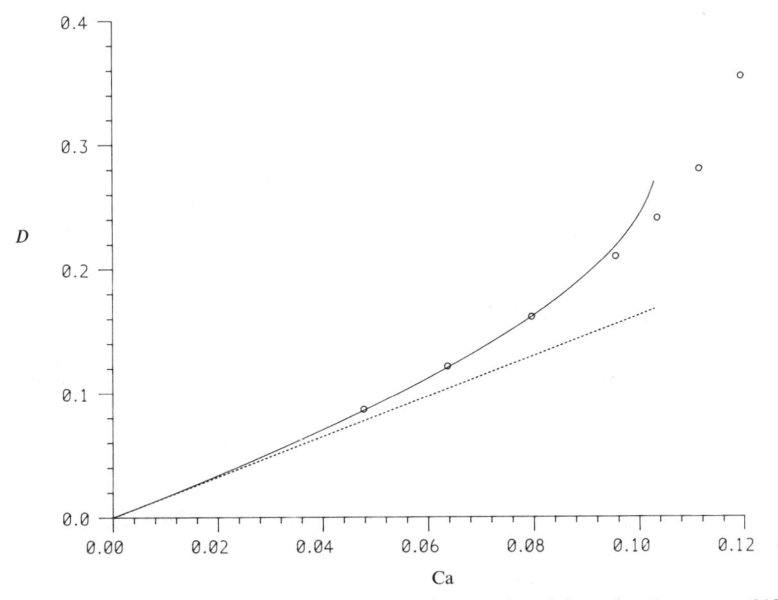

Figure 6 Variation of D with Ca for axisymmetric extensional flow. $\lambda = 1$. ---- $O(\text{Ca})$ theory of Taylor (1932); ——— $O(\text{Ca}^2)$ theory of Barthès-Biesel & Acrivos (1973a); ○ numerical computations of B. R. Duffy & J. R. Blundell (unpublished).

axisymmetric pure straining flow, where the drop cross section is circular, and the position of its axis is known a priori (Taylor 1964, Buckmaster 1972, 1973, Acrivos & Lo 1978, Hinch 1980). The drop surface is then specified in cylindrical coordinates by $r = R(z)$, $-\ell \leq z \leq \ell$.

The flow *interior* to the drop is an almost rectilinear Poiseuille flow, forced by the exterior straining motion Ca z at the surface, so that

$$u_z = \frac{1}{\lambda}\frac{\partial p}{\partial z}\frac{R^2 - r^2}{4} + \text{Ca}\, z \quad \text{for} \quad r \leq R(z),$$

where $\partial p/\partial z$ is the internal pressure gradient. As regards the *exterior* flow, the drop may be replaced by a distribution of sources $Q(z)$ along its axis, of strengths to be determined by the normal stress balance (7). Specifically, we have

$$\underbrace{-\text{Ca} + \frac{Q}{\pi R^2}}_{\text{External stresses}} = \underbrace{\frac{1}{R}}_{\text{Surface stresses}} + \underbrace{8\lambda\,\text{Ca}\int_0^z \frac{z'\,dz'}{R^2(z')} - p_0}_{\text{Internal stresses}}$$

Ambient flow + Sources = Surface tension + Poiseuille flow + Unknown constant pressure

External stresses = Surface stresses + Internal stresses

where the unknown pressure p_0 at the drop center is determined by the volume conservation requirement

$$\int_{-\ell}^{\ell} \pi R^2(z)\, dz = \tfrac{4}{3}\pi,$$

and $z = \pm\ell$ are the points at which $R(z) = 0$. The kinematic condition (6) then gives the shape evolution as

$$\frac{\partial R}{\partial t} + \mathrm{Ca}\, z \frac{\partial R}{\partial z} = \mathrm{Ca}\left[4\lambda \int_0^z \frac{z'\, dz'}{R^2(z')} - 1\right] R + \tfrac{1}{2} p_0 R - \tfrac{1}{2}. \tag{16}$$

For an inviscid drop $\lambda = 0$, an infinite set of solutions of (16) is given by

$$R = \frac{1}{2\nu\,\mathrm{Ca}}\left[1 - \left(\frac{z}{\ell}\right)^{\nu}\right] \quad \text{provided} \quad \ell = \tfrac{4}{3}(\nu+1)(2\nu+1)\,\mathrm{Ca}^2,$$

and the internal pressure $p_0 = 2\,\mathrm{Ca}\,(\nu+1)$, giving a slenderness ratio of order $\mathrm{Ca}^3 \gg 1$ for applicability of the analysis. When ν is not an even integer the shape is not analytic at $z = 0$ (Buckmaster 1972), and, as shown by Acrivos & Lo (1978) and Brady (1981), no rational matching scheme is successful near $z = 0$. Nevertheless, there remain a countable infinity of smooth steady shapes. By using the unsteady equation (16), the same authors were able to show that only $\nu = 2$ gives a *stable* solution, so that only this appears in practice. The same choice was made by Taylor (1964) without justification. Hinch (1980) has taken a nonlinear stability analysis further, and demonstrated that an initial drop shape with a significant waist can never reach a steady state, with the probable conclusion that an inviscid drop will split into as many fragments as there are more maxima than one in the initial shape.

For small nonzero λ, similar considerations arise and the unique stable steady solution is

$$R = \tfrac{1}{2}\sqrt{5}\,\frac{\lambda^{1/6}}{\alpha^{1/2}}\left[1 - \left(\frac{z\lambda^{1/3}}{\alpha}\right)^2\right], \qquad \ell = \alpha/\lambda^{1/3},$$

in which α is defined implicitly by

$$\mathrm{Ca}\,\lambda^{1/6} = \frac{1}{\sqrt{20}}\,\frac{\alpha^{1/2}}{1 + 4/3\alpha^3}.$$

There is no solution of this equation for α when $\mathrm{Ca} > \mathrm{Ca}_c = 0.148\lambda^{-1/6}$, and hence the drop must burst.

Examination of the equations above shows that the burst mechanism is as follows:

1. As the external shear rate increases, fluid flow along the drop surface is more rapid.
2. A consequently larger internal pressure gradient must be built up to push internal fluid back from the tip to the center of the drop.
3. The surface-tension force must therefore rise near the ends, so that the radius must fall.
4. To conserve drop volume the length must increase, and so the drop extends into regions of even more rapid flow. If $\lambda = 0$, no internal pressure gradient is needed, and the drop does not break.

For other types of shear flow, analogous calculations are possible but with the additional complications that the cross-sectional shape is no longer circular and the center line is not straight. Hinch & Acrivos (1979) have discussed plane hyperbolic flows and found an elliptical cross section and a breakup criterion

$$\mathrm{Ca_c} = 0.145\lambda^{-1/6},$$

which, as may be seen from Figure 3, is in excellent agreement, at any rate as far as the exponent of λ is concerned, with experiment.

The closeness of this criterion to the one above for axisymmetric extension is remarkable, and reflects the fact that only the drop cross-sectional *area* and not its shape affects the breakup criterion. This suggests the approximation of using a circular cross section for more complex flows.

For simple shear (Hinch & Acrivos 1980), even with the circle approximation, the complications are even greater in that the center line is found to be "S"-shaped and source doublets as well as sources are needed to generate the external flow. The analysis gives $\mathrm{Ca_c} = 0.054\ \lambda^{-2/3}$, again in fair agreement with experiment (Figure 5). In addition, the results of a time-dependent simulation show that the drop tip spirals toward its equilibrium position as seen in experiment, and it is also found that a subcritical flow rate Ca can nevertheless cause drop breakup if it is imposed too rapidly (cf. the observations of Section 3).

We note in passing that for these almost rectilinear flows past slender drops, calculations have been made for the effects of small and moderate amounts of inertia in either the inner or the outer fluid (Acrivos & Lo 1978, Brady & Acrivos 1982). In particular, values of $\mathrm{Ca_c}$ have been computed. This work falls outside the scope of this review; the interested reader is referred to the paper by Brady & Acrivos (1982) and references therein.

Numerical Theories for Intermediate Deformations

For deformations intermediate between near-spheres and slender bodies, numerical methods have been used. In order to examine the full range of physically realized behaviors without encountering the difficulties of nonuniqueness of the steady drop shape mentioned in the previous section, most authors have solved time-dependent problems where the shape evolves in time from a specified initial condition under prescribed flow until either an equilibrium or a burst is reached.

Rather than solve for the fluid velocity at all points in space, most studies, following Youngren & Acrivos (1976), have used a boundary-integral method by which the Stokes equations inside and outside the drop (3) are cast into an integral form that involves only quantities evaluated on the drop surface (see Ladyzhenskaya 1969). It may then be shown (Rallison & Acrivos 1978) that at a point $\mathbf{x}$ of the drop surface S at time t the fluid velocity is given by

$$\tfrac{1}{2}(1+\lambda)u_i(\mathbf{x}) - \frac{3}{4\pi}(1-\lambda)\int_{S_y} \frac{r_i r_j r_k}{r^5}\, u_j(\mathbf{y})\mathbf{n}_k(\mathbf{y})\, dS_y$$

$$= \mathrm{Ca}\,(e+\omega)_{ij}x_j - \frac{1}{8\pi}\int_{S_y}\left(\frac{\delta_{ij}}{r}+\frac{r_i r_j}{r^3}\right)n_j(y)\kappa(y)\, dS_y \quad (17)$$

in which $\mathbf{r} = \mathbf{x}-\mathbf{y}$. Hence, if S is represented by a number of collocation points $\mathbf{x}_1, \ldots, \mathbf{x}_N$, $\mathbf{u}(\mathbf{x}_i)$ may be determined by matrix inversion at any instant of time, and thus $S(t+\Delta t)$ can be computed. Two complications arise in such a computation. First, there are eigensolutions of (17) when $\lambda = 0$ or $\lambda = \infty$ (Ladyzhenskaya 1969), and so the matrix to be inverted will be singular. Fortunately, away from the extremes the inversion is straightforward.. Second, the kernel functions on both the left- and right-hand sides of (17) involve integrable singularities of order $1/r$ as $r \to 0$; these can be removed for computational purposes and dealt with analytically (Youngren & Acrivos 1976).

For acceptable accuracy in representing a general drop shape, N must be large (100, say), and hence for general λ a (100×100) matrix has to be inverted at each time step to solve (17). For an axisymmetric shape, however, the drop surface is specified by a curve, and a more modest value of N (20, say) is possible. Rallison & Acrivos (1978) have produced results for D and Ca_c for an axisymmetric straining flow for general λ, which show good agreement with the linear result of Taylor (1932), even for modest Ca. These results have since been superseded (for $\lambda = 1$) by numerical work of higher accuracy by B. R. Duffy & J. R. Blundell (unpublished), which agrees well with the quadratic theory of Barthès-Biesel & Acrivos (1973a) over a larger range of Ca (see Figure 6).

A second case where a marked numerical simplification is possible is when $\lambda = 1$. Equation (17) then becomes

$$u_i(x) = \mathrm{Ca}\,(e+\omega)_{ij}x_j - \frac{1}{8\pi}\int_S\left(\frac{\delta_{ij}}{r}+\frac{r_i r_j}{r^3}\right)n_j\kappa\, dS, \quad (18)$$

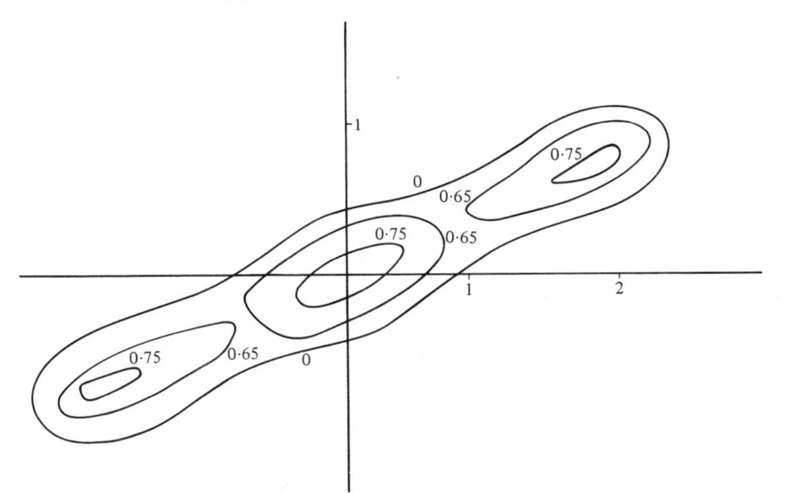

Figure 7 Incipient drop burst in simple shear; contours of constant elevation of drop surface. $\lambda = 1$, Ca just supercritical (from numerical solution by Rallison 1981).

which is, in fact, valid at all points of space, not just on S. In this case, no matrix inversion is needed to determine $\mathbf{u}$, and so nonaxisymmetric shapes can be handled with moderate amounts of computer time and storage. Rallison (1981) has exploited this simplification to study a range of nonaxisymmetric flows for $\lambda = 1$, in particular simple shear, and plane flows with varying strengths of vorticity. The principal conclusions of the study regarding drop deformation are given in the following section. For simple shear, the two burst modes mentioned in Section 3 as reported by Torza et al. (1972) were found; in Figure 7 we show the drop shape just before it bursts into two large and (at least) one small fragment for a shear rate that is just supercritical.

5. SUMMARY

With the exception of the tip-streaming phenomenon, our understanding of low-Reynolds-number drop deformation and burst in unbounded shear flows now seems fairly complete from both a theoretical and an experimental point of view. The gross features may be summarized as follows:

1. Drop deformation and burst is promoted primarily by the straining motion in the external shear. It is inhibited by the vorticity in the outer flow.
2. For globular drop shapes, surface tension acts as a restoring force resisting burst, but if the drop becomes elongated, then surface tension may promote burst.

3. Low-viscosity drops can attain highly extended stable shapes and require very strong flows to break them.
4. High-viscosity drops are pulled part by modest extensional flows (though they take a long time to break), but cannot be broken by flows with significant vorticity, however strong.
5. The history of the applied flow can be important in determining whether or not a drop breaks: rapid changes in flow strength can provoke subcritical bursts. The size and number of satellite drops produced by a burst depend on both the flow type and its history.

6. RELATED AREAS OF STUDY

Finally, we mention briefly (and incompletely) other areas where the theoretical techniques reviewed in earlier sections are being or could be applied. There are, of course, developments possible where inertia is included, but the low-Reynolds-number assumption lies at the heart of most of the work described previously, and we mention only extensions of those methods here.

First, the constitutive behavior of the interface between the two fluids can be changed so as to endow it with some (visco-) elastic response (e.g. in consequence of the presence of surfactants; Levich 1962, Rumscheidt & Mason 1961). Barthès-Biesel (1980) has coined the term "capsule" to describe a drop of fluid enclosed by an elastic membrane and immersed in a second fluid. The applications in mind here are to the deformation and burst of red blood cells (see, for example, Goldsmith & Skalak 1975), and to emulsions stabilized by interfacial cross-linking polymerization (Brédimas et al. 1983). The central theoretical difference between an elastic membrane and a surface-tension interface is that particle motions within the surface S generate elastic stresses, and so both the dynamic (5) and the kinematic (6) boundary conditions have to be modified. Small deformation analyses for capsules have been performed by Barthès-Biesel (1980), Barthès-Biesel & Rallison (1981), and by Brunn (1980, 1983), and numerical work for stronger flows has been performed by Hasan (1983).

A second natural extension of the work described here is to include the effect of solid boundaries on the drop (at any rate where the fluid in the drop does not wet the solid boundary so that contact-line problems do not arise). The practical problem of interest here is that of oil recovery from porous rocks, where oil drops suspended in water are pushed through pores of varying cross section (see Payatakes 1982). Similar low-Reynolds-number, free-surface problems arise too where film draining or drying is involved.

Third, and as yet largely unexplored, are cases where more than one drop is present in the flow (Bartok & Mason 1959). Drop coalescence promoted by shear is an important phenomenon in emulsion formation and complements the shear-breakup problem reviewed here; but, perhaps because of the additional complications of interparticle forces, this problem has, as yet, received much less attention.

Literature Cited

Acrivos, A., Lo, T. S. 1978. Deformation and breakup of a single slender drop in an extensional flow. *J. Fluid Mech.* 86:641–72

Barthès-Biesel, D. 1980. Motion of a spherical microcapsule freely suspended in a linear shear flow. *J. Fluid Mech.* 100:831–53

Barthès-Biesel, D., Acrivos, A. 1973a. Deformation and burst of a liquid droplet freely suspended in a linear shear field, *J. Fluid Mech.* 61:1–21

Barthès-Biesel, D., Acrivos, A. 1973b. The rheology of suspensions and its relation to phenomenological theories for non-Newtonian fluids. *Int. J. Multiphase Flow* 1:1–24

Barthès-Biesel, D., Rallison, J. M. 1981. The time-dependent deformation of a capsule freely suspended in a linear shear flow. *J. Fluid Mech.* 113:251–67

Bartok, W., Mason, S. G. 1959. Particle motions in sheared suspensions. VIII. Singlets and doublets of fluid spheres. *J. Colloid Sci.* 14:13–26

Batchelor, G. K. 1967. *An Introduction to Fluid Dynamics.* Cambridge: Cambridge Univ. Press

Brady, J. F. 1981. *Inertial effects in closed cavity flows and their influence in drop breakup.* PhD dissertation. Stanford Univ.

Brady, J. F., Acrivos, A. 1982. The deformation and breakup of a slender drop in an extensional flow. *J. Fluid Mech.* 115:443–51

Brédimas, M., Veyssie, M., Barthès-Biesel, D., Chhim, V. 1983. Model suspension of spherical capsules: physical and rheological properties. *J. Colloid Interface Sci.* In press

Brunn, P. O. 1980. On the rheology of viscous drops surrounded by an elastic shell. *Biorheology* 17:419–30

Brunn, P. O. 1983. The deformation of a viscous particle surrounded by an elastic shell in a general time-dependent flow field. *J. Fluid Mech.* 126:533–44

Buckmaster, J. D. 1972. Pointed bubbles in slow viscous flow. *J. Fluid Mech.* 55:385–400

Buckmaster, J. D. 1973. The bursting of pointed drops in slow viscous flow. *J. Appl. Mech.* 40:18–24

Buckmaster, J. D., Flaherty, J. E. 1973. The bursting of two-dimensional drops in slow viscous flow. *J. Fluid Mech.* 60:625–39

Cox, R. G. 1969. The deformation of a drop in a general time-dependent fluid flow. *J. Fluid Mech.* 37:601–23

Frankel, N. A., Acrivos, A. 1970. The constitutive equation for a dilute emulsion. *J. Fluid Mech.* 44:65–78

Goldsmith, H. L., Skalak, R. 1975. Hemodynamics. *Ann. Rev. Fluid Mech.* 7:213–47

Grace, H. P. 1971. Dispersion phenomena in high viscosity immiscible fluid systems and application of static mixers as dispersion devices in such systems. *Eng. Found., Res. Conf. Mixing, 3rd, Andover, N.H.* Republished 1982 in *Chem. Eng. Commun.* 14:225–77

Hadamard, J. S. 1911. Mouvement permanent lent d'une sphère liquide dans un liquide visqueux. *C.R. Acad. Sci. Paris* 152:1735–38; 154:109

Hakimi, F. S., Schowalter, W. R. 1980. The effects of shear and vorticity on deformation of a drop. *J. Fluid Mech.* 98:635–45

Hasan, A. 1983. *Modélisations analytique et numérique du mouvement et de la déformation d'une capsule en suspension libre dans un écoulement. Application au globule rouge humain.* Doct. Ing. dissertation. Univ. Technol. Compiègne

Hinch, E. J. 1980. The evolution of slender inviscid drops in an axisymmetric straining flow. *J. Fluid Mech.* 101:545–53

Hinch, E. J., Acrivos, A. 1979. Steady long slender droplets in two-dimensional straining motion. *J. Fluid Mech.* 91:401–14

Hinch, E. J., Acrivos, A. 1980. Long slender drops in a simple shear flow. *J. Fluid Mech.* 98:305–28

Ladyzhenskaya, O. 1969. *The Mathematical Theory of Viscous Incompressible Flow.* New York: Gordon & Breach. 2nd ed.

Lamb, H. 1932. *Hydrodynamics.* 6th ed. Cambridge Univ. Press

Levich, V. G. 1962. *Physicochemical Hydrodynamics.* Englewood Cliffs, N.J.: Prentice Hall

Mikami, T., Cox, R. G., Mason, S. G. 1975. Breakup of extending liquid threads. *Int. J. Multiphase Flow* 2:113–38

Olbricht, W. L., Rallison, J. M., Leal, L. G.

1982. Strong flow criteria based on microstructure deformation. *J. Non-Newtonian Fluid Mech.* 10:291–318

Payatakes, A. C. 1982. Dynamics of oil ganglia during immiscible displacement in water-wet porous media. *Ann. Rev. Fluid Mech.* 14:365–93

Rallison, J. M. 1980. Note on the time-dependent deformation of a viscous drop which is almost spherical. *J. Fluid Mech.* 98:625–33

Rallison, J. M. 1981. A numerical study of the deformation and burst of a viscous drop in general shear flows. *J. Fluid Mech.* 109:465–82

Rallison, J. M., Acrivos, A. 1978. A numerical study of the deformation and burst of a viscous drop in an extensional flow. *J. Fluid Mech.* 89:191–200

Richardson, S. 1968. Two-dimensional bubbles in slow viscous flows. *J. Fluid Mech.* 33:475–93

Richardson, S. 1973. Two-dimensional bubbles in slow viscous flows. Part 2. *J. Fluid Mech.* 58:115–27

Rumscheidt, F. D., Mason, S. G. 1961. Particle motions in sheared suspensions. XII. Deformation and burst of fluid drops in shear and hyperbolic flow. *J. Colloid Sci.* 16:238–61

Rybczyński, W. 1911. Über die fortschreitende Bewegung einer flüssigen Kugel in einem zähen Medium. *Bull. Int. Acad. Sci. Cracovie Cl. Sci. Math. Nat. Ser. A* 1911:40–46

Sherwood, J. D. 1981. Spindle-shaped drops in a viscous extensional flow. *Math. Proc. Cambridge Philos. Soc.* 90:529–36

Taylor, G. I. 1932. The viscosity of a fluid containing small drops of another fluid. *Proc. R. Soc. A* 138:41–48

Taylor, G. I. 1934. The formation of emulsions in definable fields of flow. *Proc. R. Soc. A* 146:501–23

Taylor, G. I. 1964. Conical free surfaces and fluid interfaces. *Proc. Int. Congr. Appl. Mech., 11th, Munich*, pp. 790–96

Torza, S., Cox, R. G., Mason, S. G. 1972. Particle motions in sheared suspensions. XXVII. Transient and steady deformation and burst of liquid drops. *J. Colloid Interface Sci.* 38:395–411

Tomotika, S. 1936. Breaking up of a drop of viscous liquid immersed in another viscous fluid which is extending at a uniform rate. *Proc. R. Soc. A* 153:302–18

Wegener, P. P., Parlange, J.-Y. 1973. Spherical-cap bubbles. *Ann. Rev. Fluid Mech.* 5:79–100

Youngren, G. K., Acrivos, A. 1976. On the shape of a gas bubble in a viscous extensional flow. *J. Fluid Mech.* 76:433–42

Ann. Rev. Fluid Mech. 1984. 16: 67–97

NUMERICAL SOLUTION OF THE NONLINEAR BOLTZMANN EQUATION FOR NONEQUILIBRIUM GAS FLOW PROBLEMS

S. M. Yen

Aeronautical and Astronautical Engineering Department
and Coordinated Science Laboratory, University of Illinois,
Urbana, Illinois 61801

1. INTRODUCTION

This review concerns the numerical solution of the nonlinear Boltzmann equation for a gas flow under conditions far from thermal equilibrium. The rarefied-gas flow problem, which is characterized by a large global parameter, the Knudsen number, is often thought to be the only non-equilibrium problem. An appropriate measure of the local departure from equilibrium is the local Knudsen number, which may be defined in terms of the local property gradient. Nonequilibrium conditions characterized by large property gradients do occur in certain regions in continuum-flow problems: a shock wave is a familiar example. Since equilibrium conditions exist in the upstream and downstream regions of the shock wave and since the relations between the upstream and downstream properties are known, the internal shock structure is not needed for the solution of such continuum-flow problems. Another example, which is less familiar, is the Knudsen layer next to an evaporation or a condensation interface. In contrast to the shock wave, the condition of the vapor at the interface is far from equilibrium. Neither this nonequilibrium condition nor its relation with the downstream equilibrium condition is known. This Knudsen-layer problem, therefore, cannot be treated by using a continuum approach, even though the flow characteristics in this layer may not be of interest. One of

0066-4189/84/0115-0067$02.00

the objectives of this paper is to show that the treatment and solution of nonequilibrium problems are of interest to the general fluid-mechanics community in solving not only rarefied-gas flow problems, but also certain continuum-flow problems.

The concept of transport coefficients is no longer valid in the formulation of a nonequilibrium flow problem; therefore, such a problem requires kinetic-theory treatment, which consists of solving the Boltzmann equation. The Boltzmann equation is an integro-differential equation that contains the collision integral, the microscopic transport property basic in determining the macroscopic nonequilibrium characteristics. The collision integral has often been considered intractable, thus making the solution of the Boltzmann equation impossible. The advent of digital computers has spurred interest in the numerical solution of many nonlinear differential equations in fluid mechanics; the study of the numerical solution of the nonlinear Boltzmann equation started almost as soon as the first modern digital computer became available. Numerical methods have been developed to solve the Boltzmann equation, and solutions have been obtained for several basic one-dimensional problems, as well as more-complex problems.

The distinctive feature of numerical methods for solving the Boltzmann equation is to embed in a method of integration a Monte Carlo method for evaluating the collision integral. The numerical method consists of two explicit stages: the evaluation of the collision integral, and the integration of the differential equation. In the Monte Carlo stage of calculation, there are two basic requirements: simulating the intermolecular collision phenomenon according to the Boltzmann formulation, and controlling the statistical as well as systematic errors in order to obtain fidelity in the calculation. The strategy in performing the integration and in choosing an appropriate integration scheme is the same as that for the governing equations of continuum gas flows. The consideration of velocity space in the kinetic-theory treatment requires more computation time and storage. However, significant advances have been made in increasing the computation efficiency such that Boltzmann solutions can be obtained using currently available computers.

Direct-simulation techniques have been developed to obtain the solution of the Boltzmann equation. The basic difference between the direct-simulation technique and the numerical method to solve directly the Boltzmann equation lies in the Monte Carlo stage of calculation. The validity of the direct-simulation technique depends on the degree of agreement between the Monte Carlo simulation model and the Boltzmann formulation.

We review in this paper the formulation of nonequilibrium flow problems using the kinetic-theory approach, the development of numerical methods to solve the nonlinear Boltzmann equation, and the significance of numerical solutions obtained for several basic one-dimensional problems. In the last section, we make several summary remarks.

2. KINETIC-THEORY TREATMENT

For a gas without internal degrees of freedom and in the absence of any external force, the Boltzmann equation may be written in the form

$$\frac{\partial f}{\partial t} + \mathbf{v}\frac{\partial f}{\partial \mathbf{r}} = (a - bf), \tag{1}$$

where

f = distribution function,
$(a-bf)$ = collision integral,
a = gain term of the collision integral,
= rate of scattering of molecules into $d\mathbf{v}$.
bf = loss term of the collision integral,
= rate of scattering of molecules out of $d\mathbf{v}$.

Direct numerical solution of the Boltzmann equation involves the calculation of the distribution function f as well as the collision integral $(a - bf)$ at each velocity point in a quantized velocity space. This must be done for each of the selected positions in physical space using an intermolecular collision model and a gas-surface interaction model appropriate for a given problem. From the moments of the distribution function and the collision integral, we may calculate the macroscopic properties of interest. The moments of the distribution function and the collision integral are

$$\mathscr{M}[\Phi(\mathbf{v})] = \int f\Phi(\mathbf{v})\,d\mathbf{v}, \tag{2}$$

$$I[\Phi(\mathbf{v})] = \int (a-bf)\Phi(\mathbf{v})\,d\mathbf{v}, \tag{3}$$

in which $\Phi(\mathbf{v})$ is any function of velocity. It would be of interest to study the moments and their functions that show distinct nonequilibrium behavior.

We discuss here the method of computation of flow properties of common interest for one-dimensional flow problems. For one-dimensional steady flows, the Boltzmann equation and the Boltzmann transport equation for any function of velocity Φ_k become, respectively,

$$v_x\frac{df}{dx} = (a-bf), \tag{4}$$

and

$$\frac{d}{dx}(\mathscr{M}_k) = I(\Phi_k/v_x), \tag{5}$$

in which

$$\mathscr{M}_k = \int f\Phi_k \, d\mathbf{v}, \tag{6}$$

and

$$I(\Phi_k/v_x) = \int \Phi_k(a - bf) \, d\mathbf{v}/v_x. \tag{7}$$

The ordinary macroscopic properties of a nonequilibrium gas can be determined from six moments of the velocity distribution function f (Hicks et al. 1972). The six moments are $\mathscr{M}_1$, $\mathscr{M}_2$, $\mathscr{M}_3$, $\mathscr{M}_4$, $\mathscr{M}_6$, and $\mathscr{M}_9$; the corresponding functions of velocity are $\Phi_1 = 1$, $\Phi_2 = v_x$, $\Phi_3 = v_x^2$, $\Phi_4 = v_x v^2$, $\Phi_6 = v_x^3$, and $\Phi_9 = v_\perp^2$. The moments $\mathscr{M}_2$, $\mathscr{M}_3$, and $\mathscr{M}_4$ are invariants, while $\mathscr{M}_1$, $\mathscr{M}_6$, and $\mathscr{M}_9$ are noninvariants. The gradients of $\mathscr{M}_1$, $\mathscr{M}_6$, and $\mathscr{M}_9$ are the moments of the collision integrals $I(1/v_x)$, $I(v_x^2)$, and $I(v_\perp^2/v_x)$. The moments of both the distribution function and the collision integral are of interest in studying the nonequilibrium characteristics in a gas flow.

3. NUMERICAL METHODS

There are two aspects in the development of a numerical method for solving the nonlinear Boltzmann equation: (*a*) simulating accurately the intermolecular collision phenomenon, and (*b*) integrating numerically the differential equation. The numerical methods have two explicit steps: (*a*) evaluating the Boltzmann collision integral using a Monte Carlo method, and (*b*) integrating the Boltzmann equation using a numerical scheme under specified boundary conditions. In the direct-simulation technique, the Boltzmann equation is not solved directly; however, it also has two explicit steps in dealing with, respectively, the intermolecular collisions and the motion of the simulated particles. If the simulation of the collision phenomenon is consistent with the Boltzmann formulation, then the technique can also yield the Boltzmann solution.

We divide this review of numerical methods into five sections. In the first section, the Monte Carlo methods for evaluating the Boltzmann collision integral are briefly discussed. The embedding of the Monte Carlo calculation in a numerical scheme for integrating the Boltzmann equation for one-dimensional problems is presented in the second section. The Monte Carlo direct-simulation techniques are briefly reviewed in the third section. The final two sections deal with methods for obtaining Boltzmann solutions for multidimensional and continuum-flow problems, respectively.

3.1 *Evaluation of the Boltzmann Collision Integral*

In 1955, Nordsieck (Nordsieck & Hicks 1967) devised a Monte Carlo method of evaluating the collision integral. A similar method was developed by Cheremisin (1969, 1970). A method based on polynomial approximation of the collision integral was used by Lima (1973). We describe briefly here Nordsieck's method because the details of this method and its developments are more widely known.

Nordsieck's method is a statistical sampling technique closely resembling the statistical collision phenomenon in the gas. In this method, the collision integral is replaced by an integral over a finite region of velocity space. The finite region is taken large enough so that it includes most of the molecules, say 99%. The average of the integrand over this volume and over all values of the line-of-centers vector is then approximated by the average of a large and fair sample of N values of the integrand. A Monte Carlo estimate of the value of the collision integral is given by the product of this average value with the volume.

Nordsieck's method was first applied to flows possessing axial symmetry, with the velocity vector represented by the cylindrical components—v_x and $v_\perp$—and an azimuth angle. For these flows, the velocity distribution function does not depend on the azimuth angle. The corresponding velocity space $(v_x, v_\perp)$ is quantized by covering the semicircular region with 226 fixed cells, as shown in Figure 1. Sixty-four cells are included in each sweep of azimuthal angle from 0 to 2π. The sampling algorithm is devised so that random numbers are generated to produce successive and independent collisions.

During a period of ten years, Nordsieck's method was systematically tested and refined (Hicks & Smith 1968, Hicks et al. 1969). In 1964, by which time the dominant error in computing the collision integral had been reduced to 2%, the method was applied to the solution of the Boltzmann equation. By 1967, this error was further reduced to less than 1%. Several techniques were used to reduce the systematic as well as random errors. It is of interest to review here some of the major techniques.

The quadrature error produces bias in the computed values of the two parts of the collision integral. A least-squares technique is used to make corrections to the 226 values of the collision integral to insure that the conservation equations are satisfied.

A technique found to be successful in increasing the accuracy of the calculation under a near-equilibrium condition amounts to correcting the calculated value of both parts of the collision integral in such a way that the collision integral would equal zero for a gas in equilibrium at the same values of density, temperature, and gas velocity.

The fairness of sampling used in Nordsieck's method was ascertained on the basis of comparative studies of the number of hits per velocity cell. The random errors of the calculations of a, bf, $a-bf$, and their moments were studied separately. In addition, the results of $a-bf$ and their moments were compared with certain analytical calculations and were found to be in good agreement (Hicks & Yen 1971, Yen & Ng 1974).

Nordsieck's method has also been applied to a three-dimensional velocity space and to binary gas mixtures for which two separate velocity spaces are used, one for each gas (Yen et al. 1974).

Computed functions of a, bf, and $a-bf$ can be displayed graphically in order to study their characteristics in the velocity space (Hicks & Yen 1971, Hicks et al. 1972, Yen & Tcheremissine 1981). These characteristics would be of interest in studying distinct nonequilibrium behavior in a gas flow. The graphical display would also be useful in studying the differences between the Boltzmann collision integrals and several proposed approximations.

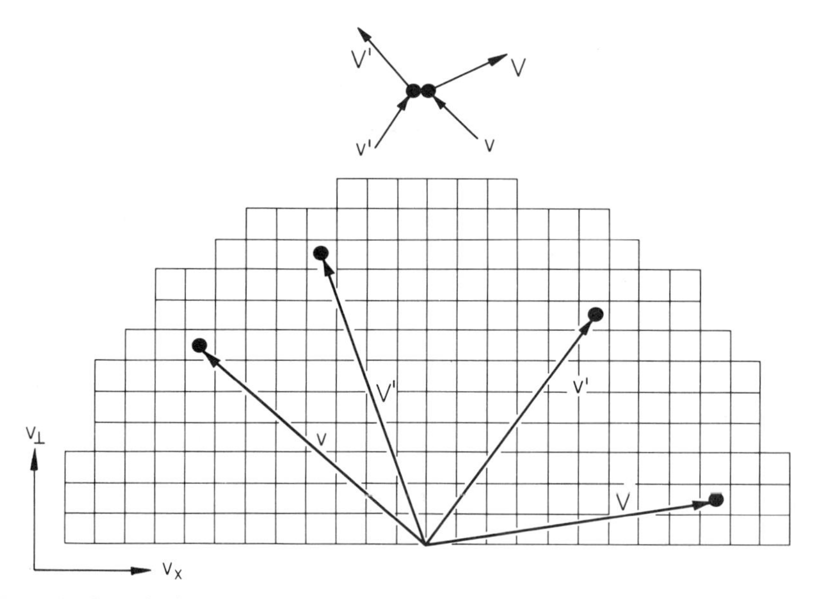

Figure 1 Quantized $(v_x, v_\perp)$ velocity space showing 226 cells; **v** and **v**′ are the velocities before collision, and **V** and **V**′ are the corresponding velocities after collision.

3.2 *Integration of the Boltzmann Equation*

In order to simplify the presentation and discussion of the integration scheme, we consider the one-dimensional Boltzmann equation

$$\frac{\partial f}{\partial t} + v_x \frac{\partial f}{\partial x} = (a - bf). \tag{8}$$

If both a and b are known, we can apply the same computation strategies and numerical schemes as those that have been developed and used to integrate the governing equations for continuum gas-dynamics problems.

There are two strategies in integrating the Boltzmann equation to obtain the steady-state solution: iterative and time dependent. Both strategies were considered by Nordsieck. He chose the iterative scheme because he felt that its computational difficulties were easier to deal with. We discuss first Nordsieck's iterative scheme (Nordsieck & Hicks 1967). The computational mesh for one-dimensional flow problems is shown schematically in Figure 2. The mesh system consists of a velocity space placed at each chosen spatial position. In this figure, three velocity spaces are shown at three positions with $\hat{x}$ = normalized spatial variable = 0, 0.5, and 1. The velocity space at each value of $\hat{x}$ is divided into two quadrants. The quadrant for which

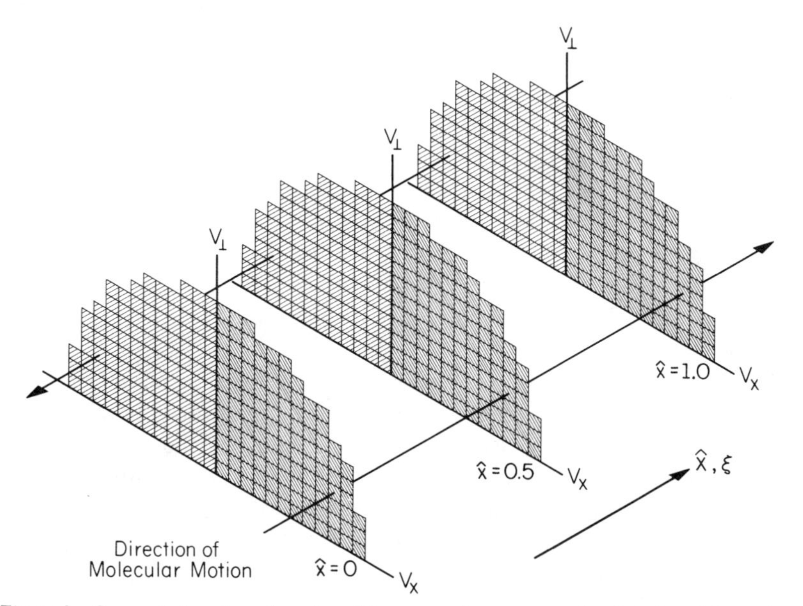

Figure 2 Computational mesh system. Three velocity spaces at $\hat{x} = 0$, 0.5, and 1 are shown. Integration of the Boltzmann equation is performed for the velocity point in each cell in the direction of the molecular motion.

$v_x > 0$ is for molecules moving in the positive direction, and that for which $v_x < 0$ is for molecules moving in the negative direction. Each quadrant contains 113 velocity points. One collision integral is associated with each velocity point; therefore, this entire set of integrals must be evaluated at each value of $\hat{x}$. The 113 Boltzmann equations for $v_x > 0$ are integrated starting at $\hat{x} = 0$ and moving to $\hat{x} = 1$, and those for $v_x < 0$ are integrated starting at $\hat{x} = 1$ and moving to $\hat{x} = 0$.

Nordsieck's iterative scheme is explicit and is based on central differencing for the spatial derivative. For molecules with $v_x > 0$, the Boltzmann equation is written as

$$v_x \left[\frac{\partial f^I}{\partial x}\right]_{i+1/2} = a_{i+1/2}^{I-1} - (bf)_{i+1/2}^{I-1}, \tag{9}$$

in which I denotes the number of iterations, and i the spatial position. The terms in Equation (9) are evaluated as follows:

$$\left[\frac{\partial f^I}{\partial x}\right]_{i+1/2} = \frac{f_{i+1}^I - f_i^I}{\Delta x}, \tag{10}$$

$$a_{i+1/2}^{I-1} = (a_{i+1}^{I-1} + a_i^{I-1})/2, \tag{11}$$

$$(bf)_{i+1/2}^{I-1} = [(bf)_{i+1}^{I-1} + (bf)_i^{I-1}]/2. \tag{12}$$

Expressions similar to Equations (10)–(12) are written for molecules with $v_x < 0$.

The iterative scheme is initiated by assuming a velocity distribution function at each x-position. The collision integrals are evaluated for the assumed distribution function. The integration for the Boltzmann equations is performed separately for molecules with $v_x > 0$ and for those with $v_x < 0$. For molecules with $v_x > 0$, the integration starts at $\hat{x} = 0$ with a specified boundary condition there. For molecules with $v_x < 0$, the integration starts at $\hat{x} = 1$ with a specified boundary condition at this position.

The convergence of the iterative process is monitored by plotting the rms difference of successive iterates, which is a measure of the residual δf. The uniqueness of the solution is ascertained by studying the rms difference between solutions with different zeroth iterates of the distribution functions. Detailed studies were made on the effects of errors due to (*a*) small discrepancies in matching the equilibrium distribution functions at the boundaries for the shock-wave problem (boundary error), (*b*) the use of an explicit instead of an implicit scheme, and (*c*) the truncation error. These studies have led to several improvements in the scheme.

Since the most time-consuming part of the computation in the numerical solution is the evaluation of collision integrals, the overall computational efficiency may be increased significantly if the number of spatial nodes can be reduced. As is discussed below, such a reduction is possible for certain numerical methods.

Aristov & Cheremisin (1976, 1980) and Cheremisin (1977) chose the time-dependent approach and proposed a splitting method. The integration is performed in two explicit steps corresponding to two successive stages of motion: a spatially uniform relaxation and a collisionless free motion. We rewrite the one-dimensional time-dependent Boltzmann equation in the following form to facilitate the presentation of their numerical schemes:

$$\frac{\partial f}{\partial t} + v_x \left[\frac{\partial f}{\partial x} \right]_i^j = (a - bf)_i^j, \tag{13}$$

in which i and j denote the indexes in the x- and t-directions, respectively. The solution of Equation (13) is replaced by a sequence of two solutions defined by the following two equations:

$$\frac{\partial f}{\partial t} = (a - bf)_i^j, \tag{14}$$

$$\frac{\partial f}{\partial t} + v_x \left(\frac{\partial f}{\partial x} \right)_i^j = 0. \tag{15}$$

The solution of Equation (14) provides the initial condition for Equation (15) at each node x_i, and the solution of Equation (15) provides the initial condition for Equation (14) at the next time step.

An advantage of using the splitting method to construct conservative difference schemes is that the conservation conditions, which should be satisfied exactly in the relaxation stage, can be imposed at each node x_i. A correction method on the distribution function is used to insure conservation. Let f_i be the solution of the relaxation problem. Then the corrected solution $\tilde{f}_i$ is

$$\tilde{f}_i = f_i[1 + P(\mathbf{v})], \tag{16}$$

in which $P(\mathbf{v})$ is a polynomial in velocity. The coefficients of the polynomial are determined to insure that the invariant moments are conserved before and after the relaxation.

The integration scheme used by Aristov & Cheremisin has three steps. The first is to find the solution $\hat{f}_i^j$ of Equation (14) using one of the following methods:

Explicit:

$$\frac{\hat{f}_i^j - f_i^j}{\Delta t} = (a - bf)_i^j. \tag{17}$$

Implicit:

$$\frac{\hat{f}_i^j - f_i^j}{\Delta t} = a_i^j - b_i^j \hat{f}_i^j. \tag{18}$$

Exponent:

$$\hat{f}_i^j = f_i^j[\exp(-\Delta t\, b_i^j)] + \frac{a_i^j}{b_i^j}[1 - \exp(-\Delta t\, b_i^j)]. \tag{19}$$

The intermediate solution $\hat{f}_i^j$ is corrected to insure conservation by using Equation (16); therefore, we have

$$\tilde{f}_i = \hat{f}_i[1 + P(\mathbf{v})]. \tag{20}$$

The final step is to obtain the solution f_i^{j+1} of the free motion of the following difference equations:

$$\begin{aligned} &\frac{f_i^{j+1} - \tilde{f}_i^j}{\Delta t} + v_x \frac{\tilde{f}_i^j - \tilde{f}_{i-1}^j}{\Delta x} = 0, \qquad v_x \geq 0, \\ &\frac{f_i^{j+1} - \tilde{f}_i^j}{\Delta t} + v_x \frac{\tilde{f}_{i+1}^j - \tilde{f}_i^j}{\Delta x} = 0, \qquad v_x \leq 0. \end{aligned} \tag{21}$$

The stability of the scheme given by Equation (21) requires that the Courant condition $|v_x|_{\max} \Delta t \leq \Delta x$ be satisfied.

3.3 *Direct-Simulation Technique*

Eaton et al. (1977) classified direct-simulation techniques into four types. The tracer method, first developed by Bird (1963, 1976), dynamically follows a subset of the molecules, and interactions between the molecules occur only within the randomly selected subset. The field method, used by, for example, Haviland (1963, 1965), follows a single test molecule that interacts with other molecules in the flow field. The tracer-field method dynamically follows a subset of the molecules, which interact with other molecules in the flow field. The full simulation method dynamically follows all the molecules in the flow. We consider in this review only the most widely used tracer method.

Just like other Lagrangian methods, there are two explicit steps in the calculation: to evaluate the interactions among the fluid particles, and to move them. In the case of molecular flows, these two steps correspond to the solution of a spatially uniform relaxation and that of a molecular motion. This splitting of the calculations into two steps is the same as that used by Aristov & Cheremisin to solve directly the Boltzmann equation, as discussed above. Recently, Nanbu (1980a,b,c, 1981, 1982a,b,c, Watanabe & Nanbu 1982) has not only developed an accurate direct-simulation technique to obtain the solution of the Boltzmann equation, but he has also analyzed the basic aspects of the direct-simulation technique in producing the Boltzmann solution. We present briefly his review and analysis of the tracer method.

There are five tracer methods, which were developed by Bird (1963, 1976), Koura (Koura & Kondo 1969, Koura 1970), Belotserkovskii & Yanitskii (1975), Deshpande (1978), and Nanbu (1980a). These methods differ from each other in the simulation scheme used in each of the two stages. Belotserkovskii & Yanitskii use a random-walk model in the molecular-motion stage, while all others consider the motion to be free. Nanbu's treatment of the molecular interaction in the relaxation stage is based on the Boltzmann equation, while that of others is based on the Kac equation (Kac 1959), which is closely related to the Boltzmann equation. As for the simulation in the relaxation stage, Bird's, Koura's, and Deshpande's methods are modified versions of the method of Belotserkovskii & Yanitskii.

Nanbu's analysis of the direct-simulation technique is based on the concept of decoupling by splitting the operators in the Boltzmann equation, as done by Aristov & Cheremisin. Let us rewrite the Boltzmann equation as

$$\partial f/\partial t = -Df + Jf, \tag{22}$$

in which $Df = \mathbf{v}(\partial f/\partial \mathbf{r})$ and $Jf = a - bf$. In terms of the operators D and J, we may write the solution of $f(\mathbf{v}, \mathbf{r}, \Delta t)$ as

$$f(\mathbf{v}, \mathbf{r}, \Delta t) = (1 - \Delta t D + \Delta t J) f(\mathbf{v}, \mathbf{r}, \mathbf{0}). \tag{23}$$

Neglecting the second-order term, we may write Equation (23) as

$$f(\mathbf{v}, \mathbf{r}, \Delta t) = (1 - \Delta t D)(1 + \Delta t J) f(\mathbf{v}, \mathbf{r}, \mathbf{0}). \tag{24}$$

The splitting of the two operators leads to two successive stages of the simulation process. The first-stage process is the spatial relaxation due to the operator $(1 + \Delta t J)$. In the second stage, the gas system undergoes a collisionless free motion due to the operator $(1 - \Delta t D)$. The step size Δx should be chosen so that the gas may be considered to be spatially uniform in this step, and the interval Δt should be selected so that the decoupling of the process into two stages is justified. Of course, there are other computational parameters to be considered in the application of the direct-simulation technique. Nanbu (1982a) has studied the effects of sample sizes, number of molecules, step size, and cutoff angle upon simulation results.

Nanbu (1980a) has developed a simulation technique on the basis of the Boltzmann formulation. The technique still takes a fairly long computation time except for Maxwellian molecules. The shock-wave solution obtained by Nanbu for the Maxwellian molecules is being compared with those obtained by Yen (Yen & Ng 1974).

Nanbu's work came to the author's attention during the preparation of this paper. Space does not allow for a more detailed presentation of his technique or his complete analysis of other direct-simulation techniques. A separate review by Nanbu of his analysis of direct-simulation techniques would be of interest.

The heart of any numerical method for solving the Boltzmann equation lies in the accurate simulation of the intermolecular-collision phenomenon according to the Boltzmann formulation. The question raised by Nanbu (1980a) concerning the disagreement with the Boltzmann formulation of the simulation model used in some current direct-simulation techniques is a pertinent one. If this disagreement is significant, then there is a simulation error at each time step. Since the direct-simulation technique is a time-dependent approach, there is a cumulative effect of this error on the solution. This simulation error plus the large statistical scatter usually encountered in the calculation using a direct-simulation technique may seriously impair its validity. Recently, the accumulation of error using Bird's method was analyzed by Belotserkovskii et al. (1980). Since accurate Boltzmann solutions have been obtained using the direct numerical method for several basic one-dimensional flow problems, it would still be desirable to make a detailed comparison with the corresponding results using the direct-simulation technique in order to assess the accuracy of the latter.

Bird (1978) reviewed the Monte Carlo direct-simulation method, but included Nordsieck's method in his review. Several remarks concerning this method were, in the author's opinion, inaccurate and misleading and should be clarified. Bird referred to the choice of cutoff boundary in the velocity space as a source of difficulty. The problem of finite cutoff in the velocity space has to be faced either explicitly or implicitly in the implementation of any Monte Carlo technique for simulating the intermolecular collisions. Furthermore, the effects of finite cutoffs on the calculation have been studied by Nordsieck. Bird also remarked in his paper that Nordsieck's method shared with Haviland's test-particle method the serious disadvantages of requiring an initial estimate of the flow and of depending on the convergence of a subsequent iteration. First, both iterative and time-dependent approaches require initialization, and this can hardly be considered a disadvantage. Second, the uniqueness of Nordsieck's iterative method has been studied by using different initial guesses for the distribution function. Third, the crucial part of Nordsieck's method lies in the accurate evaluation of the collision integral. The collision integral obtained could be embedded in either an iterative scheme or a time-dependent technique. Finally, the iterative scheme is, in general, preferable to the time-dependent scheme for obtaining steady-state solutions in many problems because of error accumulation. Bird also commented that

Nordsieck's method suffers from the same computational disadvantage as the test-particle method, namely that its application has been restricted to relatively simple flow situations. It should be pointed out that Cheremisin (1973, 1982) used the direct numerical method to solve the Boltzmann equation for multidimensional nonequilibrium as well as near-equilibrium problems. The reason that Nordsieck's method has not been applied to more-complex problems is the lack of support of basic Boltzmann research by United States government agencies.

3.4 *Methods for Multi-Dimensional Problems*

Cheremisin (1973, 1982) developed methods to solve the Boltzmann equation for two-dimensional nonequilibrium flow problems. He considered both the steady iteration and the time-dependent splitting schemes. We review here the aspects of Cheremisin's integration schemes.

The steady iteration scheme of the two-dimensional Boltzmann equation is given by

$$v_x\, \partial f^I/\partial x + v_y\, \partial f^I/\partial y = a^{I-1} - b^{I-1} f^I, \tag{25}$$

in which the superscript I denotes the iteration number. The time-splitting scheme consists of two alternating problems:

$$\begin{aligned} \partial f/\partial t &= (a - bf), \\ \partial f/\partial t + v_x\, \partial f/\partial x + v_y\, \partial f/\partial y &= 0. \end{aligned} \tag{26}$$

In both schemes, the finite-difference approximation in divergence form is used for the transport operator for each computational cell. For the steady-iteration method, an implicit difference scheme is obtained from

$$\int (\mathbf{v}\cdot\mathbf{n}) f^I\, dl = \iint (a^{I-1} - b^{I-1} f^I)\, dA, \tag{27}$$

and for the time-splitting scheme, the distribution function f is explicitly updated by

$$\int \partial f/\partial t\; dA + \int (\mathbf{v}\cdot\mathbf{n}) f\, dl = 0, \tag{28}$$

in which l is the boundary surface of the cell, $\mathbf{n}$ is a vector normal to dl, and A is the area of the cell. Equation (27), in its discretized form, implies a flux balance of the distribution function f with $\mathbf{v}$ with a source term due to molecular collisions. If we choose N velocity vectors, there are N equations to be solved for N distribution functions. The distinct feature in applying the integration scheme to nonequilibrium problems is the need to evaluate the source term, which is the collision integral, at each integration step. The price we pay to solve the nonequilibrium flow problem is the evaluation of the Boltzmann collision integral; however, we do get the benefit of implementing a simple set of governing equations with only the distribution function as the dependent variable.

It should be noted that the quantities a^{I-1} and b^{I-1} in Equation (25) are to be evaluated at each node at each iterative step. For the time-dependent scheme, the solution of the relaxation problem is to be obtained at each time step.

The other features requiring consideration in applying the method to multidimensional problems are the choice of a fair sample of velocity vectors, the design of an optimum discretization for favorable error distribution, and the consistent implementation of the appropriate boundary conditions. The direct solution of the nonlinear Boltzmann equation for multidimensional problems using Cheremisin's methods appears to be possible with the computers currently available.

3.5 *Methods for Continuum-Flow Problems*

It would be of interest to use the Boltzmann equation for continuum problems for two reasons. First, it would alleviate certain computational difficulties encountered in the continuum formulation, especially for a flow phenomenon with a large departure from equilibrium, such as that of a shock wave. Second, it would allow the use of a single numerical method to solve a problem in which different length scales prevail.

As discussed by Aristov & Cheremisin (1982), the use of a single method is more desirable than a hybrid method in which the continuum and nonequilibrium regions are dealt with separately using two different approaches, and in which the solutions are matched at chosen interface boundaries. The difficulty in implementing a hybrid method lies in smoothly matching the solutions of different equations at proper locations to keep the error uniformly distributed over the entire computational domain.

Recently, Aristov & Cheremisin (1982) proposed a method for solving the Boltzmann equation for both equilibrium and near-equilibrium flows and applied it to several problems to study its validity in obtaining solutions of the Euler and Navier-Stokes equations. Their scheme is basically the same as the splitting technique developed by them for nonequilibrium flow problems. The difference lies in the details of the implementation of two of the three steps. The first step—solving the spatially uniform relaxation problem—is replaced by either of the following two equations:

$$\hat{f}_i^j = F_{\mathrm{M}}(f_i^j) \quad \text{for equilibrium flow,} \tag{29}$$

$$\hat{f}_i^j = F_{\text{N-S}}(f_i^j) \quad \text{for near-equilibrium flow,} \tag{30}$$

in which F_M is the Maxwellian distribution function with the density, velocity, and temperature determined from f_i^j (the solution of the free-motion equation), and $F_{\text{N-S}}$ is the Navier-Stokes distribution function with the density, velocity, temperature, and their gradients determined from f_i^j. The second step is to correct the distribution function from $\hat{f}_i^j$ to $\tilde{f}_i^j$ to insure conservation. The final step is to calculate the parameters n_i^{j+1}, u_i^{j+1}, and T_i^{j+1} from the moment equations obtained from Equation (21). For example, the continuity equation becomes

$$(n_i^{j+1} - \tilde{n}_i^j)/\Delta t + (\tilde{\phi}_i^{+j} - \tilde{\phi}_{i-1}^{+j})/\Delta x + (\tilde{\phi}_{i+1}^{-j} - \tilde{\phi}_i^{-j})/\Delta x = 0 \tag{31}$$

in which

$$\tilde{\phi}_i^{+j} = \sum_{v_x > 0} v_x \tilde{f}_i^j \quad \text{and} \quad \tilde{\phi}_i^{-j} = \sum_{v_x < 0} v_x \tilde{f}_i^j. \tag{31a}$$

Aristov & Cheremisin (1982) discussed the disadvantage of using an iterative scheme to solve the kinetic equation at small Knudsen number. Such a method has slow convergence, since the changes of the distribution function and thus the flow quantities during iteration may be of the same order as the truncation error.

Pullin (1980), Reitz (1981), and Deshpande (Deshpande & Raul 1982) also applied kinetic theory approaches to equilibrium and near-equilibrium flows. These approaches are similar in principle to that of Aristov & Cheremisin.

4. NUMERICAL SOLUTIONS

Accurate numerical solutions of the nonlinear Boltzmann equation have been obtained for several steady one-dimensional nonequilibrium problems. These solutions include velocity distribution functions, collision integrals, and their moments, and they are of great interest for several reasons. First, some of the microscopic and macroscopic characteristics exhibit distinct and often unforeseen nonequilibrium behavior. Second, more details of the flow properties may be obtained from them for future comparison with other calculations and with experiment.

Also of interest are the features of the numerical method developed to deal with several computational difficulties and to increase the computational efficiency.

4.1 *Formulation of Steady One-Dimensional Flow Problems*

We may formulate the one-dimensional steady-flow problem according to the kinetic-theory approach by referring again to Figure 2, which shows schematically the computational mesh. In the mesh system, the spatial positions $\hat{x}$ of 0 and 1 correspond to the two boundaries of the flow field under consideration. The position $\hat{x} = 0.5$ is within the flow field. As discussed in Section 3, the integration is performed in two directions, one from $\hat{x} = 0$ to $\hat{x} = 1$ for molecules with $v_x > 0$ and the other from $\hat{x} = 1$ to $\hat{x} = 0$ for molecules with $v_x < 0$. The boundary conditions are given at the positions where the integration starts: $\hat{x} = 0$ for molecules with $v_x > 0$ and $\hat{x} = 1$ for molecules with $v_x < 0$. The kinetic-theory formulation exemplifies the basic treatment of the physics of the nonequilibrium flow development. The shock wave, for example, is formed by the interactions of molecules moving upstream relative to the shock and also to those moving downstream. The heat transfer in a gas depends on the interactions between the molecules moving in one direction and those moving in the other direction. In fact, the kinetic-theory treatments of the molecular flow of a shock wave and that of a heat- and mass-transfer problem are the same. The difference between the two problems lies in the boundary conditions.

The treatment of the boundary condition of the molecules moving downstream ($v_x > 0$) is the same as that of those moving upstream ($v_x < 0$). In each case, the distribution function at the initial position is known or specified; the distribution function at each of the terminal positions is unknown. The specification of the distribution function at a solid surface depends on the gas-surface interaction and the net mass flux due to, for example, evaporation or condensation.

Nonequilibrium gas-flow problems are characterized by a length parameter and a gradient parameter. The length parameter is a measure of the number of mean free paths in the nonequilibrium region. The gradient parameter is a measure of the property gradient in each mean free path. The Knudsen number Kn, defined as the ratio of the mean free path to the length of the nonequilibrium region, is commonly used as the length parameter. It is thus the reciprocal of the number of mean free paths. The gradient parameter is not commonly used. For a shock wave, Mach number is used as the gradient parameter. For heat- and mass-transfer problems, however, a gradient parameter could be defined.

The detailed nonequilibrium behavior depends on the intermolecular collisions and the gas-surface interactions. The differential collision cross section depends only on the gas, whereas the gas-surface interaction depends both on the gas and the surface.

There are two one-dimensional problems of basic interest: the shock-wave problem and the heat- and mass-transfer problem. The shock-wave problem is simpler in that it has no length parameter and its structure depends only on the intermolecular collisions. For the heat- and mass-transfer problem, the effect of both parameters must be considered, and the physics of the molecular flow depends on the intermolecular collisions as well as the gas-surface interaction.

4.2 *Shock-Wave Problem*

We describe here one of the principal features of the iterative scheme used by Nordsieck to solve the shock-wave problem (Hicks et al. 1972): the use of density as the independent variable instead of x, the spatial distance. The Boltzmann equation and the transformation function are

$$v_x(df/d\xi)(d\xi/dx) = (a-bf), \tag{32}$$

$$d\xi/dx = dn/dx = \int (a-bf)\, d\mathbf{v}/v_x. \tag{33}$$

Since dn/dx is a moment of the collision integral $I(1/v_x)$, it is evaluated from the calculation of $(a-bf)$, which changes after each iteration. There are two major reasons for choosing the density n as the independent variable. One is to stabilize the shock by eliminating the artificial origin on the x-axis. The other is to use an optimum node distribution. Since f is nearly a linear function of the density, a small number of nodes is needed. This requirement of a smaller number of nodes increases significantly the computational efficiency, since the time-consuming calculations of the collision integral are reduced. Aristov & Cheremisin (1980) demonstrated the possibility of using a very coarse mesh for the shock-wave problem by applying their conservative splitting technique.

The boundary conditions are set as follows. The distribution function for molecules moving downstream starts with that of the upstream equilibrium condition, and for molecules moving upstream, with that of the downstream equilibrium condition. The distribution functions obtained from the solution at the terminating positions are not exactly the same as those of the equilibrium conditions at these positions; therefore, there is a boundary error. This error was studied and found to be insignificant.

We present in this section the significant results obtained from the accurate Boltzmann solutions for the shock-wave problem by Hicks, Yen, Reilly, and Ng (Hicks et al. 1972, Yen & Ng 1974) for a wide range of Mach numbers (1.1–10) and for two intermolecular potentials.

A familiar way to describe the internal shock-wave structure is to show the macroscopic properties as functions of the distance x. Of particular interest is the density profile, since it can be measured experimentally and the results may be compared with theoretical calculations. However, there are several reasons why it is, in general, difficult to compare the shock-wave structure on the basis of the density profile. First, the asymmetry is not easy to see. Second, the choice of origin is arbitrary. Finally, the important characteristics such as the density gradient cannot be accurately determined, especially when there is a large statistical scatter in either the Monte Carlo calculation or the experiment. Such a statistical scatter makes

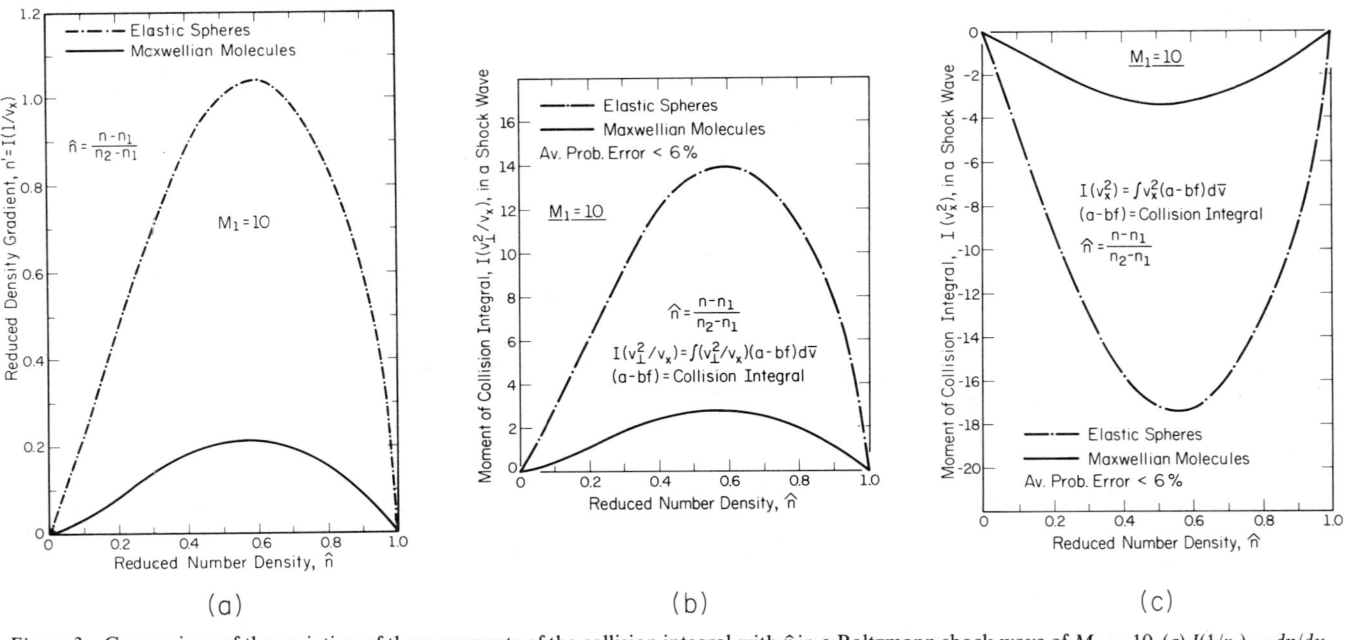

Figure 3 Comparison of the variation of three moments of the collision integral with $\hat{n}$ in a Boltzmann shock wave of $M_1 = 10$. (*a*) $I(1/v_x) = dn/dx$ = density gradient; (*b*) $I(v_\perp^2/v_x)$; (*c*) $I(v_x^2)$. The moments $I(v_\perp^2/v_x)$ and $I(v_x^2)$ are also related to the gradients of macroscopic properties in a shock wave.

a meaningful comparison difficult, since the scatter may be larger than the accuracy in determining the differential collision cross section for a given gas.

The moments of the collision integral, which are directly related to the gradients of macroscopic properties, were found to be more useful than those of the distribution function in accurately evaluating the non-equilibrium characteristics and the effect of intermolecular collision laws on them. The moments of the collision integral—$I(1/v_x)$ = the density gradient, $I(v_\perp^2/v_x)$, and $I(v_x^2)$—for a shock wave of $M_1 = 10$ for two intermolecular collision laws are shown in Figure 3. We observe from this figure not only the effect of the intermolecular collision law on the relaxation rate, but also the nature of the relaxation in the upstream and the downstream wings. Calculations of the ratios of the three integrals for the elastic spheres with the corresponding results for the Maxwellian molecules indicate that these ratios are nearly constant in the interior of the shock [$\hat{n} = (n-n_1)/(n_2-n_1) = 0.25$–$0.75$]. This constancy implies that the nature of the relaxation in the interior of a shock wave is similar and depends only weakly on the collision law.

The Boltzmann calculations show that each of the five noninvariant moments of f ($\mathcal{M}_1$, $\mathcal{M}_6$, $\mathcal{M}_9$, and two other moments) is nearly a linear function of the density. We thus conclude that the moments of the distribution function that determine the ordinary macroscopic properties are strongly coupled, and that this coupling depends only weakly on the intermolecular collision law. We might, therefore, expect that the function of macroscopic properties with respect to the density and to each other would be nearly the same for a gas whose intermolecular-collision law lies between the two extreme cases of elastic spheres and Maxwellian molecules. An example of the strong coupling of one macroscopic property with respect to the other is given in Figure 4, which shows the reduced temperature $\hat{T}$ as a function of reduced density $\hat{n}$ for the shock wave at $M_1 = 4$ and two collision laws. We observe that the effect of the collision law on the function is small, as expected. However, there are significant, albeit small, differences in the upstream and downstream wings of the shock wave. In the downstream wing ($\hat{n} = 0.75$–1) of the shock wave for Maxwellian molecules, there is a small overshoot ($\hat{T} = 1.0092$) at the position $\hat{n} = 0.9375$. The overshoot appears in the solution of the shock wave for Maxwellian molecules for $M_1 > 4$ and for elastic spheres at $M_1 = 10$. Elliott & Baganoff (1974) studied the overshoot of temperature in the downstream wing of a shock wave, as well as other characteristics in both wings.

Several measures of the departure from equilibrium in a shock wave may be calculated from the Boltzmann solutions. Some of these measures

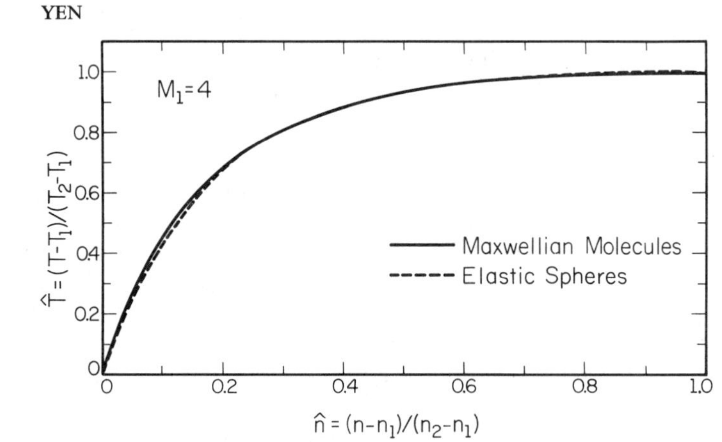

Figure 4 Comparison of variation of reduced temperature $\hat{T}$ with reduced density $\hat{n}$ in a Boltzmann shock wave for $M_1 = 4$. There are small but significant differences in the upstream wing ($\hat{n} = 0$–0.25) and the downstream wing ($\hat{n} = 0.75$–1).

exhibit some unforeseen nonequilibrium behavior. For example, a global measure of the departure from equilibrium obtained from the collision integral indicates that the largest departure occurs near the cold side of the shock for Mach numbers above 1.2 (Hicks et al. 1972).

The study of the velocity distribution function and the collision integral has led to an understanding of the physics of shock-wave formation. For example, the molecules that move upstream relative to the shock and that are being rapidly produced by the collisions play an important role in creating both the nonequilibrium behavior near the cold side of the shock and the upstream equilibrium distribution function.

4.3 *Heat- and Mass-Transfer Problems*

Figure 5 shows schematically a typical heat- and mass-transfer problem. The two plates are located a distance d apart. The hot plate, with temperature T_2, emits a gas of Maxwellian distribution either at temperature T_2 for complete accommodation or at a temperature less than T_2 for incomplete accommodation. The gas reaching the cold plate, with temperature T_1, is partially absorbed, and the gas emitted from this plate has a Maxwellian distribution of either temperature T_1 or a temperature greater than T_1. The problem, first considered by Cheremisin (1972), is referred to as the emission-absorption problem. If there is no net mass flux at each plate, then we have the familiar heat-transfer problem.

Four flow parameters are needed to characterize the heat- and mass-transfer problem. The length parameter Kn may be defined as follows:

$$\mathrm{Kn} = \sqrt{(2kT_1/m)} \Big/ \int_0^d b_{eq}\, dx, \tag{34}$$

in which k is the Boltzmann constant, m the molecular mass, and b_{eq} the average local collision frequency. We define the gradient parameter M_T as follows:

$$M_T = [2(T_2 - T_1)/(T_2 + T_1)]\ \mathrm{Kn}, \tag{35}$$

which is a measure of the temperature gradient in each mean free path. A more familiar parameter related to M_T is the temperature ratio T_1/T_2. The mass-transfer parameter is

$$\beta = \dot{m}^+(0)/\dot{m}^-(0), \tag{36}$$

in which $\dot{m}^-(0)$ is the molecular mass flux reaching the cold plate, and $\dot{m}^+(0)$ the molecular mass flux emitted from the cold plate. Since β is defined at the cold plate, it is an absorption coefficient. The parameter α is commonly used as the energy accommodation coefficient and is defined as

$$\alpha = (T_i - T_r)/(T_i - T_w), \tag{37}$$

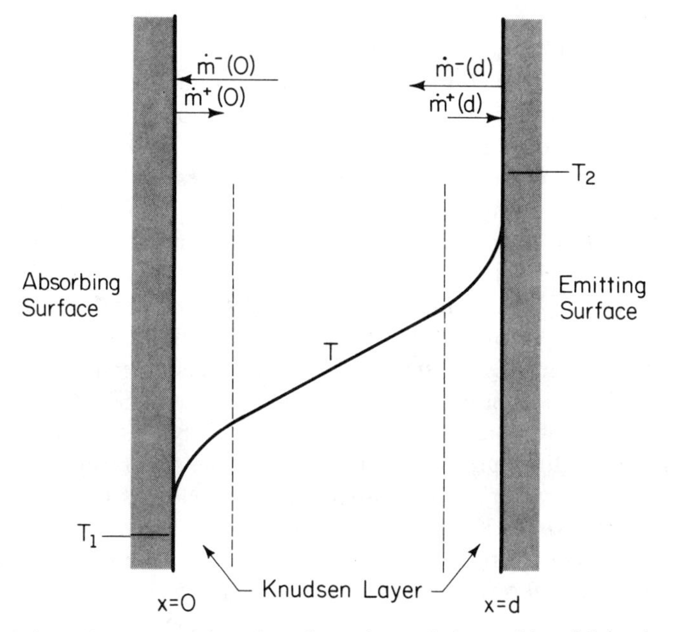

Figure 5 Schematic representation of an absorption-emission problem. Molecules emitted from the plate with temperature T_2 at $x = d$ are absorbed by the plate at temperature T_1 at $x = 0$.

in which T_i and T_r are the temperatures of the incident and reflected molecules, respectively, and T_w is the plate temperature.

The independent variable x in the Boltzmann equation is transformed to ξ as follows:

$$d\xi/dx = \text{Kn}\sqrt{(2/\pi)}nT^{1/2}. \tag{38}$$

The numerical method for integrating the Boltzmann equation for the heat and mass transfer problem differs from that for the shock wave only in the transformation function and the boundary conditions.

During the iterative process, the mass-transfer parameter β is monitored, and the distribution function of the emitted molecules is adjusted to insure that the value β remains constant at each iteration. The uniqueness of the solution was tested by comparing solutions with a different initial distribution function.

We present in this section the significant results obtained from the accurate Boltzmann solutions obtained by Yen (1971a,b, 1973, 1974, 1981, 1982) and Yen & Akai (1977) for several heat- and mass-transfer problems. These solutions include detailed calculations of the distribution functions, the collision integrals, their moments, and the functions of the moments for a wide range of parameters (Kn = 0.05–100, $T_1/T_2 = 0.05$–0.95, $\beta = 0$–100, $\alpha = 0.5$–0.9) and for two intermolecular collision laws.

The nonequilibrium condition occurs in the heat- and mass-transfer problem because of the boundary conditions imposed at the solid surfaces. Under these boundary conditions, the distribution function of the molecules emitted and/or reflected from the surface differs from that of the molecules reaching the surface. Of principal interest is the determination of the change in nonequilibrium condition as a function of the distance from the surface. For large Kn, there are very few intermolecular collisions between the plates, and the nonequilibrium condition persists in the entire region. For small Kn, the intermolecular collisions that take place produce relaxation toward equilibrium as the distance from the surface increases. If the length parameter Kn is small enough, then the molecule may reach the local equilibrium condition at a location between the two plates. This situation may be considered as the simulation of two half-space problems, one for the emitting surface and the other for the absorbing surface.

The basic characteristic of a nonequilibrium flow is the existence of a Knudsen layer. In the Knudsen layer, the gas relaxes from a condition that departs significantly from equilibrium to an equilibrium condition or one close to that. The shock wave may be considered to consist of two back-to-back Knudsen layers, one from the cold (supersonic) side to the location of maximum departure from equilibrium and the other from this location to the hot (subsonic) side. These two Knudsen layers share the boundary at which the maximum departure from equilibrium occurs. In the gas between the two plates in the heat- and mass-transfer problem, the existence of two Knudsen layers is more apparent.

In our study of the nonequilibrium behavior in a shock wave, we investigate the relaxation behavior as well as the asymmetry. In a sense, we look at the characteristics of two Knudsen layers. The asymmetry and the difference in nonequilibrium behavior between the two Knudsen layers in the heat- and mass-transfer problem are more pronounced, especially since there are more parameters to influence the flow behavior.

The moments of the distribution function within the shock wave are strongly coupled; therefore, the function of one macroscopic property with respect to any other is practically independent of the intermolecular-collision law. This independence was also found to exist in the molecular flow in the heat- and mass-transfer problem and has an even greater impact in simplifying the method of application of the kinetic-theory approach to practical problems.

The profiles in macroscopic properties such as temperature and density in the Knudsen layer in the heat-transfer problem do not often exhibit the distinct nonequilibrium relaxation behavior. On the other hand, the corresponding moments of the collision integral do show sharply the nonequilibrium characteristics in the Knudsen layers in most cases. We display in Figure 6 the density profile and the profile of the moment of the collision integral $I(v_{\perp}^2/v_x)$ for the solution with Kn $= 0.2493$, $T_1/T_2 = 0.783$, and $\alpha = 0.826$. These parameters were chosen to match those of the experiment by Teagen & Springer (1968). The Knudsen layers are seen to be more sharply defined near the plates in the profile of the moment of collision integral $I(v_{\perp}^2/v_x)$. This moment is directly related to the gradient of the moment $\mathscr{M}(v_{\perp}^2)$, which determines most of the macroscopic properties of interest. The density profile obtained from the Boltzmann solution shows a small nonlinearity in the Knudsen layer. The spatial variation of the density determined from the experiment is almost linear; therefore, the existence of Knudsen layers was not detected by the experiment.

For problems with mass transfer, we present results for two cases corresponding to different applications. The mass-transfer problem with a small value of β, i.e. large absorption, may be used to make a one-dimensional simulation of the outgassing flow-field problem near a spacecraft, since the near-vacuum condition at large distances from the spacecraft may be implemented as large absorption by a solid surface. Figure 7 shows the density profiles for two cases, $\beta = 0$ and 0.1, with Kn $= 0.1$ and $T_1/T_2 = 0.25$. We discuss first the case for $\beta = 0$. The profiles for both collision laws are qualitatively similar. The distribution function of the molecules near the absorbing surface resembles a Gaussian function but

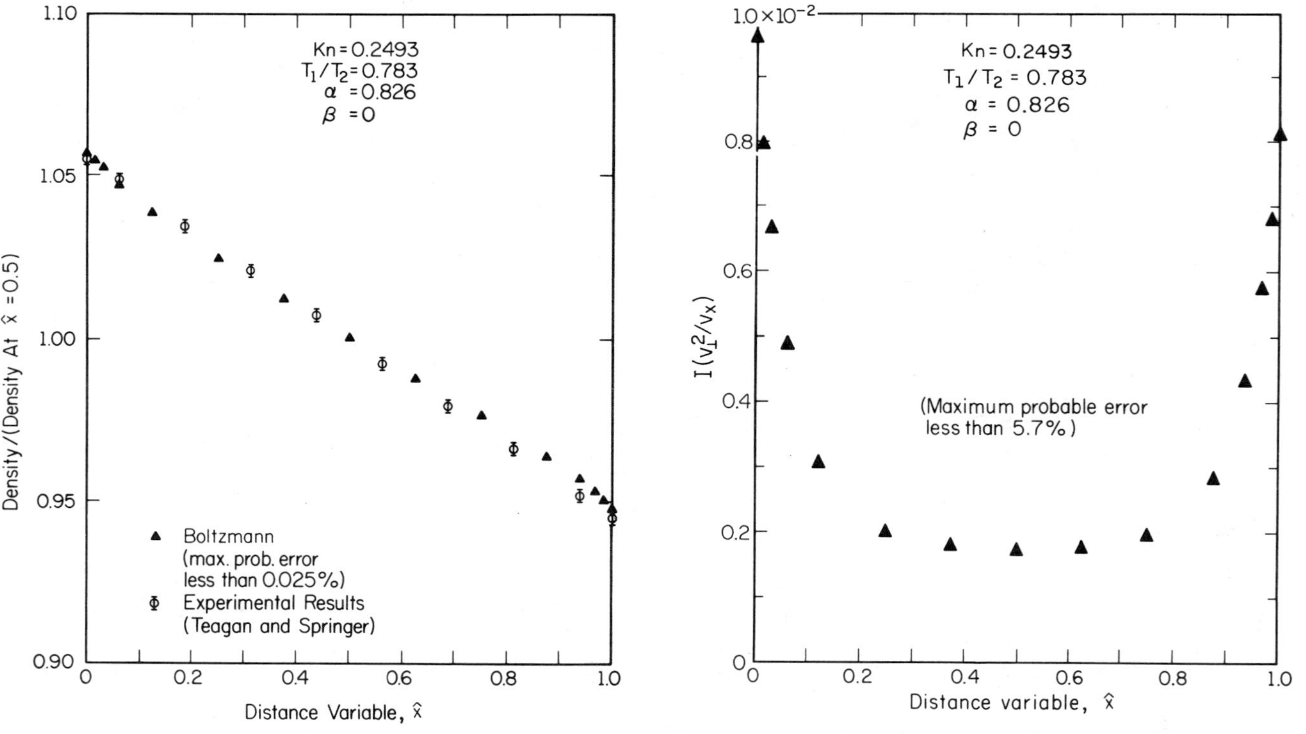

Figure 6 Knudsen-layer characteristics exhibited by the density profile and the profile of a moment of the collision integral, $I(v_\perp^2/v_x)$, obtained from the Boltzmann solution for a heat-transfer problem with $\mathrm{Kn} = 0.2493$, $T_1/T_2 = 0.783$, $\alpha = 0.826$, and $\beta = 0$. Also shown are the experimental results on the density profile by Teagen & Springer (1968).

deviates considerably from the local Maxwellian (Yen 1974). For outgassing problems, both the distribution functions and the flow properties, such as density, velocity, and temperature, at a large distance from the spacecraft are unknown and may be determined from the solution of a kinetic equation.

The corresponding results for the case of $\beta = 0.1$ show that the molecules emitted from the absorbing surface, even though few in number, play a role in altering significantly the nonequilibrium behavior near the absorbing surface. The relaxation phenomenon for the elastic spheres near the absorbing surface differs appreciably from that of the Maxwellian molecules. The Knudsen-layer characteristics near the emitting surface are qualitatively similar to those for the case of $\beta = 0$ but quantitatively different.

The study of the flow characteristics of the simple emission and absorption problem is directly relevant to that of the outgassing problem of more-complex geometry. We observe that the gas at a large distance from the outgassing surface is not in thermal equilibrium, and its condition

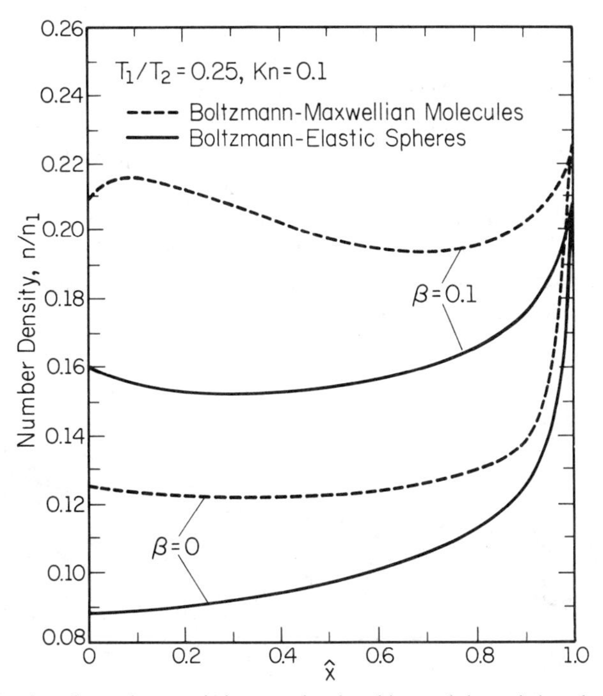

Figure 7 Knudsen-layer characteristics near the absorbing and the emitting plates exhibited by the density profile obtained from the Boltzmann solution for two cases of absorption ($\beta = 0$ and 0.1).

depends on the ambient condition. An accurate kinetic-theory treatment is thus necessary for the outgassing problem in order to implement properly the downstream boundary condition.

We next consider the problem of evaporation and condensation. Basically, this is the same as the emission-absorption problem if we consider the absorption surface as the condensation interface and the emission surface as the evaporation interface. We discuss here the difference between the Knudsen-layer characteristics near an evaporation interface and those near a condensation interface. The substance parameter may be redefined for the evaporation and condensation problem (Ytrehus 1982) as follows:

$$\hat{\beta} = \ln\left[P_2^+(d)/P_1^-(0)\right]/(1-T_1/T_2), \tag{39}$$

in which $P_2^+(d)$ is the pressure of the gas arriving at the emitting wall, and $P_1^-(0)$ is the pressure of the gas arriving at the absorbing wall. The substance parameter is so named for the condensation problem because it may be shown that it is related to the latent heat and the temperature of the liquid (Ytrehus & Alvestad 1981).

Figure 8 shows the Knudsen-layer characteristics exhibited by the temperature profiles for the evaporation and condensation problem for two values of the mass-transfer parameter, $\hat{\beta} = 2.68$ and 10.21. The solutions for these two cases were obtained using the following parameter values: (*a*) $\hat{\beta} = 2.68$: $\mathrm{Kn} = 0.1$, $\beta = 0.1$, $T_1/T_2 = 0.5$; and (*b*) $\hat{\beta} = 10.21$: $\mathrm{Kn} = 0.1$, $\beta = 0.1$, $T_1/T_2 = 0.75$. For the case $\hat{\beta} = 2.68$, the temperature profile is nearly antisymmetrical, with the temperature at the midposition ($\hat{x} = 0.5$) equal to the average value ($\hat{T} = 0.5$); therefore, the two Knudsen layers are similar in nonequilibrium behavior. For the case $\hat{\beta} = 10.21$, the temperature profile is no longer antisymmetrical. In fact, the evaporation Knudsen

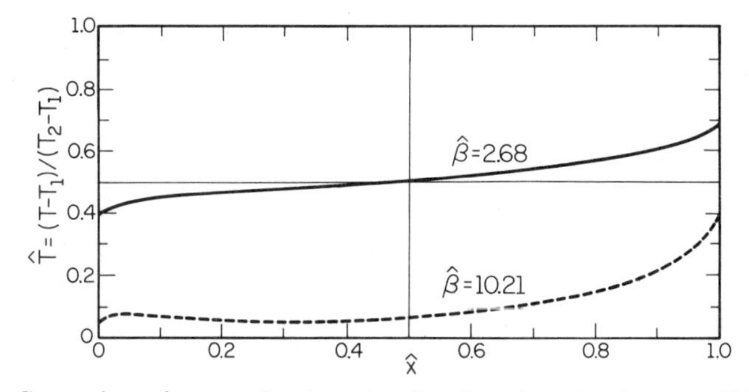

Figure 8 Comparison of evaporation (near $\hat{x} = 1$) and condensation (near $\hat{x} = 0$) Knudsen-layer characteristics exhibited by the temperature profile obtained from the Boltzmann solution for two substance parameters ($\hat{\beta} = 2.68$ and 10.21).

layer dominates the flow region. The condensation Knudsen layer is very small. The temperature at the midposition is only slightly above that of the condensation interface. Ytrehus (1982) discussed in detail the effect of the mass-transfer parameters on the Knudsen-layer characteristics.

Figure 9 shows the Knudsen-layer characteristics exhibited by two moments of the collision integral, $I(1/v_x)$ and $I(v_\perp^2/v_x)$. These moments are directly related to the gradients of temperature and several other macroscopic properties in the Knudsen layer. We observe that the mass-transfer parameter $\hat{\beta}$ affects $I(1/v_x)$, the density gradient, much more than it does $I(v_\perp^2/v_x)$. The difference in the Knudsen-layer characteristics exhibited by the temperature profiles in Figure 8 is, therefore, entirely due to the difference in the density profile. The effect of other moments of the collision integral on the Knudsen-layer characteristics has also been studied by Yen (1982).

For evaporation and condensation problems, we are mostly concerned with the effect of flow parameters on the mass flux and the jump conditions across the Knudsen layer. Since there are more parameters for these problems, it would be expensive to solve the Boltzmann equation in order to study their flow characteristics. The finding that the flow property as a function of any other property depends only weakly on the collision law has led to success in solving these problems using much simpler approaches, e.g. that of solving the Krook equation (Yen & Akai 1977, Yen 1981) and that of

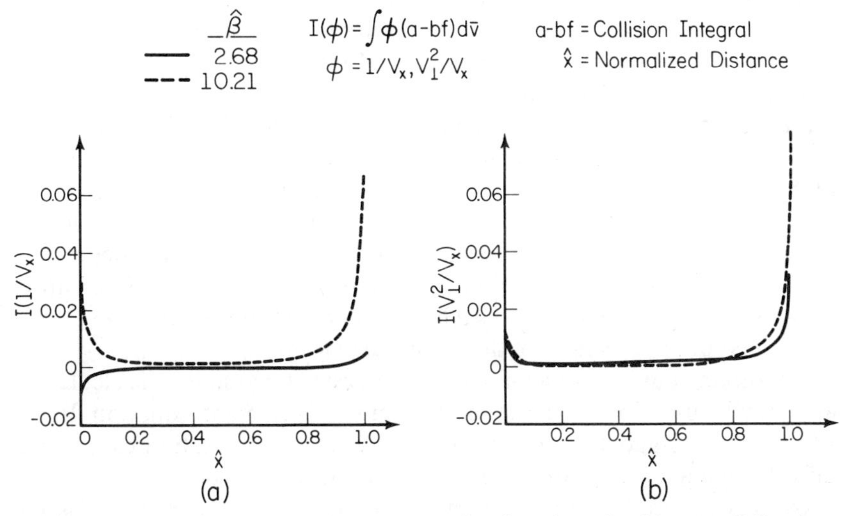

Figure 9 Comparison of evaporation (near $\hat{x} = 1$) and condensation (near $\hat{x} = 0$) Knudsen-layer characteristics exhibited by two moments of the collision integral, $I(1/v_x)$ = density gradient and $I(v_\perp^2/v_x)$, obtained from the Boltzmann solution for two substance parameters ($\hat{\beta} = 2.68$ and 10.21).

solving the Boltzmann transport equations using a more simply implemented collision law (Ytrehus 1977, Ytrehus & Alvestad 1981). The moment method of solving the Boltzmann transport equations may also be used together with a conventional numerical method to solve a continuum problem in which either an evaporation or a condensation Knudsen layer occurs.

5. SUMMARY REMARKS

The nonequilibrium condition occurs in continuum as well as rarefied-gas flow problems. The kinetic-theory treatment, which consists of solving the nonlinear Boltzmann equation, is necessary for a gas flow under nonequilibrium conditions, since the concept of transport coefficients is no longer valid.

The principal feature of a direct numerical method for solving the nonlinear Boltzmann equation for nonequilibrium flow problems is to embed in an integration scheme a Monte Carlo method that simulates the intermolecular-collision phenomenon in a gas. The Monte Carlo simulation should be in accord with the Boltzmann formulation, and the fidelity of the solutions depends on the control of statistical and systematic errors of the method. The considerations in the choice of an appropriate integration scheme for the Boltzmann equation are the same as those for the governing equations of continuum flow.

Direct numerical solution of the Boltzmann equation requires the evaluation of the collision integral in the Monte Carlo stage of calculation. The collision integral, which expresses the microscopic transport property, is directly relevant to the study of nonequilibrium flow problems, and it, as well as its functions, exhibits more distinctly the nonequilibrium behavior of the gas. Monte Carlo methods used to evaluate the nonlinear Boltzmann collision integral have been developed and may be embedded in any integration scheme. Either an iterative scheme or a time-dependent approach may be used to obtain the steady-state solution. The integration of the Boltzmann equation for each velocity point in the velocity space is performed in the direction of the molecular motion.

The direct-simulation technique also consists of two stages of calculation and may be used to obtain the solution of the Boltzmann equation. In the Monte Carlo stage, the simulation scheme for the relaxation calculation must be in agreement with the Boltzmann formulation.

Of basic interest in studying any nonequilibrium flow problem is the determination of the characteristics in the Knudsen layer, where the gas relaxes from a highly nonequilibrium condition to one of equilibrium or near-equilibrium. The moments of the distribution function, as well as the collision integral obtained from the Boltzmann solutions, exhibit distinct and often unforeseen nonequilibrium behavior of the gas in the Knudsen layer.

Analyses of the direct numerical solutions of the Boltzmann equation for several basic one-dimensional nonequilibrium flow problems have led to significant findings. Different moments of the distribution function and the collision integral exhibit distinct aspects of the complex nonequilibrium characteristics in different regions of the Knudsen layer. The moments of the distribution function in the Knudsen layer are strongly coupled; therefore, one macroscopic property as a function of any other property is almost independent of the collision law. Problems in which only the function of one property with respect to the other properties is of interest, such as evaporation and condensation, may be solved by using any collision law, and in particular, one which is simple to implement.

The nonlinear Boltzmann equation can be solved for multidimensional and more-complex problems by using the presently developed numerical methods on currently available computers. Furthermore, the kinetic-theory approach could also be used to solve continuum problems. For these problems, the Monte Carlo stage of calculation is replaced by one that requires that the distribution function is equal to that of either the Maxwellian for the inviscid flow or that of the Navier-Stokes for the viscous flow.

ACKNOWLEDGMENT

I wish to express my appreciation to Kenichi Nanbu for sending me his comments on the direct simulation techniques.

Literature Cited

Aristov, V. V., Cheremisin, F. G. 1976. Separation of the inhomogeneous operator in the Boltzmann equation. *Sov. Phys.-Dokl.* 21:601–2

Aristov, V. V., Cheremisin, F. G. 1980. The conservative splitting method for solving Boltzmann equation. *USSR Comput. Maths. Math. Phys.* 20(1):208–25

Aristov, V. V., Tcheremissine (Cheremisin), F. G. 1982. The kinetic numerical method for rarefied and continuum gas flows. *Proc. Int. Symp. Rarefied Gas Dyn., 13th.* In press

Belotserkovskii, O. M., Yanitskii, V. E. 1975. The statistical particles-in-cells method for solving rarefied gas dynamics problems. *USSR Comput. Maths. Math. Phys.* 15(5):101–14

Belotserkovskii, O. M., Erofeev, A. I., Yanitskii, V. E. 1980. A nonstationary method of direct statistical modeling of rarefied gas flows. *USSR Comput. Maths. Math. Phys.* 20:82–112

Bird, G. A. 1963. Approach to translational equilibrium in a rigid sphere gas. *Phys. Fluids* 6:1518–19

Bird, G. A. 1976. *Molecular Gas Dynamics*. London: Oxford Univ. Press. 238 pp.

Bird, G. A. 1978. Monte Carlo simulation of gas flows. *Ann. Rev. Fluid Mech.* 10:11–31

Cheremisin, F. G. 1969. A method for direct numerical integration of Boltzmann equation. In *Numerical Methods in the Theory of Gases*, ed. V. P. Shidlovskii. Moscow: Comput. Cent., USSR Acad. Sci. Transl. in *NASA TT F-638*, pp. 43–63. 188 pp.

Cheremisin, F. G. 1970. Numerical solution

of a kinetic Boltzmann equation for homogeneous stationary gas flows. *USSR Comput. Maths. Math. Phys.* 10(3): 125–37

Cheremisin, F. G. 1972. Motion of rarefied gas between infinite parallel emitting and absorbing plane surfaces. *Fluid Dyn. (Izv. Akad. Nauk USSR)* 7(2): 351–53

Cheremisin, F. G. 1973. Solution of the plane problem of the aerodynamics of a rarefied gas on the basis of Boltzmann's kinetic equation. *Sov. Phys.-Dokl* 18: 203–4

Cheremisin, F. G. See Tcheremissine

Deshpande, S. M. 1978. An unbiased and consistent Monte Carlo game simulating the Boltzmann equation. *Rep. 78 FM 4*, Dept. Aerosp. Eng., Indian Inst. Sci., Bangalore. 19 pp.

Deshpande, S. M., Raul, R. 1982. Kinetic theory based fluid-in-cell method for Eulerian fluid dynamics. *Rep. 82 FM 14*, Dept. Aerosp. Eng., Indian Inst. Sci., Bangalore. 67 pp.

Eaton, R. R., Fox, R. L., Touryan, K. J. 1977. Isotope enrichment by aerodynamic means: a review and some theoretical considerations. *J. Energy* 1: 229–36

Elliott, J. P., Baganoff, D. 1974. Solution of the Boltzmann equation at the upstream and downstream singular points in a shock wave. *J. Fluid Mech.* 65: 603–24

Haviland, J. K. 1963. Determination of shock wave thickness by the Monte Carlo method. In *Rarefied Gas Dynamics*, ed. J. Laurmann, 1: 274–96. New York: Academic. 541 pp.

Haviland, J. K. 1965. The solution of two molecular flow problems by the Monte Carlo method. In *Methods of Computational Physics*, ed. B. Alder, S. Fernbeck, M. Rotenberg, 4: 109–209. New York: Academic. 385 pp.

Hicks, B. L., Smith, M. A. 1968. On the accuracy of Monte Carlo solution of the nonlinear Boltzmann equation. *J. Comput. Phys.* 3: 58–79

Hicks, B. L., Yen, S. M. 1971. Collision integrals for rarefied gas flow problems. In *Rarefied Gas Dynamics*, ed. D. Dini, pp. 845–52. Pisa: Edit. Tech. Sci. 1270 pp.

Hicks, B. L., Yen, S. M., Reilly, B. J. 1969. Numerical studies of the nonlinear Boltzmann equation. Part II: Studies of new techniques of error reduction. *Coord. Sci. Lab. Rep. R-412*, Univ. Ill., Urbana. 21 pp.

Hicks, B. L., Yen, S. M., Reilly, B. J. 1972. The internal shock wave structure. *J. Fluid Mech.* 53: 85–111

Kac, M. 1959. *Probability and Related Topics in Physical Sciences*. London: Interscience. 266 pp.

Koura, K. 1970. Transient Couette flow of rarefied gas mixtures. *Phys. Fluids* 13: 1457–66

Koura, K., Kondo, K. 1969. Solutions of unsteady nonlinear flow problems by the Monte Carlo method. In *Rarefied Gas Dynamics*, ed. L. Trilling, H. Y. Wachman, pp. 181–84. New York: Academic. 907 pp.

Lima, E. G. 1973. Method for numerical solution of the Boltzmann equation. *USSR Comput. Maths. Math. Phys.* 13(6): 246–54

Nanbu, K. 1980a. Direct simulation scheme derived from the Boltzmann equation. I. Monocomponent gases. *J. Phys. Soc. Jpn.* 49: 2042–49

Nanbu, K. 1980b. Direct simulation scheme derived from the Boltzmann equation. II. Multicomponent gas mixtures. *J. Phys. Soc. Jpn.* 49: 2050–54

Nanbu, K. 1980c. Direct simulation scheme derived from the Boltzmann equation. III. Rough sphere gases. *J. Phys. Soc. Jpn.* 49: 2055–58

Nanbu, K. 1981. Direct simulation scheme derived from the Boltzmann equation. IV. Correlation of velocity. *J. Phys. Soc. Jpn.* 50: 2829–36

Nanbu, K. 1982a. Direct simulation scheme derived from the Boltzmann equation. V. Effects of sample size, number of molecules, step size, and cut-off angle upon simulation data. *Rep. Inst. High Speed Mech., Tohoku Univ, Jpn.* 45: 19–41

Nanbu, K. 1982b. Direct simulation scheme derived from the Boltzmann equation. VI. Velocity correlation in a model cell. *J. Phys. Soc. Jpn.* 51: 59–62

Nanbu, K. 1982c. Direct simulation scheme derived from the Boltzmann equation. VIII. Velocity correlation relevant to boundary conditions. *J. Phys. Soc. Jpn.* 51: 1763–68

Nordsieck, A., Hicks, B. L. 1967. Monte Carlo evaluation of the Boltzmann collision integral. In *Rarefied Gas Dynamics*, ed. C. L. Brundin, pp. 695–710. New York: Academic. 879 pp.

Pullin, D. I. 1980. Direct simulation methods for compressible inviscid ideal-gas flow. *J. Comput. Phys.* 34: 231–44

Reitz, R. D. 1981. One-dimensional compressible gas dynamics calculations using the Boltzmann equation. *J. Comput. Phys.* 42: 108–23

Tcheremissine (Cheremisin), F. G. 1977. Methods of solution of the Boltzmann kinetic equation. In *Topics in Numerical Analysis*, ed. J. J. H. Miller, 3: 417–28. New York: Academic. 477 pp.

Tcheremissine (Cheremisin), F. G. 1982. Solution of the Boltzmann equation for plane rarefied gas dynamic problems. *Proc. Int. Symp. Rarefied Gas Dyn., 13th.* In press

Teagen, W. P., Springer, G. S. 1968. Heat transfer and density distribution measurements between parallel plates in the transition region. *Phys. Fluids* 11: 497–506

Watanabe, Y., Nanbu, K. 1982. Direct simulation scheme derived from the Boltzmann equation. VII. Ergodicity of simulation solutions. *Rep. Inst. High Speed Mech., Tohoku Univ., Jpn.* 45: 43–76

Yen, S. M. 1971a. Solutions of the Boltzmann and Krook equations for heat transfer problems with Maxwell and evaporating boundary conditions. In *Rarefied Gas Dynamics*, ed. D. Dini, pp. 853–60. Pisa: Edit. Tech. Sci. 1270 pp.

Yen, S. M. 1971b. Monte Carlo solutions of nonlinear Boltzmann equation for problems of heat transfer in rarefied gases. *Int. J. Heat Mass Transfer* 14: 1865–69

Yen, S. M. 1973. Numerical solutions of nonlinear kinetic equations for a one-dimensional evaporation-condensation problem. *Comput. Fluids* 1: 367–77

Yen, S. M. 1974. Solutions of kinetic equations for the nonequilibrium gas flow between emitting and absorbing surfaces. In *Rarefied Gas Dynamics*, ed. M. Becker, M. Fiebig, pp. A15.1–10. Porz-Wahn, West Germany: DFVLR Press. 1305 pp.

Yen, S. M. 1981. Numerical solution of the Boltzmann and Krook equations for a condensation problem. In *Rarefied Gas Dynamics*, ed. S. S. Fisher. *Prog. Astronaut. Aeronaut.* 74: 356–62

Yen, S. M. 1982. Nonequilibrium behavior of the Knudsen layer near a condensing surface. *Proc. Int. Symp. Rarefied Gas Dyn., 13th.* In press

Yen, S. M., Akai, T. J. 1977. Nonlinear numerical solutions for an evaporation-effusion problem. In *Rarefied Gas Dynamics*, ed. J. L. Potter. *Prog. Astronaut. Aeronaut.* 51: 1175–83

Yen, S. M., Ng, W. 1974. Shock wave structure and intermolecular collision laws. *J. Fluid Mech.* 65: 127–44

Yen, S. M., Tcheremissine (Cheremisin), F. G. 1981. Monte Carlo solution of the nonlinear Boltzmann equation. In *Rarefied Gas Dynamics*, ed. S. S. Fisher. *Prog. Astronaut. Aeronaut.* 74: 287–304

Yen, S. M., Hicks, B. L., Osten, R. M. 1974. Further developments of a Monte Carlo method for evaluation of Boltzmann collision integrals. In *Rarefied Gas Dynamics*, ed. M. Becker, M. Fiebig, pp. A12.1–10. Porz-Wahn, West Germany: DFVLR Press. 1305 pp.

Ytrehus, T. 1977. Theory and experiments of gas kinetics in evaporation. In *Rarefied Gas Dynamics*, ed. J. L. Potter. *Prog. Astronaut. Aeronaut.* 51: 1197–1212

Ytrehus, T. 1982. Gaskinetics and thermodynamic aspects in evaporation and condensation Knudsen layers. *Tech. Rep. AAE 82-2, UILU-82-0502*, Univ. Ill., Urbana. 51 pp.

Ytrehus, T., Alvestad, J. 1981. A Mott-Smith solution for nonlinear condensation. In *Rarefied Gas Dynamics*, ed. S. S. Fisher. *Prog. Astronaut. Aeronaut.* 74: 330–45

Ann. Rev. Fluid Mech. 1984. 16: 99–137

NUMERICAL SIMULATION OF TURBULENT FLOWS[1]

Robert S. Rogallo and Parviz Moin

Computational Fluid Dynamics Branch, NASA Ames Research Center, Moffett Field, California 94035

1. INTRODUCTION

A century has passed since O. Reynolds demonstrated that fluid flow changes from an orderly and predictable state to a chaotic and unpredictable one when a certain nondimensional parameter exceeds a critical value. The chaotic state, turbulence, is the more common one in most flows at engineering and geophysical scales, and its practical significance, as well as the purely intellectual challenge of the problem, has attracted the attention of some of the best minds in the fields of physics, engineering, and mathematics. Progress toward a rigorous analytic theory has been prevented by the fact that turbulence dynamics is stochastic (often having underlying organized structures) and strongly nonlinear. There are, however, rigorous kinematic results that stem from tensor analysis and the linear constraint of continuity, and these allow a reduction of variables in the statistical description of the velocity field in certain cases, especially for isotropic turbulence. Rigorous dynamical results are available only for limiting cases where the governing equations can be linearized, and although the required limits are seldom approached in practice, the linear analysis provides guidance for model development. In spite of the dearth of rigorous nonlinear results, we have accumulated over the years a surprisingly good qualitative understanding of turbulence and its effects. Indeed, the gems of turbulence lore are the scaling laws for particular domains (either in physical or wave space), which result from the recognition of the essential variables and the constraint of dimensional invariance. In particular, the Kolmogorov law, and the law of the wall, are so well established that compatability with them is required of any theory or simulation.

All attempts at a statistical theory of turbulence have ultimately been faced with the problem of closure, that is, the specification at some order of a statistical quantity for which no governing equation exists. The success of the closure model depends not only on the flow configuration, but also on the statistical order at which results are desired. When the closure model is inadequate for accurate determination of the desired statistics, the model must be improved, or closure postponed to yet higher order. Most of the closures attempted to date may be classified as either one-point or two-point, depending on the number of spatial points appearing in the desired statistical results. Reviews of the many one-point closures are given by Reynolds (1976) and Lumley (1980). The much more complicated two-point closures [the Direct Interaction Approximation (DIA) and the related Test Field Model (TFM) of Kraichnan (see Leslie 1973), and the Eddy Damped Quasi-Normal Markovian (EDQNM) model of Orszag (1970), Lesieur & Schertzer (1978), and Cambon et al. (1981)] have been limited to homogeneous (usually isotropic) flows, where symmetry allows a reduction of variables.

Progress in the experimental study of turbulence has not been as difficult as that of analysis, but it has required great ingenuity in the collection of data and often in setting up the flows themselves. The results are usually of two kinds: statistical and visual. The velocity statistics are collected by use of hot-wire probes and, more recently, also by use of the laser Doppler velocimeter. Flow visualization has been particularly useful in aiding the interpretation of statistical data and identifying persistent flow structures. The primary difficulty with experimental turbulence data is the lack of it; the theoretician needs a number of statistical quantities, some of which (for example, those involving the pressure) are difficult to measure. A secondary problem is the isolation of the effect of a single parameter; for example, the effect of rotation on the decay of turbulence generated by a screen in a wind tunnel must be separated from the effect of rotation on the turbulence-generation process itself. Modern electronic recording and computing equipment has increased the quantity and quality of available data and has led to more-sophisticated analysis techniques (for example, conditional averages and pattern recognition).

The turbulence problem is so challenging that any research tool found successful in even remotely similar problems is quickly brought to bear. The two-point closures are such examples, as are the concepts of "critical phenomena," "strange attractor," and "renormalization." The high-speed digital computer is another recently developed tool with obvious application to the problem. The computer is used in other ways in fluid dynamics (see Van Dyke's article in this volume), but its most straightforward use is for "brute force" numerical simulation.

The numerical simulation of turbulence as we know it today rests largely on foundations laid down by the meteorologists at the National Center for Atmospheric Research (NCAR); their early work is reviewed by Fox & Lilly (1972). Since that time, computer capacity has increased by over an order of magnitude as has the number of workers in the field. Although some progress has been made in the efficiency and accuracy of computational algorithms, particularly in the adaptation of spectral methods, the primary pacing item determining our ability to simulate turbulence is the speed and memory size of the computing hardware (Chapman 1979, 1981).

The choice between simulation and experiment for a specific flow reduces to two questions: can the desired data be obtained at the required accuracy, and if so, how much will it cost? At the present time, simulation can provide detailed information only about the large scales of flows in simple geometries, and is advantageous when many flow quantities at a single instant are needed (especially quantities involving pressure) or where the experimental conditions are hard to control or are expensive or hazardous. Simulation cannot provide statistics that require a very large sample or that remain sensitive to Reynolds number even at high Reynolds number. It is particularly advantageous to use both simulation and experiment for delicate questions involving stability or sensitivity to external influences.

Turbulence consists of chaotic motion, and often persistent organized motions as well, at a range of scales that increases rapidly with Reynolds number. This restricts complete numerical resolution to low Reynolds number. When the scale range exceeds that allowed by computer capacity, some scales must be discarded, and the influence of these discarded scales upon the retained scales must be modeled. We shall distinguish between completely resolved and partly resolved simulations by referring to them as "direct" and "large-eddy" (LES), respectively, although these terms are often used interchangeably in the literature to indicate both without distinction. The descriptor "large-eddy" is misleading when the important flow structures to be resolved are extremely small as are those near solid boundaries, and at the dissipation scale at high Reynolds number. The attraction of direct simulation is that it eliminates the need for ad hoc models, and the justification often advanced is that the statistics of the large scales vary little with Reynolds number and can be found at the low Reynolds numbers required for complete numerical resolution. This approach has been successful for unbounded flows where viscosity serves mainly to set the scale of the dissipative eddies, but it has not been successful for wall-bounded flows, such as the channel flow, where computational capacity has so far not allowed a Reynolds number at which turbulence can be maintained. This is typical of many flows of engineering interest and forces the development of the LES approach.

The basic philosophy of LES is to explicitly compute only the large-scale motions that are directly affected by the boundary conditions and are therefore peculiar to the problem at hand. The small-scale motion is assumed to be more nearly universal, that is, its statistics and their effect upon the large scales can be specified by a small number of parameters. We hope that convergence of the method with increasing resolution will be rapid, because our ability to parameterize the sub-grid scale (SGS) effects should improve as the SGS length and time scales become disparate from those at energetic scales, and also simply because the SGS effects are proportional to the reduced SGS energy. The LES approach lies between the extremes of direct simulation, in which all fluctuations are resolved and no model is required, and the classical approach of O. Reynolds, in which only mean values are calculated and all fluctuations are modeled.

The numerical simulation of turbulence requires judgments with respect to the governing equations, initial and boundary conditions, and numerical resolution and methods. In the following sections we discuss some of the available choices and the results that follow from them.

2. GOVERNING EQUATIONS

We limit our discussion of simulation technique to flows of incompressible Newtonian fluids governed by the Navier-Stokes equations. Effects of buoyancy, compressibility, density stratification, magnetic forces, and passive scalar transport introduce new physical phenomena but increase the simulation difficulty in degree rather than kind. The convective terms of the equations produce a range of scales limited by molecular diffusion, so that with sufficiently low Reynolds number the entire range can be numerically resolved and no modification of the governing equations is required. When computer capacity does not allow complete resolution and the equations are not modified to take this into account, the computed values may have no relation to fluid physics. The numerical algorithm may become unstable as the smallest computed scales accumulate energy or, when energy-conserving numerical approximations are used, the energy may reach a nonphysical equilibrium distribution among the finite degrees of freedom. Orszag (see Fox & Lilly 1972) has demonstrated that energy-conserving numerics in inviscid isotropic flow lead to energy equipartition among the degrees of freedom, and this is often used as a check of algorithms and programming in simulation codes. When the viscosity is not zero but is too small to allow accurate resolution of the dissipation scales, an energy-conserving algorithm collects energy at the smallest computed scales until the dissipation and cascade rates reach equilibrium. Kwak et al. (1975) show that this excess energy, trapped at the mesh scale rather than cascading to the Kolmogorov scale, produces too rapid an energy transfer from large scales. This would be expected if the small scales act on the large scales as an eddy viscosity with a value, proportional to the length and velocity scales of the trapped energy, that is increased by the entrapment. One of the most important modifications of the Navier-Stokes equations is a mechanism for removal of energy from the computed scales that mimics as closely as possible the physical cascade process. The first step in an LES is then to define the variables that can be resolved and their governing equations.

The values at discrete mesh points of a simulation represent flow variables only in some average sense, and one way to define this sense is to find the differential equations that are exactly equivalent to the discrete approximation of the Navier-Stokes equations (Warming & Hyett 1974). The popular second-order central difference formula for the derivative of a continuous variable, for example, gives exactly the derivative of a second continuous variable that is an average of the first one:

$$\frac{u(x+h)-u(x-h)}{2h}=\frac{d}{dx}\left\{\frac{1}{2h}\int_{x-h}^{x+h}u(\xi)\,d\xi\right\}. \tag{1}$$

This shows how a discrete operator filters out scales less than the mesh size h. The direct use of such operators on the terms of the Navier-Stokes equations then introduces a different averaged variable for each term, depending on the derivative and discrete operator involved. This direct approach is therefore limited to completely resolved flows where the averages cause no information loss and all such averages give the same value. When the Reynolds number is too high for the direct approach, the range of scales can be limited to a resolvable size by explicitly filtering the Navier-Stokes equations. This formally defines the averaging process that separates resolvable from subgrid scales and the SGS stresses that must be modeled. When the smallest scale, $O(\Delta)$, allowed by the filter and the SGS model is sufficiently greater than the smallest scale, $O(h)$, resolved by the mesh, the results of the computation are independent of the choice of numerical algorithm and depend only on the filter and SGS model. Complete separation of physics from numerics is very costly in an LES, where mesh doubling in three directions increases the cost by an order of magnitude or more and in practice $\Delta = O(h)$ in each direction. Thus, resolution of the smallest computed scales is often marginal, and care is required to insure that the truncation crror is less than the physical SGS effects. Leonard (1974) applies the homogeneous filter

$$\bar{f} = \int_{-\infty}^{+\infty} G(\mathbf{x}-\xi) f(\xi)\, d\xi, \qquad f = \bar{f} + f' \tag{2}$$

to the Navier-Stokes equations to obtain the "resolvable-scale" equations

$$\begin{aligned} &\frac{\partial \bar{u}_i}{\partial t} + \frac{\partial}{\partial x_j} \overline{u_i u_j} + \frac{1}{\rho} \frac{\partial \bar{p}}{\partial x_i} = \nu \nabla^2 \bar{u}_i, \\ &\frac{\partial \bar{u}_i}{\partial x_i} = 0. \end{aligned} \tag{3}$$

Here, and throughout the paper, an overbar denotes a resolvable scale quantity and a prime denotes an SGS quantity. The convective fluxes are

$$\overline{u_i u_j} = \overline{\bar{u}_i \bar{u}_j} + Q_{ij}, \qquad Q_{ij} = \overline{\bar{u}_i u'_j} + \overline{u'_i \bar{u}_j} + \overline{u'_i u'_j}. \tag{4}$$

The equations must be closed by specifying these fluxes as functionals of the resolved variables. The terms containing u' must be modeled, but only the deviation from isotropy has dynamic effect. The $\overline{\bar{u}_i \bar{u}_j}$ term may be computed directly from resolved variables. When the average is uniform over an unbounded homogeneous dimension (space or time) or is a statistical (ensemble) average, the postulates of O. Reynolds lead to $\overline{u_i u_j} = \bar{u}_i \bar{u}_j + \overline{u'_i u'_j}$, but the postulates do not apply to averages over bounded domains (Monin & Yaglom 1971, Leonard 1974). The convolution (2) simplifies to $\bar{f}(\mathbf{k}) = G(\mathbf{k}) f(\mathbf{k})$ in wave space, from which it follows that $\overline{f'} = G(1-G)f$ is zero only when G is piecewise constant at values of 0 or 1. Reynolds' average is equivalent to $G(\mathbf{k}) = 0$ for $|\mathbf{k}| > 0$, and Fourier spectral methods implicitly filter with $G(\mathbf{k}) = 0$ for $|\mathbf{k}| > k_{\max}$. In the latter case $\overline{\bar{u}_i \bar{u}_j} = \bar{u}_i \bar{u}_j$ for resolved $\mathbf{k}$, but $\overline{\bar{u}_i u'_j} \neq 0$ there.

An alternative derivation of the resolvable scale equations by Schumann (1975) averages the equations over the cell volumes of a fixed mesh. This leads directly to the integral form of the Navier-Stokes equations in which time derivatives of cell-volume velocity averages are related to differences of cell-surface average stress and momentum flux. The various surface averages of momentum flux are decomposed as in (4) assuming Reynolds postulates, and the required surface averages of velocity and its gradient are related to volume averages of velocity by Taylor-series expansion. There is an inconsistency between the assumption of piecewise constant velocity required for validity of the Reynolds postulates and the use of Taylor-series expansions, but the resulting equations, except for the SGS model, are precisely those obtained by Deardorff (1970) using the continuous averaging process (1) and second-order numerics on a staggered uniform mesh.

The cell-volume averaging used by Deardorff and Schumann does not satisfy Reynolds postulates, and the difference $\overline{u_i u_j} - \bar{u}_i \bar{u}_j$ is modeled. The part of this, $\overline{\bar{u}_i \bar{u}_j} - \bar{u}_i \bar{u}_j$, that can be computed directly from resolved variables is known as the Leonard term. Leonard (1974) shows that this term removes significant energy from the computed scales and should probably not be lumped with the SGS terms. If direct calculation of the term is difficult he proposes a simple model, based on its Taylor-series expansion:

$$\overline{\bar{u}_i \bar{u}_j} \sim \bar{u}_i \bar{u}_j + \frac{\gamma}{2} \nabla^2 (\bar{u}_i \bar{u}_j) + \cdots, \qquad \gamma = \int_{-\infty}^{+\infty} |\xi|^2 G(\xi)\, d\xi. \tag{5}$$

At low Reynolds number Clark et al. (1979) find this form to be quite accurate when compared with values from a direct simulation. Shaanan et al. (1975) used a numerical operator for the divergence of the flux tensor in the Navier-Stokes equations that has lowest-order truncation error of nearly the form (5), thereby implicitly capturing the Leonard term. Most subsequent authors who explicitly filter the equations simply compute $\overline{\bar{u}_i \bar{u}_j}$ (Mansour et al. 1979). Clark et al. (1977) also find that the measured "cross" terms $C_{ij} = \overline{\bar{u}_i u'_j} + \overline{u'_i \bar{u}_j}$ drain significant energy from the resolved scales. Again, part of the effects can be captured by a Taylor-series expansion of the *resolved scale* velocity:

$$\overline{u'_i \bar{u}_j} \sim \overline{u'_i}\, \bar{u}_j + \frac{\Delta^2}{12} \frac{\partial \overline{u'_i}}{\partial x_k} \frac{\partial \bar{u}_j}{\partial x_k} + \cdots, \tag{6}$$

where $\overline{u'_i} = \bar{u}_i - \bar{\bar{u}}_i$, and we have used a Gaussian filter, $G(\xi) = \sqrt{(6/\pi\Delta)}\, e^{-6\xi^2/\Delta^2}$. Clark et al. (1977) propose a different model for the cross terms, but its derivation involves the Taylor-series expansion of the SGS velocity field. The dependence of the modeled terms in (4) upon the filter (for example, the vanishing Leonard term for sharp filters in wave space) suggests that simulation accuracy might be improved by a particular choice. Deardorff (1970) and Schumann (1975) use cell-volume averages related as in (1) to their finite-difference operators, and Chollet & Lesieur (1981) use the sharp filter implied by their Fourier spectral methods. When the choice of filter is divorced from the numerical algorithm, and this can only occur for $\Delta \gg h$, the Gaussian filter (Kwak et al. 1975, Shaanan et al. 1975, Mansour et al. 1978, Moin & Kim 1982) is usually used for homogeneous dimensions because it provides a smooth transition between resolved and subgrid scales and is positive definite (in fact Gaussian) in both physical and wave space. The optimum choice is of course the combination of filter and model that minimizes the total simulation error. The ratio of filter to mesh resolution, Δ/h, serves primarily to control numerical error, while the form of the filter and the form of the closure model determine the modeling error. The dependence of the model on the filter is studied, in isotropic flow within the TFM framework, by Leslie & Quarini (1979) and, for solutions of the Burgers equation, by Love (1980).

The averaged Navier-Stokes equations (3, 4) provide a conceptual framework for the discussion of modeling. The practical value of explicitly filtering the convective terms is a matter of current debate. The Leonard term is $O(\Delta^2)$, so it seems pointless to compute it separately in simulations using second-order numerics with error of $O(h^2)$ unless $\Delta/h \gg 1$ and the filtered field is well resolved. When the Leonard term is not swamped by numerical error, the filter, SGS stresses, and velocity field are related by (4), and the filter and model, $M(u)$, should in principle be selected together to minimize in some sense the modeling error $\overline{u_i u_j} - \overline{\bar{u}_i \bar{u}_j} - M_{ij}(\bar{u})$; the filtered convection $\overline{\bar{u}_i \bar{u}_j}$ is then computed directly. Kwak et al. (1975), for example, assume a Gaussian filter and a Smagorinsky (1963) SGS model and optimize the filter width and model constant by matching decay rate and spectral shape from the LES with experimental data for isotropic turbulence. A general study of filter and model forms has not yet been attempted. But the true filter is always uncertain because of the inherent inability of SGS models to exactly satisfy (4), so that the Leonard term cannot be found without error. An argument against separate treatment of the Leonard term is advanced by Antonopoulos-Domis (1981), who finds that in his LES calculations it moved energy from the small resolved scales to the large ones, rather than to the subgrid scales as predicted by Leonard. Leonard & Patterson (unpublished) point out that in isotropic turbulence the transfer spectrum $T(k)$ associated with the flux $\bar{u}_i \bar{u}_j$ is negative at small k, positive at large k, and is conservative. The transfer spectrum associated with the filtered flux $\overline{\bar{u}_i \bar{u}_j}$ is simply $G(k)T(k)$ and can reasonably be expected to remove energy from the resolved scales. The proper way to determine the effect of the Leonard term is to measure the energy transfer associated with the filtered convective term in an *accurately resolved field.* Studies of this kind by Leonard & Patterson, Clark et al. (1979), and Leslie & Quarini (1979) have verified the energy drain but at a lower magnitude than Leonard's original estimate. Antonopoulos-Domis draws his conclusions from simulations with no viscous or modeled turbulent terms. His results do indicate that the approximate form (5) alone is not sufficient to stabilize the calculation, but they do not indicate the effect of the Leonard term in a well-resolved calculation. A more general problem with explicit filters is the difficulty of extending them to inhomogeneous dimensions, where differentiation and filtering do not in general commute, but this does not seem insurmountable.

The equations of LES are then essentially the original Navier-Stokes equations written for averaged variables, with a filtered convection term and additional terms to model the effects of the unresolved scales. The only change from the original analysis of O. Reynolds is the use of averages over bounded domains, which requires the convective term to be filtered. The crux of the problem remains the closure model.

3. MODELS

Statistical homogeneity in space or time reduces the dimensions of the Reynolds-averaged problem, and all of the effects of fluctuations in the missing dimensions must be accounted for by the model. The variation of correlations in the remaining inhomogeneous dimensions is peculiar to the specific problem and cannot be modeled in a universal way. In an LES the equations are averaged over only small scales and retain all space-time dimensions. The averaging process is chosen to resolve numerically the physical features of interest, and the desired statistics are measured directly from the computed scales. The role of the model is not to provide these statistics directly, but to prevent the omission of the unwanted scales from spoiling the calculation of scales from which statistics are taken.

It is apparent from the LES work to date that the most important contribution of the model is to provide, or at least allow, energy transfer between the resolved and subgrid scales at roughly the correct magnitude. This transfer is usually from resolved to subgrid scales but may be reversed near solid boundaries, where the small productive eddies are not resolved and the SGS model must account for the lost production. Models can be tested either by directly comparing the modeled quantity with the model itself, using data from a reliable source (theory, experiment, or direct simulation), or by using the model in an LES and comparing results with those from a reliable source. The detailed information required for the former test can be supplied only by theory or simulation, and in practice the latter procedure is the more common. This is consistent with the LES philosophy; the model is not required to supply detailed information about the subgrid scales. But there is frequently a need to improve the model's description of physical detail and thus allow increased reliance on the model and lower computation cost. The sequence of model complexity could follow the same path as for the Reynolds-averaged equations, with the introduction of separate equations for the SGS stress or energy (Deardorff 1973). But in an LES the SGS length scales are given by the filter width, and velocity scales can be estimated from the smallest resolved scales. Bardina et al. (1980) suggest that the SGS stresses themselves be modeled by an extrapolation of the computed stresses at the smallest resolved scales. The simplest model, $\overline{u'_i u'_j} \sim C\overline{u'_i}\,\overline{u'_j} = C(\bar{u}_i - \bar{\bar{u}}_i)(\bar{u}_j - \bar{\bar{u}}_j)$, has been tested by McMillan et al. (1980) using data from direct simulations. The model correlates much better with the data than does a typical eddy-viscosity model, but Bardina et al. find that it is not sufficiently dissipative to stabilize an LES.

The effects of discarded scales on computed ones consist of "local" contributions, which diminish rapidly as the interacting scales are separated, and "nonlocal" contributions, which are significant even for widely separated scales. The interaction between scales of similar size retains the full complexity of the original turbulence problem, so there is little hope of modeling the local effects well. On the other hand, interaction of disparate scales is easier to analyze, so that nonlocal effects can be modeled with greater confidence.

The modern statistical theories of isotropic turbulence (DIA, TFM, EDQNM) provide models in which the roles of the various scales can be determined. Kraichnan (1976) and Leslie & Quarini (1979) evaluate the transfer spectrum within the TFM model, showing explicitly the local and nonlocal (in wave space) effects of the truncated scales on the energy flow within the resolved scales. The transfer spectrum is of the form $T(k) = -2\nu(k)\sqrt{E(k_m)/k_m}\,k^2 E(k) + U(k)$, where k is the wave-number magnitude, k_m is the limit of wave-number resolution, $\nu(k)$ is a nondimensional eddy viscosity, and E is the three-dimensional energy spectrum. The first term arises from stresses like $\overline{\bar{u}_i u'_j}$, while the "backscatter" term $U(k)$ arises from the $\overline{u'_i u'_j}$ stresses. The forms $\nu(k)$ and $U(k)$ depend upon both the filter and the energy spectrum; Kraichnan considers a sharp k filter in an infinite inertial subrange, and Leslie & Quarini extend these results to a Gaussian filter and more-realistic spectra. Kraichnan finds that the local effects are confined to scales within an octave of k_m and are characterized by a rapid rise in transfer as k approaches k_m. The net energy flow across k_m is dominated by this local transfer as described by Tennekes & Lumley (1972). Below this local range, $k < k_m/2$, the viscosity is independent of k [but depends on time through $E(k_m)$], and the backscatter decays like k^4 (Lesieur & Schertzer 1978). This backscatter might be important in unbounded flows, where length scales grow indefinitely and, as Leslie & Quarini note, its form is not well represented by an eddy-diffusion model because neither its magnitude nor anisotropy level are set by the large scales. Their results indicate that a Gaussian filter damps the SGS contribution to the local cascade too severely and broadens its range; this suggests that a sharper filter might be found in which the Leonard term carries the entire local transfer and leaves only the nonlocal effects to be modeled. Chollet & Lesieur (1981) achieve the same end using Kraichnan's effective eddy viscosity to successfully close both EDQNM and LES calculations. Chollet (1982) closes an LES by coupling it to an EDQNM calculation for the effective eddy viscosity, thus avoiding an assumed SGS energy spectrum. This is a rather elaborate "one-equation" model. The extension of EDQNM to homogeneous anisotropic flows by Cambon et al. (1981) allows application of this approach to less-restricted SGS stresses, but at a great increase in complexity. Yoshizawa (1979, 1982) relates these statistical closures in wave space to the gradient-diffusion closures in physical space by a formal multiscale expansion. The assumption that the SGS time scale, as well as space scales, is disparate from those of the resolved scales leads to SGS stresses that are locally isotropic at lowest order and of gradient-diffusion form (scalar eddy viscosity) at next order. The more interesting limit of commensurate time scales, leading to homogeneous but anisotropic SGS turbulence at lowest order, is prevented by the resulting complexity of the required DIA closure.

The gradient-diffusion model for SGS stresses is usually postulated with appeal to the similar stresses produced by molecular motion. But it is well known (Tennekes & Lumley 1972, Corrsin 1974) that the required scale separation, present in the case of molecular diffusion, does not occur between all of the scales of turbulence. In the Reynolds-averaged equations for flows having a single length and time scale, the gradient-diffusion form is required by dimensional analysis but the model cannot handle multiple scales (Tennekes & Lumley 1972). The eddy-viscosity model of the SGS stress tensor is

$$\tau_{ij} = Q_{ij} - \tfrac{1}{3}Q_{kk}\,\delta_{ij} \sim -2\nu_T S_{ij}, \tag{7}$$

where ν_T is the eddy viscosity, and $S_{ij} = \frac{1}{2}(\partial \bar{u}_i/\partial x_j + \partial \bar{u}_j/\partial x_i)$ is the strain-rate tensor of the resolved scales; the SGS energy $\frac{1}{3}Q_{kk}$ can be combined with the pressure and has no dynamic effect.

Smagorinsky (1963) proposes an eddy-viscosity coefficient proportional to the local large-scale velocity gradient:

$$\nu_T = (C_S\Delta)^2|\mathrm{S}|. \tag{8}$$

Here, C_S is a constant, the filter width Δ is the characteristic length scale of the smallest resolved eddies, and $|S| = \sqrt{S_{ij}S_{ij}}$. This model and its variants have been used in numerical simulations with considerable success. Assuming that scales of $O(\Delta)$ are within an inertial subrange so that $|S|$ can be found from Kolmogorov's spectrum, the analysis of Lilly (1966), with a Kolmogorov constant of 1.5, gives values of C_S from 0.17 to 0.21, depending on the numerical approximation for S_{ij}. Subsequent investigators determine C_S in an empirical manner. In large-eddy simulations of decaying isotropic turbulence, Kwak et al. (1975), Shaanan et al. (1975), Ferziger et al. (1977), and Antonopoulos-Domis (1981) obtain C_S by matching the computed energy-decay rate to the experimental data of

Comte-Bellot & Corrsin (1971). For several computational grid volumes and different filters they find C_S to be in the range 0.19–0.24. None of these calculations extends to an inertial subrange, and different treatments of the Leonard stresses and numerical methods are used; thus, the small variation of C_S indicates its insensitivity to the details of the energy-transfer mechanism in isotropic turbulence.

In a simulation of high-Reynolds-number turbulent channel flow, Deardorff (1970) finds that the use of the value of C_S estimated by Lilly causes excessive damping of SGS intensities, but that a value of 0.1 gives energy levels close to those measured by Laufer (1951). Deardorff (1971) attributes this difference in C_S to the presence of mean shear, which is not accounted for in Lilly's analysis. In the calculation of inhomogeneous flows without mean shear, where buoyancy is the primary driving mechanism, Deardorff (1971) finds $C_S = 0.21$ appropriate. Lower values lead to excessive accumulation of energy in one-dimensional energy spectra near the cutoff wave number.

Using flow fields generated by direct numerical simulation of decaying isotropic turbulence at low Reynolds number, Clark et al. (1977, 1979) and McMillan & Ferziger (1979) tested the accuracy of Smagorinsky's model and calculated C_S. They give values of C_S comparable to those obtained empirically in the large-eddy simulations. McMillan et al. (1980), using data from direct simulations of strained homogeneous turbulence, find that C_S decreases with increasing strain rate, which confirms the conclusions of Deardorff (1971). With the mean strain rate removed from the computation of the model, C_S is nearly independent of the mean strain, a highly desirable property for the model. Fox & Lilly (1972) point out that the removal of the mean shear might have allowed Deardorff (1970) to use the higher C_S value of Lilly.

In addition to calculating model parameters, direct simulations are also used to determine how well the forms of the SGS models represent "exact" SGS stresses. For isotropic turbulence, the results show that the stresses predicted by Smagorinsky's model (and other eddy-viscosity models) are poorly correlated with the exact stresses. The model performance is worse still in homogeneous flows with mean strain or shear. The notable success of calculations using the Smagorinsky model seems to reflect the ability of this model to stabilize the calculations, and also shows that low-order statistics of the large scales are rather insensitive, in the flows considered, to the details of the SGS motions.

Several alterations and extensions to Smagorinsky's model have been proposed. A modification consistent with the classical two-point closures replaces the local magnitude of the strain-rate tensor, $|S|$, in (8) with its ensemble average $\langle S \rangle$ (Leslie & Quarini 1979). Although in numerical solutions of the Burgers equation (Love & Leslie 1979) this modification improves the results, direct testing in isotropic flow by McMillan & Ferziger (1979) shows only a slight improvement. For free-shear flows, Kwak et al. (1975) suggest that it is appropriate to use the magnitude of vorticity $|\omega|$, rather than $|S|$, in (8), because the former vanishes in an irrotational flow. For isotropic turbulence this modification does not cause significant differences in large-scale statistics, but a substantial disparity is reported in small-scale statistics such as the velocity derivative flatness (Ferziger et al. 1977), which indicates the sensitivity of the smallest resolved scales to the SGS model. To account for mean shear in an LES of turbulent channel flow, Schumann (1975) introduces a two-part eddy-viscosity model. One part models the SGS stress fluctuations, and the other part, which reduces to Prandtl's mixing-length model for very coarse grids, accounts explicitly for the contribution of the mean shear.

When the grid resolution near a solid boundary is inadequate, the SGS flow field includes highly dynamic anisotropic eddies that contribute a significant portion of the total turbulence production and do not take a passive and dissipative role. Moin & Kim (1982), like Schumann, use a two-part eddy-viscosity model to account fully for the contributions to energy production by the finely spaced high- and low-speed streaks near the wall (see Section 4 and Kline et al. 1967) that are not adequately resolved in the spanwise direction:

$$\tau_{ij} = -\nu_T(S_{ij} - \langle S_{ij} \rangle) - \nu_T^*(y)\langle S_{ij} \rangle. \tag{9}$$

Here $\langle\ \rangle$ indicates an average over planes parallel to the walls. The first term in (9), the Smagorinsky model with mean shear removed, has essentially dissipative and diffusive effects on the resolvable scale turbulence intensities, $\sqrt{\langle(\bar{u}_i - \langle \bar{u}_i \rangle)^2\rangle}$. The second term accounts for the SGS energy production corresponding to SGS dissipation of mean kinetic energy $\langle \bar{u} \rangle^2$ but, in contrast to the first term, does not contribute to the dissipation of resolvable-scale turbulent kinetic energy. It does, however, indirectly enhance resolvable-scale energy production by representing the effect of the SGS stresses on the mean-velocity profile. Indeed, when Moin & Kim (1982) excluded the second term of (9) the computed flow did not transfer sufficient mean energy to the turbulence to sustain it against *molecular* dissipation. The characteristic length scale associated with ν_T^* is Δ_3, the filter width in the spanwise direction (normal to the mean flow and parallel to the wall), multiplied by an appropriate wall-damping factor to account for the expected y^3 or y^4 behavior of the Reynolds shear stress near the wall ($y = 0$). The influence of ν_T^* diminishes as the resolution of the spanwise direction is increased and the wall-layer streaks are better resolved.

Eddy-viscosity models of the type described above implicitly assume that the SGS turbulence is in equilibrium with the large eddies and adjusts itself instantaneously to changes of the large-scale velocity gradients. It may be desirable (certainly in transitional flows) to allow a response time for the SGS eddies to adjust to the changes in the resolvable flow field. Following Prandtl, Lilly (1966) assumes an eddy viscosity proportional to the SGS kinetic energy q^2, i.e. $\nu_T = c\Delta q$. The equation for q^2, derived formally from the Navier-Stokes equations, contains several terms that must be modeled. Schumann (1975) successfully uses this model for the fluctuating SGS stresses in his calculation of turbulent flows in channels and annuli. Grotzbach & Schumann (1979) extend the model to lower Reynolds numbers. In addition to dividing the SGS stresses into mean and fluctuating parts, a noteworthy feature of Schumann's formulation is its explicit allowance for anisotropic grids. Different characteristic length scales and dimensionless coefficients determined by grid geometry appear in the representation of the various surface-averaged SGS stresses. The utility of the model is demonstrated by its ability to simulate turbulent flow in an annulus, with relatively high grid anisotropy, by changing only the mesh-geometry parameters. Parameters of a physical nature retained the values used in the channel flow calculations (Schumann 1975).

Deardorff (see Fox & Deardorff 1972) finds that the Smagorinsky model smears out the mean temperature gradient that occurs when buoyant convection is terminated by a stably stratified overlayer. For a more realistic model, Deardorff (1973) resorts to transport equations for the SGS stresses. This involves 10 additional partial differential equations. The closure models in these equations are essentially analogous to the corresponding models in the Reynolds-averaged equations. These models may not be appropriate because the behavior and relative importance of the various correlations involving only small scales are different than those involving the total turbulence (Ferziger 1982). Although the transport model does lead to improved results, the prospect of such a complex treatment of the SGS stresses is less attractive to us than a judicious distribution of mesh points and the possibility of extracting more-accurate models directly from information carried at the resolved scales.

In the discussion of the equations and models for LES we have considered flows in which the statistics of interest are determined by the large scales. This is appropriate for engineering purposes, but there are also very fundamental and interesting questions about the small scales to be answered. These are concerned with intermittency and structure at small scale, and the implications for Kolmogorov's universal equilibrium hypothesis and its later modifications. Siggia (1981) outlines a conceptual procedure analogous to LES in which the large scales are modeled and the small scales are computed. The model for the missing large scales appears as a forcing term in the equations for the small scales. Siggia argues that if the large-scale effects depend on a small number of parameters (the dissipation rate is an obvious one) and the model is accurate enough, the limited scale range of the simulation might represent the intermittency achieved by the larger range of scales occurring at high Reynolds numbers. Unfortunately this is not possible in a calculation based on a periodic field in a fixed mesh, because the small-scale spatial intermittency that can be represented is directly limited by the number of mesh points and this geometric constraint cannot be modeled away. The vortex method of Leonard (1980) is not grid limited and is a more natural way to describe the intermittent vorticity fields occurring at high Reynolds numbers.

4. RESOLUTION REQUIREMENTS

Over two decades ago Corrsin (1961) demonstrated that the direct numerical simulation of high-Reynolds-number flows places an overwhelming demand on computer memory and speed. [See Chapman (1979) for a comprehensive study of the grid requirements for computational aerodynamics.] In direct simulations the number of spatial grid points is determined by two constraints: first, the size of the computational domain must be large enough to accommodate the largest turbulence scales (or the scale of the apparatus), and second, the grid spacing must be sufficiently fine to resolve the dissipation length scale, which is on the order of the Kolmogorov scale, $\eta = (\nu^3/\varepsilon)^{1/4}$. The ratio of these two scales (cubed) provides an estimate for the total number, N, of mesh points. In turbulent channel flow, for example, macroscales in the directions parallel to the walls determined from the two-point correlation measurements of Comte-Bellot (1963) and the average dissipation rate $\varepsilon = u_\tau^2 U_m/\delta$ give $N \simeq (6\mathrm{Re}_m)^{9/4}$ (Moin 1982); here Re_m is the Reynolds number based on the channel half-width, δ, and the average flow speed, U_m; and $u_\tau = \sqrt{\tau_w/\rho}$ is the wall shear velocity determined by the shear stress at the wall, τ_w, and the fluid density ρ. It is assumed that four grid points in each direction are required to resolve an eddy, and that $U_m/u_\tau \simeq 20$. Temporal resolution of the smallest computed eddies requires the time step Δt to be on the order of $(\nu/\varepsilon)^{1/2} = (\delta/u_\tau)\,\mathrm{Re}_m^{-1/2}$. At the moderate Reynolds number $\mathrm{Re}_m = 10^4$, roughly 5×10^{10} grid points and 2×10^3 time steps are necessary for the flow to reach a statistically steady state (a total flow time of $100\delta/U_m$). Such a computation is beyond the capabilities of presently available computers. However, if the bulk of the dissipation occurs at scales larger than 10η rather than η, direct simulation of channel or pipe flow may be possible in the near future at the lowest Reynolds numbers studied experimentally ($\mathrm{Re}_m \sim 2500$; see Eckelmann 1974).

In contrast to wall-bounded turbulent shear flows, which cannot be sustained below a critical Reynolds number, homogeneous and free-shear flows remain turbulent at Reynolds numbers for which all scales of motion can be resolved. The large-scale features of these flows are nearly independent of Reynolds number, and statistics determined from them are relevant at higher Reynolds numbers. However, in the simulation of unbounded shear flows such as turbulent jets and mixing layers (especially at low Reynolds numbers), the computational domain must be large enough to allow development of the long wavelength instability typical of these flows.

In LES the resolution requirements are determined directly by the range of scales contributing to the desired statistics and indirectly by the accuracy of the model. The less accurate the model, the further the modeled scales must be separated from the scales of interest. In engineering calculations the important scales contain the dynamic physical events responsible for turbulent transport of heat and matter and the production of turbulent energy. Near walls the principal flow structures are high- and low-speed streaks, which are finely spaced in the spanwise direction (Kline et al. 1967) and provide most of the turbulence energy production. The mean spanwise spacing of the streaks is about 100 wall units ($100\nu/u_\tau$), but streaks as narrow as 20 wall units probably occur and would require $h_3^+ \sim 5$ for complete spanwise resolution (Moin 1982). Similar considerations in the streamwise direction lead to $h_1^+ = 20$ to 30. Using 64 grid points normal to the wall (with nonuniform spacing to resolve the viscous sublayer and outer layers) and with computational periods in directions parallel to the walls chosen in accordance with two-point correlation measurements, the total number of grid points is estimated to be $N \sim 0.06\ \mathrm{Re}_m^2$. Although at high Reynolds numbers this is prohibitively large, detailed simulation of the important large eddies can be performed at low Reynolds numbers ($\mathrm{Re}_m \sim 5000$) with presently available computers. The Reynolds number of resolvable flows can be significantly increased when a fine mesh in the lateral directions is embedded near the walls (Chapman 1979), but for practical applications much computer power is still needed to calculate the flow in this extremely thin layer. If the wall-layer dynamics can be replaced by reliable outer-flow boundary conditions (see Section 5.2), the number of grid points becomes low enough to use LES for engineering computations on current computers (Chapman 1981).

Another practical difficulty in both direct and large-eddy simulations is the cost of obtaining an adequate sample for the flow statistics. The various scales of motion are not equally sampled; the scale sample is inversely proportional to the scale volume. With appeal to the ergodic hypothesis, ensemble averages can be replaced by averages over homogeneous space-time dimensions. For low-order velocity statistics a sample of 10^3 nodes appears adequate, but much larger samples are required for statistics of intermittent velocity derivatives and this problem increases with Reynolds number (Fox & Lilly 1972, Ferziger et al. 1977, Siggia 1981). When the homogeneous dimensions (there is usually at least one) do not provide an adequate sample, the statistics can be collected from an ensemble of flows evolving from independent initial conditions, but this is very costly and poses a serious problem for simulation of inhomogeneous flows.

5. NUMERICAL METHODS

Numerical implementation of the governing equations consists of four main issues: numerical approximation of spatial derivatives, initial and boundary conditions, time-advancement algorithm, and computer implementation and organization. In each category there are options available, and the choice of the overall algorithm depends on the problem under consideration, the cost, and the computer architecture.

5.1 *Spatial Representation*

Second- and fourth-order finite differences and spectral methods are used to approximate spatial derivatives. Since turbulent flows involve strong interaction among various scales of motion, special care should be taken that numerical representation of derivatives be faithful to the governing equations and the underlying physical mechanisms. For example, approximations with appreciable artificial viscosity, such as upwind difference schemes, significantly lower the effective Reynolds number of the calculation, and their dissipative mechanism distorts the physical representation and dynamics of large as well as small eddies. The formal order of accuracy associated with a difference method, which defines the asymptotic error for infinite resolution, may be less important than the accuracy of the method at the coarse resolution applied at the smallest computed scales. The accuracy of a method at various scales is illustrated by its ability to approximate the derivative of a single Fourier mode e^{ikx} (Mansour et al. 1979). For a given number of grid points, all difference schemes are inaccurate for values of wave number k near π/h, the highest wave number that can be represented on the grid. However, for intermediate values of k some schemes are significantly more accurate than others having the same formal order of accuracy.

The spectral method (Gottlieb & Orszag 1977) is a very accurate numerical differentiator at high k values. In this method the flow variables are represented by a weighted sum of eigenfunctions, with weights determined using the orthogonality properties of the eigenfunctions. The derivatives are obtained from term-by-term differentiation of the series or by using the appropriate recursion relationships (Fox & Parker 1968). The choice of eigenfunctions depends on the problem and the boundary geometry and conditions. For problems with periodic boundary conditions Fourier series are the natural choice, but for arbitrary boundary conditions orthogonal polynomials that are related to the eigenfunctions of singular Sturm-Liouville problems should be used (Gottlieb & Orszag 1977). Expansions based on these polynomials do not impose parasitic boundary conditions on higher derivatives, and for smooth functions they provide rapid convergence independent of the boundary conditions.

The difference between spectral and "pseudo-spectral" approximations is in the way products are computed (Orszag 1972). The advantage of the more expensive spectral method is the exact removal of aliasing errors (Orszag 1972); Patterson & Orszag (1971) give efficient techniques for handling aliasing errors arising in bilinear products. These errors are not peculiar to pseudo-spectral methods; finite-difference approximations of products also contain aliasing errors, but the errors are less severe owing to the damping at high k of the difference approximations. Aliasing errors usually increase with the order of accuracy of difference schemes (Orszag 1971).

One serious consequence of aliasing errors is the violation of the invariance properties of the Navier-Stokes equations. It is easily shown that in the absence of viscous terms and time-differencing errors, the governing equations conserve mass, momentum, energy, and circulation. Aliasing errors can violate these invariance properties and lead to nonlinear numerical instabilities (Phillips 1959). Lilly (1964) demonstrates that the staggered-mesh difference scheme (see Harlow & Welch 1965) preserves these invariance properties. When the nonlinear terms in the momentum equations are cast in the rotational form, $\boldsymbol{\omega} \times \mathbf{u} + \nabla(\mathbf{u}^2/2)$, properly invariant numerical solutions are obtained with pseudo-spectral and most finite-difference methods (Mansour et al. 1979).

For sufficiently smooth functions, spectral methods are more accurate than difference schemes having the same number of nodes. In contrast to higher-order difference methods, which require special treatment near the boundaries, spectral methods allow proper imposition of the boundary conditions. However, for the flow field to be sufficiently smooth, the smallest scale of motion present should be well resolved on the computational grid; otherwise, the rapid convergence of spectral methods is badly degraded. Cost constraints usually prohibit thorough resolution of the small scales; in direct simulations this means a mesh too coarse to capture the dissipation scales and in LES calculations means a filter or SGS model that does not remove sufficient energy from the small scales. In simulations of "two-dimensional turbulence" in a periodic box, Herring et al. (1974) find that the accuracy of spectral calculations is comparable to that of second-order (conservative but aliased) difference calculations having approximately twice the number of grid points in each direction. The advantage of the spectral method as an accurate differentiator is limited by the error that arises from truncation of small scales produced by the nonlinear terms.

A very important attribute of spectral methods is their self-diagnosis property. Inadequate grid resolution is reflected in excessive values of high-order expansion coefficients (Herring et al. 1974, Moin 1982). Fourier analysis of finite-difference solutions can also reveal poor resolution (Grotzbach 1981), but damping at high wave numbers masks its detection until the computational grid is insufficient to represent even the larger scales of motion.

5.2 *Boundary and Initial Conditions*

In turbulence simulations, the major difficulty with specification of boundary conditions occurs at open boundaries where the flow is turbulent. The flow variables at these boundaries depend on the unknown flow outside the domain. The unavoidably ad hoc conditions specified at these boundaries should be designed to minimize the propagation of boundary errors. Periodic boundary conditions are generally used for directions in which the flow is statistically homogeneous, but this implies that quantities at opposite faces of the computational box are perfectly correlated. If the periodic solution obtained is to represent turbulence, the period must be significantly greater than the separation at which two-point correlations vanish. The computed two-point correlation functions then serve as a good check of the adequacy of the size of the period.

Periodic boundary conditions for homogeneous turbulence subjected to uniform deformation may be applied only in a coordinate system moving with the (linear) mean flow. In this system the mean convection relative to the mesh vanishes, and the equations do not refer explicitly to the space variables. However, the computational grid is being continuously deformed, and the calculations must be stopped when the domain becomes so distorted that the flow cannot be resolved in all directions (Roy 1982). In the case of uniform shear, a convenient remeshing procedure (Rogallo 1981, Shirani et al. 1981) allows the computations to continue until the scale of the largest resolved eddies becomes bounded by the period. A clever implementation of the procedure for a finite-difference calculation by Baron (1982) uses shifting boundary values on a fixed mesh. The problem of length-scale growth is common to both experiments and computations. In homogeneous flows or unbounded inhomogeneous flows, the macroscales of turbulence grow until they reach the dimensions of the wind tunnel or the size of the computational box. When this occurs, meaningful statistics cannot be obtained from the large scales. To study the evolution of the flow for longer times it is tempting to use a coordinate transformation that continuously expands the computational box in time, but such a transformation reintroduces explicit spatial dependence in the governing equations. On the other hand, the calculation can be interrupted and the mesh rescaled to cover a new range of larger scales. The interpolation of the existing field to the new mesh causes some information loss; to minimize the damage the process should be carried out while the two-point correlations still show a significant uncorrelated range.

One of the more challenging, and virtually untouched, problems is that of turbulent inflow and outflow boundary conditions in nonhomogeneous directions. The inflow problem appears to be more troublesome, since in most cases the influence of the upstream conditions persists for large distances downstream. Of course, one way to avoid the problem is to prescribe a small orderly perturbation on an incoming laminar flow and follow the flow through transition to turbulence. However, in addition to more stringent requirements on the treatment of the small-scale motions in transitional flows, the required length of the computational box for the entire process is prohibitively large in some cases. The use of turbulent inflow and outflow conditions appears to be a practical necessity for flows such as boundary layers, where linear-stability theory predicts a long transitional zone.

The implementation of inflow and outflow conditions in simulations of free turbulent shear flows has so far been avoided by use of the "frozen turbulence" approximation. The physical problem, which is homogeneous in time but not in the mean-flow direction, is replaced by a computational problem that is homogeneous in the flow direction but not in time. The inflow condition is replaced by an initial condition, and periodic boundary conditions in the mean-flow direction are applied. Although the time-developing approximation of the "real flow" has most of its features, important differences remain. In a spatially developing turbulent mixing layer, for example, the mean streamlines within the layer are inclined to the direction of the flow outside the layer, but those in the time-developing flow are not.

Two approaches have been taken for implementing irrotational free-stream conditions in free-shear flows. Orszag & Pao (1974), Mansour et al. (1978), and Riley & Metcalfe (1980a) use a finite computational domain with stress-free boundary conditions in which the normal velocity and the normal derivative of the tangential velocities are zero. The turbulence field is confined to the central region of the domain and is surrounded by irrotational flow that extends to the boundaries. The subsequent use of

Fourier series implies the existence of image flows above and below the computational box that influence the dynamics of the flow inside. A better approach (Cain et al. 1981) maps the infinite domain into a finite computational box and applies the free-stream (or no-stress) boundary conditions at the boundaries of the transformed domain. The coordinate transformation used by Cain et al. allows a fairly simple use of Fourier spectral methods.

The specification of boundary conditions at smooth solid boundaries does not pose any difficulty; the velocity at the wall is the wall velocity. In the vicinity of the wall, the flow field is composed of small, energetic eddies associated with large mean-velocity gradients (see Section 4). For practical applications it is desirable to avoid the high cost of resolving this wall region by replacing flow near the wall with boundary conditions applied somewhat away from the wall. In simulations of high-Reynolds-number turbulent channel flow, Deardorff (1970) and later Schumann (1975) modeled the flow near the wall by applying such boundary conditions in the logarithmic layer. Once again it is not clear how to specify boundary conditions within a turbulent flow. For example, Schumann (1975) assumes that the fluctuations of wall shear stress, τ_w, are perfectly correlated with those of the streamwise velocity one mesh cell from the wall. Space-time correlation and joint probability density measurements of τ_w and u by Rajagopalan & Antonia (1979) support this assumption very close to the wall provided that a (sizable) time delay between these two quantities is introduced (see also Eckelmann 1974). The accuracy of this assumption degrades as the point of application of boundary conditions moves away from the wall; the normalized correlation is unity at the wall but is only about 0.5 in the logarithmic layer at $y^+ = 40$ ($y/\delta = 0.031$) (Rajagopalan & Antonia 1979). However, Robinson (1982) reports a correlation as high as 0.7 at $y^+ = 300$ ($y/\delta = 0.03$) in experiments at an order of magnitude higher Reynolds number ($\mathrm{Re}_\theta = 32{,}800$). These experimental results indicate that the u-τ_w correlation at a fixed y^+ improves with increasing Reynolds number, but at least part of this apparent improvement results from inadequate probe resolution at high Reynolds numbers. Robinson's wire length extends 200 wall units in the spanwise direction. Nevertheless, Schumann's assumption of u-τ_w correlation is reasonable and can be improved by including a space-time shift. Chapman & Kuhn (1981) propose a two-dimensional wall-layer structure retaining only the transverse spatial variation. They use detailed experimental data to set the length scales and phase relations of the velocity at the outer edge of the layer and obtain good agreement with experiment for the internal layer structure. Their wall-layer edge conditions have not yet been used as boundary conditions for the outer flow. The detailed pressure-velocity data provided by simulation (Moin & Kim 1982, Kim 1983) should be useful for the formulation of wall-layer edge conditions of the kind proposed by Chapman & Kuhn.

A three-dimensional velocity field satisfying the continuity equation and boundary conditions must be specified to initialize the calculation. Within these constraints, a random fluctuating velocity field is superimposed on a prescribed mean-velocity profile. Although the initial turbulence field can be defined with the desired intensity profiles and energy spectra, its higher-order statistics become physically realistic only after an adjustment period (see Orszag & Patterson 1972, Riley & Metcalfe 1980a). For example, the velocity derivative skewness is initially zero but quickly rises to a realistic value. The evolution of time-developing flows (those that never reach a statistically steady state) is often quite sensitive to the initial conditions for the large scales.

5.3 *Time Advancement*

Starting from initial conditions, the governing equations are advanced in time subject to the incompressibility constraint. We discuss time-advancing algorithms as they are applied to the incompressible Navier-Stokes equations. The additional SGS terms in the LES equations pose little additional numerical difficulty, and virtually identical numerical methods are used.

Time advancement may be done either explicitly or implicitly; explicit schemes are much easier to implement and have a much lower cost per step. The popular second-order explicit Adams Bashforth and Leapfrog schemes require only one evaluation of the time derivatives per step, but they do require retention of variables at step $n-1$ in order to advance from step n to $n+1$. The self-starting Runge-Kutta schemes (second, third, and fourth order) cost more per step but have better stability properties and therefore allow larger steps. The multiple evaluations of nonlinear terms required by Runge-Kutta methods can be used to reduce the cost of controlling aliasing errors in Fourier spectral calculations (Rogallo 1981).

When using explicit methods, the incompressibility constraint at each time step is usually enforced by solving a Poisson equation for pressure rather than by direct use of the continuity equation. To satisfy the discrete continuity constraint, the discrete Poisson problem must be derived using the same differencing operators used in the discrete momentum and continuity equations (Kwak et al. 1975). The staggered-grid difference scheme (see Harlow & Welch 1965) leads to a particularly simple Laplacian operator, whereas with standard centered-difference methods the operator is less compact and causes spatial pressure oscillations due to the uncoupling of even and odd points.

The choice of proper boundary conditions for the pressure equation is ambiguous (Moin & Kim 1980). Usually the Neumann boundary condition obtained from the normal momentum equation is used, but a Dirichlet boundary condition can also be derived from the tangential momentum equations. When spectral methods are used with explicit time advancement, the fact that both conditions cannot be simultaneously enforced implies the inability to impose complete velocity boundary conditions (Moin & Kim 1980). With the second-order staggered finite-difference scheme, the need for pressure boundary conditions does not arise. The continuity equation at the interior cells, together with the momentum equations (at the interior grid points) and the velocity boundary conditions, leads to a closed system of algebraic equations for pressure.

The root of this difficulty with spectral methods is that explicit methods treat the governing equations as an initial-value problem rather than as a boundary-value problem. Implicit methods require the solution of a boundary-value problem at each time step, thus allowing the natural imposition of velocity boundary conditions. Moreover, in simulations of wall-bounded flows, implicit treatment of the viscous terms overcomes the severe restriction on time step that arises from the small grid spacing normal to the wall. For these reasons all the calculations that extend to the wall use semi-implicit time-advancement algorithms (Orszag & Kells 1980, Moin & Kim 1980, 1982, Kleiser & Schumann 1979). In these calculations the nonlinear terms are advanced by the Adams Bashforth method. Fourier expansions are used in homogeneous dimensions, and either Chebyshev polynomial expansions or second-order difference methods are used in the direction normal to the wall. Recently, Leonard & Wray (1982) have developed a semi-implicit spectral method based on expansion in divergence-free vector functions. In this representation of the velocity, each term satisfies the continuity equation as well as the boundary conditions. Since the continuity constraint is satisfied by the expansion functions, pressure does not appear and only two velocity components are required to define the velocity field; this significantly reduces computer memory requirements. In wall-bounded flows the time step required for accurate resolution (see Section 4) is much larger than that required for convective stability, which suggests that advancement of the convective terms by implicit methods may be advantageous. Deardorff (1970) and Schumann (1975) translate the coordinate system at constant speed, reducing the mean convection velocity relative to the mesh to allow increased time steps. Alternatively, convection by the mean velocity can be handled implicitly; this is much simpler than a complete implicit treatment.

For problems in general geometries the computational complexity of spectral algorithms is not appreciably greater than that of difference algorithms when the boundary conditions allow use of explicit time advancement and the physical domain can be analytically mapped to a simple computational domain. But the linear convective stability criterion for the explicit advancement is more severe (by a factor of π for second-order central differences). With fully or partially implicit time advancement the computational complexity of spectral algorithms is much greater than that of difference algorithms. The nonconstant coefficients that arise when a complicated physical domain is mapped to a simple computational domain lead to dense matrix equations for the spectral coefficients. It is impractical to solve these equations by direct techniques; only iterative procedures appear to be feasible (Orszag 1980), and the accuracy and efficiency of the method depend on the number of iterations required to obtain the converged solution at the next step.

6. RESULTS

The flows simulated to date fall into one of three classes: homogeneous (unbounded), unbounded inhomogeneous, and wall bounded. The emphasis of the work can be classified as fundamental physics, development of simulation technique, and application to real problems. In the preceding sections we have discussed some of the work on technique. In this section we present typical fundamental results for three simple shear flows: homogeneous turbulence in uniform shear, the evolution of a turbulent mixing layer, and turbulent channel flow. These flows exhibit many of the complications found in real engineering problems. The homogeneous shear flow introduces anisotropy and production at large scales, the mixing layer adds turbulent diffusion and intermittence at the large scales, and the channel introduces solid boundaries near which all of these complications occur at small scales as well. These three flows are well documented by high-quality experimental data and have been simulated using a variety of numerical methods and a range of resolution.

Turbulence in uniform shear exhibits growing length scales, $O(L)$, and velocity scales, $O(q)$, which appear to approach fixed ratios as the flow evolves (Harris et al. 1977), and the characteristic time of the turbulence, $O(L/q)$, locks on to the characteristic time of the shear, $O(S^{-1})$. It is plausible that the turbulence ultimately attains a self-similar structure with exponential growth of length and velocity scales (Rogallo 1981). The early evolution of isotropic turbulence subjected to uniform shear is predicted well by the linear theory of rapid distortion (Deissler 1961, 1972). Although this theory incorrectly predicts ultimate turbulence decay, its prediction of the Reynolds-stress anisotropy and two-point correlations is surprisingly accurate (Townsend 1976). The first simulation of homogeneous shear was

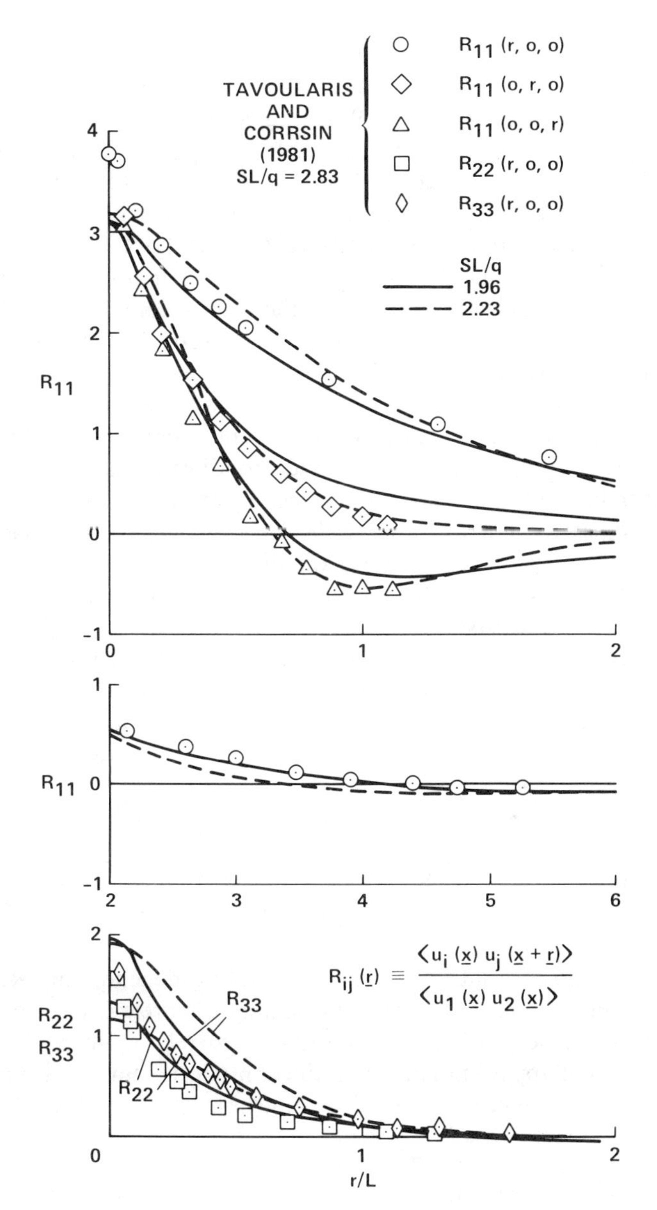

Figure 1 Self-similarity of the autocorrelations in homogeneous shear turbulence (from Rogallo 1981).

the $16 \times 16 \times 16$ finite-difference LES by Shaanan et al. (1975). Their results agree qualitatively with the experimental data, even though periodic boundary conditions were applied on a fixed mesh (see Section 5). More details of the flow are obtained in the $64 \times 64 \times 64$ direct spectral simulations of Feiereisen et al. (1981) and Shirani et al. (1981) in which compressibility effects and passive scalar transport, respectively, are studied. The results of Rogallo's (1981) $128 \times 128 \times 128$ direct spectral simulation indicate that even at a macroscale Reynolds number an order of magnitude below that of Tavoularis & Corrsin (1981), the large-scale statistics of the experiment can be reproduced. A major difficulty is the definition of a characteristic length for the energy-containing scales. The integral scale depends strongly on the largest computed scales for which the statistical sample is poor. In Figure 1 the computed correlations for two simulations are compared with the data of Tavoularis & Corrsin. The correlations are normalized by the turbulent shear stress rather than normal stresses, and the separation is made nondimensional by reference to the longitudinal integral scale in the mean-flow direction. This scaling should collapse the correlations of the large scales; the correlations of streamwise velocity collapse well for the different Reynolds numbers and characteristic times ratios, SL/q, but collapse for the transverse velocity components is less satisfying.

The calculated flow fields can be used as detailed data for the development and testing of closure models. As an example, the tensor sum of the pressure-strain correlation (the "slow" term) and the deviator of dissipation,

$$\varepsilon\phi_{ij} = -2\overline{pS_{ij}} + 2(\varepsilon_{ij} - \tfrac{1}{3}\varepsilon\,\delta_{ij}), \qquad \varepsilon_{ij} = \nu\overline{\frac{\partial u_i}{\partial x_k}\frac{\partial u_j}{\partial x_k}}, \quad \varepsilon = \varepsilon_{ii} \tag{10}$$

is usually modeled in a Reynolds-stress closure (Lumley 1980) by a scalar multiple of the Reynolds-stress anisotropy tensor, $\phi_{ij} \sim \beta b_{ij}$, where $b_{ij} = \overline{u_i u_j}/\overline{u_k u_k} - \frac{1}{3}\delta_{ij}$.

Lumley proposes that the scalar coefficient depends on Reynolds number, the invariants of the stress tensor, and other relevant scalars of the flow. The two tensors (Figure 2*a*) are indeed correlated in the calculated fields, and the collapse obtained by the linear model (Figure 2*b*) supports its use (but its performance in other anisotropic homogeneous flows does not;

Figure 2 Lumley's (1980) linear model of pressure-strain correlation and dissipation anisotropy. (*a*) Dependence of modeled tensor on Reynolds-stress anisotropy tensor; (*b*) comparison of modeled and measured values; (*c*) variation of model coefficient with Reynolds number. The data points represent independent flow fields at a wide range of parameters (from Rogallo 1981).

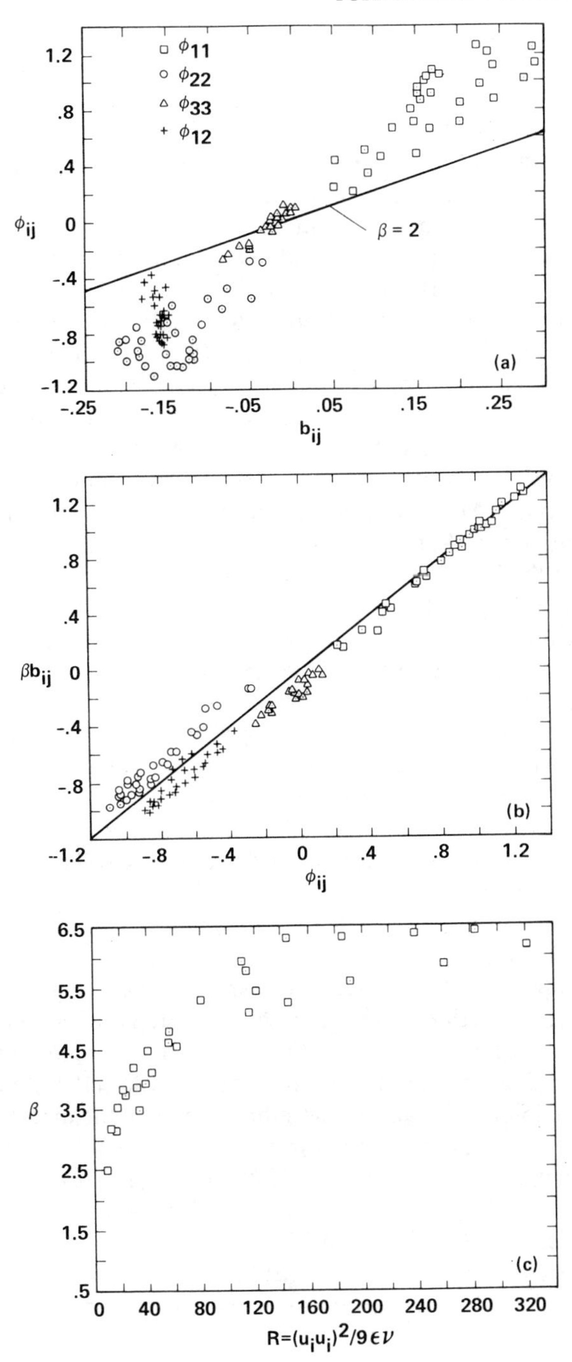

ϕ_{11}
ϕ_{22}
ϕ_{33}
ϕ_{12}
$\beta = 2$
(a)
ϕ_{ij}
b_{ij}
1.2
.8
.4
0
-.4
-.8
-1.2
-.25
-.15
-.05
.05
.15
.25
(b)
βb_{ij}
ϕ_{ij}
-1.2
-.8
-.4
0
.4
.8
1.2
(c)
β
6.5
5.5
4.5
3.5
2.5
1.5
.5
0
40
80
120
160
200
240
280
320
$R=(u_iu_i)^2/9\epsilon\nu$

see Rogallo 1981). The orderly nature of the small remaining error suggests the possibility of higher-order model terms. The increase of the model coefficient, β, with Reynolds number (Figure 2*c*) has been predicted by Lumley, but it should be noted that other scalar attributes of the flow, particularly the ratio of shear and turbulence time scales, are important and they are also varying among the data shown.

The mixing layer separating two uniform streams of differing speed has been studied analytically, experimentally, and recently by simulation. Much of the recent work is concerned with the observed organized vortical structures that result from Kelvin-Helmholtz instability in turbulent layers and their downstream growth by pairing (Roshko 1976). It is found experimentally that the evolution of the layer is strongly influenced by imposed perturbations, and the simulations indicate an analogous sensitivity to initial conditions. Simulations of the LES type have been performed at low resolution by Mansour et al. (1978) and Cain et al. (1981). In the calculations of Mansour et al., the roll-up stage of the flow is inhibited by a mesh domain too short to support unstable waves. When vortex cores are included in the initial field the eddy-viscosity model, in the presence of the mean shear, prevents the proper growth of energy and length scales. The problem appears to be simply one of inadequate resolution. The mesh of Cain et al., on the other hand, is scaled to include the fundamental instability wave and its subharmonic. Roll up of the layer occurs, with the resulting vortices meandering in the spanwise direction and pairing locally to form a network of vortex tubes. Riley & Metcalfe (1980b), using a $32 \times 32 \times 32$ direct spectral simulation, show (as do Cain et al.) that the presence of an energetic two-dimensional instability wave modulates the layer growth; the early growth is more rapid, but once roll up has occurred growth is delayed until the vortices approach each other (by turbulent diffusion, convection by a subharmonic, spanwise variations in proximity, etc.) closely enough for pairing to occur. The spanwise vorticity field of a turbulent mixing layer (Figure 3*a*) clearly shows coherent structures, even though the layer growth is statistically self-similar. The structures are not simple two-dimensional vortices however, as the vorticity at another spanwise plane (Figure 3*b*) indicates. Metcalfe & Riley (1981) increase the computational domain to capture the subharmonic of the instability wave. These $64 \times 64 \times 64$ mesh results confirm their earlier results, and the larger domain eliminates a spurious growth of turbulence intensity found there. This flow illustrates the importance of not constraining potentially important scales, in this case the instability scale.

The most extensive application of LES has been the calculation of fully developed turbulent channel flow. In the first realistic numerical simulation

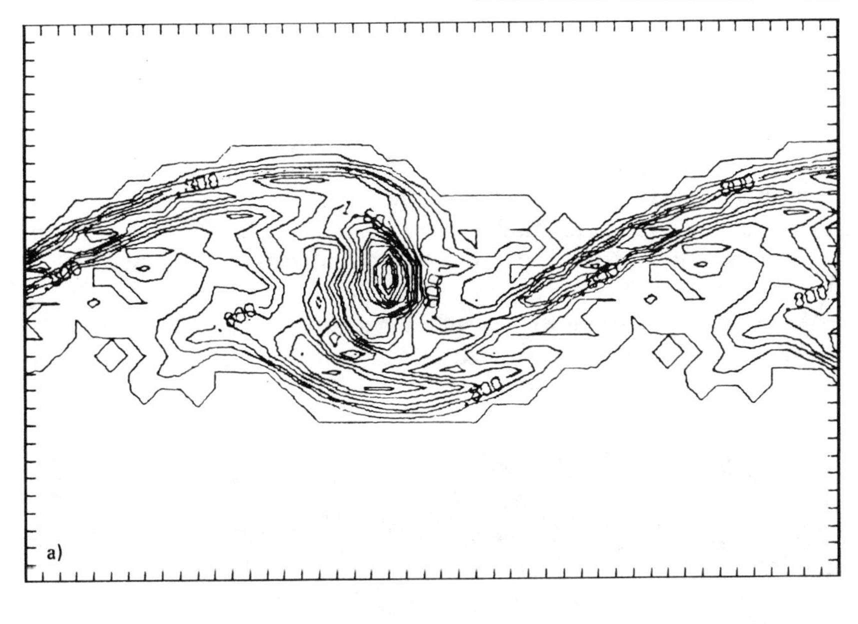

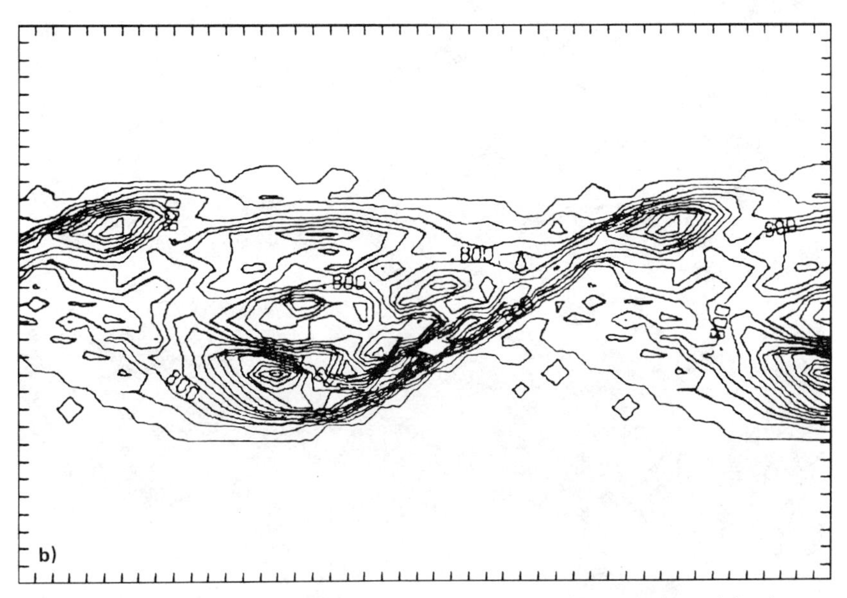

Figure 3 Distribution of the spanwise vorticity component in a turbulent mixing layer as viewed in the spanwise direction. The distribution is shown in two planes separated by half the computational period (from Riley & Metcalfe 1980b).

of turbulence, Deardorff (1970) calculated this flow at a very high Reynolds number using only 6720 grid points. Schumann (1975) and Grotzbach & Schumann (1979) used up to 65,536 grid points, included temperature fluctuations and heat transfer, and considered a range of moderate Reynolds numbers (Re $> 10^4$), but like Deardorff, they *modeled* the wall-layer dynamics. In these calculations the mean-velocity profile, turbulent intensities, and pressure statistics are in good agreement with the experimental data. Moin & Kim (1982) calculated the channel flow at Re $= 13{,}800$ (based on channel half-width δ and centerline velocity), and extended the calculations to the wall using a nonuniform mesh with total of 516,096 grid points. The computed velocity and pressure field was used to study the time-dependent structure of the flow and its relationship to

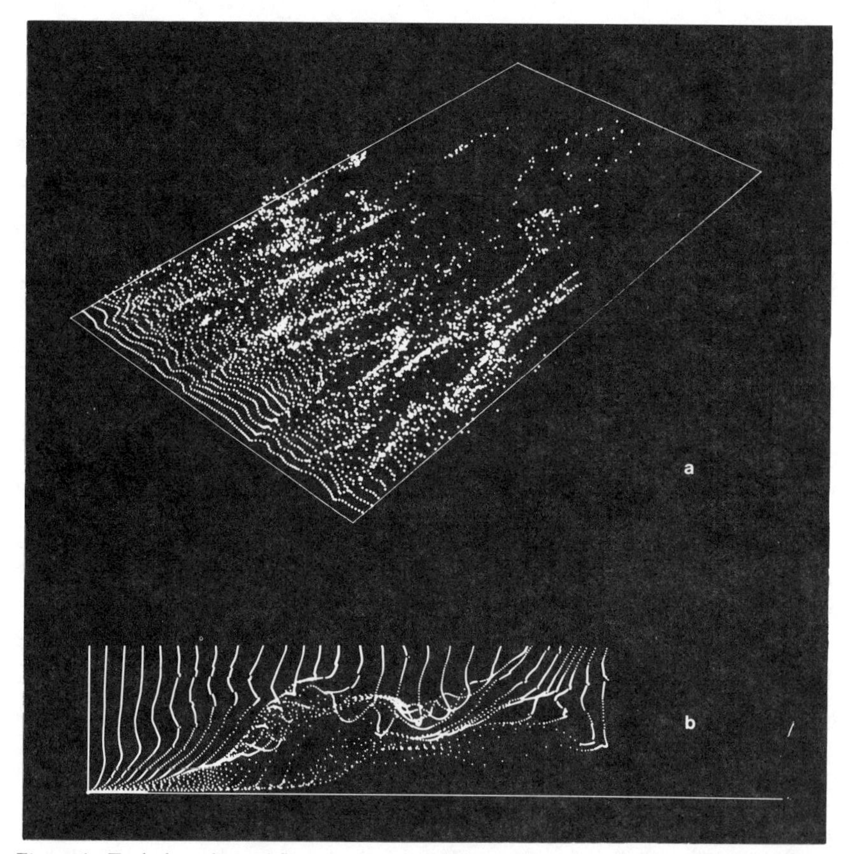

Figure 4 Turbulent channel flow visualized by fluid markers (simulated hydrogen bubbles). (*a*) Markers introduced on a line in the spanwise direction at $y^+ = 6$; (*b*) markers introduced on a line normal to the wall; view extends to $y^+ = 240$ (from data of Moin & Kim 1982).

various flow statistics (including those appearing in the time-averaged Reynolds-stress equations). The detailed flow field was analyzed with contour plots of the instantaneous velocity, pressure, and vorticity fluctuations; with higher-order statistical correlations; and with tracking of passive particles in the flow. In particular, a motion picture was made simulating hydrogen-bubble flow-visualization experiments (see Kim et al. 1971; Kline et al. 1967). In Figure 4 two typical frames from this film show the paths of bubbles generated near the wall ($y^+ \simeq 6$) along a line in the spanwise direction and of bubbles generated along a line normal to the channel wall. Various distinct flow features, including the wall-layer streaks (Figure 4*a*), and the formation of profiles with multiple inflection points and ejection of fluid from the wall region (Figure 4*b*), are in accordance with laboratory observations.

The contours of wall-pressure fluctuations from the turbulent channel flow simulations of Grotzbach & Schumann (1979) are shown in Figure 5. In agreement with experimental measurements (Bull 1967, Willmarth 1975), the large-scale pressure fluctuations are correlated at considerably greater distances in the lateral direction than they are in the mean-flow direction. This feature is reproduced in the calculations of Moin & Kim (1982), where localized regions of high pressure intensity are also observed. The two-point pressure correlations of Moin & Kim (1982) indicate that the spanwise elongation of pressure eddies persists across the entire channel. Figure 6 shows the two-point velocity and pressure correlations in the vicinity of the wall ($y/\delta = 0.06$, $y^+ = 38$). The pressure correlation is negative for large streamwise separations but is always positive for

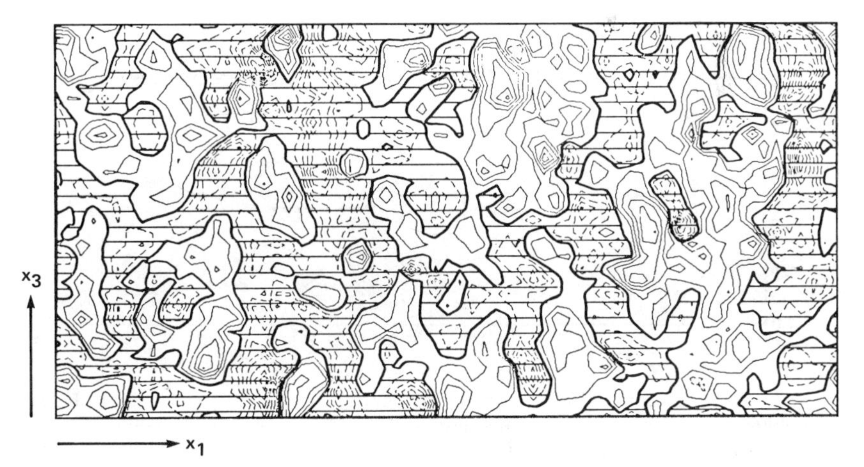

Figure 5 Pressure distribution at the wall in turbulent channel flow (from Grotzbach & Schumann 1979).

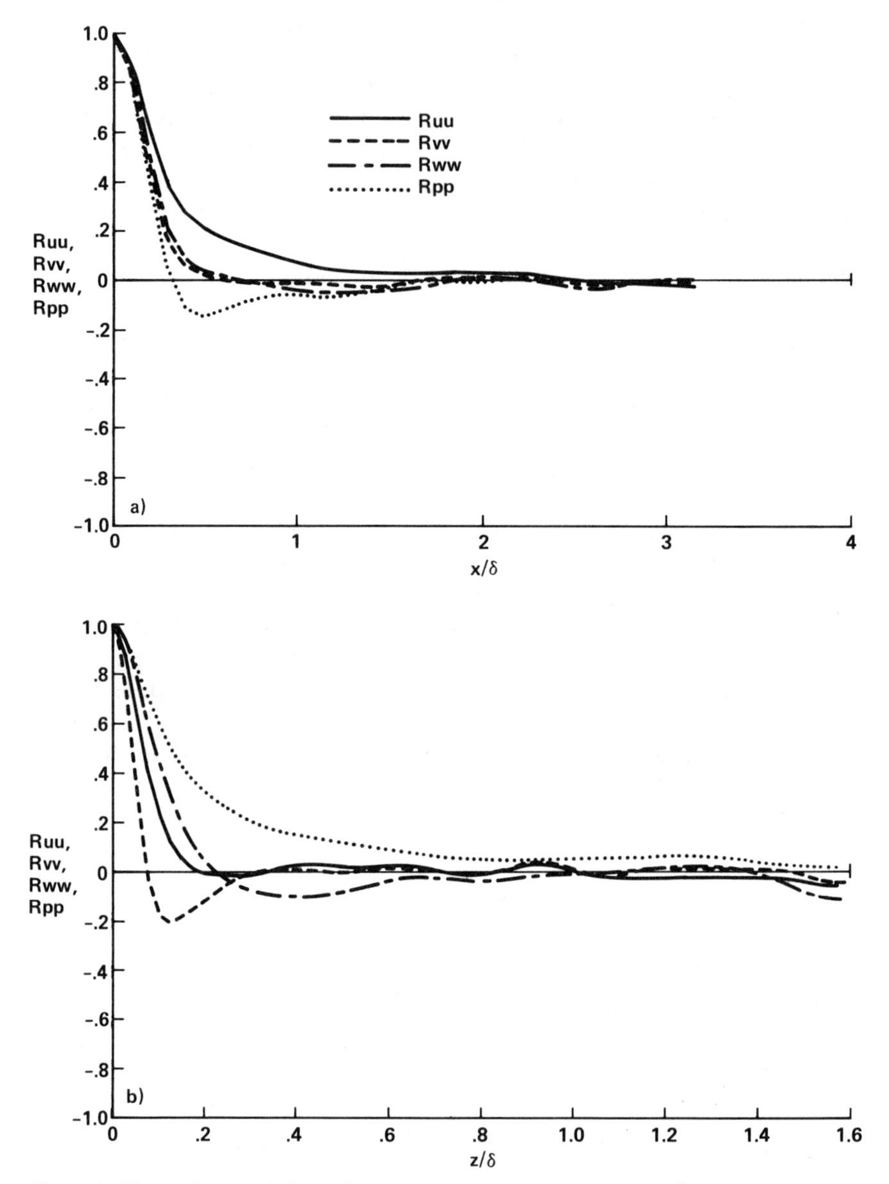

Figure 6 Two-point correlations of pressure and velocity near the wall ($y^+ = 38$) in turbulent channel flow. (*a*) Points separated in streamwise direction; (*b*) points separated in spanwise direction (from data of Moin & Kim 1982).

spanwise separations. The same characteristics are exhibited by experimentally measured wall pressure correlations (Bull 1967). Thus, the fluctuating pressure gradients driving the flow are stronger in the streamwise direction than in the spanwise direction. In the vicinity of the wall the pressure fluctuations are correlated over larger lateral distances than are the velocity components, but in the streamwise direction it is the velocity fluctuations (particularly the streamwise component) that are correlated over larger distances.

Recently, Kim (1983) has further studied the spatial structure of the wall layer by applying a conditional sampling technique to the "data" generated by Moin & Kim (1982). Figure 7 shows the signatures of the pressure and streamwise velocity component, during a "bursting event," obtained using a variant of the VITA conditional sampling technique of Blackwelder & Kaplan (1976). The velocity signatures are remarkably similar to the experimental results. The pressure signatures (which can be obtained experimentally only at the walls) indicate localized peaks during the detected bursting event, with adverse pressure gradient associated with flow deceleration. The pressure signature persists at significantly larger

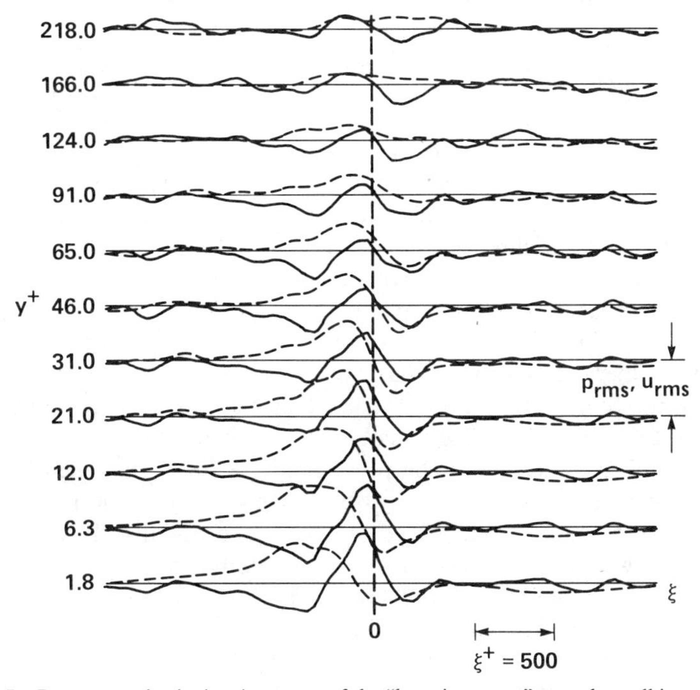

Figure 7 Pressure and velocity signatures of the "bursting event" near the wall in turbulent channel flow. ----- streamwise velocity; ——— pressure (from Kim 1983).

distances normal to the wall than does the velocity signature; this suggests that the fluctuating pressure gradient driving the wall layer is imposed by the outer flow. The conditionally averaged transverse velocity components and streamwise vorticity component, displayed in planes normal to the flow direction, show a distinct pair of counter-rotating vortical structures associated with the bursting process.

In addition to the fundamental studies outlined above, LES has also been used in practical engineering applications, where it can be more cost effective than the multiple transport equation statistical models (Schumann et al. 1980). For example, in a problem related to nuclear-reactor safety, Grotzbach (1979) used a very-coarse-grid ($16 \times 16 \times 8$) LES to investigate the effect of buoyancy on flow mixing in the downcomer of a reactor. He found that buoyancy enhances mixing of the entering hot and cold fluid streams and prevents a "hot chimney" from developing along the length of downcomer adjacent to the reactor core. These results were later confirmed by experimental measurements. LES appears to be the only viable predictive computational tool in applications that involve aerodynamic noise, reduction of turbulent skin friction (for example, flow over compliant boundaries), and other applications in which the details of turbulence dynamics play a dominant role.

7. SUMMARY

Numerical simulation has become a viable complement to experiment in both fundamental and applied turbulence research. Its growing popularity reflects both its promise of realistic answers to a difficult problem and the continuing rapid decline in computing costs. We expect this trend to continue. In addition to the advances in computer capacity of the last decade, less easily measured progress has been made in simulation technique and in the utilization of simulation results. A notable development in numerical algorithms has been the use of spectral methods for direct simulations in simple geometries. This method is not very attractive at present for complex LES calculations involving mesh mapping and implicit time advancement. The premise of LES, that turbulence calculations can be closed more easily by truncating scales of motion rather than statistical moments, is supported by results, especially those in wall-bounded flows. But the hope that very simple eddy-viscosity models would be sufficient has not proved correct for the anisotropic SGS stresses caused by high mean field gradients, at least with a reasonable number of mesh points. Anisotropic meshes, which cause ambiguity in the definition of SGS length scales, and moderate Reynolds numbers, at which the roles of the various scales overlap, introduce additional modeling difficulties. The decomposition of SGS stress into mean and fluctuations, which essentially models separately the SGS energy transfer from the mean flow and that from the remainder of the resolved scales, provides a workable closure for the wall-bounded cases reported. Despite the ad hoc nature of the model, it demonstrates the ability of an LES to base the SGS model on subsets of the resolved set of scales. The explicit calculation of the Leonard and "cross" terms, and the related modeling ideas of Bardina et al. (1980), also directly utilize more information from the resolved scales to reduce model error.

The nature of the flow near walls requires the expensive resolution of very small scales. The cost of resolution can be reduced by embedding a fine mesh only near the wall. However, the scale disparity between "wall" and "wake" layers, the presence of the overlap "log" layer, and the known form of the organized eddies near the wall strongly suggest that in some practical applications the wall layer can be replaced by a boundary condition for the wake layer that is imposed in the log layer. This situation is analogous to the separation at high Reynolds number of the energy-containing scales and the dissipative scales by an inertial range, and we certainly believe closure is possible in the inertial range in that case.

Inflow and outflow boundary conditions present a major obstacle in the calculation of complex engineering flows. In self-similar cases (wakes, jets, mixing layers, etc.) the use of periodic boundary conditions in the appropriate similarity coordinates seems natural, but in the more general case it will be necessary to measure the sensitivity of computed values to the inflow and outflow conditions used.

The future of turbulence simulation appears bright indeed. While there remains much work to be done on simulation technique, modeling, and numerical methods, we have already reached the point of being able to generate more information than we are able to digest. One can imagine in the near future a researcher at a graphics terminal with access to computed turbulent flow fields of high resolution. He will be able to display any desired quantity computed from the field (for example, statistical averages or three-dimensional visualizations of fluid motion and eddy structure). The computer can answer any question about the fields it holds, and the researcher can devote his time to the really difficult effort of finding the right question to ask. The experimentalist must arrange his experiment and gather the specific data needed to answer his questions; if these answers suggest other questions, the experiment must frequently be rerun to collect new data. The use of stored simulation results places fewer constraints on the questions that can be answered, and allows rapid interactive display of results. An experimentalist with access to such a data base would be able to evaluate the choices of data to be taken from the experiment; for example, he could tune a conditional sampling strategy to capture more precisely the events of interest. A person developing turbulence models could use the data base to evaluate proposed models. Furthermore, the flow-field data base can be shared by other researchers who do not have the computer power required to generate the fields, but do have enough power to probe them. The development of hardware and software tools for interactive probing of simulation results, the availability of the computed flow fields (in computer-readable form), and the advancement in computer capacity will ultimately determine the degree to which simulation enhances our understanding and ability to control turbulence.

Literature Cited

Antonopoulos-Domis, M. 1981. Large-eddy simulation of a passive scalar in isotropic turbulence. *J. Fluid Mech.* 104:55–79

Bardina, J., Ferziger, J. H., Reynolds, W. C. 1980. Improved subgrid-scale models for large-eddy simulation. *AIAA Pap. 80-1357*

Baron, F. 1982. *Three-dimensional large-eddy simulation of turbulent shear flows.* Doctoral thesis. Univ. Pierre Marie Curie, Paris

Blackwelder, R. F., Kaplan, R. E. 1976. On the wall structure of the turbulent boundary layer. *J. Fluid Mech.* 76:89–112

Bull, M. K. 1967. Wall pressure fluctuations associated with subsonic turbulent boundary layer flow. *J. Fluid Mech.* 28:719–54

Cain, A. B., Reynolds, W. C., Ferziger, J. H. 1981. A three-dimensional simulation of transition and early turbulence in a time-developing mixing layer. *Rep. TF-14,* Dept. Mech. Eng., Stanford Univ.

Cambon, C., Jeandel, D., Mathieu, J. 1981. Spectral modelling of homogeneous non-isotropic turbulence. *J. Fluid Mech.* 104: 247–62

Chapman, D. R. 1979. Computational aerodynamics development and outlook. *AIAA J.* 17:1293–1313

Chapman, D. R. 1981. Trends and pacing items in computational aerodynamics. See Reynolds & MacCormack 1981, pp. 1–11

Chapman, D. R., Kuhn, G. D. 1981. Two-component Navier-Stokes computational model of viscous sublayer turbulence. *AIAA Pap. 81-1024*

Chollet, J. 1982. Two-point closures as a subgrid scale modeling for large eddy simulations. *NCAR Preprint 0901/82-16.* Presented at Symp. Turbul. Shear Flows, Karlsruhe, West Ger., Sept. 1983

Chollet, J., Lesieur, M. 1981. Parameterization of small scales of three-dimensional isotropic turbulence utilizing spectral closures. *J. Atmos. Sci.* 38:2747–57

Clark, R. A., Ferziger, J. H., Reynolds, W. C. 1977. Evaluation of subgrid-scale turbulence models using a fully simulated turbulent flow. *Rep. TF-9,* Dept. Mech. Eng., Stanford Univ.

Clark, R. A., Ferziger, J. H., Reynolds, W. C. 1979. Evaluation of subgrid-scale models using an accurately simulated turbulent flow. *J. Fluid Mech.* 91:1–16

Comte-Bellot, G. 1963. *Contribution a l'etude de la turbulence de conduite.* Doctoral thesis. Univ. Grenoble

Comte-Bellot, G., Corrsin, S. 1971. Simple Eulerian time correlation of full- and narrow-band velocity signals in grid-generated "isotropic" turbulence. *J. Fluid Mech.* 48:273–337

Corrsin, S. 1961. Turbulent flow. *Am. Sci.* 49:300–25

Corrsin, S. 1974. Limitations of gradient transport models in random walks and in turbulence. See Frenkiel & Munn 1974, pp. 25–60

Deardorff, J. W. 1970. A numerical study of three-dimensional turbulent channel flow at large Reynolds numbers. *J. Fluid Mech.* 41:453–80

Deardorff, J. W. 1971. On the magnitude of the subgrid scale eddy coefficient. *J. Comput. Phys.* 7:120–33

Deardorff, J. W. 1973. The use of subgrid transport equations in a three-dimensional model of atmospheric turbulence. *J. Fluids Eng.* 95:429–38

Deissler, R. G. 1961. Effects of inhomogeneity and of shear flow in weak turbulence fields. *Phys. Fluids* 4:1187–98

Deissler, R. G. 1972. Growth of turbulence in the presence of shear. *Phys. Fluids* 15: 1918–20

Durst, F., Launder, B. E., Schmidt, F. W., Whitelaw, J. H., eds. 1979. *Turbulent Shear Flows I.* Berlin: Springer. 413 pp.

Eckelmann, H. 1974. The structure of the

viscous sublayer and the adjacent wall region in a turbulent channel flow. *J. Fluid Mech.* 65:439–59

Feiereisen, W. J., Reynolds, W. C., Ferziger, J. H. 1981. Numerical simulation of compressible, homogeneous, turbulent shear flow. *Rep. TF-13*, Dept. Mech. Eng., Stanford Univ.

Ferziger, J. H. 1982. State of the art in subgrid scale modeling. In *Numerical and Physical Aspects of Aerodynamic Flows*, ed. T. Cebici, pp. 53–68. New York: Springer. 636 pp.

Ferziger, J. H., Mehta, U. B., Reynolds, W. C. 1977. *Large eddy simulation of homogeneous isotropic turbulence.* Presented at Symp. Turbul. Shear Flows, Penn. State Univ., University Park

Fox, D. G., Deardorff, J. W. 1972. Computer methods for simulation of multidimensional, nonlinear, subsonic, incompressible flows. *J. Heat Transfer* 94:337–46

Fox, D. G., Lilly, D. K. 1972. Numerical simulation of turbulent flows. *Rev. Geophys. Space Phys.* 10:51–72

Fox, L., Parker, I. 1968. Chebyshev polynomials in numerical analysis. London: Oxford Univ. Press

Frenkiel, F. N., Munn, R. E., eds 1974. *Proc. Symp. Turbul. Diffus. Environ. Pollut., Charlottesville, 1973. Advances in Geophysics*, Vol. 18A. New York: Academic. 462 pp.

Gottlieb, D., Orszag, S. A. 1977. Numerical analysis of spectral methods: theory and application. *CBMS-NSF Reg. Conf. Ser. Appl. Math.*, Vol. 26. Philadelphia: SIAM. 170 pp.

Grotzbach, G. 1979. Numerical investigation of radial mixing capabilities in strongly buoyancy-influenced vertical, turbulent channel flows. *Nucl. Eng. Des.* 54:49–66

Grotzbach, G. 1981. Spatial resolution requirements for numerical simulation of internally heated fluid layers. In *Numerical Methods in Laminar and Turbulent Flow*, ed. C. Taylor, B. A. Schrefler, pp. 593–604. Swansea, UK: Pineridge

Grotzbach, G., Schumann, U. 1979. Direct numerical simulation of turbulent velocity-, pressure-, and temperature-fields in channel flows. See Durst et al. 1979, pp. 370–85

Harlow, F. H., Welch, J. E. 1965. Numerical calculation of time-dependent viscous incompressible flow of fluid with free surface. *Phys. Fluids* 8:2182–89

Harris, V. G., Graham, J. A. H., Corrsin, S. 1977. Further experiments in nearly homogeneous turbulent shear flow. *J. Fluid Mech.* 81:657–87

Herring, J. R., Orszag, S. A., Kraichnan, R. H., Fox, D. G. 1974. Decay of two-dimensional homogeneous turbulence. *J. Fluid Mech.* 66:417–44

Kim, H. T., Kline, S. J., Reynolds, W. C. 1971. The production of turbulence near a smooth wall in a turbulent boundary layer. *J. Fluid Mech.* 50:133–60

Kim, J. 1983. On the structure of wall-bounded turbulent flows. *NASA TM-84313.* Also *Phys. Fluids* 1983. In press

Kleiser, L., Schumann, U. 1979. Treatment of incompressibility and boundary conditions in 3-D numerical spectral simulations of plane channel flows. *Proc. GAMM Conf. Numer. Meth. Fluid Mech., 3rd*, pp. 165–73. Braunscheig/Wiesbaden: Friedr. Vieweg & Sohn

Kline, S. J., Reynolds, W. C., Schraub, F. A., Runstadler, P. W. 1967. The structure of turbulent boundary layers. *J. Fluid Mech.* 30:741–73

Kollmann, W., ed. 1980. *Prediction Methods for Turbulent Flows.* New York: Hemisphere. 468 pp.

Kraichnan, R. H. 1976. Eddy viscosity in two and three dimensions. *J. Atmos. Sci.* 33:1521–36

Krause, E., ed. 1982. *Proc. Int. Conf. Numer. Methods Fluid Dyn., 8th, Aachen. Lecture Notes in Physics*, Vol. 170. New York: Springer. 569 pp.

Kwak, D., Reynolds, W. C., Ferziger, J. H. 1975. Three-dimensional time-dependent computation of turbulent flow. *Rep. TF-5*, Dept. Mech. Eng., Stanford Univ.

Laufer, J. 1951. Investigation of turbulent flow in a two-dimensional channel. *NACA Rep. 1053*

Leonard, A. 1974. Energy cascade in large-eddy simulations of turbulent fluid flows. See Frenkiel & Munn 1974, pp. 237–48

Leonard, A. 1980. Vortex methods for flow simulation. *J. Comput. Phys.* 37:289–335

Leonard, A., Wray, A. 1982. A new numerical method for the simulation of three-dimensional flow in a pipe. See Krause 1982, pp. 335–42

Lesieur, M., Schertzer, D. 1978. Amortissement autosimilaire d'une turbulence a grand nombre de Reynolds. *J. Méc.* 17:610–46

Leslie, D. C. 1973. *Developments in the Theory of Turbulence.* Oxford: Clarendon. 368 pp.

Leslie, D. C., Quarini, G. L. 1979. The application of turbulence theory to the formulation of subgrid modelling procedures. *J. Fluid Mech.* 91:65–91

Lilly, D. K. 1964. Numerical solutions for the shape-preserving two-dimensional thermal convection element. *J. Atmos. Sci.* 21:83–98

Lilly, D. K. 1966. On the application of the eddy viscosity concept in the inertial subrange of turbulence. *NCAR Manuscr. 123*

Love, M. D. 1980. Subgrid modelling studies with Burgers' equation. *J. Fluid Mech.* 100:87–110

Love, M. D., Leslie, D. C. 1979. Studies of sub-grid modelling with classical closures and Burgers' equation. See Durst et al. 1979, pp. 353–69

Lumley, J. L. 1980. Second order modelling of turbulent flows. See Kollmann 1980, pp. 1–32

Mansour, N. N., Ferziger, J. H., Reynolds, W. C. 1978. Large-eddy simulation of a turbulent mixing layer. *Rep. TF-11*, Dept. Mech. Eng., Stanford Univ.

Mansour, N. N., Moin, P., Reynolds, W. C., Ferziger, J. H. 1979. Improved methods for large eddy simulations of turbulence. See Durst et al. 1979, pp. 386–401

McMillan, O. J., Ferziger, J. H. 1979. Direct testing of subgrid-scale models. *AIAA J.* 17:1340–46

McMillan, O. J., Ferziger, J. H., Rogallo, R. S. 1980. Tests of new subgrid-scale models in strained turbulence. *AIAA Pap. 80-1339*

Metcalfe, R. W., Riley, J. J. 1981. Direct numerical simulations of turbulent shear flows. See Reynolds & MacCormack 1981, pp. 279–84

Moin, P. 1982. Numerical simulation of wall-bounded turbulent shear flows. See Krause 1982, pp. 53–76

Moin, P., Kim, J. 1980. On the numerical solution of time-dependent viscous incompressible fluid flows involving solid boundaries. *J. Comput. Phys.* 35:381–92

Moin, P., Kim, J. 1982. Numerical investigation of turbulent channel flow. *J. Fluid Mech.* 118:341–77

Monin, A. S., Yaglom, A. M. 1971. *Statistical Fluid Mechanics*, Vol. 1. Cambridge, Mass: MIT Press. 769 pp.

Orszag, S. A. 1970. Analytical theories of turbulence. *J. Fluid Mech.* 41:363–86

Orszag, S. A. 1971. Numerical simulation of incompressible flows within simple boundaries: accuracy. *J. Fluid Mech.* 49:75–112

Orszag, S. A. 1972. Comparison of pseudospectral and spectral approximation. *Stud. Appl. Math.* 51:253–59

Orszag, S. A. 1980. Spectral methods for problems in complex geometries. *J. Comput. Phys.* 37:70–92

Orszag, S. A., Kells, L. C. 1980. Transition to turbulence in plane Poiseuille and plane Couette flow. *J. Fluid Mech.* 96:159–205

Orszag, S. A., Pao, Y. 1974. Numerical computation of turbulent shear flows. See Frenkiel & Munn 1974, pp. 225–36

Orszag, S. A., Patterson, G. S. 1972. Numerical simulation of three-dimensional homogeneous isotropic turbulence. *Phys. Rev. Lett.* 28:76–79

Patterson, G. S., Orszag, S. A. 1971. Spectral calculations of isotropic turbulence: efficient removal of aliasing interactions. *Phys. Fluids.* 14:2538–41

Phillips, N. A. 1959. An example of nonlinear computational instability. In *The Atmosphere and Sea in Motion*, ed. B. Bolin, pp. 501–4. New York: Rockefeller Inst. Press

Rajagopalan, S., Antonia, R. A. 1979. Some properties of the large structure in a fully developed turbulent duct flow. *Phys. Fluids* 22:614–22

Reynolds, W. C. 1976. Computation of turbulent flows. *Ann. Rev. Fluid Mech.* 8:183–208

Reynolds, W. C., MacCormack, R. W., eds. 1981. *Proc. Int. Conf. Numer. Methods Fluid Dyn., 7th, Stanford/NASA-Ames, 1980. Lecture Notes in Physics*, Vol. 141. New York: Springer. 485 pp.

Riley, J. J., Metcalfe, R. W. 1980a. Direct numerical simulations of the turbulent wake of an axisymmetric body. In *Turbulent Shear Flows II*, eds. J. S. Bradbury, F. Durst, B. E. Launder, F. W. Schmidt, J. H. Whitelaw, pp. 78–93. Berlin: Springer. 480 pp.

Riley, J. J., Metcalfe, R. W. 1980b. Direct numerical simulation of a perturbed, turbulent mixing layer. *AIAA Pap. 80-0274*

Robinson, S. K. 1982. An experimental search for near-wall boundary conditions for large eddy simulation. *AIAA Pap. 82-0963*

Rogallo, R. S. 1981. Numerical experiments in homogeneous turbulence. *NASA TM-81315*

Roshko, A. 1976. Structure of turbulent shear flows: a new look. *AIAA J.* 14:1349–57

Roy, P. 1982. Numerical simulation of homogeneous anisotropic turbulence. See Krause 1982, pp. 440–47

Schumann, U. 1975. Subgrid scale model for finite difference simulations of turbulent flows in plane channels and annuli. *J. Comput. Phys.* 18:376–404

Schumann, U., Grotzbach, G., Kleiser, L. 1980. Direct numerical simulation of turbulence. See Kollmann 1980, pp. 124–258

Shaanan, S., Ferziger, J. H., Reynolds, W. C. 1975. Numerical simulation of turbulence in the presence of shear. *Rep. TF-6*, Dept. Mech. Eng., Stanford Univ.

Shirani, E., Ferziger, J. H., Reynolds, W. C. 1981. Mixing of a passive scalar in isotropic and sheared homogeneous turbulence. *Rep. TF-15*, Dept. Mech. Eng., Stanford Univ.

Siggia, E. D. 1981. Numerical study of small-scale intermittency in three-dimensional turbulence. *J. Fluid Mech.* 107:375–406

Smagorinsky, J. 1963. General circulation experiments with the primitive equations. I. The basic experiment. *Mon. Weather Rev.* 91:99–164

Tavoularis, S., Corrsin, S. 1981. Experiments in nearly homogeneous turbulent shear flow with a uniform mean temperature gradient. Part 1. *J. Fluid Mech.* 104:311–47

Tennekes, H., Lumley, J. L. 1972. *A First Course in Turbulence*. Cambridge, Mass: MIT Press. 300 pp.

Townsend, A. A. 1976. *The Structure of Turbulent Shear Flow*. Cambridge Univ. Press. 429 pp. 2nd ed.

Warming, R. F., Hyett, B. J. 1974. The modified equation approach to the stability and accuracy analysis of finite-difference methods. *J. Comput. Phys.* 14:159–79

Willmarth, W. W. 1975. Pressure fluctuations beneath turbulent boundary layers. *Ann. Rev. Fluid Mech.* 7:13–38

Yoshizawa, A. 1979. A statistical investigation upon the eddy viscosity in incompressible turbulence. *J. Phys. Soc. Jpn.* 47:1665–69

Yoshizawa, A. 1982. A statistically-derived subgrid model for the large-eddy simulation of turbulence. *Phys. Fluids.* 25: 1532–38

Ann. Rev. Fluid Mech. 1984. 16: 139–77

NONLINEAR INTERACTIONS IN THE FLUID MECHANICS OF HELIUM II[1]

H. W. Liepmann

Graduate Aeronautical Laboratories, California Institute of Technology, Pasadena, California 91125

G. A. Laguna

Los Alamos National Laboratory, Los Alamos, New Mexico 87545

INTRODUCTION

Besides its practical importance in a host of technical applications, fluid mechanics retains its intrinsic interest as a physical discipline. The governing equations are nonlinear, and hence the motion of fluids demonstrates the complexities of solution of a nonlinear field theory, a fact that has been appreciated more and more in recent times. The most striking manifestations of this nonlinearity are shock waves and turbulence, corresponding to nonlinear wave and vortex interactions, respectively.

Several years ago a study of liquid helium was initiated at GALCIT[2] to arrive at a better understanding of the fluid-mechanical aspects of He II, in particular to demonstrate or disprove the validity of the Landau-London two-fluid equations *beyond* the linear approximation. It was hoped that our extensive experience with "classical" turbulence and shock-wave motion could help in making a contribution to He II dynamics.

In this article the liquid is considered strictly as a fluid described by a set of thermodynamic and dynamic variables interrelated by the two-fluid equations of London and Landau.

The macroscopic quantum effects of liquid helium below the λ transition

[2] Graduate Aeronautical Laboratories, California Institute of Technology.

appear in the form of an additional pair of conjugate thermodynamic state variables: $\xi = \rho_n/\rho$, the fractional ratio of normal fluid density to density, and w, the so-called counterflow velocity. Thus, the thermodynamics of He II is contained in a chemical potential μ, which is a function $\mu = \mu(p, T, w)$ of the variables p, T, and w, and which satisfies a corresponding thermodynamic identity $d\mu = -SdT + vdp - \xi wdw$.

For small w (small in a sense that will become clearer later), the expression

$$\frac{\partial \mu}{\partial w} = -\xi w \tag{1}$$

can be integrated and μ written as

$$\mu = \mu_0(p, T) - \tfrac{1}{2}\xi_0(p, T)w^2. \tag{2}$$

The corresponding equations of motion—the equivalent equations to the Euler and Navier-Stokes equations of fluid dynamics—can be written either in terms of two densities ρ_n and ρ_s and two velocity components, respectively, or else (and preferably) in terms of p and ξ and two velocities $\rho\mathbf{v} = \rho_n\mathbf{v}_n + \rho_s\mathbf{v}_s$ and $\mathbf{w} = \mathbf{v}_n - \mathbf{v}_s$.

In the following, only occasional use is made of the very complex dissipation terms in the equations of motion. Consequently, the equations are written down in Figure 1 without the dissipative terms. This paper

---

$$\frac{\partial \rho}{\partial t} + \nabla \cdot (\rho \mathbf{v}) = 0,$$

$$\frac{\partial(\rho \mathbf{v})}{\partial t} + \nabla \cdot [\rho \mathbf{v}\mathbf{v} + \xi(1-\xi)\rho \mathbf{w}\mathbf{w} + p\tilde{I}] = 0,$$

$$\frac{\partial(\mathbf{v} - \xi \mathbf{w})}{\partial t} + \nabla[\mu + \tfrac{1}{2}v^2 - \xi \mathbf{w}\cdot\mathbf{v} + \tfrac{1}{2}\xi^2 w^2] = 0,$$

$$\frac{\partial E}{\partial t} + \nabla \cdot \mathbf{Q} = 0,$$

where

$$E = \rho(e + \tfrac{1}{2}v^2) + \tfrac{1}{2}\xi(1-\xi)\rho w^2$$

$$\mathbf{Q} = (h + \tfrac{1}{2}v^2)\rho\mathbf{v} + sT(1-\xi)\rho\mathbf{w} + \tfrac{1}{2}\xi(1-\xi)\rho\mathbf{v}w^2$$
$$+ \xi(1-\xi)[\mathbf{w}\cdot\mathbf{v} + (1-\xi)w^2]\rho\mathbf{w}$$

Figure 1 Two-fluid equations of motion written in terms of the bulk velocity $\mathbf{v}$ and relative velocity $\mathbf{w}$.

emphasizes the manifestations of the nonlinearities of these equations, which are analogous to the nonlinearities in ordinary fluid dynamics but more complex.

In classical fluid mechanics, the study of certain simple flows has provided a disproportionate share of our present-day knowledge. Jets, free-shear layers, flat-plate boundary layers, and shock waves still appear frequently in the fluid-mechanical literature. With liquid helium, on the other hand, most experimental investigations have chosen very restrictive geometries, such as flow through bundles of narrow capillaries or toroids packed with fine powder. While these experiments are extremely useful in elucidating some of the striking quantum-mechanical aspects of liquid helium, many of the more basic fluid-mechanical properties have remained largely unexplored.

Appropriate boundary conditions for fluid motion past real, solid boundaries, such as in capillaries, are often not at all trivial. Indeed, Lin (1959) attempted to construct a theory of He II motion based almost exclusively on altered fluid-solid boundary conditions. Consequently, the research at GALCIT was concentrated on geometrically simple flows for which the fluid-solid interactions were of minor importance, such as fluid jets and shock-wave motion.

This article presents, thus, a review of recent work on jets and shock-wave motion in liquid helium II; the emphasis is on experiments done by our group at GALCIT, because we are aware of very few other investigations that approach the subject from a similar point of view.

We cannot possibly review completely the very large literature on "turbulence" and critical velocities in liquid helium. We do, however, deal with turbulence in the context of counterflow jets and demonstrate and discuss the unique contribution that shock-wave experiments can make to the study of the breakdown of superfluidity.

Most articles on liquid helium start out with a detailed presentation of Landau's two-fluid equations. We feel that this has been done so often that we can omit this discussion. Instead, we refer the reader to the extensive treatments of two-fluid hydrodynamics that are available in the review by Roberts & Donnelly (1974) and in the books by Putterman (1974) and Donnelly (1967). For the fundamentals of He II, the book by London (1954) and the appropriate chapters in the Landau & Lifshitz series (1959) are still unsurpassed.

COUNTERFLOW JETS

In this section we review several experiments on the counterflow jet (Figure 2) that have been done in our laboratory. P. Kapitza (1941) was the first to report observations of a well-defined jet emerging from the mouth of a counterflow channel. The two-fluid model identifies the jet as normal fluid, while the orifice represents a sink for the superfluid. By studying the deflection of a vane, Kapitza verified that the jet spread at a rate appropriate for a jet in the viscous normal fluid, apparently uninfluenced by the presence of a superfluid component.

If we take a closer look at the flow field of Figure 2, we see that the counterflow jet in the two-fluid picture produces a large relative velocity between the superfluid and normal fluid, but without the confinement of walls. In the classical picture of sink flow, $v_s \approx 0$ several diameters away from the orifice. Using Kapitza's jet-spreading measurements, the relative velocity is then $w = v_n - v_s \approx v_n$. Thus, the counterflow jet affords an opportunity to study a two-fluid flow field at high relative velocity but without perturbing boundaries. It is exactly for this reason that jets, wakes, and free-shear layers have remained such enduring structures in fluid-dynamics research. It is also true that even after years of extensive study these classical flows are not fully understood. The relatively recent discovery by Brown & Roshko (1974) of large coherent structures in a shear layer is a case in point.

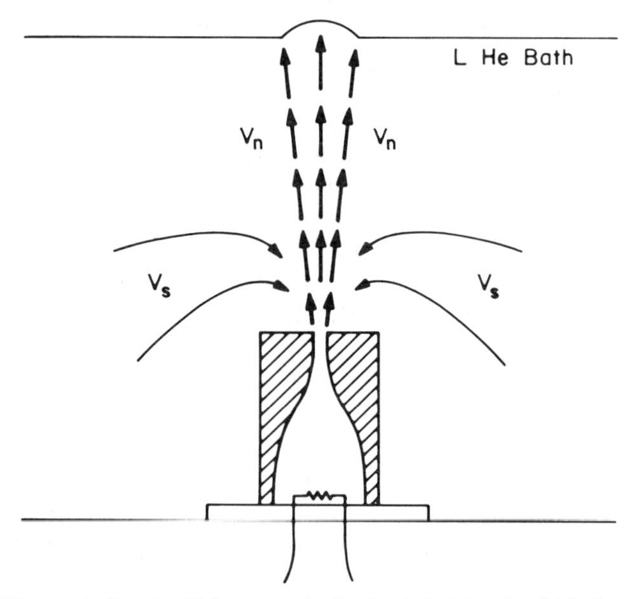

Figure 2 The counterflow jet. This represents the classical picture in which the normal fluid forms a jet and the orifice is a sink for the inviscid superfluid.

Mutual Friction and Gorter-Mellink Force

In both ordinary fluids and liquid helium the greatest challenge to the hydrodynamicist is in the characterization and understanding of turbulence. Laminar jets are quite well understood: the analytic solution to the equations of motion is given in most textbooks on elementary fluid dynamics. Turbulent jets, however, remain subjects of current research. Since the experiments to be described later are on turbulent counterflow jets, we present some results of earlier studies on turbulent pipe counterflow in liquid helium.

Consider ordinary counterflow in a tube of circular or rectangular cross section. The two-fluid equations predict a small temperature gradient because of the normal-fluid viscosity:

$$\frac{\partial T}{\partial x} = -\frac{G\eta_n}{(\rho s)^2 T} q_x = -a(T) q_x, \tag{3}$$

where G is a geometry-dependent constant, and $q = \rho s T v_n$ is the heat flux. Equation (3) can be integrated, providing $a(T)$ is assumed to be constant (small temperature difference), to give $\Delta T = -alq$ for the temperature difference along a pipe of length l. This expression has been verified for small heat fluxes by Keesom & Keesom (1936). As the heat flux is increased, a critical heat flux is reached beyond which the experimentally determined relation between the temperature difference and the heat flux is $\Delta T = -l(aq + bq^3)$.

To account for this cubic heat-flux dependence, Gorter & Mellink (1949) proposed appending a mutual-friction term to the two-fluid equations. The frictional force was proportional to the cube of the relative velocity of the superfluid and normal fluid, and given by the expression $\mathbf{F}_{sn} = -A\rho_s\rho_n w^2 \mathbf{w}$. The equations of motion for supercritical heat flux become

$$\frac{\partial \mathbf{v}_s}{\partial t} + \nabla\left(\frac{v_s^2}{2} + \mu\right) = A\rho_n w^2 \mathbf{w}, \tag{4}$$

$$\frac{\partial \mathbf{v}_n}{\partial t} + (\mathbf{v}_n \cdot \mathrm{grad})\mathbf{v}_n + \frac{1}{\rho}\nabla p = \frac{\eta_n}{\rho_n}\nabla^2 \mathbf{v}_n - \frac{\rho_s}{\rho_n} s\nabla T - A\rho_s w^2 \mathbf{w}. \tag{5}$$

For one-dimensional counterflow in a tube, these equations give a temperature gradient in the supercritical regime

$$\nabla T = -aq - b(T)q^3, \tag{6}$$

$$b(T) = \frac{A\rho_n}{s(\rho_s s T)^3}.$$

Note that $b(t)$ is independent of the size or shape of the channel. This law seems to apply to wide channels over a broad range of heat fluxes and channel dimensions. The coefficient A was measured by Vinen (1957a) and found to be temperature dependent, but only very weakly dependent on the size of the channel.

On the basis of his experimental observations, Vinen (1957b) proposed a mechanism for the mutual-friction force. According to Vinen, the critical relative velocity is representative of a transition to a turbulent state of the superfluid. This turbulent state is characterized by a tangled mass of quantized vortex lines, and is roughly analogous to the random fluctuations in the vorticity of a turbulent classical fluid, but with circulation around each vortex core quantized as suggested by Onsager (1949) and Feynman (1955). Mutual friction results from the interaction of excitations, which make up the normal fluid, with the tangled mass of vortices. Using dimensional analysis and frequent appeals to classical fluid mechanics, Vinen was able to extract the essential dependence of the Gorter-Mellink force on relative velocity.

Although intuitively appealing, the idea of mutual friction does not appear applicable in all situations where there is a relative velocity. Temperature-difference measurements in a channel where $\mathbf{v}_s$ and $\mathbf{v}_n$ can be varied independently (Kramers 1966, Van Der Heijden et al. 1972) show a more complicated dependence on the velocities than can be expressed as a simple function of $\mathbf{w}$. To explain these and other experiments, more friction terms were added to the equations of motion; however, the predictive value of Vinen's model is then lost and its physical basis becomes somewhat open to question, since it relies heavily on geometry independence. One cannot tell a priori which mutual-friction terms should be added to describe a flow situation. Still, the Gorter-Mellink law provides results in excellent agreement with pure counterflow in wide channels.

The experiments presented in the following sections further test the idea of mutual friction as it might apply to the counterflow jet. The first two test the prediction of Vinen's theory that there must be a temperature gradient and an additional attenuation of second sound wherever there is a large $\mathbf{w}$, independent of geometry. The last two experiments are flow-field measurements made by a Doppler-velocimetry technique and by flow visualization using hydrogen and deuterium particles.

Temperature Gradient and Attenuation Measurements in the Jet

In this section, we review two experiments designed to test the predictions of the mutual-friction theory for the free jet. As previously remarked, the jet is a localized region of high relative velocity. If mutual friction is indeed a volume force, then there should be a temperature gradient extending from the orifice all the way to the free surface.

Dimotakis & Broadwell (1973) carried out an experiment to measure this temperature gradient by traversing a small carbon thermometer along the axis of the jet. A diagram of this experiment is shown in Figure 3. The jets were formed from counterflow channels designed in such a way that the only area of high relative velocity was far from the heater. The thermometers used in these measurements were cubes of graphite, less than 0.025 cm on edge so as not to obstruct the flow. The startling result of these measurements is shown in Figure 4. A temperature gradient, $\nabla T = -b(T)q^3$, exists only in the channel but disappears within one channel diameter in the free jet.

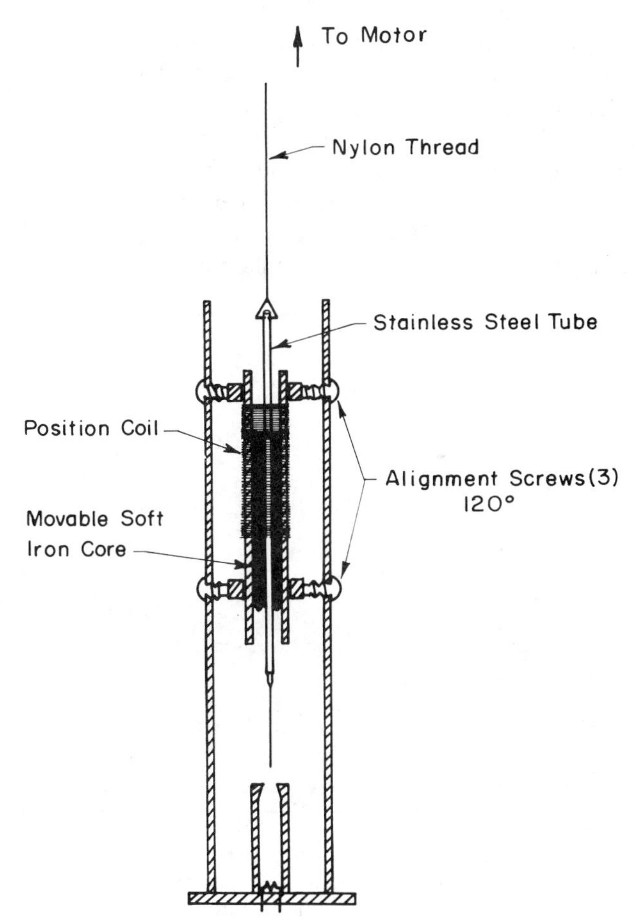

Figure 3 Apparatus for axial temperature measurements in a counterflow jet.

In another experiment, Laguna (1975) measured the attenuation of second sound in the free jet. The counterflow jet was directed through a plane parallel resonator for high-frequency (100–500 kHz) second sound. Any additional attenuation in the jet decreases the value of Q for the cavity. The attenuation in the free jet was found to be a factor of 15 less than the predicted value, and the power dependence was $q^{1.5}$ rather than q^2. We show in the next section that the results of the second-sound attenuation measurements can be explained by a simple geometrical-optics calculation of the attenuation of a beam traversing a turbulent boundary layer.

The two measurements—second-sound attenuation and temperature gradient—indicate a severe problem with our understanding of turbulent counterflow jets. One possible explanation for these measurements is that the Gorter-Mellink force is applicable only to channels. Hence, mutual friction is not a volume force but instead relies on walls for the interaction. An alternative explanation is that the superfluid is entrained by the normal fluid, thus making the relative velocity zero in the free jet. This alternative was rejected at the time of the experiments because the superfluid flow field would require a stagnation point at the orifice, which would have to remain stationary while the relative velocity varied by two orders of magnitude (see Keesom & Keesom 1936). For reasons that are presented below, we now favor the entrainment theory despite its unusual flow field.

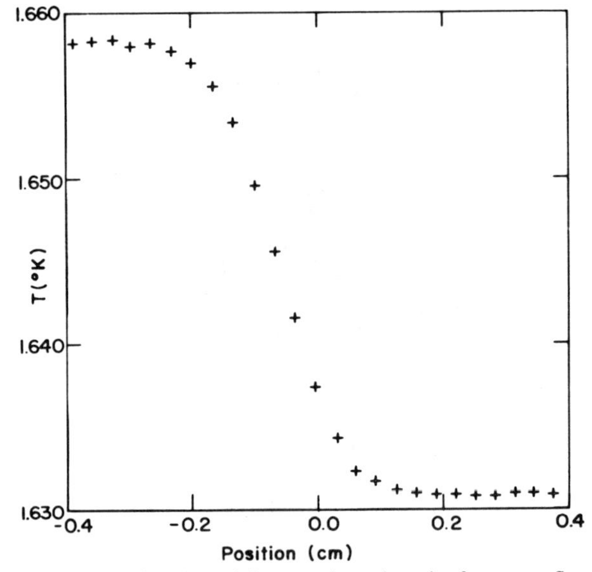

Figure 4 Temperature as a function of distance along the axis of a counterflow jet similar to that shown in Figure 2. The temperature gradient extends only over the region of high heat flux confined by the channel walls.

Calculation of the Attenuation of Second Sound in a Counterflow Jet

A simple explanation exists for the measured additional attenuation of second sound in a counterflow jet. It is well known that a beam of light traversing a turbulent boundary layer is scattered by fluctuations in the index of refraction. An analogous situation must hold for any wave. Specifically, if we attribute an index of refraction to second sound ($n = c_{2,0}/c_2$), then fluctuations of n will reduce the intensity of second sound at the receiver. We present this calculation here for the first time because it leads to several conclusions concerning the counterflow jet. The success of a simple, classical calculation warns that one need not always invoke detailed quantum properties in order to explain an observation in liquid helium.

The attenuation at a given temperature, heat flux, and position was observed to be independent of frequency. In an ordinary fluid, frequency-independent attenuation of sound is observed when the wavelength is shorter than the microscale of the turbulence. The extent to which this is satisfied is discussed later, but the observed frequency independence suggests a geometrical-optics calculation.

For a plane turbulent jet in an ordinary fluid, the mean-velocity profile becomes self-preserving around $x/d = 40$–50 (Tennekes & Lumley 1972). Until the turbulence is self-preserving, the flow has a memory of the nozzle geometry. Even though this calculation is concerned with $x/d < 20$, it will be assumed that the small scale of turbulence has already reached its asymptotic state of local isotropy.

MEAN-SQUARE DEFLECTION ANGLE A simple calculation of the attenuation uses geometrical acoustics and is based on similar computations by one of us (Liepmann 1952) for the diffusion of a light ray passing through a turbulent boundary layer. Let x be the streamwise direction, y the cross-stream direction, and δ the thickness of the boundary layer. The index of refraction is in the form $n = n_0(y)(1+n')$, with $n' \ll 1$ representing the random fluctuating part of the index of refraction due to the turbulence, $\langle n' \rangle = 0$ (brackets $\langle\ \rangle$ indicate time averaging).

The mean-square angle of deflection is given by

$$\varepsilon^2(\delta) = \frac{2}{n_0^2(\delta)} \int\int n_0(y) n_0(\xi) \left(\frac{\partial n'}{\partial x}\right)^2 R(|y-\xi|)\, dy\, d\xi, \tag{7}$$

provided that the fluctuations are homogeneous over regions larger than the correlation length. In Equation (7) R is a correlation function, i.e.

$$\left(\frac{\partial n'}{\partial x}\right)_y \left(\frac{\partial n'}{\partial x}\right)_\xi = \left(\frac{\partial n'}{\partial x}\right)^2 R(|y-\xi|). \tag{8}$$

INDEX OF REFRACTION OF SECOND SOUND Khalatnikov (1956) has shown that for second sound in the presence of a flow, $c_2 = c_{2,0} + v_y + \gamma w_y$, where v_y and w_y are the components of the mass flow velocity and the relative velocity in the direction of second-sound propagation, and γ is a monotonically decreasing function of temperature. For the jet, w_y is taken to be zero. The calculation has been tried with $v_y = 0$, $w_y \neq 0$, but a different temperature dependence results. The index of refraction is

$$n = \frac{c_{2,0}}{c_2} = 1 - \frac{v_y}{c_2}. \tag{9}$$

Let $v_y = \langle v_y \rangle + v_y'$, where $\langle v_y \rangle$ is the mean velocity and v_y' is the turbulence component, i.e. $\langle v_y' \rangle = 0$. The index of refraction can now be written in the form

$$n = \left(\frac{\langle v_y \rangle}{c_{2,0}}\right)\left(1 - \frac{v_y'}{c_{2,0}}\right). \tag{10}$$

It will become necessary to choose a correlation length to complete the calculation. Dimotakis (1974) has suggested a dimensionless group to scale the onset of mutual friction in a channel. For a critical heat flux q_c, the relation

$$\frac{Aq_c l}{sT} = 1 \tag{11}$$

seems to hold, where l is the channel diameter and A is the coefficient in the force of mutual friction. The correlation length proposed here is

$$\frac{1}{\Lambda} = a\frac{Aq}{sT}, \tag{12}$$

where a is a number of order unity.

For the jet, δ is outside the region of high velocity so one can reasonably take $n_0(\delta) = 1$. If the correlation function falls to zero rapidly for increasing $|y-\xi|$, then one can replace $R(|y-\xi|)$ with $\Lambda\delta(|y-\xi|)$, where Λ is the scale of fluctuations and $\delta(|y-\xi|)$ is the Dirac δ-function:

$$\varepsilon^2(\delta) = 2\Lambda \int_0^\delta n_0^2(y)\,\frac{\partial n'^2}{\partial x}\,dy. \tag{13}$$

Applying the usual relations for a turbulent jet (Tennekes & Lumley 1972, Schlichting 1968) yields finally

$$\varepsilon^2(\delta) = 0.156\frac{\delta d}{\Lambda x}\left(\frac{U_J}{c_{2,0}}\right)^2. \tag{14}$$

CONCLUSION The attenuation is related to the mean-square deflection by the formula $\alpha = 2\langle\varepsilon^2\rangle^{1/2}/\mathbf{L}$, where L is the size of the beam. Finally, we find after substitutions for Λ and U_J that $\alpha = f(T,\delta)q^{3/2}/\mathbf{L}$. Note that δ, the width of the jet, is a function of the distance downstream from the orifice. The function $f(T,\delta)/L$ is plotted in Figure 5 for a position 20 diameters from the jet exit, along with the attenuation measured at this location. The constant in Λ has been chosen to be 1/3.

It is necessary to examine the assumption that Λ is greater than the wavelength. With the above choice of the constant, Λ ranges from 0.18 to 0.41 mm, while the second-sound wavelength goes from 0.2 to 0.15 mm for a 100-kHz frequency in the temperature range 1.50–2.05 K. Thus the condition on the coherence length is at least approximately satisfied.

The result of this calculation is extremely suggestive, and the following may be inferred:

1. The superfluid must move with the normal fluid. Using a relative velocity instead of v gives the wrong temperature dependence. This supports the entrainment hypothesis.
2. The jet is turbulent. A laminar profile cannot give attenuation of the magnitude observed. Using the spreading relations for a turbulent jet removes most of the geometry dependence of the attenuation.

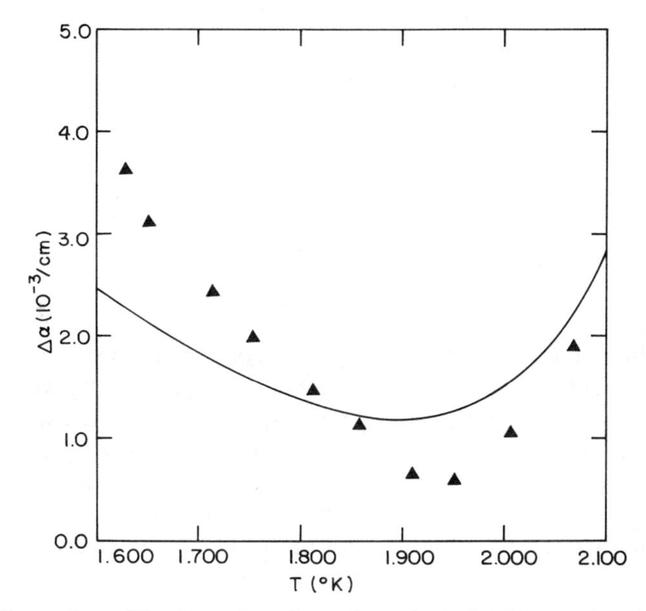

Figure 5 Comparison of the attenuation of second sound calculated from geometrical optics with the measured attenuation. Experimental results shown as triangles.

3. The coherence length in the jet is the Gorter-Mellink length scale. This suggests that the mutual friction acts away from solid boundaries but in a previously unsuspected way. Mutual friction, as a source of entropy, assumes a role something like viscous dissipation in an ordinary fluid.

Velocity-Sensitive Phase Measurements in the Counterflow Jet

The acoustic properties of second sound provide an elegant way to measure velocity fields in liquid helium. Khalatnikov's equation for the speed of second sound is $c_2 = c_{2,0}(p, T) + v_k + \gamma(p, T)w_k$, where $c_{2,0}$ is the speed of second sound in quiescent fluid, and v_k and w_k are the components of bulk and relative velocity in the direction of the second-sound wave vector **K**. The thermodynamic coefficient,

$$\gamma(p, T) = 2\frac{\rho_s}{\rho} - \frac{TS}{c_p \rho_n}\left(\frac{\partial \rho_n}{\partial T}\right), \tag{15}$$

has been measured in pure counterflow ($v = 0$) by Johnson & Hildebrandt (1969). If one considers a plane second-sound wave running between an emitter and detector, the phase of the wave at the detector is given by

$$\varphi = \int_0^D \mathbf{K} \cdot d\mathbf{x}. \tag{16}$$

To first order in the velocity perturbations, this will be

$$\varphi = \varphi_0 - \frac{\omega L}{c_{2,0}^2}\left(\langle v_k \rangle + \gamma(p, T)\langle w_k \rangle + \frac{\partial c_{2,0}}{\partial T}\langle \delta T \rangle\right). \tag{17}$$

The quantities $\langle v_k \rangle$, $\langle w_k \rangle$, and $\langle \delta T \rangle$ are averages of fluctuating quantities in the perturbed region, and L is the distance that the second-sound beam must travel through the perturbed region.

The geometry used by Dimotakis & Laguna (1977) for phase-sensitive velocimetry on the counterflow jet is shown in Figure 6*a*. Ordinary attenuation of second sound at frequencies of 500 kHz insured no standing waves between the emitter and detector. The experiments were done by modulating the jet and measuring the change in phase of the second sound relative to a local oscillator. In Figure 6*b* we see one series of measurements where the second sound traveled at an angle of 60° with respect to the jet axis. There is both a mean phase shift due to the steady-flow component and a fluctuating phase shift due to turbulent velocity fluctuations. When the second-sound beam is directed 90° to the jet axis, only a fluctuating component is observed because $\langle \mathbf{v} \rangle \cdot \mathbf{K} = 0$.

The observed phase shift versus heat flux at a given temperature was

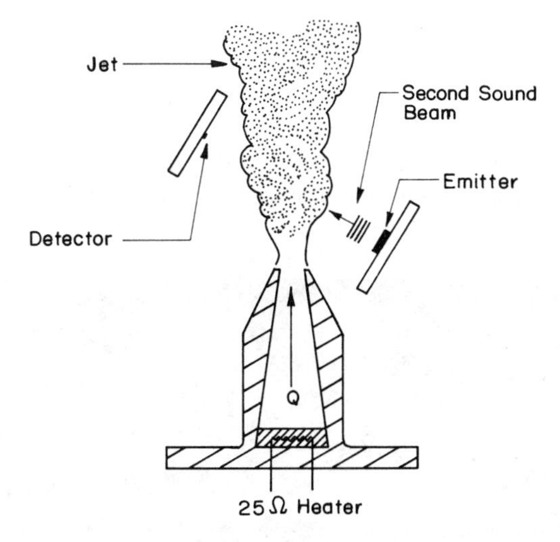

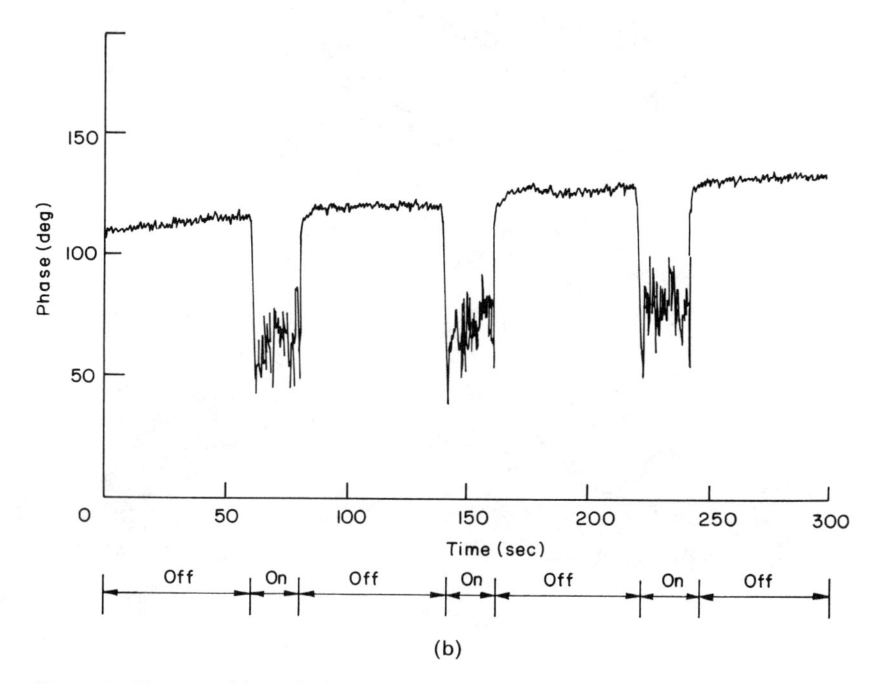

Figure 6 Phase-sensitive velocimetry in the counterflow jet. (*a*) Geometry of the phase-sensitive detection; (*b*) fluctuating velocity component superimposed on a mean velocity when the jet heater current is periodically modulated.

compared with two limiting cases, $\langle v_k \rangle = 0$ or $\langle w_k \rangle = 0$. Although not totally conclusive, the data are in better agreement with $w_k = 0$, providing some direct evidence for entrainment of superfluid by the normal-fluid jet. The limiting case $v_k = 0$ is not even qualitatively similar to the observations. Many assumptions had to be made in this comparison, so it is hoped that an ambitious theorist will attempt a more realistic calculation of the flow field.

Figure 7 The counterflow-jet flow field made visible by a neutrally buoyant mixture of hydrogen and deuterium particles.

Photographs of the Jet

As a final experiment, we present in Figure 7 a photograph of the jet by Dimotakis & Smith (1981).[3] The flow-visualization technique involved the injection of a neutrally buoyant mixture of hydrogen and deuterium through a specially constructed transfer tube. The vortex near the orifice has never been observed in a classical fluid.

Future Studies on Counterflow Jets

A consistent picture of the counterflow jet seems to be emerging: the normal-fluid jet entrains the superfluid. Since $w = 0$ in this flow field, one would not expect to see the consequences of mutual friction, and this explains the absence of a temperature gradient or additional second-sound attentuation. On the other hand, we do not as yet have a mechanism for superfluid entrainment. It has been suggested that the mechanism of entrainment is the same as in mutual friction, but this leads us back to the original question of the role of walls in the Gorter-Mellink force. Clearly, much more theoretical work is needed on counterflow jets.

From the experimental side, there is an enormous amount of information in the counterflow jet that we have only begun to use. The second-sound velocimetry technique of Dimotakis & Laguna (1977) can certainly be used in the jet to analyze correlations and fluctuation spectra as in ordinary fluids. It could also be profitable to look at other classical flows such as the free shear layer. Any theory of turbulence in liquid helium will have to stand the test of application to these flows.

SHOCK WAVES

General Properties

Shock waves are an essential element of the fluid mechanics of any compressible medium. In the absence of viscosity and heat conductivity, such as for an ideal fluid, there exist interfaces at which the normal velocity component of the flow is discontinuous and all thermodynamic variables change discontinuously from one equilibrium state to another. Viscosity and heat-conductivity effects alter the discontinuities to narrow transition layers with thicknesses that are, under ordinary conditions, small compared with practically any characteristic length parameter of the flow geometry. Consequently, in the vast majority of flow problems, shock waves can be considered as discontinuity surfaces separating two thermodynamic equilibrium regions in the flow.

[3] Senior thesis, California Institute of Technology; private communication

These obviously nonlinear waves travel with a wave velocity c that can be determined from the so-called jump conditions of the equations of motion—the integral relations between conditions just upstream and downstream of the discontinuity surface. In the limit of a negligible jump in the thermodynamic variables, the solution reduces to a linear wave with velocity determined by the thermodynamic state, the discontinuity surface becoming one of the characteristic surfaces of the equation of motion. Conversely, to a set of characteristics or waves there corresponds a set of possible shock waves.

In He II there exist two sets of characteristics corresponding to "first" and "second" sound, and hence two different types of shock waves: pressure shocks, which are similar to shock waves in any classical fluid, and thermal shocks, which correspond to nonlinear "second-sound waves" and are a unique feature of He II. Consequently, the study of shock waves in He II is important for two reasons. First, since shocks are nonlinear waves, the correct prediction of their properties from the equations of motion represents a test of their validity. A test of the correctness of the two-fluid equations of liquid helium including their nonlinear terms is not at all trivial. Furthermore, shock waves separate two different equilibrium states, and hence shocks can be used to create a state uniquely defined by the wave velocity c and the upstream conditions without manipulating solid boundaries. Indeed, this state is rather independent from the influence of or the interaction with solid boundaries. This feature of shock-wave motion has been used for a long time in chemistry, chemical kinetics, and related fields, where often thermodynamic states difficult to achieve any other way are required, e.g. states with very high temperatures.

Using shock waves to produce a well-defined state has a particular appeal for studies of He II because it makes it possible to investigate conditions largely independent of the fluid-solid wall interactions. These interactions are certainly not well enough understood to take their effects for granted. In particular, vortex interactions with solid walls are not even a simple matter in ordinary fluids, let alone in He II. In the subsequent sections, studies of shock-wave motion in liquid helium are discussed. A brief section on shock waves in gaseous helium is included to elucidate the method used to propagate strong pressure shocks in liquid helium. Very interesting phenomena occur when a strong shock wave impinges on the gas-liquid interface, raising both temperature and pressure almost instantaneously at the interface and hence producing both pressure and temperature waves in the liquid.

Characteristics, Jump Conditions

The two-fluid equations written for one space variable and time are four first-order partial differential equations of the form

$$A_{ij}\frac{\partial u_j}{\partial t}+B_{ij}\frac{\partial u_j}{\partial x}=0. \tag{18}$$

The corresponding equation for the roots of the characteristic determinant is then obtained in the usual way from the coefficients of the derivatives. The roots determine the characteristic directions in the x, t-plane, i.e. the phase velocities of the corresponding waves. The four equations of motion thus lead to the following fourth-degree algebraic equation for the wave speeds:

$$c^4-(a^2+\gamma b^2)c^2+a^2b^2=0, \tag{19}$$

$$a^2=\left(\frac{\partial p}{\partial \rho}\right)_{\rm s}\equiv c_{1,0}^2,\qquad b^2=\frac{\rho_{\rm s}}{\rho_{\rm n}}\frac{S^2T}{c_{\rm p}}\equiv c_{2,0}^2.$$

The small but finite coefficient of thermal expansion or the related difference from unity in the ratio of the specific heats couples the resulting pressure and temperature waves (i.e. first and second sound are not exactly independent wave motions, but the coupling is quite weak).

The jump conditions for shock waves are obtained from the equations of motion, written (in conservation form) formally by

$$\frac{\partial}{\partial t}\rightarrow -c[\;],\qquad \mathrm{div}\rightarrow \mathbf{n}\cdot[\;],$$

$\mathbf{c}=c\mathbf{n}$ wave velocity.

They are written out in Figure 8. This set of algebraic equations relating the thermodynamic variables ahead and behind the shock wave was apparently first given by Khalatnikov (1965). The relations are too involved to permit an explicit, analytical solution. Khalatnikov gave, in addition, the approximation for weak shocks, which includes the first-order corrections to the phase velocity of first and second sound:

$$c_1=c_{1,0}\left(1+\frac{\Delta p}{2}\frac{\partial}{\partial p}\ln(\rho c_{1,0})\right), \tag{20}$$

$$c_2=c_{2,0}\left(1+\frac{\Delta T}{2}\frac{\partial}{\partial T}\ln\left(\frac{\partial s}{\partial T}c_{2,0}^3\right)\right). \tag{21}$$

More recently, Sturtevant (1976) and Sturtevant & Moody (private communication, 1983) have used a computer to solve the exact jump conditions. The results are quite useful in checking the approximations in the weak-shock solution, but are hampered by the lack of thermodynamic data for the dependence of the variables of state on the counterflow velocity, w. In discussing shock waves, it is usually convenient to define shock strengths in terms of a "Mach number," the ratio of the actual wave speed to the corresponding linear wave velocity. The application of Mach number as the nondimensional parameter is not quite as useful in liquids as in gases. In gases, the ratio of fluid to signal velocity and the ratio of directed to random energy are proportional to M and M^2, respectively. In liquids, the internal energy or enthalpy is not directly related to the speed of sound and hence to the Mach number. Thus, a shock Mach number close to unity may correspond to a large pressure ratio across the shock.

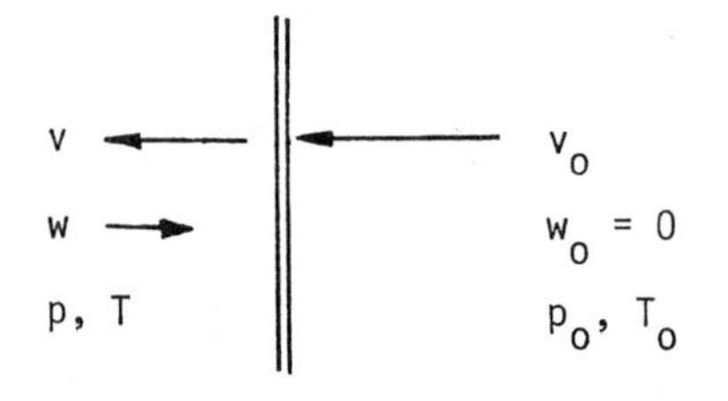

$$\rho_0 v_0 = \rho v$$

$$p_0 + \rho_0 v_0^{\,2} = p + \rho v^2 + \xi(1-\xi)\,\rho w^2$$

$$\mu_0 + \frac{1}{2} v_0^{\,2} = \mu + \frac{1}{2}(v-\xi w)^2$$

$$\rho_0\, s_0\, T_0\, v_0 = \rho s T\,[v + (1-\xi)w] + \xi\rho w\,[v + (1-\xi)w]^2$$

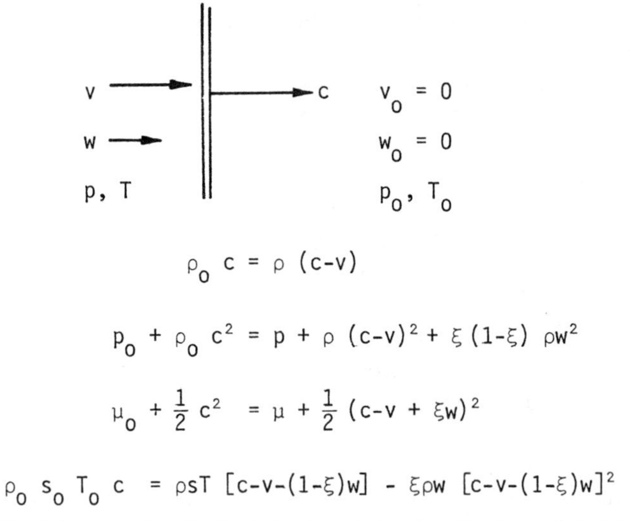

Figure 8 Shock-jump relations in shock-fixed and laboratory-fixed coordinates, respectively.

In the following sections we review pertinent experiments on shock-wave propagation in liquid helium. Both pressure and thermal shocks were studied in our laboratory, including quite recently the propagation of thermal shock waves into rotating helium.

Shock Tubes

A cryogenic shock tube for liquid-helium research was developed by Liepmann et al. (1973) and Cummings (1974). A shock wave is produced in gaseous helium by slight modifications of well-known shock-tube techniques and is permitted to impinge on a liquid-helium surface. The shock tube, in a somewhat later version, is shown in Figure 9. The differences from the usual shock tubes are obvious from the figure. Note in particular the diaphragm technique, which makes it possible to produce a sequence of some 60 shocks without the need to open the tube; this feature was suggested by D. Coles, and makes operation at cryogenic temperatures feasible.

The governing equations for shock tubes are found in most modern texts on gasdynamics, e.g. Liepmann & Roshko (1957). A few conclusions from shock-tube theory particularly pertinent to the present work are repeated here. The gas in both the driver and driven section is helium, a monatomic gas that under all working conditions encountered is very nearly a perfect gas with a ratio of specific heats of 5/3. Simple shock-tube theory relates the shock Mach number to the pressure and temperature ratios across the diaphragm. Figure 10 shows the "ideal shock-tube theory" compared with measured shock Mach numbers. The agreement at cryogenic temperatures in the test section is excellent, and indeed much better than corresponding measurements at elevated temperatures. This fact, important for our applications, is simply due to the temperature dependence of the viscosity and density, which results in a comparatively large Reynolds number based on the tube diameter and hence a drastic reduction in wall-friction effects. For the present task of producing shock waves in liquid helium by impinging a strong shock wave from the gas on the liquid-gas interface, pressure and temperature in the test section are evidently connected by the vapor-pressure-temperature relation of helium. Consequently, for a given driver temperature of, say, 300 K, there exists a specific set of states in the p, T-diagram of helium that can be reached. The final states behind such a shock reflected from a solid surface are shown in Figure 11. The gas-liquid interface does not correspond exactly to a solid interface but the difference is relatively slight, and hence Figure 11 presents a rather accurate picture of what can be done with such a device.

The pressure and temperature corresponding to these end states behind the reflected gasdynamical shock are thus nearly instantaneously impressed on the liquid-helium surface and hence form the initial conditions for shock-wave propagation into the liquid. Evidently it is possible with this technique to have interesting initial conditions, corresponding, for example, to a supercritical state of the helium, and thus a disappearance of a true gas-liquid interface.

The method using a gasdynamical shock tube to propagate shock waves into liquid helium leads naturally to a combination of pressure and temperature shocks in He II. To study temperature shock waves in the

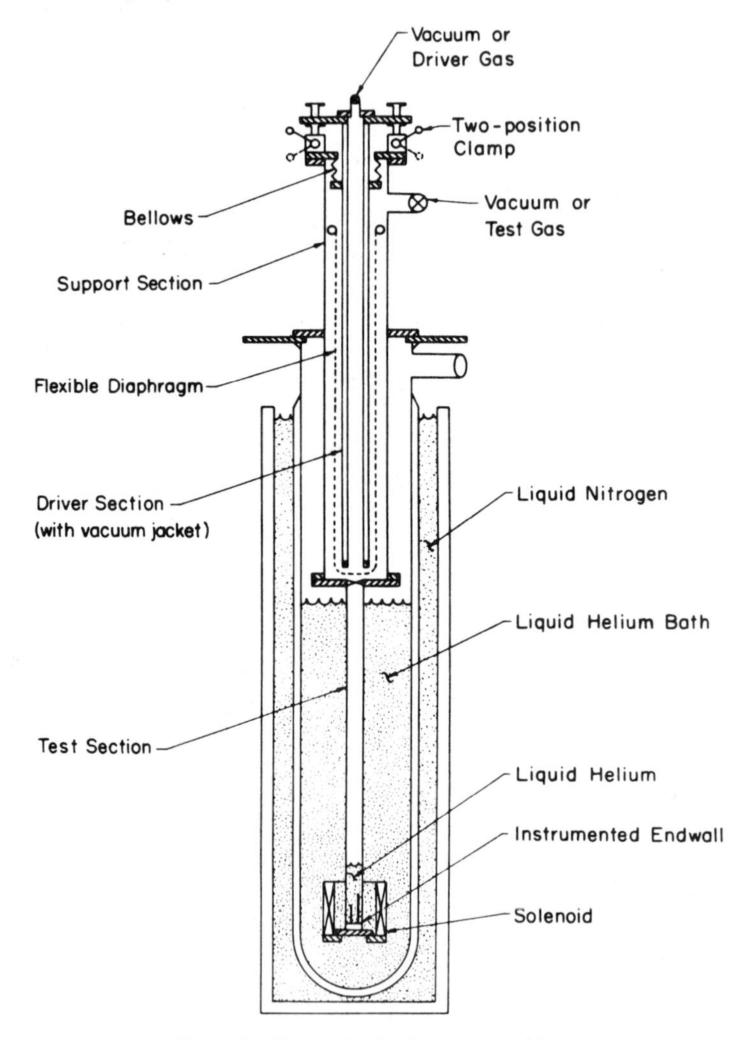

Figure 9 Cryogenic shock-tube assembly.

absence of significant pressure waves, a different technique has to be used. It is possible to produce almost instantaneous steps in the temperature by the rapid heating of a thin conducting film. Corresponding shock tubes for thermal shocks have been constructed on this basis by Cummings (1976), Cummings et al. (1978), Turner (1979), and Torczynski (1981). Figure 12 shows one of these later versions of a thermal shock tube.

A device to produce pressure shock waves without significant temperature waves has apparently not yet been developed. It is in principle possible to build such a device, e.g. by using an instantaneously moved piston or membrane, but the technical difficulties are significant. Shock arrival times

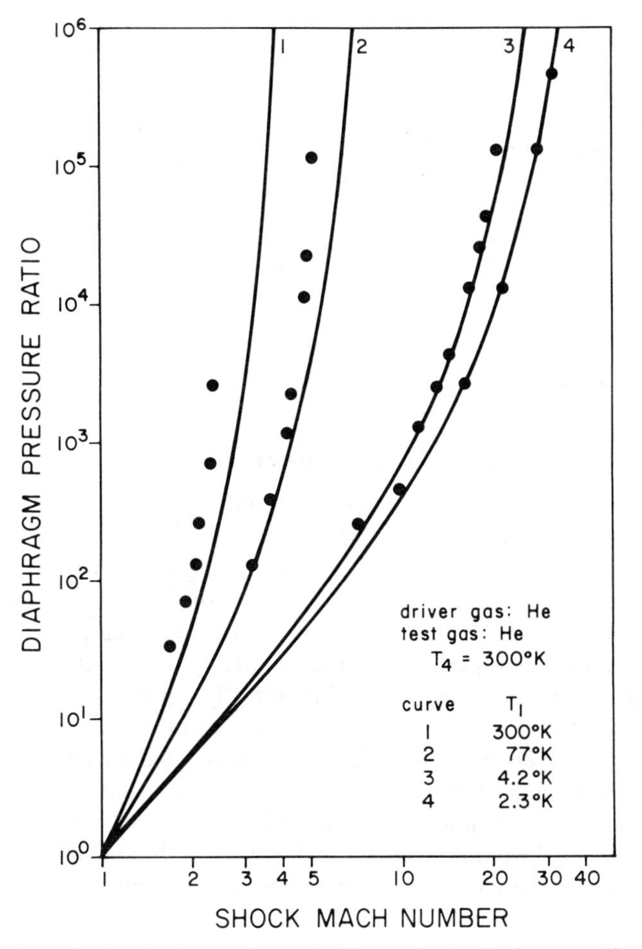

Figure 10 Ideal shock-tube theory compared with experiments in the cryogenic shock tube. Experimental results shown as circles.

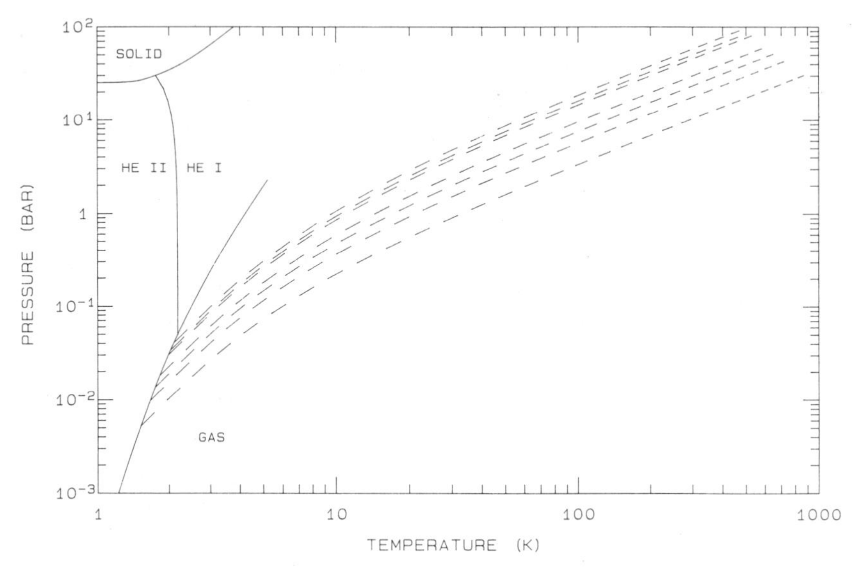

Figure 11 Final thermodynamic states at the gas-liquid interface obtainable by shock-wave reflection. Each line segment corresponds to an increase in diaphragm pressure ratio by a factor of $2^{1/4}$.

and shock strength were measured using superconducting thin-film sensors operated in the transition region and magnetically biased to vary the transition temperature.

Shock Waves in He II, Experiments

The two-fluid equations lead to the definite jump conditions for He II shock waves shown in Figure 8. The pertinent results of the numerical solution for the waves can be seen in Figure 13. The first task of experiments is to verify the qualitative and quantitative prediction of the theory. The existence of the two shock waves—a pressure wave with little temperature change across it and a temperature wave at nearly constant pressure—are strikingly demonstrated by impinging a gasdynamical shock on a He II surface and observing the wave propagation into the liquid. This was first done by Cummings (1976), and with improved techniques was continued by Wise (1979). Figure 14*a* shows a typical result. As usual, the traces of the wave motion are exhibited in a nondimensional x, t (χ, τ) diagram. The incoming and reflected gasdynamical shocks and the transmitted pressure and temperature shocks are evident. Repeating the experiment at slightly higher temperatures, such that the end condition behind the pressure shock corresponds to He I, leads to the result shown in Figure 14*b*. As expected, the temperature wave is missing: He I behaves like a classical fluid. Note

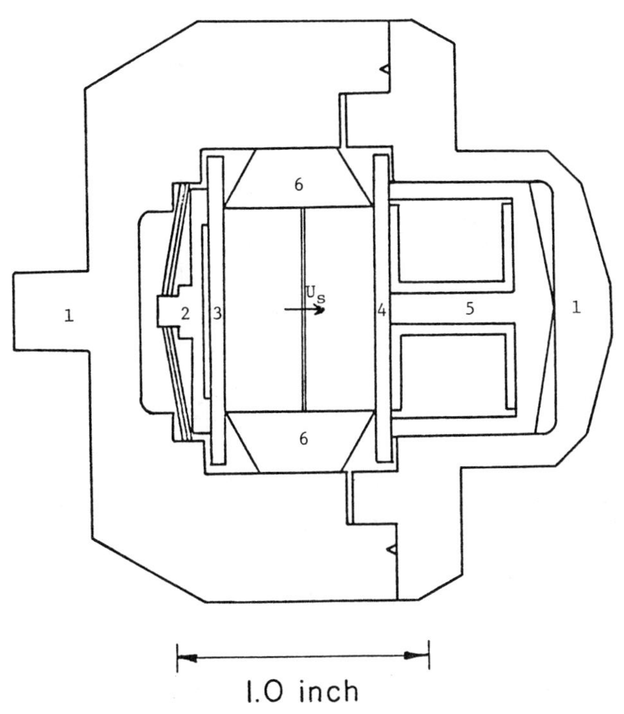

Figure 12 Device to generate and study temperature shock waves in He II. 1. Brass housing; 2. Spring loading for heater; 3. Quartz substrate of heater; 4. Quartz substrate of sensor; 5. Sensor-biasing magnet (superconducting); 6. Teflon tube, down which the shock propagates.

that at the conditions shown, the lambda transition occurs within the shock transition zone. The qualitative results of the experiments are thus in complete agreement with the two-fluid equations. A typical quantitative comparison of theory and experiments based on the shock-jump conditions in gaseous and liquid helium is presented in Figure 15. The trace of the pressure shock in He II comes out very accurately, but there remain small, significant differences in the speeds of the reflected shock and the temperature shock. These differences, already noted by Cummings and by Wise, are almost certainly connected with the heat flow and evaporation in region 5 of the wave diagram: in this region the temperature has to drop from the high value behind the reflected shock to the initial temperature of He II. In one particular experiment, this drop was from some 200 K to 1.5 K. Matching the heat flux by conduction in He I to the heat flux by counterflow in He II leads to extremely large temperature gradients in He I,

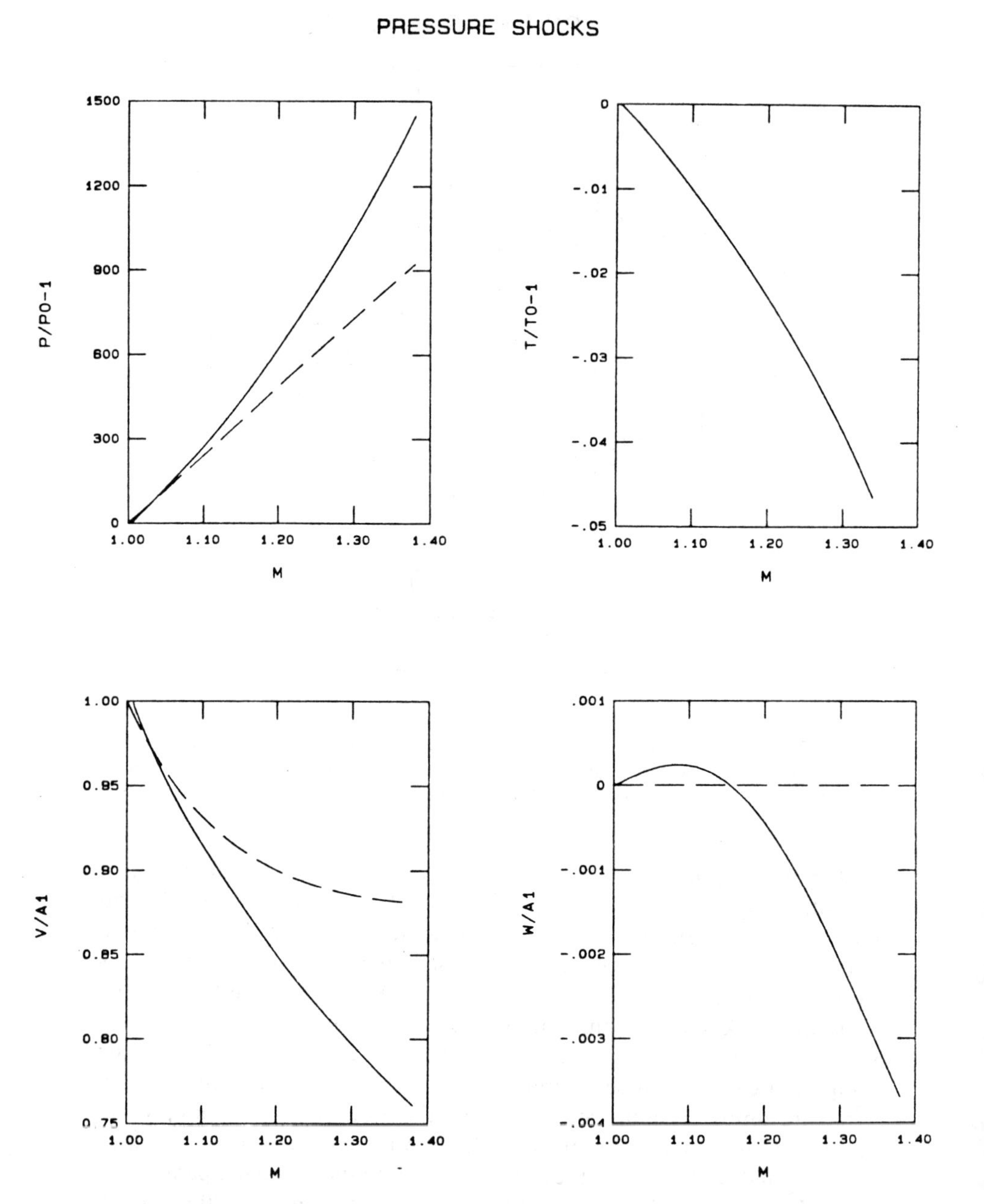

Figure 13a Results of the numerical solution of the shock-jump relations for He II. Pressure shocks for $T_0 = 1.80$ K and $p_0 =$ SVP. ——— Numerical results; ------ Khalatnikov approximation.

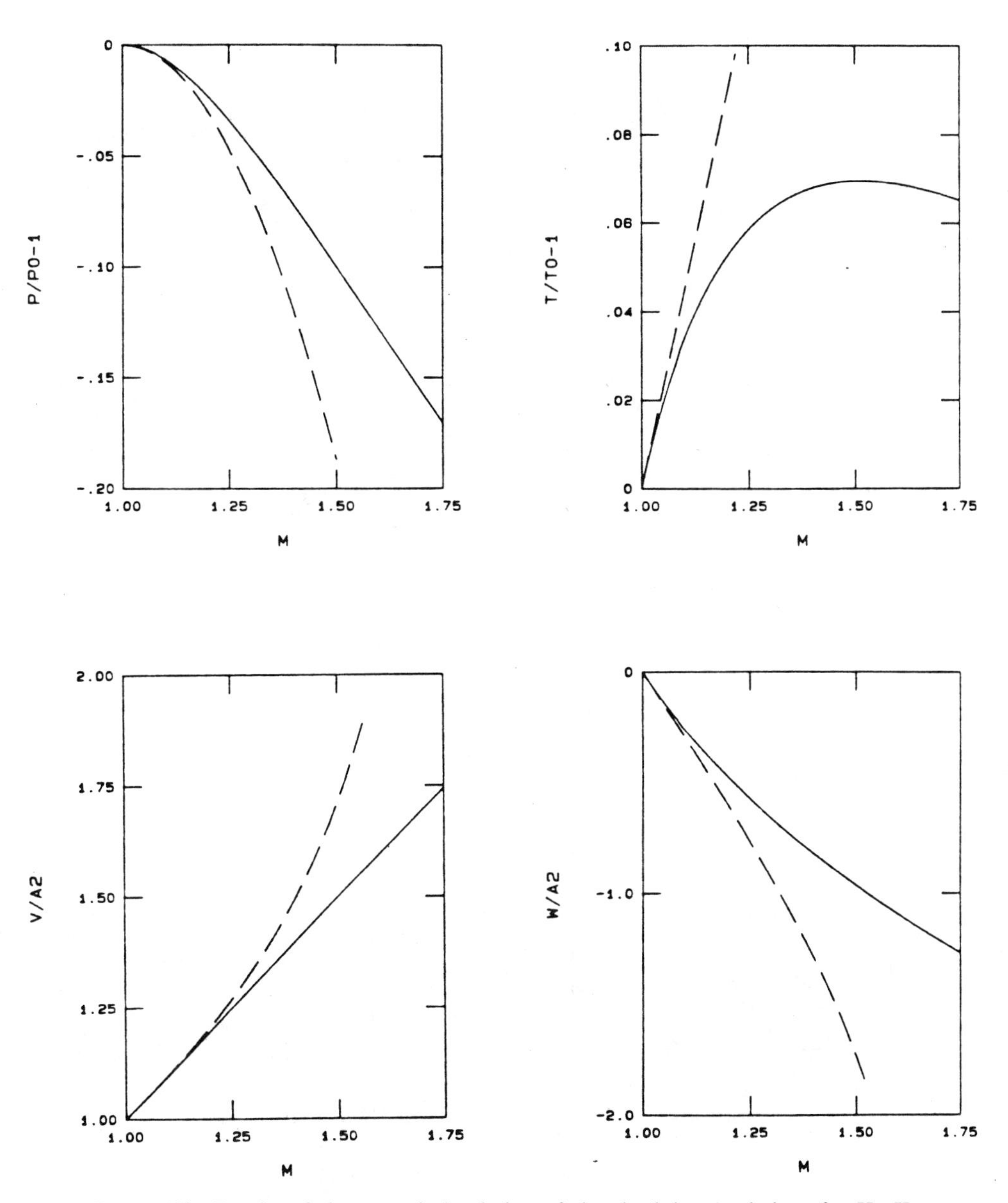

Figure 13b Results of the numerical solution of the shock-jump relations for He II. Temperature shocks for $T_0 = 1.60$ K and $p_0 = 1$ bar. ———— Numerical results; ------ Khalatnikov approximation.

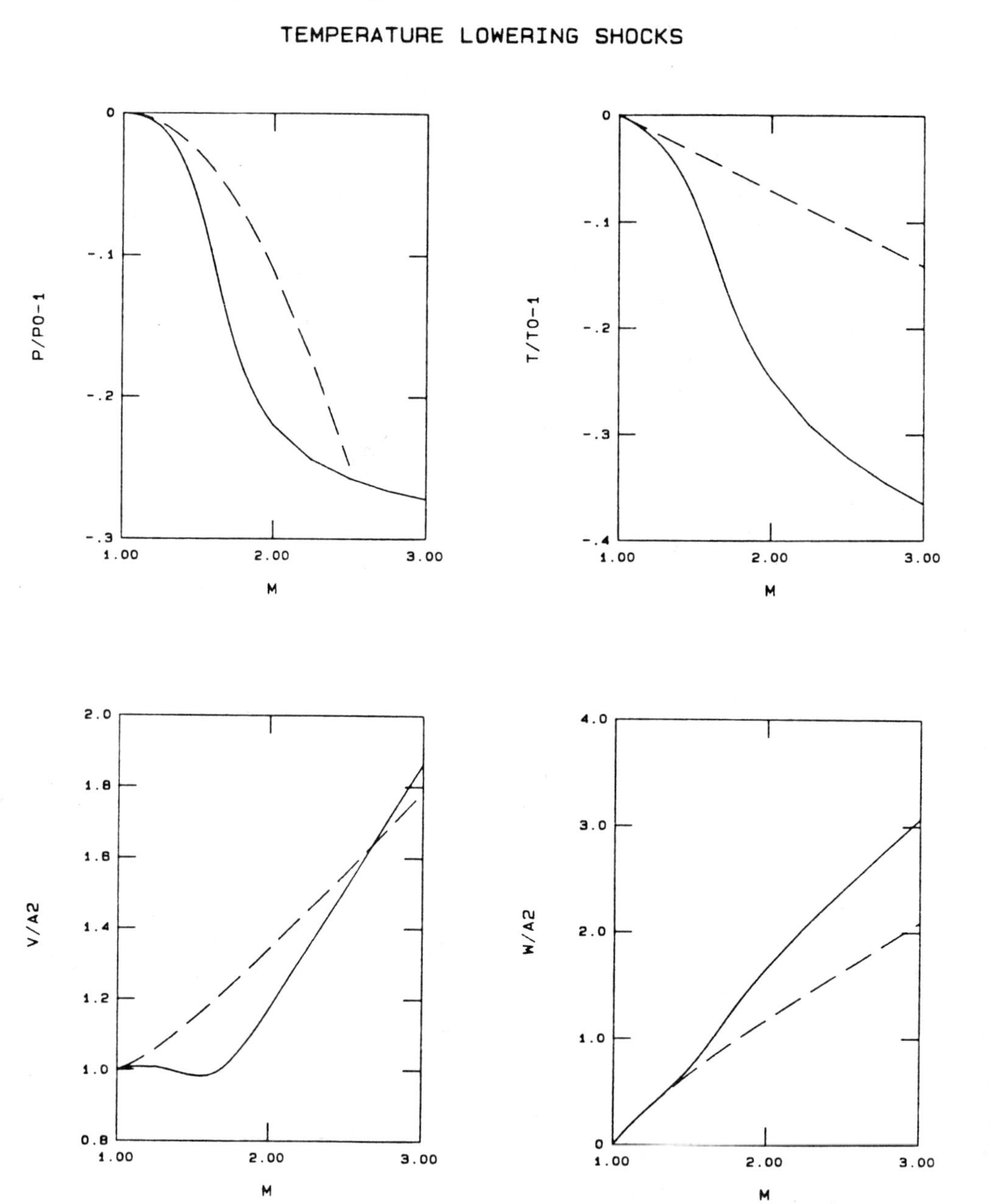

Figure 13c Results of the numerical solution of the shock-jump relations for He II. Temperature shocks for $T_0 = 2.10$ K and $p_0 = 1$ bar. ——— Numerical results; ------ Khalatnikov approximation.

since the temperature drop occurs over distances of the order of ten microns or so. These recent computations by Moody (1983) show quite clearly that the heat flow after the shock reflection and the formation of a second-sound shock under these conditions are still not completely understood.

The numerical computations shown in Figure 15 take into account the small temperature drop in the passage through a pressure shock in He II. For the conditions of experiment II, these computations yield a temperature in region 6 of 2.017 K and a pressure of 11.74 bar. The Khalatnikov equations used by Wise in his evaluation of the same data give 2.095 K. Consequently, according to the numerical solution the fluid is still He II, while according to the Khalatnikov approximation the lambda line has been passed and the fluid is He I. In the experiments no temperature shock was observed, and this fact was considered conclusive evidence that the fluid in region 6 was indeed He I. This earlier conclusion is, however, not completely safe. Near the lambda line the temperature shocks are temperature lowering. Therefore, the rise in temperature downstream of the pressure shock is not discontinuous and may not have been registered. On the other hand, the numerical solution has to use thermodynamic data, which are good to second order only. Consequently, the computed temperature may well be slightly off. It would have been a relatively easy matter to clarify this point while the experiments were still in progress; unfortunately, the problem was not realized at the time.

Temperature shock waves were studied earlier by Osborne (1951) and by Dessler & Fairbank (1956). In the latter experiments the increased velocity of a small temperature pulse riding on top of a larger one was measured and found to agree well with the Khalatnikov second-order theory. Osborne demonstrated a particularly interesting phenomenon for temperature shocks: the steepening coefficient changes sign. For $0.5\ \mathrm{K} < T < 0.95\ \mathrm{K}$ and $1.88\ \mathrm{K} < T < T_\lambda$ the coefficient is negative, and hence shock waves form toward the rear of a positive temperature pulse, i.e. the temperature decreases discontinuously. Osborne demonstrated this predicted behavior. More recently, Turner (1979) showed both theoretically and experimentally that the same effect must lead to a double-shock formation in the neighborhood of the temperature $T = 1.88$ K, where the steepening coefficient vanishes. His results, which demonstrate this typically nonlinear result of the two-fluid theory, are shown in Figure 16. Some schlieren photographs showing finite temperature waves were obtained early on by Gulyaev (1960, 1967) and these demonstrate the use of schlieren observations for shock-tube studies in He II.

These experiments demonstrate qualitative agreement with the predictions of the two-fluid theory, and for weak waves there is quantitative agreement as well. The latter is not true for strong waves. The experiments

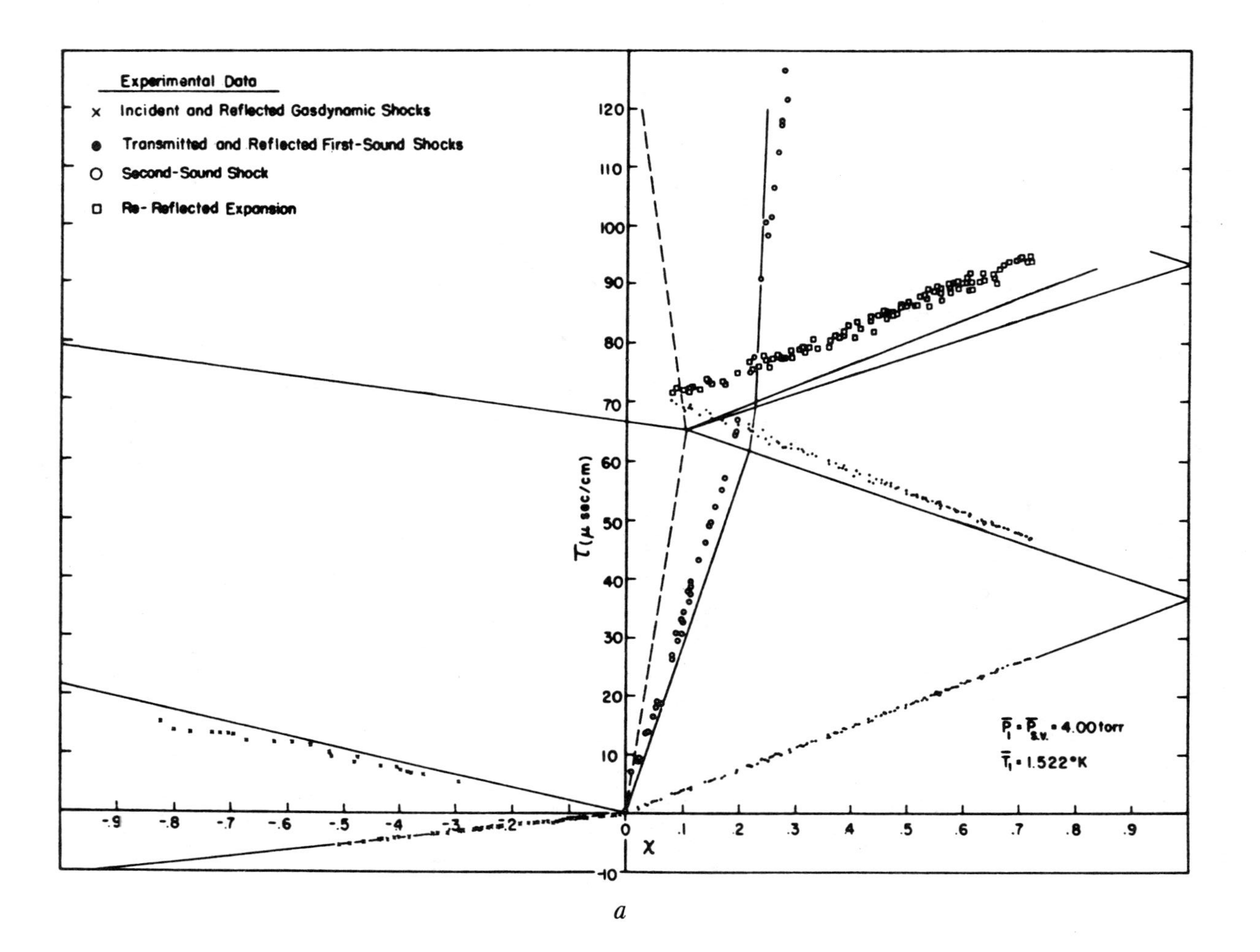

Experimental Data
× Incident and Reflected Gasdynamic Shocks
● Transmitted and Reflected First-Sound Shocks
○ Second-Sound Shock
□ Re-Reflected Expansion
τ (μ sec/cm)
X
$\bar{p}_1 = \bar{p}_{S.V.} = 4.00$ torr
$\bar{T}_1 = 1.522$°K

a

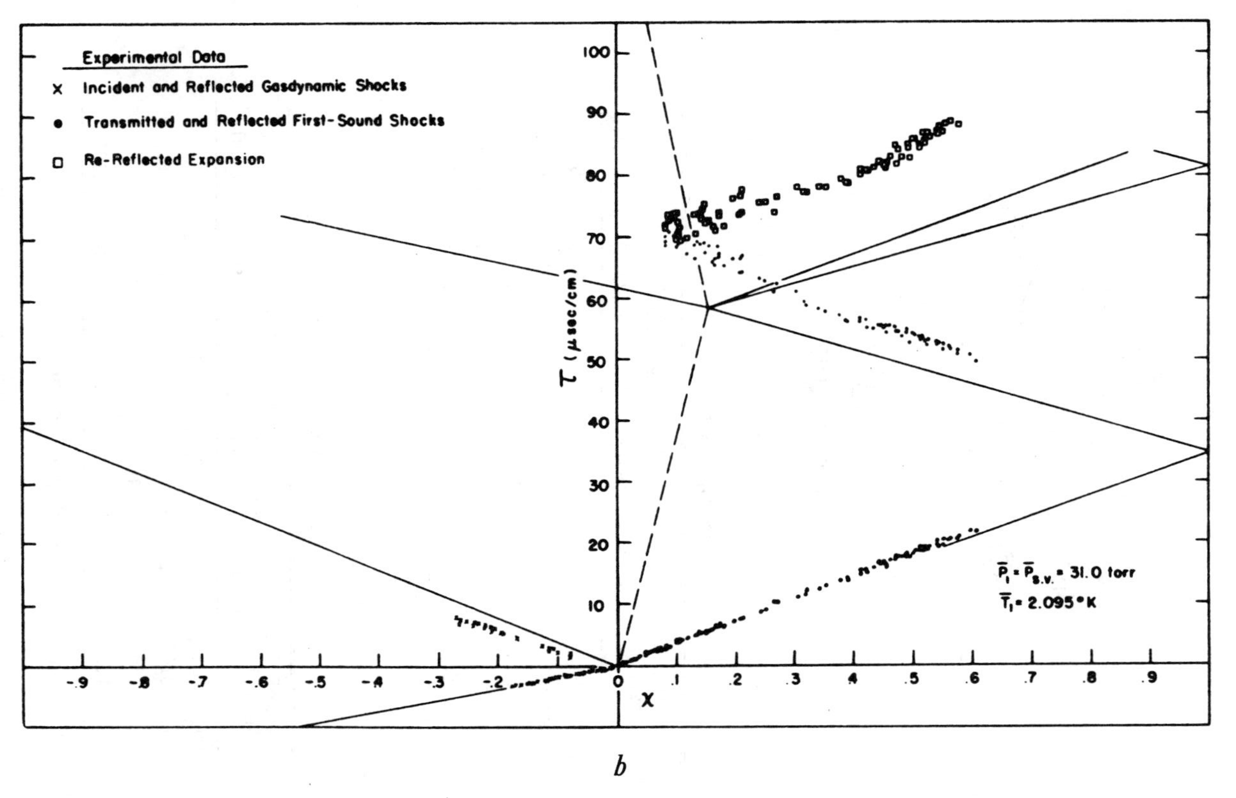

b

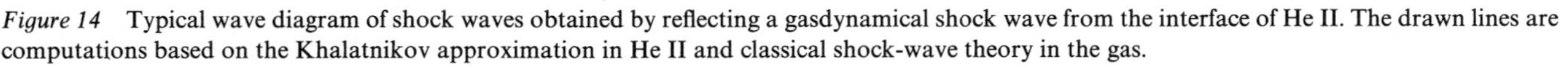

Figure 14 Typical wave diagram of shock waves obtained by reflecting a gasdynamical shock wave from the interface of He II. The drawn lines are computations based on the Khalatnikov approximation in He II and classical shock-wave theory in the gas.

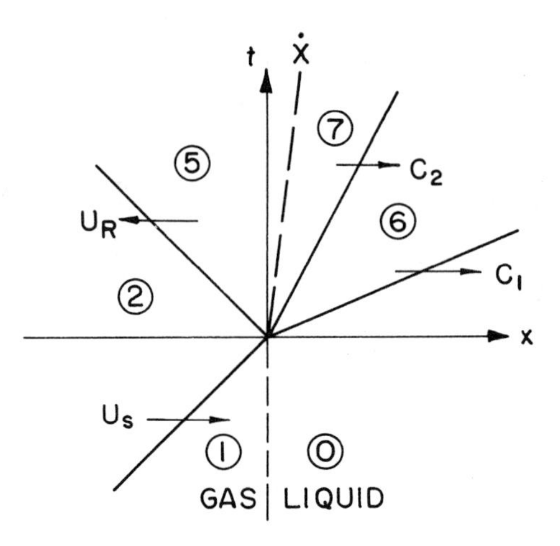

EXPERIMENT I

WAVE SPEEDS (m/sec)	REGION	COMPUTED FLOW STATES p(bar)	T(K)	v(m/sec)	w(m/sec)
U_S = 940.	1	5.33×10^{-3}	1.522	0	0
U_R = 542.	2	1.13	82.2	701.	0
c_1 = 269.	5	7.37	208.	-59.0	0
c_2 = 31.2	6	7.11	1.504	18.17	0.264
$\dot{X}$ = 10.6	7	7.10	1.557	18.18	4.992

EXPERIMENT II[a]

WAVE SPEEDS (m/sec)	REGION	COMPUTED FLOW STATES p(bar)	T(K)	v(m/sec)	w(m/sec)
U_S = 556.	1	4.13×10^{-2}	2.095	0	0
U_R = 314.	2	2.32	31.4	408.	0
c_1 = 276.	5	13.1	72.4	-18.7	0
$\dot{X}$ = 11.2	6	11.74	2.017	29.04	-0.184

[a] No temperature shock observed.

Figure 15 Numerical evaluation of the variables of state based on the experiments of Figure 14.

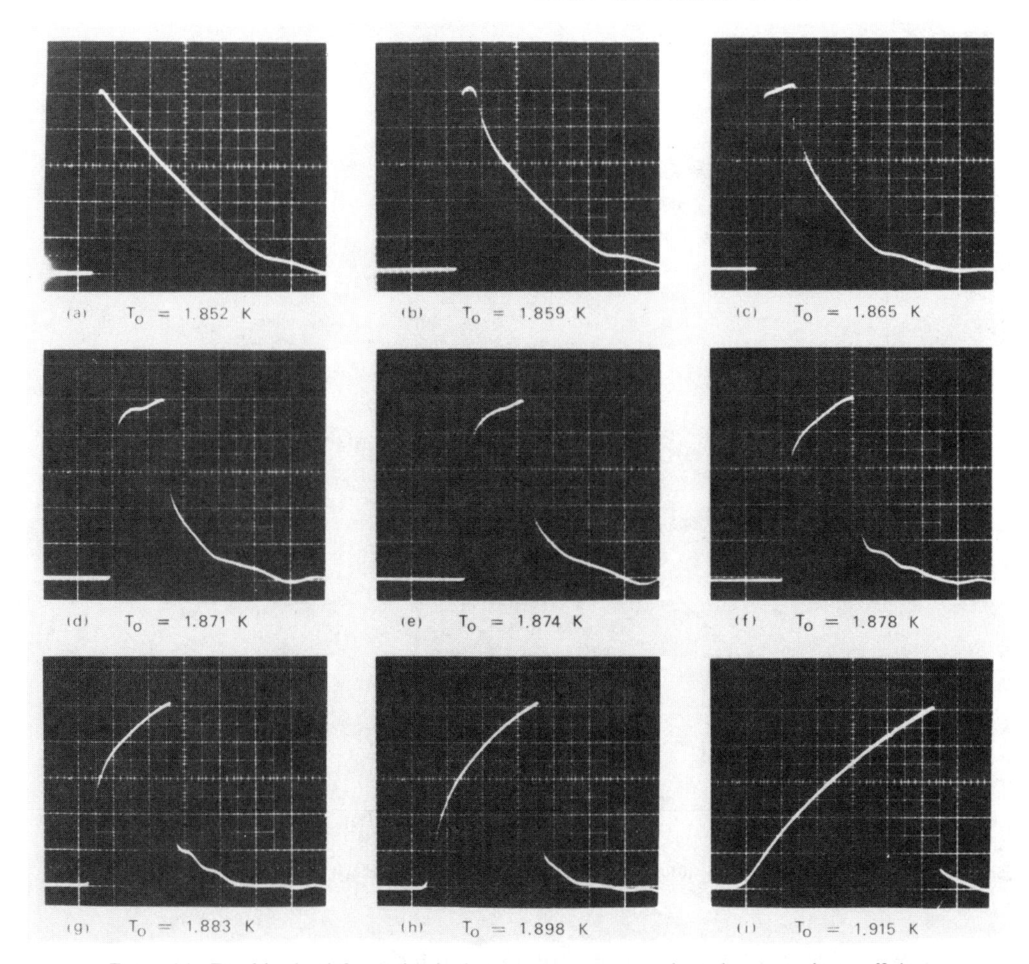

Figure 16 Double-shock formation in the temperature range where the steepening coefficient changes sign.

by Cummings et al. (1978), Turner (1979), and Rogers (1979) on temperature shocks used heat fluxes that exceeded the ones used in the earlier experiments, and they found definite deviations from the Khalatnikov result. Figure 17 shows a set of measurements due to Torczynski. The excellent agreement with the second-order theory, as well as the striking deviations for stronger shocks, is evident. This disagreement of theory and experiment can be traced to the breakdown of superfluidity and is discussed in the next section. The comparison between theory and experiment in the Cummings-Wise experiments (with a combination of temperature and pressure shock

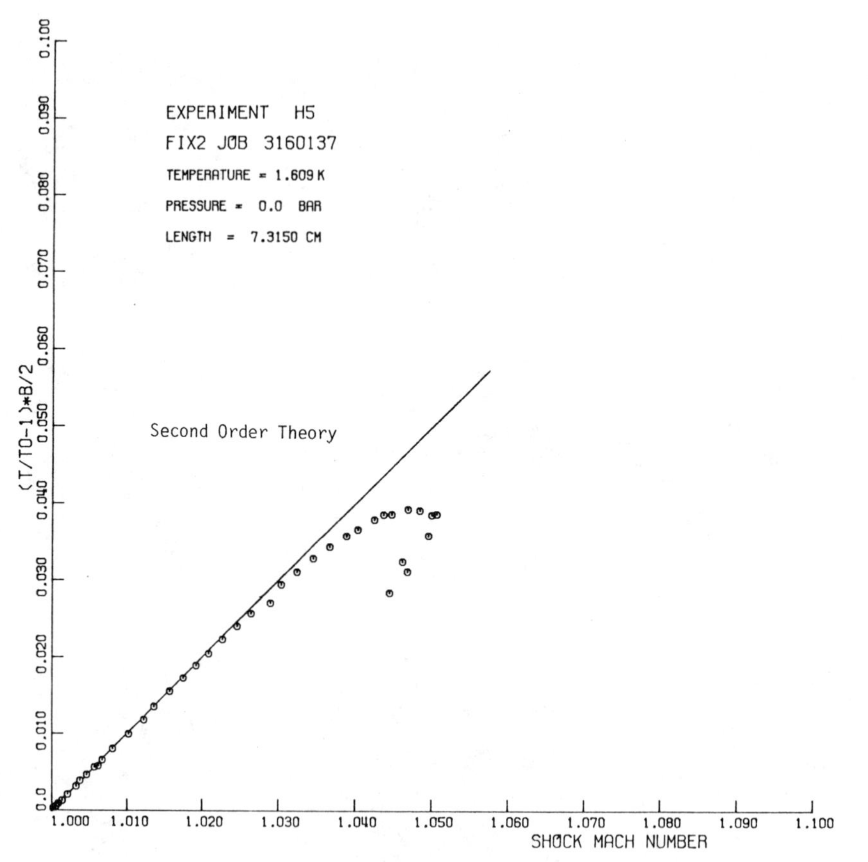

Figure 17 Temperature shock-wave measurements compared with Khalatnikov's second-order theory.

waves) very probably involves liquid-gas interface problems, which are briefly discussed further on.

Shock Waves and the Breakdown of Superfluidity

A shock wave propagating into fluid at rest changes the variables of state to new equilibrium values, and a uniform mass flow with velocity v is induced. For a classical fluid, two variables, such as entropy and pressure, determine the thermodynamic state downstream of the shock. In He II, the third state variable, the counterflow velocity w, is altered as well by the passage of a shock wave. For a pressure shock the jump in w is usually quite small, but for thermal shocks it is quite significant. It is known from heat-flow measurements in channels and porous media that superfluidity ceases to exist once a certain value for w is exceeded. This breakdown of superfluidity at an "intrinsic" critical value of w corresponds more to a thermodynamic phase transition than to, say, the laminar-turbulent transition in fluid mechanics to which it is sometimes compared. Thermal shock waves are thus a unique tool to produce well-defined states of He II including the third state variable, w. Observations of thermal shocks of increasing strength should therefore result in a determination of the intrinsic critical velocity in the absence of noticeable wall effects.

At the suggestion of one of us, this problem was investigated by Turner (1979), who measured the development of an initially rectangular heat pulse as a function of the temperature rise in the pulse (the shock strength). The comparison of the measured wave velocities with Khalatnikov's second-order theory discussed earlier showed that systematic deviations occurred as the shock strength exceeded a certain limit. These deviations are not fully explained by the neglected higher-order terms in Khalatnikov's equations,

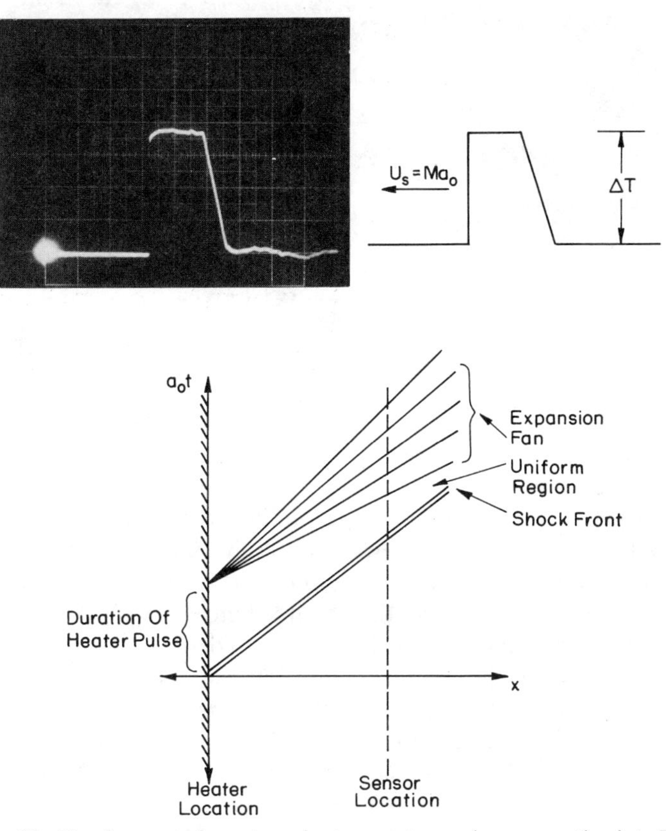

Figure 18 Development of a rectangular temperature pulse propagating into He II.

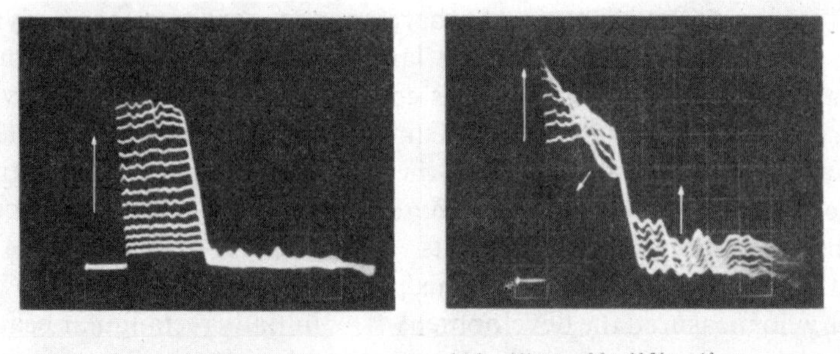

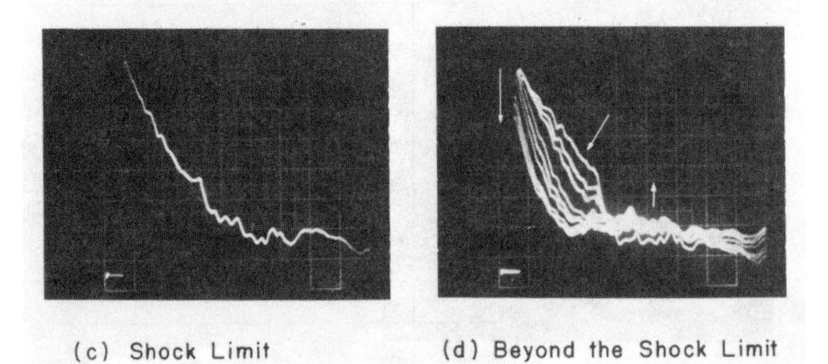

Figure 19 Measurements of the modification of heat pulses in He II with increasing temperature loading.

as demonstrated by a comparison with the numerical computations of Sturtevant & Moody.

The original characteristic shape of the pulse consists of a sudden increase, followed by a constant section and a final slow decrease, quite similar to a pressure pulse in a classical fluid (Figure 18). Beyond a certain temperature loading, however, the picture changes, finally reaching a state that Turner terms the "shock limit," a rapid increase followed by an exponential "warm" tail (Figure 19). The corresponding counterflow velocity that was found independent of the geometry is thus a "fundamental critical velocity." Schlieren photographs of a thermal shock wave, shown in Figure 20, confirm that the shock is quite planar and that the downstream state is therefore uniform. Turner's and Torczynski's critical velocities are summarized in Figure 21.

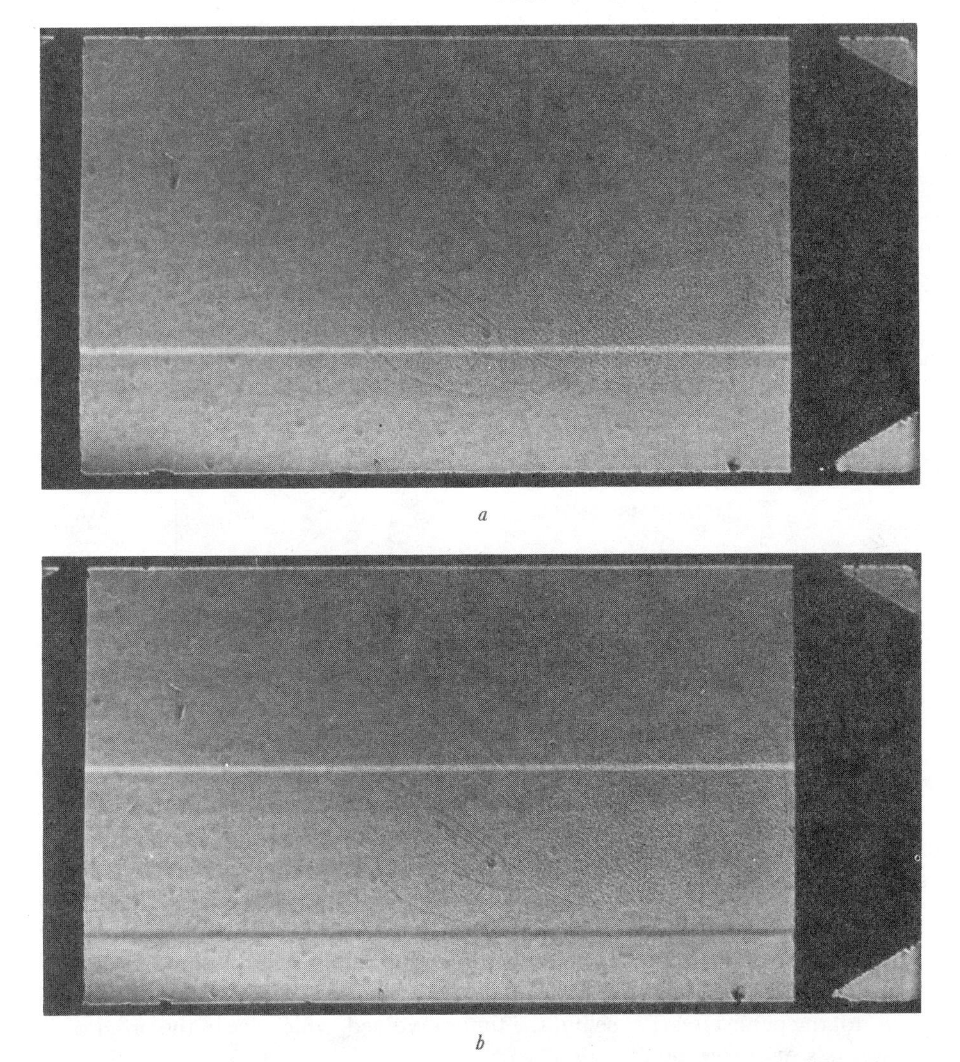

a

b

Figure 20 Schlieren photographs of a single (*a*) and a double (*b*) shock wave propagating into He II. Note that the waves are quite planar and intersect the sidewalls without any observable curvature.

The measured velocities are evidently much higher than corresponding results of steady counterflow in a channel or tube of the same diameter. These values are even higher than the critical velocities measured by Notarys (1969) with which they are compared in Figure 21; we do not imply, however, that these two measurements refer to exactly the same physical state. One of the most interesting observations in these shock-

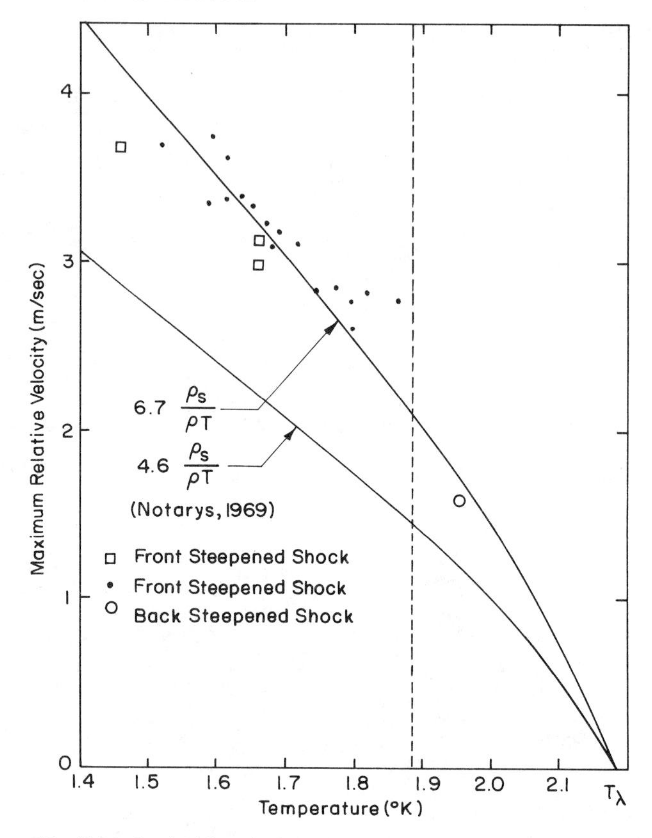

Figure 21 Critical velocities obtained from temperature-shock measurements.

propagation studies is the long-lasting effect the passage of a shock wave has on a subsequent one. This effect, which leads to the "shock limit" for a weaker subsequent shock, demonstrates the persistence of some perturbations left in the wake of a strong shock (Figure 22). The surprising fact is not the persistence per se but the times involved, which are of the order of seconds or even minutes.

It is true that the very low viscosity of the normal fluid implies a very slow decay of any vortex motion, but it is not evident what length scales determine the corresponding times. In any case, perturbations are left in the wake of a shock wave near or beyond the "shock limit," possibly having vortex character. This result suggests the study of second-sound propagation into rotating He II, i.e. propagation into a well-defined vorticity field. Experiments like these have been carried out by Torczynski (1983), and the expected effect on the "shock limit" has indeed been found. A full account of these experiments, which are not yet concluded, will be published in due time.

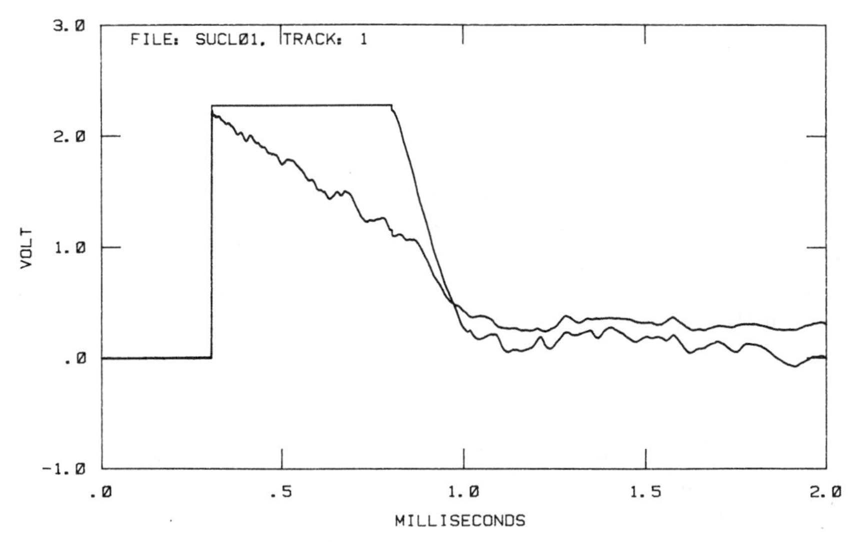

Figure 22 Perturbations left by a temperature shock affecting a subsequent shock one second later.

SUMMARY AND CONCLUSIONS

We have demonstrated some of the many interesting facets of the fluid mechanics of liquid helium. Even the limited work at GALCIT on the subject could not be reported in full here because of space limitations; a great number of fascinating avenues remain to be explored. The most important results of the experiments reported here seem to be the following: The nonlinear convection terms in the two-fluid equations are similar to the ones in the Navier-Stokes equations, and hence one expects to find both turbulence and shock waves. The jet experiments have demonstrated velocity fluctuations that are certainly qualitatively similar to classical turbulent fluctuations; furthermore, the overall behavior of the jet closely resembles the known classical character. Thus, turbulence certainly exists in the normal fluid, which is an excitation gas! Results for the interactions between the super and normal components are less conclusive. The experiments indicate an entrainment of the superfluid into the jet leading to a rather complex velocity field. A quantitative explanation on the basis of the speculative Gorter-Mellink force or Vinen's theory does not seem straightforward. Much more work on He II turbulence is needed to clarify the problem. The use of very-high-frequency second sound for Doppler velocimetry has proven to be a very useful tool for this purpose.

The observed existence and behavior of pressure and temperature shock waves is certainly in qualitative agreement with the predictions of the two-fluid equations. In particular, the double temperature shock near 1.9 K agrees qualitatively as well as quantitatively with the prediction of the two-fluid equations, and hence probably represents the best check of the correctness of the nonlinear terms in these equations. The technique of using shock waves to produce definite thermodynamic states in a fluid has proven to be as useful in He II as in classical fluid dynamics and chemical kinetics. The critical counterflow velocities in the wake of thermal shock waves of some meters/second seem to be the highest observed so far. At 1.5 K the corresponding critical heat flux is about 10^5 W m^{-2}. There remain a host of interesting problems to be explored using the shock-wave technique in both He II and in the vapor near the critical point.

ACKNOWLEDGMENTS

The authors would like to acknowledge the many helpful discussions with faculty and students at GALCIT. The assistance of D. M. Moody and J. R. Torczynski was of great value in the preparation of the manuscript. The schlieren photographs (Figure 20) were taken by Dagmar Gerthsen during her stay at GALCIT. Jacquelyn Beard's patient use of the CIT word processor was essential to the on-time completion of this paper and is gratefully acknowledged.

Literature Cited

Brown, G. L., Roshko, A. 1974. On density effects and large structure in turbulent mixing layers. *J. Fluid Mech.* 64:775–816

Cummings, J. C. 1974. Development of a high-performance cryogenic shock tube. *J. Fluid Mech.* 66:177–87

Cummings, J. C. 1976. Experimental investigations of shock waves in liquid helium I & II. *J. Fluid Mech.* 75:373–83

Cummings, J. C., Schmidt, D. W., Wagner, W. J. 1978. Experiments on second-sound shock waves in superfluid helium. *Phys. Fluids* 21:713–17

Dessler, A. J., Fairbank, W. M. 1956. Amplitude dependence of velocity of second sound. *Phys. Rev.* 104:6–10

Dimotakis, P. E. 1974. Gorter-Mellink scale and critical velocities in liquid helium II counterflow. *Phys. Rev. A* 10:1721–23

Dimotakis, P. E., Broadwell, J. E. 1973. Local temperature measurements in supercritical counterflow in liquid helium II. *Phys. Fluids* 16:787–95

Dimotakis, P. E., Laguna, G. A. 1977. Investigations of turbulence in a liquid helium II counterflow jet. *Phys. Rev. B* 15:5240–44

Donnelly, R. J. 1967. *Experimental Superfluidity*. Univ. Chicago Press. 246 pp.

Feynman, R. P. 1955. Applications of quantum mechanics to liquid helium. In *Progress in Low Temperature Physics*, ed. C. J. Gorter, pp. 17–53. Amsterdam: North-Holland. 418 pp.

Gorter, C. J., Mellink, J. H. 1949. On the irreversible processes in liquid helium II. *Physica* 15:285–304

Gulyaev, A. I. 1960. Schlieren photography of thermal pulses in liquid He 4. *Sov. Phys.-JETP* 30:34–43

Gulyaev, A. I. 1967. Visual observation of second sound by the Toepler method. *JETP Lett.* 5:325–27

Johnson, E. M., Hildebrandt, A. F. 1969. Entrainment of second sound in steady-state counterflowing normal and superfluid helium II. *Phys. Rev.* 178:292–94

Kapitza, P. L. 1941. The study of heat transfer in helium II. *Acad. Sci. USSR J. Phys. USSR* 4:181–210

Keesom, W. A., Keesom, A. P. 1936. On the heat conductivity of liquid helium. *Physica* 3:359–60

Khalatnikov, I. M. 1956. The propagation of sound in moving helium II and the effect of a thermal current upon the propagation of second sound. *Sov. Phys.-JETP* 3:649–51

Khalatnikov, I. M. 1965. Discontinuities in a

superfluid. *Introduction to the Theory of Superfluidity*, Ch. 13, pp. 81–87. New York: Benjamin. 206 pp.

Kramers, H. C. 1966. Experimental data on different types of flow in liquid helium. In *Superfluid Helium*, ed. J. F. Allen, pp. 199–214. New York: Academic. 304 pp.

Laguna, G. A. 1975. Second-sound attenuation in a supercritical counterflow jet. *Phys. Rev. B* 12:4874–81

Landau, L. D., Lifshitz, E. M. 1959. *Fluid Mechanics*. Reading, Mass: Addison-Wesley. 536 pp. (Transl. from Russian)

Liepmann, H. W. 1952. Deflection and diffusion of a light ray passing through a boundary layer. *Douglas Rep. SM14397*

Liepmann, H. W., Roshko, A. 1957. *Elements of Gasdynamics*. New York: Wiley. 439 pp.

Liepmann, H. W., Cummings, J. C., Rupert, V. C. 1973. Cryogenic shock tube. *Phys. Fluids* 16:332–33

Lin, C. C. 1959. Hydrodynamics of liquid helium II. *Phys. Rev. Lett.* 2:245–46

London, F. 1954. *Superfluids*, Vol. 2. New York: Wiley. 221 pp.

Moody, D. M. 1983. *I. Numerical solution of the superfluid helium shock jump conditions. II. Experimental investigation of the liquid He II-vapor interface*. PhD thesis. Calif. Inst. Technol., Pasadena. 88 pp.

Notarys, H. A. 1969. Pressure-driven superfluid helium flow. *Phys. Rev. Lett.* 22:1240–42

Onsager, L. 1949. Discussione & osservazioni. *Nuovo Cimento* 6:249–50

Osborne, D. V. 1951. Second sound in liquid helium II. *Proc. Phys. Soc. London Ser. A* 64:114–23

Putterman, S. J. 1974. *Superfluid Hydrodynamics*. Amsterdam: North Holland. 442 pp.

Roberts, P. H., Donnelly, R. J. 1974. Superfluid mechanics. *Ann. Rev. Fluid Mech.* 6:179–225

Rogers, P. L. 1979. *Experimental investigation of second sound shock waves in liquid helium II*. AE thesis. Calif. Inst. Technol., Pasadena. 46 pp.

Schlichting, H. 1968. *Boundary-Layer Theory*, p. 697. New York: McGraw-Hill. 747 pp.

Sturtevant, B. 1976. Shock waves in liquid helium II. *Bull. Am. Phys. Soc.* 21(10):1223 (Abstr.)

Tennekes, H., Lumley, J. L. 1972. *A First Course in Turbulence*. Cambridge, Mass: MIT Press. 300 pp.

Torczynski, J. R. 1981. Effect of quantized vortex lines on intrinsic critical velocity measurements in rotating He II. *Bull. Am. Phys. Soc.* 26:1254 (Abstr.)

Torczynski, J. R. 1983. *Second sound shock waves in rotating superfluid helium*. PhD thesis. Calif. Inst. Technol., Pasadena. 96 pp.

Turner, T. N. 1979. *Second sound shock waves and critical velocities in liquid helium II*. PhD thesis. Calif. Inst. Technol., Pasadena. 216 pp.

Van Der Heijden, G., De Voogt, W. J. P., Kramers, H. C. 1972. Forces in the flow of liquid helium II. *Physica* 59:473–97

Vinen, W. F. 1957a. Mutual friction in a heat current in liquid helium II. I. Experiments on steady heat currents. II. Experiments in transient effects. *Proc. R. Soc. London Ser. A* 240:114–27, 128–43

Vinen, W. F. 1957b. Mutual friction in a heat current in liquid helium II. III. Theory of mutual friction. *Proc. R. Soc. London Ser. A* 242:493–515

Wise, J. L. 1979. *Experimental investigation of first and second-sound shock waves in liquid helium II*. PhD thesis. Calif. Inst. Technol., Pasadena. 128 pp.

Ann. Rev. Fluid Mech. 1984. 16: 179–93

SECONDARY FLOW IN CURVED OPEN CHANNELS

Marco Falcón

Instituto de Mecánica de Fluidos, Universidad Central de Venezuela, Caracas, Venezuela

INTRODUCTION

This article concerns steady secondary flow in open-channel bends and reviews the state of the subject by reference to a few available studies of flow over rigid and erodible beds in curved channels with relatively simple planform. The secondary or transverse flow is important because (*a*) it is partly responsible for the large-scale bed topography of natural alluvial-channel bends, (*b*) it can interact dynamically with the primary flow, and (*c*) it is useful for studies of diffusion and navigation in natural waterways. The associated subject of instability of an initially straight-channel flow and its subsequent meandering has been reviewed by Callander (1968, 1978) and is not treated here. Stress-induced currents in straight channels due to turbulence anisotropy also are not considered.

As is shown in Figure 1, the radial or secondary-flow velocity component, u, occurs in planes perpendicular to the primary direction of motion (i.e. to the channel axis) and is originated by the centrifugal acceleration, v^2/r, due to channel curvature (the coordinate system shown is cylindrical). The vertical velocity component, w, not shown in the figure, is generally considered as part of the secondary-flow phenomenon. The streamwise velocity component, v, varies from zero at the bed to a maximum value at or near the water surface. Consequently, the centrifugal force is greatest near the water surface and decreases toward the bed. This force will tend to convey water radially outward and thus accumulate it in the outer portion of the channel. Therefore, a second principal radial force is due to the radial pressure gradient, $\partial p/\partial r$, otherwise manifest as the well-known transverse superelevation phenomenon; if the pressure is assumed to be distributed hydrostatically, the radial pressure gradient is linearly

0066-4189/84/0115-0179$02.00

proportional to the local radial slope of the water surface, $\partial H/\partial r$. The first of these forces is strongly variable over the depth, while the second is constant; consequently, there is a net radial force that changes sign at some elevation within the flow. It is this force imbalance that produces the secondary radial velocity, u. The difference between the centrifugal and pressure-gradient forces is mainly balanced by the vertical gradient of the radial shear stress, $\partial\tau_{zr}/\partial z$, and the convective inertial force, $\rho v\, \partial u/\partial s$ (only if the secondary flow is not fully developed), both of which depend on the distribution of u. Indeed, the u-distribution evolves along the channel axis such that the radial force balance is satisfied. For strongly curved flows or in the bank-affected regions, other terms in the radial dynamic equation may become important. Bathurst et al. (1979) mention that secondary circulation is weakest at low and high discharges and strongest at medium discharges. In the latter case, medium flows still occupy a more or less sharply curved thalweg and thus present relatively large velocities and curvatures, which

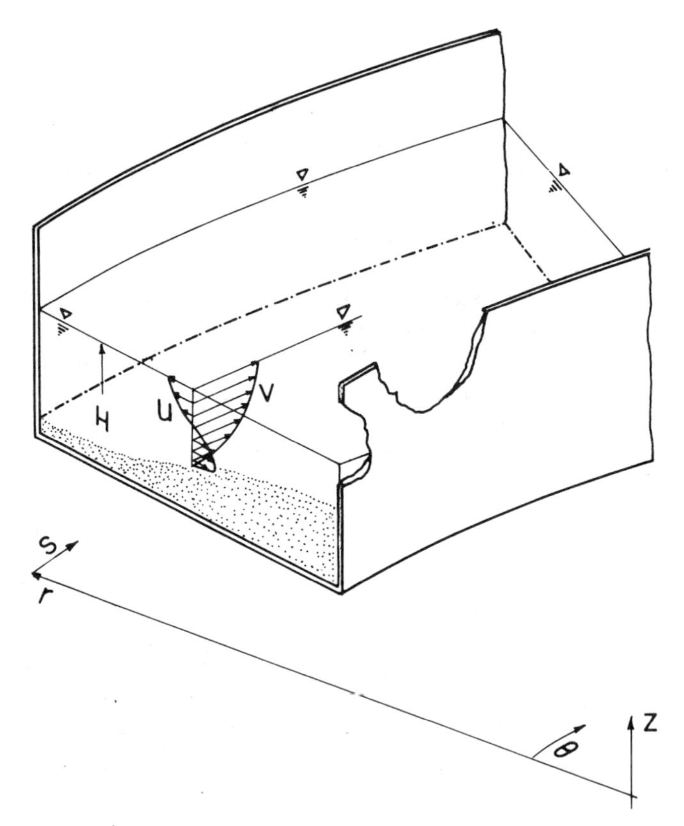

Figure 1 Definition sketch for curved channel flow.

combine to produce a large centrifugal force acting on fluid particles. The first two cases, however, have either very low velocities or a very large radius of curvature, which in both cases leads to a reduced centrifugal effect.

According to Bathurst et al., secondary flow can sometimes transfer the position of the maximum longitudinal bed shear stress from the point where the maximum longitudinal velocity occurs to the region of strong downwelling where the reverse and main secondary circulatory cells meet and where there is a significant compressing of the isovels.

The existence of constant secondary flow has been questioned by Götz (1980). The oscillations of the secondary spiral axis indicated by him suggest an interesting problem for further research.

In the case of meandering flows where the flow progresses along the channel from one meander to the next, the direction of the centrifugal force (relative to the channel axis) will be periodically reversed, so that the u-distribution in Figure 1 will have first one sense and then another. Likewise, the directions of the radial water-surface slope and radial pressure gradient will be periodically reversed. In addition, because the radial bed shear stress and both the longitudinal and radial sediment-transport rates vary periodically along the channel, the transverse bed slope also will vary periodically. The resulting depth variation requires that fluid be transported across the channel, first in one direction and then in the other. Therefore, the u-distribution may be considered to be composed of a rotational and a translational component. Due to the inertia of the flowing fluid, represented mainly by the term $\rho v\, \partial u/\partial s$, the u-distribution does not respond instantaneously to variations in the centrifugal force; instead, phase shifts exist between the channel-axis curve and the streamwise distributions of the transverse velocity component, the radial bed shear stress, and the radial water-surface and bed slopes. These phase shifts will increase as the magnitude of the inertial term, $\rho v\, \partial u/\partial s$, increases relative to that of the shear-stress (frictional) term $\partial \tau_{zr}/\partial z$. In light of these considerations, it follows that in fully developed curved flow the secondary flow is directed radially outward (inward) near the upper (lower) layers of the flow (see Figure 1), while in nonuniform flow the secondary flow pattern is variable.

Following the analysis of Zimmermann & Kennedy (1978), Falcón (1979) examined meandering-channel flow in terms of the moments exerted on it. Considering a full-width (but not including the wall-affected zones) fluid control volume extending between the bed and free surface, of angular increment $\Delta\theta$, and taking moments about a longitudinal axis located at, say, the center of gravity of the fluid within the control volume, leads to a balance of the torque of the centrifugal force, the torque of the radial shear stress on the bed, and the moment of the net momentum flux from the control volume (which is related to the $\rho v\, \partial u/\partial s$ term in the radial dynamic equation). These will be denoted as the driving, frictional, and inertial torques, respectively, and all must vary periodically along the channel. In terms of the moment-of-momentum equation, flow in sinusoidally meandering channels is analogous, at least loosely, to a forced, mass-dashpot, vibrating linear system. Through this analogy, it is again seen that if the frictional torque (dashpot) is the principal balance to the driving torque, the phase shift of the u-distribution (and related quantities, including transverse bed slope) will be small, while if the principal balance is between the driving and inertial torques, this phase shift can be expected to be quite large. This reasoning again leads to the existence of phase shifts for the various aforementioned geometric, kinematic, and dynamic variables that characterize meandering flows.

The analogy to a linear vibrating system breaks down if the secondary flow significantly interacts dynamically with the primary flow, if the convective acceleration terms $u\,\partial u/\partial r$ and $w\,\partial u/\partial z$, in addition to $v\,\partial u/\partial s$, are important, and if the ratios of the width and depth to the minimum radius of curvature are not much smaller than unity. In these strongly curved cases, higher harmonics and nonlinearities enter into the governing equations, and the phenomenon is no longer analogous to a simple harmonic system.

UNIFORM CURVED FLOW

Consider a very long curved channel with constant width and rigid vertical side walls with a common planform center of curvature. The central longitudinal channel axis at the level of the bed has a constant slope, S_0, and describes a helix in space, which if projected onto a horizontal plane becomes a circle of radius r_c. The system can be conveniently described in cylindrical coordinates. If the radial slopes of the bed and water surface are to be constant along the channel, then the local streamwise slope, $S(r)$, of both must be given by $S = S_0 r_c/r$. The flow considered herein is treated as uniform in the sense that all flow and channel properties are invariant along any helix with radius r and with slope S, as defined above.

We now review the basic tenets of some analyses for determining the transverse velocity distribution, u. These analyses start from a simplified radial dynamic equation that retains three basic terms: the radial pressure-gradient term, the centrifugal-force term, and the radial shear-force term. They can be considered weakly curved open-channel flow analyses in the sense that a logarithmic or power-law velocity distribution for the streamwise velocity component is assumed to hold (i.e. it is assumed that the distorting effect of the transverse flow on the primary velocity distribution, v, is negligible). In addition, inner-wall separation phenomena are neglected, and the analyses are assumed valid within a central portion of the channel where wall effects are considered negligible.

Vertical integration of the simplified radial dynamic equation

$$\partial H/\partial r - v^2/gr = 1/\gamma \; \partial\tau_{zr}/\partial z,$$

where g is the acceleration of gravity and γ is the fluid specific weight, under the hydrostatic assumption (i.e. $\partial p/\partial r$ is constant throughout the depth of flow and equal to $\gamma\,\partial H/\partial r$), together with the boundary condition that the radial shear stress is negligible at the free surface, yields the radial shear-stress distribution. This result, however, contains the local radial water-surface slope as an unknown. Rozovskii (1957) and Kikkawa et al. (1976) assume that for a hydraulically smooth boundary the radial shear stress at the level of the bed, τ_{0r}, is negligible and thereby calculate $\partial H/\partial r$. Then, using a constant momentum-diffusion coefficient, E (which can be obtained from primary-flow considerations), and the radial shear-stress distribution, a radial velocity distribution is obtained by integration of a simplified Boussinesq expression for the radial shear stress:

$$\tau_{zr} = \rho E \; \partial u/\partial z.$$

The integration constant is determined by imposing the condition of zero net radial flow. As a consequence of the assumption that τ_{0r} is zero, the u-distribution obtained has a finite value and a vertical tangent at the level of the bed (see Kikkawa et al.'s u-distribution in Figure 2). In the case of a hydraulically rough boundary, Rozovskii assumes that the directions of the velocity component within, and the shear stress acting on, a horizontal plane are the same at an elevation equal to the bottom roughness height.

Engelund (1974) introduces the slip-velocity concept (a finite velocity at the level of the bed) in order to justify working with a constant eddy viscosity. This in turn leads to a parabolic primary velocity distribution. The lower boundary condition used by him is similar to Rozovskii's for the hydraulically rough boundary case, except for the fact that it is applied directly at the bed. Engelund's treatment improves upon the previous analyses because it applies to an arbitrary roughness condition and thus

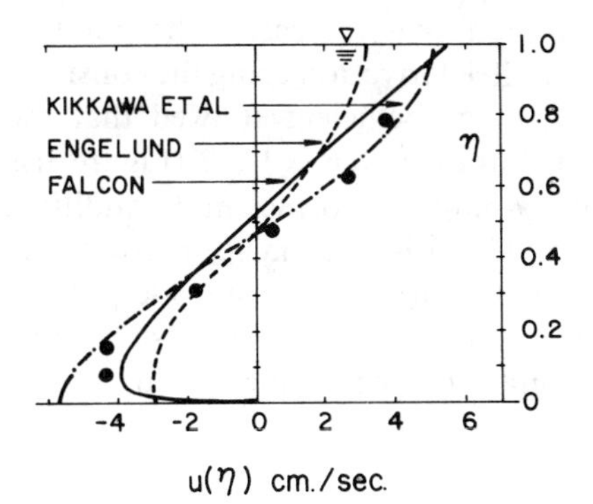

Figure 2 Comparison of theoretical predictions for the radial-velocity component by Kikkawa et al. (1976), Engelund (1974), and Falcón (1979) with measurements by Kikkawa et al.

does not assume that τ_{0r} is zero for the hydraulically smooth bed case. Naas (1977), however, found from numerous field measurements that the parabolic velocity distribution deviates significantly from the measured values over the lower 40% of the channel depth.

Following Zimmermann & Kennedy (1978), Falcón (1979) uses the power-law distribution for the primary velocity component, v, together with Nunner's (1956) law,

$$n = 1/\sqrt{f},$$

which relates the power-law exponent, $1/n$, to the Darcy-Weisbach friction factor, f, in order to calculate the u-distribution. The power-law distribution was taken because it is nonsingular at the bed and is easier to work with than the logarithmic distribution. In addition, a depth-varying, scalar, eddy-diffusion coefficient is deduced from a simplified Boussinesq expression for the primary-flow shear stress:

$$E = \tau_{z\theta}/(\rho\, \partial v/\partial z).$$

This is done by assuming a primary shear-stress distribution that varies linearly over the depth, and the velocity gradient is obtained from the above mentioned power-law distribution. Then, insertion of the variable eddy-diffusion coefficient into the simplified Boussinesq expression for the radial shear stress and integration yields the u-distribution. The constant of integration is determined from the condition that $u = 0$ at the level of the bed, and the local value of the transverse water-surface slope is obtained by imposing the condition of no net radial flow.

The dimensionless radial-velocity distribution obtained by Falcón is shown in Figure 3 for values of n equal to 2.5 (large bed roughness), 5 (intermediate bed roughness), and 7.5 (low bed roughness) (r and d are local values of the radial coordinate and depth, respectively; V_A is the depth-averaged primary velocity component). The shape of the curves is in agreement with Rozovskii's experimental findings: for smooth beds the velocity continues to increase as the bed is approached down to a very small distance from the bed; as the bed roughness is increased, the point of maximum velocity moves upward. A parallel derivation using the constant, depth-averaged value of the eddy-diffusion coefficient showed that the significant shape variation of the u-distribution with $n = 1/\sqrt{f}$ (Figure 3) is due to the vertical variability of the eddy-diffusion coefficient. In addition, use of the correct boundary condition, $u(0) = 0$, is essential for reproducing the strong variations of $u(\eta)$ near the bed for large values of n. Despite the fact that the power-law v-distribution and the Boussinesq expression for τ_{zr} are not valid in the immediate vicinity of the channel bed, the u-distributions appear to be reasonable.

Falcón deducted the effect of the net forward momentum flux, due to the secondary flow, from the along-slope weight component of the fluid and found that the longitudinal bed shear stress is reduced. Then, the corresponding decrease in the Darcy-Weisbach friction factor, f, and increase in n (the inverse of the exponent in the velocity power-law distribution) were calculated. It is thus assumed that the primary velocity distribution continues to follow the power law in spite of the distorting effects of the transverse flow. Application of the above corrective procedure led to very acceptable predictions of the experimentally obtained transverse bed profiles by Zimmermann (1974) and Hooke (1974) in curved open-channel flows with movable sand beds; this constitutes indirect evidence of the adequacy of the procedure. Because the u-distribution is sensitive to the value of n, or equivalently, to the friction factor, f, the procedure for correcting the value of f is important. However, it appears that the resistance to the primary flow, induced by the secondary flow, may be a laboratory curiosity (i.e. it was found to be negligible for all field data analyzed and not too important for the larger laboratory flows examined).

The theoretical u-distributions derived by Engelund, Kikkawa et al., and Falcón are compared with one of Kikkawa et al.'s experiments in Figure 2. In general, these data are, to the author's knowledge, the best available because the authors gave the slope, S_0, of the central longitudinal channel axis, used a relatively long curved flume, and took care to insure that the secondary flow was fully developed. It can be seen in Figure 2 that for these particular data Kikkawa et al.'s u-distribution overpredicts near the bed,

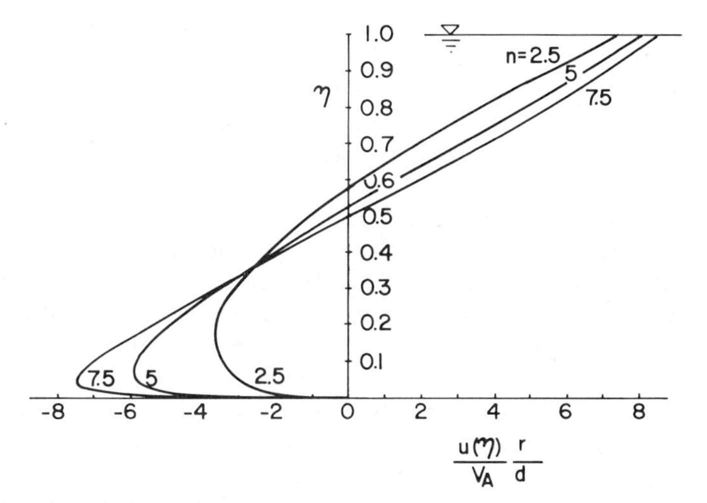

Figure 3 Dimensionless radial-velocity distribution illustrating the influence of the friction factor, $f = 1/n^2$ (Falcón 1979).

Engelund's seems to underpredict, and Falcón's u-distribution falls in between. Near the free surface, Kikkawa et al.'s distribution comes closest to the measured data. However, because the theoretical u-distributions shown in Figure 2 are the result of a highly simplified turbulence model, and because at present there are not many experimental data that conform to the previously defined uniform curved-flow condition, it would only appear possible to conclude that the distributions are reasonable and of the correct order of magnitude. These theories are all based on the existence of a Boussinesq, scalar, eddy-diffusion coefficient, and the only way to improve the derived u-distributions would be to use an improved turbulence model. Such an attempt, using a k-ε two-equation model to define a scalar turbulent momentum-diffusion coefficient, has been formulated by Leschziner & Rodi (1979) and is referred to in the next section.

Flow in curved rigid channels is a difficult three-dimensional turbulent boundary-layer problem, further complicated by the anisotropic nature of the turbulence structure near solid boundaries. In order to predict accurately the bed shear stress, it may be necessary to use a tensorial form of the momentum-diffusion coefficient. Additionally, in sediment-transporting curved channels a better formulation than those currently available is required in order to predict solid transport rates within such complex flow fields.

DeVriend (1981b) presents a numerical solution of fully developed laminar flow in a curved, rigid, rectangular open channel by solving the full Navier-Stokes equations over the total cross section of the flow. The interaction between the primary and secondary flows is investigated, and of particular interest is the prediction of a reverse circulatory cell adjacent to the outer wall that occurs when the Dean number of the flow, defined as the product of a depth Reynolds number and the square root of the ratio of depth to radius of curvature, is larger than about 50. Bathurst et al. (1979) present field confirmation of such reverse cells, not extending more than one or two depths into the flow, when the outer bank is steep, but they find no evidence of such cells when the bank is shelving. The latter observation is also confirmed by deVriend (1981a), who studied the fully developed turbulent-flow case by means of a Boussinesq, scalar, eddy-diffusion turbulence model and presented comparisons between a simplified computational model (similarity profiles for the velocity distributions over the flow cross section) and a fully three-dimensional computation, both of which were further simplified in accordance with his previous laminar-flow solution. In addition, deVriend (1981a) states that in fully developed flow, wall effects penetrate the flow cross section completely, no matter how shallow the channel may be (i.e. the secondary flow in regions far from the banks should not be calculated without considering, as Rozovskii,

Engelund, Kikkawa et al., and Falcón did, the interaction with the near-bank flow). This effect is due more to convective acceleration effects than to lateral shear stresses. According to deVriend, and on the basis of a laminar-flow model, when fluid enters a curved channel, the secondary flow develops to a nearly fully developed condition after a relatively short distance. Thereafter, the secondary flow interacts significantly with, and modifies, the transverse distribution of the primary-flow velocity. This, in turn, affects the secondary-flow distribution, and it is at this stage, over a relatively large length of curved channel, that wall effects penetrate completely into the central core of the flow. DeVriend (personal communication) indicates that simplified, central, shallow-channel models may be useful in those portions of the channel (at a relatively short distance from the curved entrance) where the secondary flow has reached its nearly fully developed condition, but where the subsequent changes mentioned above are not yet significant.

Although reasonable u-distributions can be calculated with simple turbulence models, this does not appear to be the case for the corresponding calculated transverse bed shear stress, which is an important quantity in the application of curved-flow models to erodible bed channels. Falcón (1979) calculated this stress from a simplified moment-of-momentum equation (based on the hydrostatic condition) and found it to differ significantly from that corresponding to the Boussinesq expression. Examination of the depth-integrated radial dynamic equation shows that the radial bed shear stress is a relatively small quantity, equal to the difference of two relatively large quantities: the centrifugal force and the pressure-gradient force. Therefore, a small error in the transverse water-surface slope will lead to a large error in the magnitude of τ_{0r}. This suggests that the magnitude of the radial bed shear stress may be sensitive to the particular turbulence model used, and thus until a turbulence model capable of representing curvature and anisotropy effects of the flow plus near-bed fluid-sediment interactions is formulated, it would appear that use of the moment-of-momentum approach for calculating the transverse bed shear stress is a useful alternative when the flow can be assumed to be hydrostatic. Otherwise, Naot & Rodi's (1982) work, illustrating the use of an anisotropic-turbulence model for calculating stress-induced secondary currents in straight (closed- and open-channel) ducts, might constitute a useful point of departure in order to calculate more accurately τ_{0r} in curved open-channel flows.

NONUNIFORM CURVED FLOW

Gottlieb (1976) studied both experimentally and theoretically the case of steady flow in an erodible-bed, weakly meandering channel (i.e. a nearly straight channel) in which bed load was the dominant mode of transport. The channel geometry and flow characteristics were purposely chosen so that the governing equations could be linearized. A three-dimensional calculation of the transverse velocity distribution is presented as part of an overall solution for the large-scale bed topography. It is noteworthy that very large phase shifts of the bed topography with respect to the channel apex were measured in Gottlieb's experiments. This is a consequence of the fluid inertia, represented by the term $\rho v\, \partial u/\partial s$ in the radial dynamic equation, and its effect has been referred to in the introductory section. Also, for two rigid-bed experiments, measurements of the centerline vertical distribution of the angular deviation of the velocity vector from the longitudinal direction at the channel apex demonstrate the existence of a secondary-flow velocity component that is opposite to what a uniform-flow analysis would suggest (i.e. the transverse velocity was directed radially inward near the surface and radially outward near the channel bottom). Gottlieb does not present any measured transverse velocity distributions. His analysis demonstrates that there is an important interrelationship between the transverse rate of change of the depth-averaged primary-flow velocity, the longitudinal and vertical variations of the transverse velocity component, the transverse bed shear stress, the large-scale bed topography, and the longitudinal and transverse bed load transports.

Falcón (1979) reanalyzed the above problem, using one more equation than Gottlieb (the moment-of-momentum equation), in order to solve only for various channel-centerline quantities such as the u-distribution and the transverse bed slope. In this manner, and unlike Gottlieb, he did not have to use wall boundary conditions in order to calculate the solution for the u-distribution inwardly toward the channel centerline. The transverse bed shear stress then results from satisfying simultaneously the simplified Boussinesq turbulence model and a simplified moment-of-momentum equation in which it is assumed that the integral of the vertical shear stress, τ_{rz}, over the depth is zero. Falcón predicted most of Gottlieb's experimental results with reasonable accuracy. The secondary-flow velocity distribution corresponding to one of Gottlieb's (1976) movable-bed experiments is shown in Figure 4.

Another fundamental difference between Gottlieb's and Falcón's analyses is that the latter used the mean sediment particle size in his analysis, while the former used a constant, dynamic, sediment-friction angle; in addition, Gottlieb used Engelund's (1974) slip-velocity concept for both the longitudinal and transverse velocity distributions, while Falcón imposed the condition of null velocity at the bed. These differences are important because the vertical distribution of the transverse velocity component was found to influence strongly the previously mentioned phase shifts in weakly meandering flows. This should also be taken into account when considering deVriend's (1981b) conclusions related to curved turbulent flows on the basis of a laminar-flow model.

Leschziner & Rodi (1979) present a three-dimensional numerical solution of strongly curved open-channel flow over a rigid bed with vertical walls. The dynamic equations are simplified, but all the inertial terms are retained. For appropriate boundary conditions, these equations are solved together with the continuity equation and in conjunction with two more transport equations, which involve the turbulence-model quantities k and ε. The latter define a space-varying, Boussinesq, scalar, eddy-diffusion coefficient. This turbulence model is denoted as a two-equation "k-ε model" and is described by Launder & Spalding (1974). It is assumed that the universal law of the wall applies in an appropriately selected layer whose outer edge defines a plane of finite difference nodes lying closest to the walls. A wall function is formulated that links the near-wall velocities, calculated from the momentum equations, to the wall shear stress. The model is applied to one of Rozovskii's (1957) experiments, and the predictions for the vertical distribution of the primary and secondary velocity components are

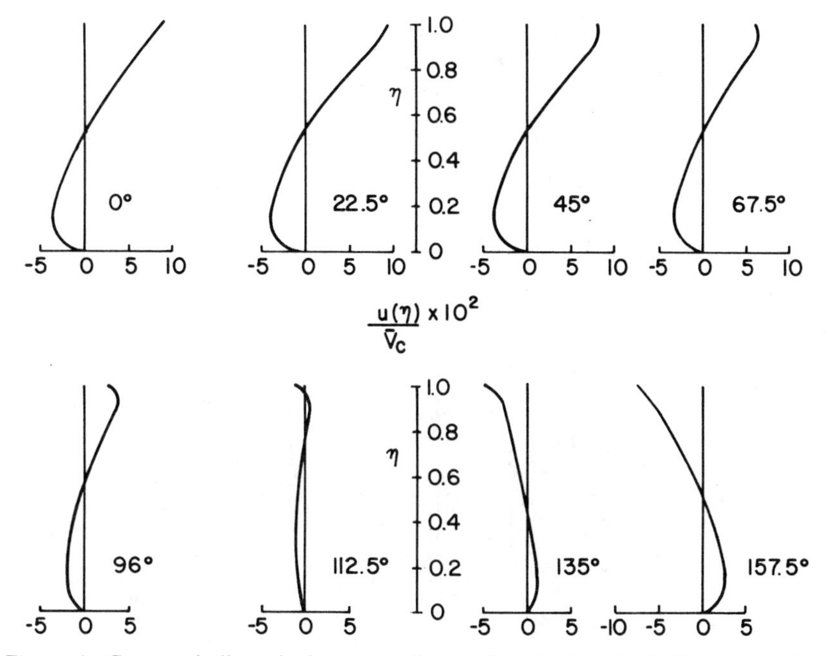

Figure 4 Computed dimensionless, centerline, radial-velocity distributions at various angular positions downstream of the channel apex for Gottlieb's (1976, pp. 56–62) run 1s in a weakly meandering channel (Falcón 1979).

shown in Figure 5. The calculated transverse water-surface slopes differ significantly from the observed ones. This will have an important effect upon the transverse bed shear stress, which is not calculated. Both Anwar & Rodger's (1981) discussion of the paper and Leschziner & Rodi's (1981) closure mention the possibility of applying the k-ε model to flows with anisotropic turbulence.

It is noteworthy that various authors have attempted to verify their uniform curved-flow analyses, which generally assume a logarithmic distribution for the primary velocity component, with Rozovskii's experiments, which are decidedly nonuniform. In these experiments the bed of the channel had no slope, and thus uniform curved flow of the type mentioned in the introduction could not exist; and, in addition, a relatively strong secondary-flow inertial resistance existed. As a result it is quite difficult to estimate a Chezy coefficient for Rozovskii's experiments. In addition, Rozovskii's curved-flow experiments are of the strongly curved type in the sense that the vertical distribution of the primary velocity component

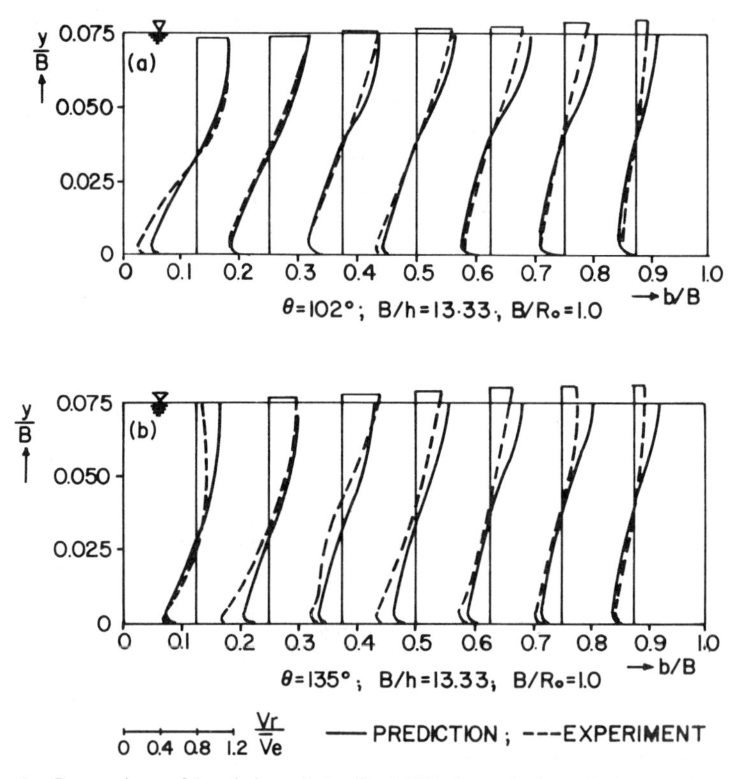

Figure 5 Comparison of Leschziner & Rodi's (1979) theoretical prediction for the radial-velocity component with measurements from an experiment by Rozovskii (1957).

deviates significantly from a logarithmic distribution. Leschziner & Rodi adequately circumvent this problem by solving for the v-distribution from a system of partial differential equations with appropriately selected boundary conditions. Further verification of their model with other author's data would seem desirable.

DeVriend (1981a) presents a numerical solution for turbulent nonuniform flow in curved channels of arbitrary, but gradually varying, depth. His approach is similar to Leschziner & Rodi's work, but simpler, because a Boussinesq turbulence model, instead of a two-equation k-ε model, is used for the sake of computational economy. The work of these authors shows the need for investigating the effects of near-bank turbulence anisotropy on the nature of the constitutive equation relating stresses and fluid deformations in order to predict bank shear stresses. Because these phenomena are intimately related to bank-recession rates, their study would be useful.

It should be noted that both observations and calculations of vertical u-distributions for meandering flows indicate that these distributions are not similar in the vertical, and also that the mean transverse velocity is not always oriented in the same direction as the radial shear stress applied to the bed (see Figure 4). The latter condition appears to have been imposed in some vertically integrated two-dimensional flow models of lakes, estuaries, etc. (see Falcón 1983).

In nonuniform curved-channel flows the resulting large-scale bed topography cannot be inferred from the secondary-flow pattern. The sediment continuity equation, reflecting the effect of spatial rates of change of the longitudinal and transverse depth-integrated sediment-transport rates upon the bed levels, governs in this case. Hooke (1980) dwells upon this point extensively.

DISCUSSION AND CLOSING REMARKS

For the case of steady uniform curved flow in open-channel bends, reasonable distributions of the secondary-flow velocity have been obtained with a simple Boussinesq turbulence model. By using a depth-varying, scalar, eddy-diffusion coefficient together with the null-velocity condition at the channel bed, Falcón (1979) shows that the transverse velocity distribution is sensitive to the friction factor of the flow, a fact suggested by Rozovskii's experiments. For strongly curved nonuniform flows, Leschziner & Rodi (1979) use a k-ε turbulence model, which defines a Boussinesq, scalar, eddy-diffusion coefficient, and present good predictions for one of Rozovskii's experiments. A basic difference between the uniform and nonuniform analyses is that in the former it is assumed that the primary velocity component follows a logarithmic or power-law distribution, while in the latter this distribution is part of the calculation.

There is some evidence that most of the interesting effects that arise due to the existence of the secondary flow and that have been calculated for, or observed in, laboratory flows may be laboratory curiosities. These effects include the large secondary-flow inertial resistance in Zimmermann's (1974) and Hooke's (1974) erodible-bed curved-flow experiments; the large phase shifts of the transverse velocity distribution with respect to the channel apex in Gottlieb's (1976) experiments; and the strong interaction of the primary and secondary flows in Rozovskii's (1957) strongly curved nonuniform flows. Falcón (1979) found the inertial resistance to the primary flow induced by the secondary flow to be negligible for all field data analyzed and not too important for the larger laboratory flows examined. This can be explained from Yen's (1965) experimental finding that the magnitude of the ratio u/v is of $O(d/r)$. Therefore, because in most rivers the magnitude of the ratio d/r is much smaller than that corresponding to typical laboratory values, the dynamic effect of the transverse velocity component on the primary flow is usually negligible in natural streams. Similarly, meandering rivers are characterized by ratios of depth to meander wavelength with magnitudes that can be several orders smaller than the laboratory values corresponding to, say, Gottlieb's experiments with weakly meandering flows. The magnitude of the longitudinal rate of change of various dynamic quantities is proportional to the magnitude of the depth-to-wavelength ratio. Therefore, when the latter is much smaller than unity, the former, which represents the nonuniformity effect, will also be negligible, and the flow tends to behave as quasi-uniform. This reasoning then suggests that the results of a uniform curved-flow analysis should be applicable to natural meandering streams, for example, at the channel apex. Some evidence of this is presented by Falcón (1979).

Although the use of simple Boussinesq turbulence models appears to yield acceptable transverse velocity distributions, this does not seem to be the case for the transverse bed shear stress, which is an important quantity for predicting riverbed topography. It seems that better turbulence models will have to be developed before bank recession and the transverse bed shear stress can be calculated accurately. In the meantime, the moment-of-momentum equation, as used by Zimmermann & Kennedy (1978) and Falcón (1979) in order to obtain the transverse shear stress at the bed within a central flow region, appears to be a useful alternative for mildly curved flows and for many natural river situations where the pressure can be assumed to be distributed hydrostatically.

Literature Cited

Anwar, H., Rodger, J. 1981. Discussion of calculation of strongly curved open-channel flow by Leschziner, M., Rodi, W. *J. Hydraul. Div. ASCE* 107 : 142–43

Bathurst, J. C., Thorne, C. R., Hey, R. D. 1979. Secondary flow and shear stress at river bends. *J. Hydraul. Div. ASCE* 105 : 1277–95

Callander, R. A. 1968. *Instability and river meanders*. PhD thesis. Univ. Auckland, New Zealand

Callander, R. A. 1978. River meandering. *Ann. Rev. Fluid Mech.* 10 : 129–58

deVriend, H. J. 1981a. Velocity redistribution in curved rectangular channels. *J. Fluid Mech.* 107 : 423–39

deVriend, H. J. 1981b. *Steady flow in shallow channel bends*. PhD thesis. Tech. Hogeschool Delft, The Netherlands

Engelund, F. 1974. Flow and bed topography in channel bends. *J. Hydraul. Div. ASCE* 100 : 1631–48

Falcón, M. 1979. *Analysis of flow in alluvial channel bends*. PhD thesis. Univ. Iowa, Iowa City

Falcón, M. 1983. Discussion of modeling circulation in depth-averaged flow by Ponce, V. M., Yabusaki, S. B. *J. Hydraul. Div. ASCE* 109 : 150–51

Gottlieb, L. 1976. Three dimensional flow pattern and bed topography in meandering channels. *Inst. Hydrodyn. Hydraul. Eng., Tech. Univ. Denmark, Ser. Pap. 11*. 79 pp.

Götz, W. 1980. Discussion of secondary flow and shear stress at river bends by Bathurst, J., Thorne, C., Hey, R. *J. Hydraul. Div. ASCE* 106 : 1710–13

Hooke, R. LeB. 1974. Distribution of sediment transport and shear stress in a meander bend. *J. Geol.* 83 : 543–65

Hooke, R. LeB. 1980. Discussion of shear stress distribution in stable channel bends by Nouh, M. A., Townsend, R. D. *J. Hydraul. Div. ASCE* 106 : 1271–72

Kikkawa, H., Ikeda, S., Kitagawa, A. 1976. Flow and bed topography in curved open channels. *J. Hydraul. Div. ASCE* 102 : 1327–42

Launder, B. E., Spalding, D. B. 1974. The numerical calculation of turbulent flows. *Comput. Meth. Appl. Mech. Eng.* 3 : 269–89

Leschziner, M. A., Rodi, W. 1979. Calculation of strongly curved open channel flow. *J. Hydraul. Div. ASCE* 105 : 1297–1314

Leschziner, M. A., Rodi, W. 1981. Closure: Calculation of strongly curved open channel flow. *J. Hydraul. Div. ASCE* 107 : 1111–12

Naas, S. L. 1977. *Flow behavior in alluvial channels*. PhD thesis. Colo. State Univ., Ft. Collins

Naot, D., Rodi, W. 1982. Calculation of secondary currents in channel flow. *J. Hydraul. Div. ASCE* 108 : 948–68

Nunner, W. 1956. Wärmübergang und Druckabfall in rauhen Rohren. *VDI Forschungsh. 455*. 39 pp.

Rozovskii, I. L. 1957. *Dvizhenie Vody na Povorote Otkrytogo Rusla*. Kiev: Izd. Akad. Nauk Ukr. SSR. 188 pp. Transl., 1961, *Flow of Water in Bends of Open Channels*. Jerusalem: Israel Program Sci. Transl. 234 pp.

Yen, B. C. 1965. *Characteristics of subcritical flow in a meandering channel*. PhD thesis. Univ. Iowa, Iowa City

Zimmermann, C. 1974. *Sohlausbildung, Reibungsfaktoren, und Sediment-Transport in gleichförmig gekrümmten und geraden Gerinnen*. PhD thesis. Inst. Hydromech., Univ. Karlsruhe, West Germany

Zimmermann, C., Kennedy, J. F. 1978. Transverse bed slopes in curved alluvial streams. *J. Hydraul. Div. ASCE*: 104 : 33–48

Ann. Rev. Fluid Mech. 1984. 16: 195–222

VORTEX SHEDDING FROM OSCILLATING BLUFF BODIES

P. W. Bearman

Department of Aeronautics, Imperial College, London SW7 2BY, England

1. INTRODUCTION

When placed in a fluid stream, some bodies generate separated flow over a substantial proportion of their surface and hence can be classified as bluff. On sharp-edged bluff bodies, separation is fixed at the salient edges, whereas on bluff bodies with continuous surface curvature the location of separation depends both on the shape of the body and the state of the boundary layer. At low Reynolds numbers, when separation first occurs, the flow around a bluff body remains stable, but as the Reynolds number is increased a critical value is reached beyond which instabilities develop. These instabilities can lead to organized unsteady wake motion, disorganized motion, or a combination of both. Regular vortex shedding, the subject of this article, is a dominant feature of two-dimensional bluff-body wakes and is present irrespective of whether the separating boundary layers are laminar or turbulent. It has been the subject of research for more than a century, and many hundreds of papers have been written. In recent years vortex shedding has been the topic of Euromech meetings reported on by Mair & Maull (1971) and Bearman & Graham (1980), and a comprehensive review has been undertaken by Berger & Wille (1972).

Vortex shedding and general wake turbulence induce fluctuating pressures on the surface of the generating bluff body, and if the body is flexible this can cause oscillations. Oscillations excited by vortex shedding are usually in a direction normal to that of the free stream, and amplitudes as large as 1.5 to 2 body diameters may be recorded. In addition to the generating body, any other bodies in its wake may be forced into oscillation. Broad-band force fluctuations, induced by turbulence produced in the flow around a bluff body, rarely lead to oscillations as severe as those caused by vortex shedding. Some form of aerodynamic instability, such that move-

0066-4189/84/0115-0195$02.00

ments of a bluff body develop exciting forces in phase with the body's velocity, may also lead to large oscillation amplitudes. Galloping is one example of an instability whereby some bluff bodies can extract energy from a fluid stream and sustain oscillations. In this review, only vibrations caused by vortex shedding are considered; for a fuller discussion of flow-induced vibrations, the work of Blevins (1977) is recommended.

Excellent surveys of vortex-induced oscillations of bluff bodies have been written by Parkinson (1974) and Sarpkaya (1979). It is hoped that the present paper will complement these reviews and give an opportunity to reconsider some of the ideas presented in them in the light of more recent developments. The majority of previous experimental work in this area is related to vortex shedding from circular cylinders. Mathematical models developed to predict vortex-induced oscillations are also mainly concerned with this body shape. This preoccupation with the circular cylinder is justified on a number of arguments; the most persuasive are that it is an important structural form, and, as convincingly reasoned by Morkovin (1964), that it presents a challenging fundamental problem for fluid mechanicists. In the present review, however, vortex shedding from vibrating bluff bodies of various forms is considered in order to identify common features. We begin with a discussion of vortex shedding from fixed bluff bodies.

2. FIXED BLUFF BODIES

Although there is no complete solution to the problem of vortex shedding, we have a reasonably clear insight into the mechanism, and models are continuously being developed to describe it mathematically. Gerrard (1966) has given an extremely useful physical description of the mechanics of the vortex-formation region. A key factor in the formation of a vortex-street wake is the mutual interaction between the two separating shear layers. It is postulated by Gerrard (1966) that a vortex continues to grow, fed by circulation from its connected shear layer, until it is strong enough to draw the opposing shear layer across the near wake. The approach of oppositely signed vorticity, in sufficient concentration, cuts off further supply of circulation to the growing vortex, which is then shed and moves off downstream.

Gerrard's vortex-formation model is illustrated in Figure 1 by a sketch showing an instantaneous filament line pattern. Entrainment plays an important role in vortex formation, and Figure 1 indicates several entrainment processes. Entrained fluid (*a*) is engulfed into the growing vortex while (*b*) finds its way into the developing shear layer. The near-wake region between the base of the body and the growing vortex oscillates in

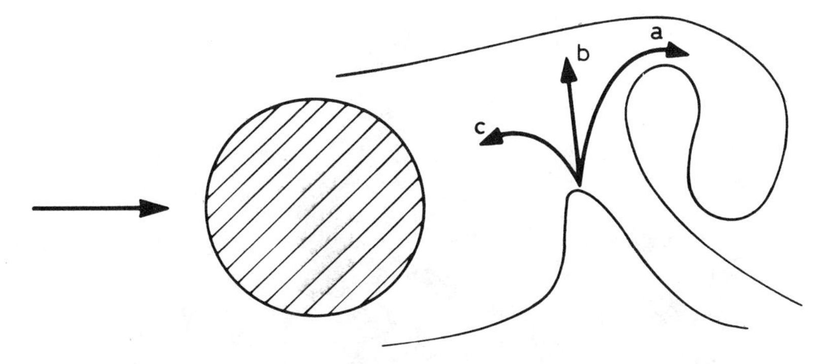

Figure 1 Vortex-formation model showing entrainment flows (Gerrard 1966).

size, and some further fluid, (*c*), is temporarily entrained into it. Entrained flow (*a*), which contains some fluid with oppositely signed vorticity to that in the growing vortex, is the largest of the three flows. The photograph in Figure 2 shows smoke filaments around a bluff body, and the interaction between one shear layer and the vortex forming on the opposite side of the wake is clearly seen.

Since vortex formation involves the mixing of flows of oppositely signed vorticity, the strengths of individual wake vortices will be less than the total circulation shed from one side of a bluff body during a shedding cycle. Fage & Johansen (1927) and Roshko (1954) estimated the rate of shedding of circulation from a bluff body by considering the mean base pressure. The rate of shedding of circulation at a separation point is given approximately by $d\Gamma/dt = \frac{1}{2}U_b^2$, where U_b is the mean velocity at the edge of the boundary layer at separation. Assuming that pressure remains constant through a shear layer, then $U_b^2 = U^2(1 - C_{p_b})$, where U is free-stream velocity and C_{p_b} is the mean base pressure coefficient. Davies (1976) has considered the additional contribution to $d\Gamma/dt$ from unsteady velocities generated at the separation point, but concluded that it can be neglected. Hence, the fraction of the original circulation that survives vortex formation is given by the expression

$$\alpha = 2S\Gamma_v/UD(1 - C_{p_b}), \tag{2.1}$$

where Γ_v is the strength of a wake vortex and S is the body Strouhal number ($S = nD/U$, where n is the vortex-shedding frequency and D is a characteristic dimension of the body, usually the body width).

In order to calculate α, an estimate must be made of the strength of the wake vortices Γ_v. Fage & Johansen (1927) and Roshko (1954) have found Γ_v by matching a measured vortex convection speed to a predicted value obtained from a suitable array of idealized point vortices. Fage & Johansen

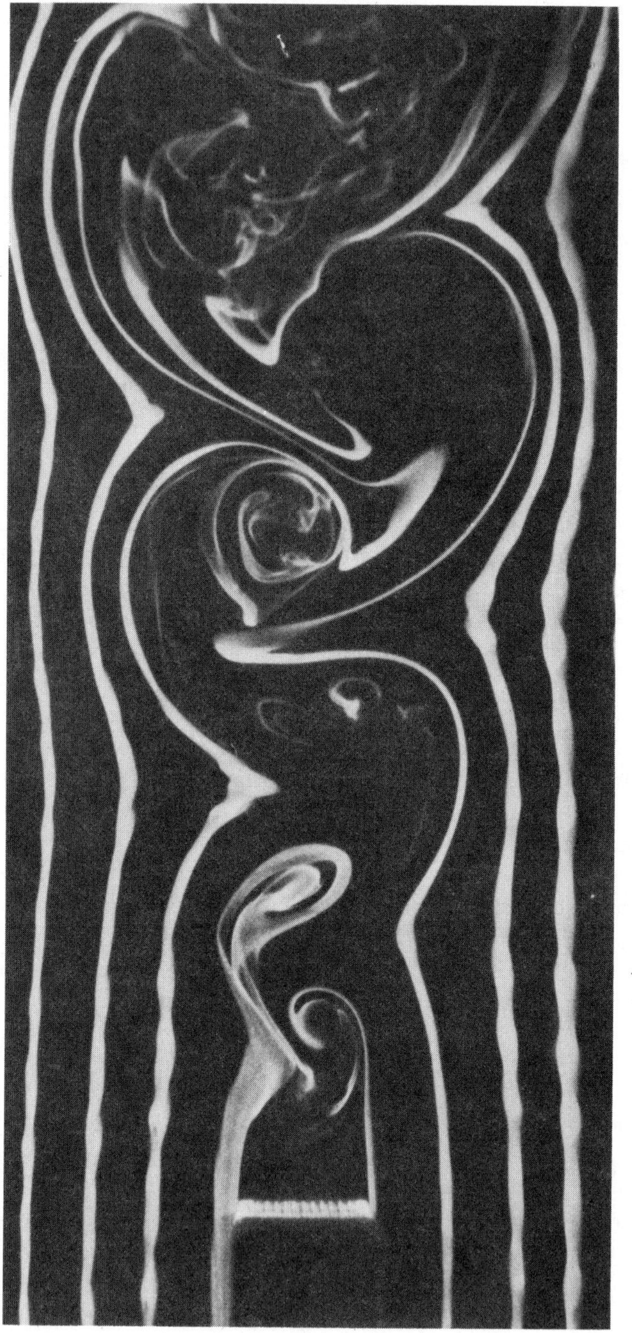

Figure 2 Visualization of the flow in a bluff-body wake.

found $\alpha = 0.6$ for a flat plate, and Roshko estimated $\alpha = 0.43$ as the average value for a circular cylinder, 90° wedge, and flat plate. More detailed analyses have been carried out by Bloor & Gerrard (1966) and Davies (1976), who used diffusing viscous vortices to represent the wake. Again by matching measured and calculated wake velocities, an estimate of Γ_v could be found. The resulting values of α are lower; for a circular cylinder Bloor & Gerrard found α to be between 0.2 and 0.3, depending on the Reynolds number, and for a *D*-shaped cylinder Davies estimated $\alpha = 0.26$.

The process of shear-layer interaction and circulation reduction is extremely well illustrated in the numerical calculations of Abernathy & Kronauer (1962). They represented two parallel shear layers of opposing signs by rows of point vortices. The rows were disturbed and calculation proceeded until clouds of vortices resembling a vortex street were formed. Point vortices of opposite sign mix within the clouds, and the value of α calculated from these numerical experiments is about 0.6. Although these calculations may not have been carried out with the utmost mathematical rigor, they represent a very important step forward in the understanding of the mechanism of vortex shedding. They show that it is the presence of two shear layers, rather than the bluff body itself, that is primarily responsible for vortex shedding. The presence of the body merely modifies the process by allowing feedback between the wake and the shedding of circulation at the separation points.

Abernathy & Kronauer's pioneering work led to the development of the discrete-vortex method for calculating vortex shedding from bluff bodies. Gerrard (1967) developed a similar method but included a circular cylinder in the flow field. Further improvements followed rapidly. A blunt-based body was studied by Clements (1973) and an inclined flat plate by Sarpkaya (1975). Circular-cylinder flow has been calculated by Deffenbaugh & Marshall (1976), Sarpkaya & Schoaff (1979a, b), Stansby (1981), and Stansby & Dixon (1982). Our intent here is not to review the achievements and shortcomings of the discrete-vortex technique, but merely to note that it has become an important method for modeling two-dimensional bluff-body flow.

Much of the experimental research into vortex shedding concerns the influence of disturbances. In addition to Reynolds number, bluff-body flows have been found to depend on factors such as body-surface finish, aspect ratio, end-plate design, flow turbulence, blockage ratio, and acoustic noise level. Another important finding has been that two-dimensional bodies in uniform flow are unlikely to shed two-dimensional vortices. The lack of two dimensionality can be assessed by investigating the spanwise variation of some unsteady quantity related to vortex shedding, such as fluctuating surface pressure, sectional lift force, or a fluctuating velocity just outside the shear layer at separation. The correlation between two points a distance z apart is given by

$$R(e, z) = \overline{e_1 e_2}/\overline{e^2}, \tag{2.2}$$

where e is the quantity under consideration and 1 and 2 are two stations a distance z apart. Since the mean flow is two dimensional, then $\sqrt{\overline{e_1^2}} = \sqrt{\overline{e_2^2}} = \sqrt{\overline{e^2}}$. The correlation length L is defined by the integral

$$L = \int_0^\infty R(e, z)\, dz. \tag{2.3}$$

In the range of Reynolds numbers between 10^4 and 10^5, L for a circular cylinder varies between about 3 and 6 cylinder diameters (see Sarpkaya 1979). Sharp-edged bodies generate vortices with correlation lengths of a similar order: for a square section, $L \approx 5.5D$ (Vickery 1966, Pocha 1971, Lee 1975, Bearman & Obasaju 1982). For a flat plate normal to a flow, Bearman & Trueman (1972) have found $L = 10D$. From these measurements it can be deduced that although the mutual interaction between two shear layers that leads to the generation of vortex shedding may be very strong, the spanwise coupling is comparatively weak.

As the aspect ratio of a constant cross-section bluff body is reduced, vortex shedding becomes weaker, although it never seems to completely disappear. Fail et al. (1959) detected vortex shedding behind flat plates of varying aspect ratio down to an aspect ratio of unity. Vortex shedding behind spheres has been observed by Achenbach (1974) and behind circular discs and cones by Calvert (1967). On low-aspect-ratio bluff bodies, the forces generated by vortex shedding are too weak to cause any significant oscillations. Wootton (1969) has observed, during wind-tunnel tests on model chimney stacks of circular cross section, that significant vortex-induced oscillations are unlikely to occur for chimney height to diameter ratios less than 5 or 6.

Comparatively small departures from strictly two-dimensional conditions can have a large influence on vortex shedding. The effect of mild taper on vortex shedding from a body of circular cross section has been studied by Gaster (1969), and the effect of mean shear on a bluff body with fixed separation has been investigated by Maull & Young (1973). In both cases the Strouhal number, based on local diameter or local velocity as appropriate, tends to the value for a cylindrical body in a uniform flow. Maull & Young found that in a shear flow the shedding occurred at a fixed frequency within cells extending a few body diameters. In both cases the vortex-shedding correlation length is greatly reduced, compared with its value in the equivalent two-dimensional problem, because of the spanwise variation in the shedding frequency.

Regular vortex shedding can be seriously disrupted if the flow separation line is not straight. Tanner (1972) has investigated the effect of various trailing-edge geometries on the drag of blunt-trailing-edge wings. He found that vortex shedding could be suppressed and base drag reduced if the trailing edge was stepped so as to break the straight separation line. Naumann et al. (1966) fixed separation on a circular cylinder by a series of staggered wires and observed no regular vortex shedding. Another example of the suppression of shedding by nonstraight separation lines is the case of a circular cylinder in the upper transition regime ($8.5 \times 10^5 < Re < 3.5 \times 10^6$) between supercritical and postcritical flow. [The nomenclature used here is identical to that suggested by Roshko (1961), except that postcritical is used in place of transcritical.] In supercritical flow the transition from laminar to turbulent flow occurs within a separation bubble, and the separation of the resulting turbulent boundary layer takes place well around the cylinder. Within the upper transition regime the bubbles break down at random points across the span, and a straight separation line is lost. At higher Reynolds numbers transition occurs within the attached boundary layer, straight separation lines are reestablished, and regular vortex shedding returns.

3. OSCILLATING BLUFF BODIES

When a bluff body is oscillating in a fluid stream, or when it is exposed to a stream with an imposed oscillation, vortex shedding can be dramatically altered. The oscillations provide a means for coupling the flow along the span of the body, and this usually results in a large increase in the correlation length. Closer inspection of the flow reveals that the body motion can take control of the instability mechanism that leads to vortex shedding. Hence, the flow generated by vortex shedding around a vibrating bluff body can have very significant differences from that around a fixed one. Controlling oscillations may be generated by the body or the flow, but usually oscillations develop from an interaction between vortex shedding and a flexibly mounted bluff body. In laboratory experiments, oscillations of bluff bodies may be vortex-induced or forced by a machine capable of generating a controlled vibratory motion. Similar interaction effects may be produced by vibrating the flow, and in a series of experiments on square-section cylinders Pocha (1971) used a gust tunnel to produce a transversely oscillating flow. Acoustic noise can develop correlated oscillatory particle velocities in the flow around a bluff body, which may lead to modifications of the vortex-shedding process that are similar to those produced by body motion. The interaction between vortex shedding and sound has been investigated by Parker (1966, 1967), Gaster (1971), Archibald (1975), and Welsh & Gibson (1979).

Practical examples of vortex-induced oscillations, such as those discussed by Naudascher & Rockwell (1980), invariably involve more complicated structures and flow environments than those treated in fundamental investigations. Wind-excited oscillations of towers and chimney stacks, for example (see Scruton 1963), are affected by the shear and turbulence in the approaching flow. Real structures have a number of possible modes of vibration, and this is an important factor to be considered when determining the vortex-induced response of long slender bluff bodies such as cables (Ramberg & Griffin 1976, Peltzer 1982). An important application area for bluff-body research is in the field of marine technology, where bodies are exposed to waves, currents, or a combination of both (Griffin 1980, Griffin & Ramberg 1982). Even in the absence of a mean velocity, when the flow is purely oscillatory, vortex shedding can still be the dominant mechanism (Bearman et al. 1981). Progress on these practical engineering problems has been made possible through the understanding gained from fundamental studies.

3.1 *Oscillations of a Flexibly Mounted Bluff Body*

Vortex shedding most commonly induces oscillations in a direction transverse to that of the stream at flow speeds where the frequency of shedding coincides with the frequency of oscillation of the body. Maximum oscillation amplitudes occur over a range of the reduced velocity U/ND, where N is the body oscillation frequency. If the vortex-shedding frequency from one side of a bluff body is n, then the reduced velocity for maximum amplitude is close to $1/S$, where S is given by nD/U. It is significant that it is not precisely equal to the inverse of the Strouhal number.

Each time a vortex is shed, a weak fluctuating drag is generated, and oscillations can be induced in-line with a fluid flow. This occurs at a reduced velocity equal to approximately $1/2S$. Another form of wake instability has been observed for circular cylinders in a range of reduced velocities less than $1/2S$. Vortices are shed in symmetric pairs and only form into the familiar staggered arrangement some distance downstream of the cylinder. This symmetric shedding is induced by the motion of the body and gives rise to a force in phase with the body velocity. In-line oscillations caused by this effect have not been observed for bodies in air; but in denser fluids such as water, where the ratio of the mass of fluid displaced to the mass of the body is substantially greater, serious oscillations can occur. In-line oscillations of this type on full-scale marine piles have been reported by Wootton et al. (1972). Comparisons between model experiments and full scale have been carried out by Hardwick & Wootton (1973) and King (1974).

A comparison carried out by King (1974) of the in-line oscillations of a model cantilevered circular cylinder at a Reynolds number of 6×10^4 and a

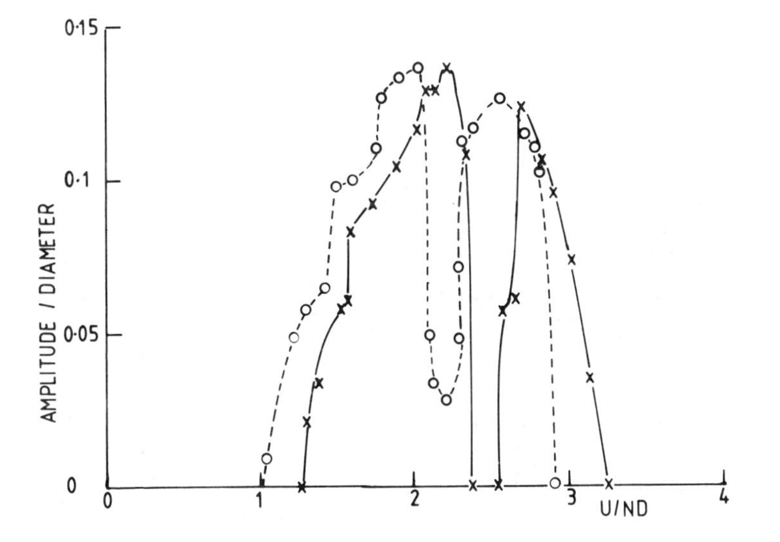

Figure 3 Vortex-excited in-line oscillations of a flexible cantilevered circular cylinder in water. ○, Re = 6×10^5; ×, Re = 6×10^4: in each case $2M\delta_s/\rho D^2 = 0.048$ (King 1974).

full-scale pile at a Reynolds number of 6×10^5 is shown in Figure 3. The two regions of instability caused by the two different shedding patterns are clearly visible. The maximum amplitudes of oscillation were less than $0.15D$ and hence much smaller than those experienced in vortex-excited cross-flow vibrations. Insufficient fundamental research has been carried out to fully explain the origin of the exciting force in the symmetric shedding regime. Hardwick & Wootton (1973) observed oscillations at Reynolds numbers of 6×10^3, but Griffin & Ramberg (1982) report that there is a lower limit of about 1.2×10^3 below which in-line oscillations are not excited in water. Interesting experiments have been carried out by Tanida et al. (1973), who forced a cylinder to oscillate in-line with a flow at reduced velocities in the range under consideration. They oscillated a circular cylinder at a fixed amplitude of $0.14D$ at a Reynolds number of 4000. No positive exciting force was found, although the fluid damping force generated by the motion of the cylinder was considerably less than that calculated assuming quasi-steady flow and using the measured drag coefficient. It is intriguing to ask whether or not an exciting force would have been measured if the oscillation amplitude had been reduced.

3.2 *Equation of Motion for a Flexibly Mounted Bluff Body*

Consider a rigid two-dimensional bluff body placed normal to a flow of velocity U and mounted on springs. The equation of motion for the body can be written as

$$M\ddot{y}+\beta\dot{y}+Ky = F(t), \tag{3.1}$$

where y is the displacement of the body, M the mass per unit span, $\beta\dot{y}$ a viscous-type damping force associated with the springs and their mounting, K the stiffness of the springs, and $F(t)$ the time-dependent fluid force. It is customary to replace β by $4\pi\delta_s MN_0$, where δ_s is the fraction of critical damping, and to let $(K/M)^{1/2}/2\pi = N_0$, the undamped natural frequency of the body. The fluctuating fluid force on the body can be expressed in terms of a coefficient C_y, where $C_y = F(t)/\frac{1}{2}\rho U^2 D$. This coefficient accounts for the component in the y direction of the total instantaneous fluid force acting on a cross section of the oscillating body and embodies elements that could be considered as fluid inertia and damping forces. In order to understand the loading caused by the complex mechanism of vortex shedding, it is more instructive to avoid any division at this stage and to keep all fluid forces on the right-hand side of Equation (3.1). Hence, Equation (3.1) can be rewritten

$$M\ddot{y}+4\pi N_0\,\delta_s M\dot{y}+4\pi^2 N_0^2 My = C_y\rho U^2 D/2. \tag{3.2}$$

Equation (3.2) is a general equation that can be used to describe the response in the transverse or in-line directions due to approaching turbulence, galloping, or vortex shedding. Much of the research into bluff-body flows has been aimed at finding a suitable form for the appropriate fluid-loading coefficient C_y. In this discussion y is assumed to be the displacement in the transverse direction and C_y the transverse force coefficient on a bluff body shedding vortices. For large-amplitude, steady-state, vortex-induced oscillations, the fluid force and the body displacement response oscillate at the same frequency n_v, which is usually close to N_0. When a bluff body is responding to vortex shedding, the fluid force must lead the excitation by some phase angle ϕ. Hence, the displacement y and the fluid force C_y can be represented by the expressions

$$y = \bar{y}\sin 2\pi n_v t,$$

$$C_y = \bar{C}_y \sin(2\pi n_v t+\phi). \tag{3.3}$$

Substitution of these forms in Equation (3.2) and equating sine and cosine terms lead to the following relationships:

$$\frac{N_0}{n_v} = \left[1-\frac{\bar{C}_y}{4\pi^2}\cos\phi\left(\frac{\rho D^2}{2M}\right)\left(\frac{U}{N_0 D}\right)^2\left(\frac{y}{D}\right)^{-1}\right]^{-1/2}, \tag{3.4}$$

$$\frac{\bar{y}}{D} = \frac{\bar{C}_y}{8\pi^2}\sin\phi\left(\frac{\rho D^2}{2M\delta_s}\right)\left(\frac{U}{N_0 D}\right)^2\frac{N_0}{n_v}. \tag{3.5}$$

By considering the order of magnitude of the various terms in Equation (3.4) (see Parkinson 1974), it can be shown that for large-amplitude oscillations of a body in air, where $\rho D^2/2M$ might be typically of order 10^{-3}, the frequency of body oscillations should be close to its natural frequency. This is confirmed by experiments. In a denser fluid such as water, where $\rho D^2/2M$ may be of order unity, the frequency of oscillation of the body can be appreciably different from its natural frequency. The steady-state response amplitude of a bluff body to vortex shedding is given by Equation (3.5). If $\rho D^2/2M$ is small such that $n_v \approx N_0$, then Equation (3.5) can be replaced by

$$\frac{\bar{y}}{D} = \frac{\bar{C}_y}{8\pi^2} \sin \phi \left(\frac{\rho D^2}{2M\delta_s}\right)\left(\frac{U}{N_0 D}\right)^2. \tag{3.6}$$

It is clear from Equations (3.4), (3.5), and (3.6) that the phase angle ϕ plays an extremely important role. The response amplitude does not depend on C_y alone but on that part of C_y in phase with the body velocity. Hence, measurements of the sectional fluctuating transverse force coefficients (hereafter referred to as the lift coefficients) on a range of stationary bluff-body shapes will give little indication of the likely amplitudes of motion of similar bodies flexibly mounted.

3.3 *Free and Forced Vibrations*

Free-vibration experiments on circular cylinders in air and water have been compared by Griffin & Ramberg (1982). Plots of oscillation amplitude versus reduced velocity are reproduced in Figure 4, where it can be seen that for similar values of the mass-damping parameter, $2M\delta_s/\rho D^2$, the maximum oscillation amplitudes are nearly the same. Two further points should be noted in Figure 4. First, the maximum responses in the two cases occur at similar values of reduced velocity. In the water experiment the reduced velocity has been formed using the oscillation frequency measured in still water. Second, away from the maximum response the amplitudes recorded are quite different in the two cases. The more lightly damped cylinder used in the air experiment responds over a narrow band of reduced velocity, whereas the more highly damped cylinder in water responds over a much broader range.

Although in the case reported by Griffin & Ramberg similar values of the mass-damping parameter lead to similar maximum responses, this does not necessarily always occur. Sarpkaya (1978) has demonstrated that for low values of the mass-damping parameter the nondimensional mass and damping affect the response separately. This can be expected from Equations (3.4) and (3.5), where low values of $2M/\rho D^2$ influence the response frequency and hence the oscillation amplitude. Griffin &

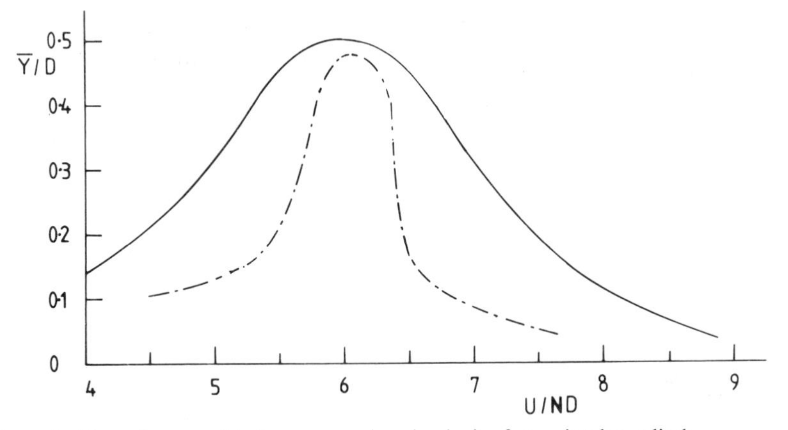

Figure 4 Cross-flow amplitude versus reduced velocity for a circular cylinder. ———, water, $2M/\rho D^2 = 7.6$, $\delta_s = 5.1 \times 10^{-2}$, $2M\delta_s/\rho D^2 = 0.39$; — · — · —, air, $2M/\rho D^2 = 68$, $\delta_s = 4.3 \times 10^{-3}$, $2M\delta_s/\rho D^2 = 0.29$ (Griffin & Ramberg 1982).

Koopmann (1977) and Sarpkaya (1978) have suggested the use of a response parameter $S_g = 8\pi^2 M\delta_s S^2/\rho D^2$. Sarpkaya (1978) has concluded from a parametric study of the separate and combined effects of the structural damping, the nondimensional mass, and the response parameter that the maximum response of a cylinder is governed by the response parameter alone only for values of S_g greater than about unity.

If the value of $\rho D^2/2M$ is sufficiently small so that the response can be predicted using Equation (3.6), and if it is assumed that the reduced velocity for maximum response is equal to the inverse of the Strouhal number, then $\bar{y}/D = \bar{C}_y \sin\phi/2S_g$. Using such a relationship, the amplitudes of oscillation of freely vibrating bluff bodies can be used to estimate the excitation $\bar{C}_y \sin\phi$, the force coefficient out of phase with the displacement. Griffin & Koopmann (1977) have gathered together estimates of $\bar{C}_y \sin\phi$ for a circular cylinder from various experiments, and these have been plotted in Figure 5 against a scaled amplitude ratio $2\bar{y}I/D$. The parameter I is the modal response factor and has been used in an attempt to correlate data from cantilevered and pivoted cylinders and cylinders moving in parallel motion. For cylinders in parallel motion, $I = 1$. The indications from Figure 5 are that the excitation increases in the range of amplitudes up to about $0.5D$ and then decreases such that there is no excitation beyond an amplitude of roughly $1.5D$.

One of the most interesting, as well as detailed, investigations into the response of freely vibrating cylinders has been carried out by Feng (1968). His best known results, those on a lightly damped circular cylinder, are shown in Figure 6. He discovered that higher amplitudes were achieved

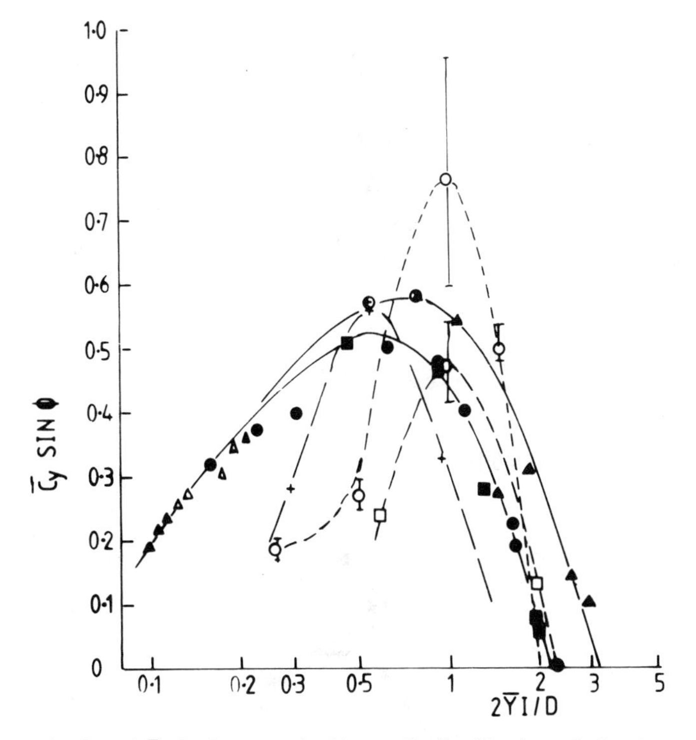

Figure 5 A plot of $\bar{C}_y$ sin ϕ versus double amplitude. Circular-cylinder data gathered together by Griffin & Koopmann (1977) and Griffin (1980). The legend for the data points is given in Griffin (1980).

when reduced velocity was increased over a certain range than when it was decreased back over the same range. This hysteresis effect has since been the subject of much discussion, but it has still to be fully explained. Feng also measured the phase angle ϕ, and it can be seen from Figure 6 that this varies with reduced velocity.

Experiments such as those of Feng (1968) demonstrate that the flow around a freely vibrating bluff body can change very rapidly with changes in reduced velocity. Since varying the reduced velocity also changes the amplitude ratio, it can be very difficult to determine the relative importance of these two effects on the resulting flow field. Hence a number of investigators have chosen to carry out forced oscillation experiments in which the amplitude and frequency of the motion can be varied at will. Each method has its advantages and disadvantages, but the forced-vibration technique has the important benefit that conditions can be closely controlled. However, due to the interaction with the structural parameters,

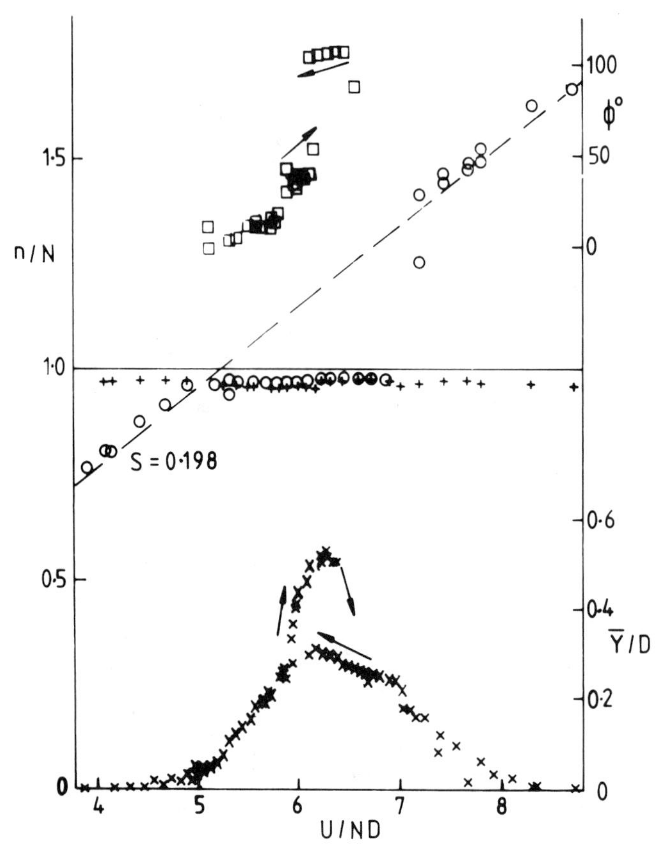

Figure 6 Oscillation characteristics for a freely vibrating circular cylinder with $2M\delta_s/\rho D^2 = 0.4$. ○, vortex-shedding frequency; +, cylinder frequency; □, phase angle; ×, oscillation amplitude (Feng 1968).

only a very limited range of the reduced velocities and amplitude ratios studied in a forced-vibration experiment is likely to be important in a freely vibrating one. For a freely suspended bluff body oscillating at a steady amplitude, it can be assumed that if the same body is forced to oscillate at a similar amplitude ratio, reduced velocity, and Reynolds number, then the flow patterns will be identical. This bold statment presumes that the precise previous history of the motion is unimportant. The available experimental evidence suggests that free and forced-vibration flows are the same.

3.4 *The Influence of Body Motion on Vortex Shedding*

Research into oscillating bluff-body flows has identified a number of significant changes that occur in the vortex-shedding behavior. Perhaps the two best-known effects are the capture of the vortex-shedding frequency by the body frequency over a range of reduced velocity, and the large increase in correlation length that occurs when the vortex-shedding frequency coincides with the body frequency. The movement of the body provides a means of synchronizing the moment of shedding along its length. The threshold oscillation amplitude required to cause large increases in the correlation length depends on the shape of the bluff body under consideration. For a circular cylinder the critical value of $\bar{y}/D$ is about 0.05, but for bodies with sharp-edged separation it may be significantly lower. Figure 7 shows an example of the increase of spanwise correlation of the flow along a square-section cylinder when it is forced into oscillation at an amplitude of $\bar{y}/D = 0.1$ and at a reduced velocity where the vortex and body frequencies coincide (Bearman & Obasaju 1982). The correlation coefficient is formed here by measuring the fluctuating pressures at two points on a side face of the cylinder a distance z apart. If the fluctuating pressure signals had been filtered so as to pass only the parts at the vortex-shedding frequency, then the correlation would have been even closer to unity across the span. At high Reynolds numbers, small-scale fluctuations caused by turbulence are unlikely to be much affected by body movement.

The "range of capture" over which the vortex-shedding frequency is locked to the body frequency is dependent on oscillation amplitude—the larger the amplitude, the larger the range of capture. In the results of Feng

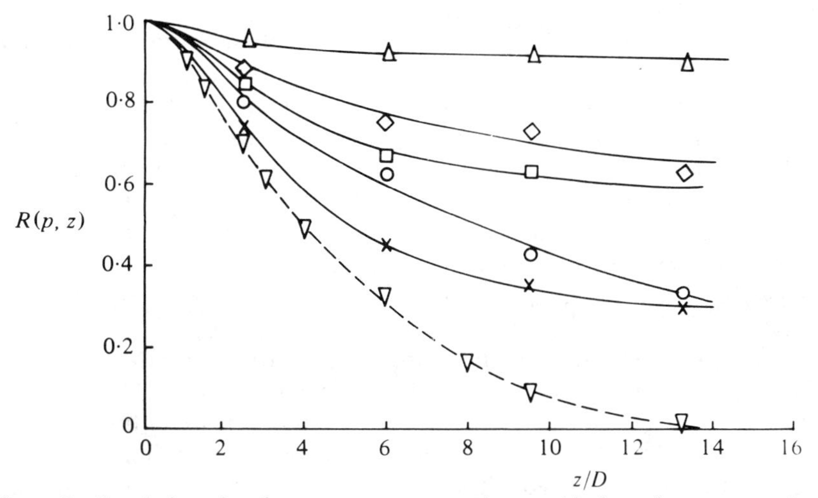

Figure 7 Correlation of surface pressures, measured on a side face of a square-section cylinder, versus spanwise separation. ∇, body stationary. Body oscillating with $\bar{y}/D = 0.1$: $\triangle$, U/ND within the lock-in range 7.3–8.5 ; $\times$, $U/ND = 6.2$; $\bigcirc$, 7.0 ; $\diamond$, 8.8 ; $\square$, 12.0 (Bearman & Obasaju 1982).

(1968), plotted in Figure 6, the range of capture is seen to extend from $U/ND = 5$ to 7.4, but in his experiment the circular cylinder is vibrating freely and hence the amplitude also varies over this range of reduced velocity. In forced-vibration experiments, where the amplitude is held constant, it is observed that the extent of the range of capture depends on the amplitude level. The range of capture always encompasses the reduced velocity, which is equal to the inverse of the Strouhal number measured for the body when stationary. This reduced velocity is referred to as the resonant point. The location of the range of capture, relative to the resonant point, depends very much on the shape of the bluff body. Feng (1968) has measured the vortex-shedding frequency behind a freely vibrating circular cylinder and a D-shaped cylinder. In the case of the circular cylinder the range of capture begins at the resonant point and extends to higher values of reduced velocity. The D-section showed a different behavior, with the high-velocity end of the range of capture just about coinciding with the resonant point.

The range of capture has also been measured in forced-vibration experiments for a variety of bluff-body shapes, but no clear pattern emerges and it must be concluded that the behavior of vortex shedding in this range depends very much on the body shape. For a circular cylinder the range of capture begins at the resonant point (Bearman & Currie 1979); for a square section (Bearman & Obasaju 1982), D-section, and flat plate the range of capture straddles the resonant point; and for a triangular section with one vertex pointing downstream the range of capture ends at the resonant point (Bearman & Davies 1977). The reasons for these variations are not understood.

The range of capture, or lock-in, is a small interval of reduced velocity over which flow conditions about a bluff body change rapidly with reduced velocity. The time period between the shedding of vortices remains constant, and hence the Strouhal number, which is equal to the inverse of the reduced velocity, changes. Rearranging Equation (2.1), the nondimensional strength of a vortex in the wake is given by

$$\Gamma_v/UD = \alpha(1-C_{p_b})U/2ND. \tag{3.7}$$

If it were supposed that the base-pressure coefficient and the circulation fraction α remain unchanged through lock-in, then Equation (3.7) indicates that the strength of the wake vortices should increase with increasing reduced velocity. This simple analysis also shows that amplification of the vortex strength, compared with its stationary-body value, will occur if the lock-in extends to reduced velocities above the resonant point. Support for the assumption that α is little affected by body motion comes from Davies (1976), who shows for a D-section cylinder that $\alpha = 0.26$ when it is stationary and $\alpha = 0.24$ at lock-in with $\bar{y}/D = 0.2$. It cannot be assumed with any certainty, however, that other bluff-body shapes will behave in a similar way.

In the vortex-formation region there is a feedback between the rolling up of the shear layers into vortices and the flow around the body, such that a small increase in vortex strength can lead to an increase in base suction and drag and hence an increase in shed vorticity at the separation points. Such a situation is clearly unstable, because the extra vorticity can lead to even stronger vortices; however, it is presumed that entrainment acts to control the near-wake flow. Hence, interfering with the vortex-shedding instability mechanism may produce large changes in vortex strength, base pressure, drag, and sectional fluctuating lift coefficient. The results of Davies (1976) for a D-section cylinder show within the lock-in range, for $\bar{y}/D = 0.2$, an increase in Γ_v/UD of 36% and a decrease in C_{p_b} of 71%, compared with the fixed-body values.

Measurements showing an increase in C_D on an oscillating circular cylinder through lock-in have been reported by Honji & Taneda (1968), Sedrak (1971), Tanida et al. (1973), and Sarpkaya (1978). Base-pressure measurements on a circular cylinder made by Stansby (1976) are shown in Figure 8, where $C_{p_b}/C_{p_{bo}}$ (the ratio of the minimum value of the base

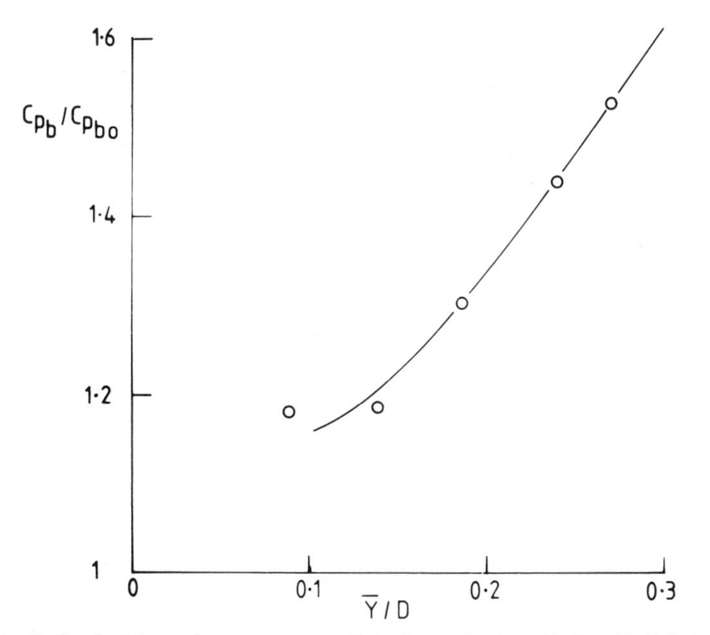

Figure 8 Ratio of minimum base-pressure coefficient on a circular cylinder at lock-in to value for a stationary cylinder versus oscillation amplitude (Stansby 1976).

pressure coefficient at lock-in to the stationary-cylinder value) is plotted against $\bar{y}/D$. Bearman & Davies (1977) have reported base-pressure measurements on a flat plate, D-shape, and triangular section with a vertex pointing downstream. The D-section and the flat plate show decreases in base pressure at lock-in. Results for the triangular section and a square section investigated by Bearman & Obasaju (1982) are plotted in Figure 9 and are seen to show a similar behavior with no amplification in base suction. At reduced velocities below lock-in, where the sections are being forced to oscillate at frequencies higher than their natural shedding frequencies, the base-suction coefficients, and hence the drag coefficients, are dramatically reduced. These results show that the shape of the afterbody, the region of a bluff body downstream of its separation points, plays an extremely important role in determining the response of the flow to body movements.

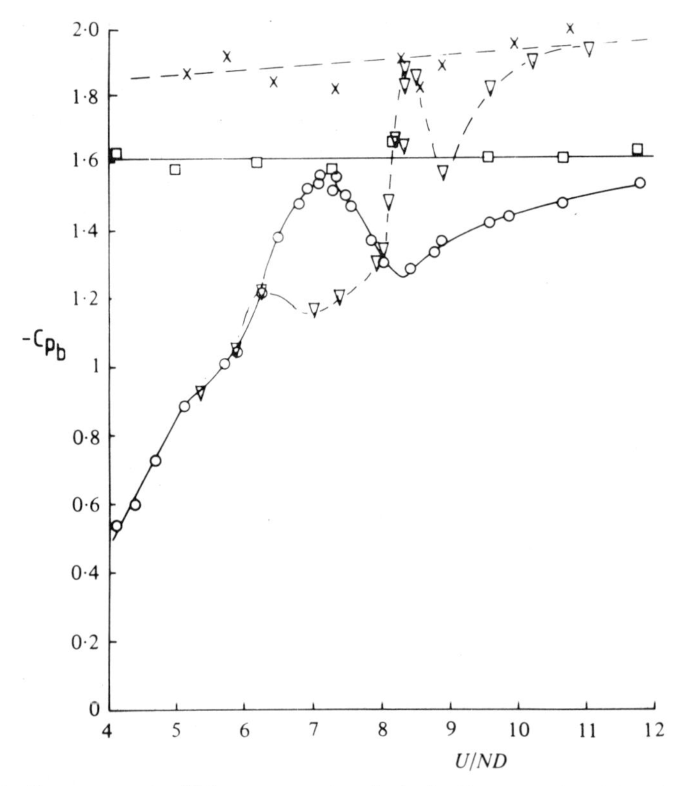

Figure 9 Base-pressure coefficient versus reduced velocity. Square section: □, stationary; ○, oscillating, $\bar{y}/D = 0.1$ (Bearman & Obasaju 1982). Triangular section: ×, stationary; ▽, oscillating, $\bar{y}/D = 0.2$ (Bearman & Davies 1977).

Within the lock-in range, all the available results show an increase in the local fluctuating lift coefficient. The improved two dimensionality of the flow at lock-in can be expected to increase the strength of shed vortices over their fixed-body values. Further increases in vortex strength, leading to higher values of $\bar{C}_y$, are caused by the direct influence of the body movement on the flow field at a section. A common feature is a reduction in the length of the vortex-formation region, with stronger vortices forming nearer the base.

Measurements of $\bar{C}_y$ on oscillating circular cylinders have been made by Bishop & Hassan (1964), Jones (1968), Sarpkaya (1978), and Staubli (1981). In these investigations, in which the body was forced to oscillate, the phase angle between lift fluctuations and body displacement was also measured. The only other bluff-body shape where several sets of $\bar{C}_y$ values are available is the square section, which has been investigated by Wilkinson (1974), Otsuki et al. (1974), Nakamura & Mizota (1975) and Bearman & Obasaju (1982).

A typical set of Sarpkaya's circular-cylinder data is shown in Figure 10 for a forced amplitude of vibration of $y/D = 0.5$. Figure 10*a* shows the in-phase component $\bar{C}_y \cos \phi$ and Figure 10*b* the out-of-phase component or excitation $\bar{C}_y \sin \phi$. These results indicate that the phase angle changes rapidly through lock-in, and that there is a substantial vortex-induced force to excite oscillations only at reduced velocities around 5. Sarpkaya also points out that the in-phase or inertia component of $\bar{C}_y$ at the point of maximum excitation, when expressed as an inertia coefficient, has a value very similar to that for a cylinder performing small-amplitude oscillations in still fluid. This result shows that if the natural frequency of oscillation of a cylinder placed in water is measured in still water, then at vortex resonance a similar frequency of oscillation will be recorded. The maximum values of the out-of-phase components measured by Sarpkaya (1978) have been included in the data plotted in Figure 5. In Sarpkaya's experiments the maximum excitation occurs at $U/ND = 5$, whereas other investigators have reported higher values of reduced velocity.

Variations in the in-phase and out-of-phase components through lock-in are caused primarily by large phase-angle changes. The phase-angle changes measured on a circular cylinder for $\bar{y}/D = 0.11$ by Feng (1968) and Bearman & Currie (1979), for free and forced vibrations, respectively, are shown in Figure 11. The phase in these experiments was found by measuring the phase difference between the suction pressure 90° from the front of the cylinder and the displacement. Feng (1968) measured the phase angle for a number of cases with different structural dampings and, hence, with different variations of $\bar{y}/D$ with U/ND. Using his data, it has been possible to find a number of measurements at $\bar{y}/D = 0.11$ for different

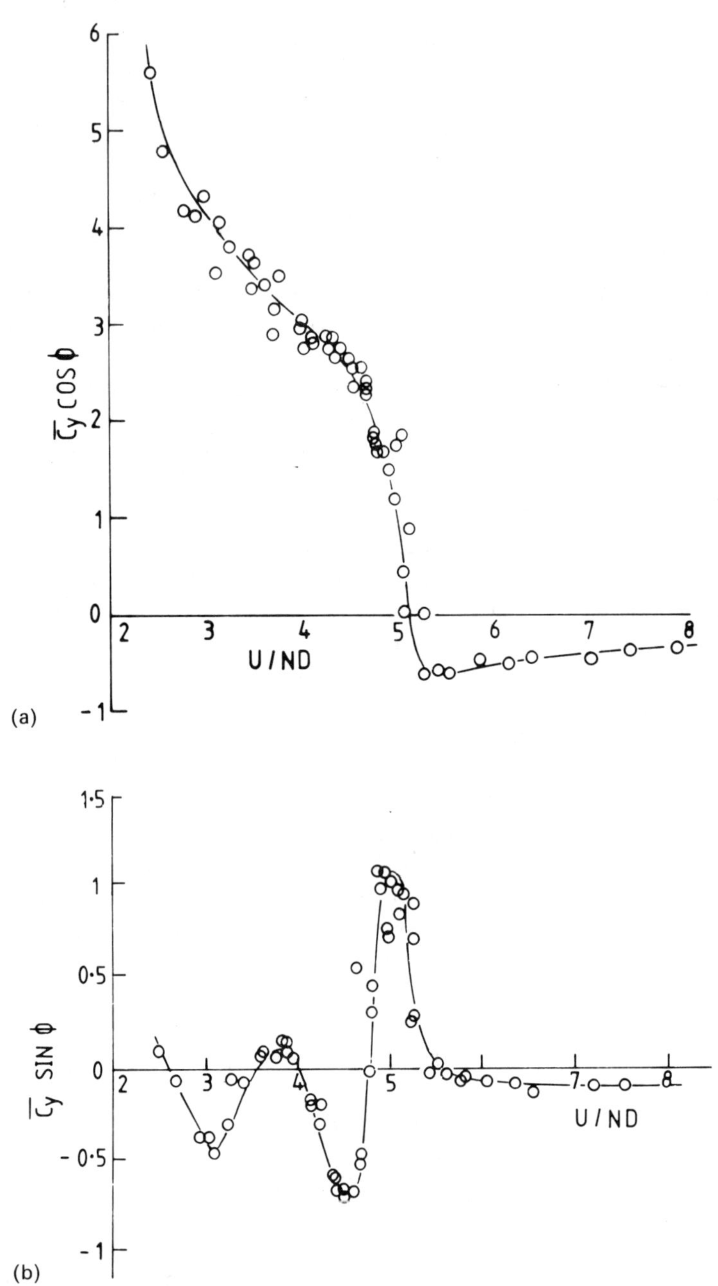

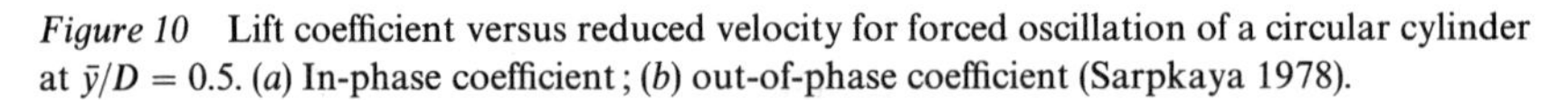

Figure 10 Lift coefficient versus reduced velocity for forced oscillation of a circular cylinder at $\bar{y}/D = 0.5$. (*a*) In-phase coefficient; (*b*) out-of-phase coefficient (Sarpkaya 1978).

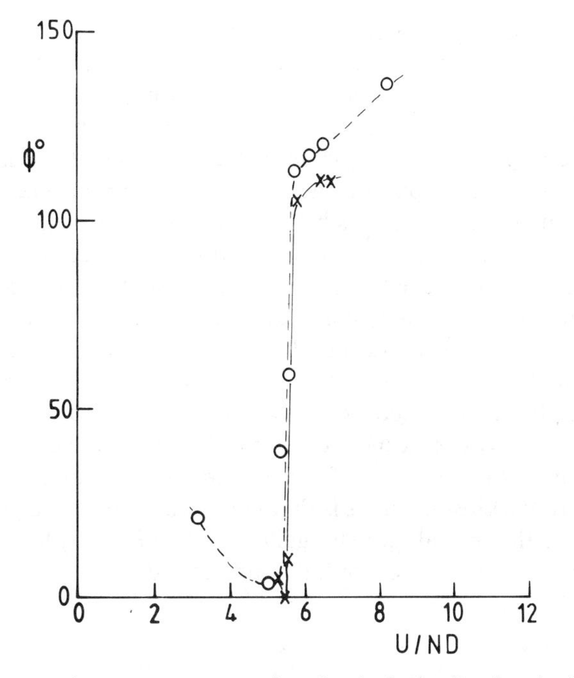

Figure 11 Phase-angle measurements on a forced and a freely vibrating circular cylinder at $\bar{y}/D = 0.11$. ○, forced vibration (Bearman & Currie 1979); ×, free vibration (Feng 1968).

reduced velocities. These changes in phase angle indicate that the point in an oscillation cycle at which a forming vortex generates its maximum lift force alters drastically with varying reduced velocity. Zdravkovich (1981), by examining the flow-visualization studies of Den Hartog (1934), Meier-Windhorst (1939), Angrilli et al. (1974), and Griffin & Ramberg (1974), has noted how the timing of vortex shedding changes through the lock-in range. His conclusions, which are mostly drawn from experiments on freely vibrating circular cylinders, suggest that there is a sudden change in phase. He notes that in the lower reduced-velocity region of the lock-in range, the vortex formed on one side of a cylinder was shed when the cylinder was near the maximum amplitude on the opposite side. With increase in reduced velocity the timing changes suddenly, such that the vortex with the same circulation as before is shed when the cylinder reaches the maximum amplitude on the same side. Controlled forced-vibration experiments carried out by Bearman & Currie (1979) at fixed amplitudes show that although the phase change occurs over a small range of reduced velocity, it is progressive and not a discontinuity.

Measurements of $C_{y\text{rms}}$ (root-mean-square values of C_y) and phase angle

ϕ for a square-section cylinder undergoing forced vibration are shown in Figures 12 and 13, respectively. Nakamura & Mizota (1975) measured lift directly on a length of cylinder about 4.5D long, whereas Bearman & Obasaju (1982) estimated C_{yrms} from fluctuating pressure measurements at one cross section. An interesting feature of the phase-angle measurements is that with increasing amplitude the reduced velocity at which the phase becomes positive increases. This behavior severely limits the amplitude of vortex-induced oscillations of square-section cylinders, since increasing amplitude at a fixed reduced velocity does not lead to increased excitation, as it can in the case of a circular cylinder. This result emphasizes the importance of a clearer physical understanding of the flow around vortex-excited bodies because, based on the values of C_{yrms} measured on fixed bluff bodies, it might be incorrectly supposed that the square section would be more susceptible to vortex-induced oscillations than the circular cylinder. A further interesting feature of the square section, as investigated by Wawzonek & Parkinson (1979), is that vortex resonance and galloping can combine when the critical speed for galloping is sufficiently low. The physics of this interaction process is not fully understood.

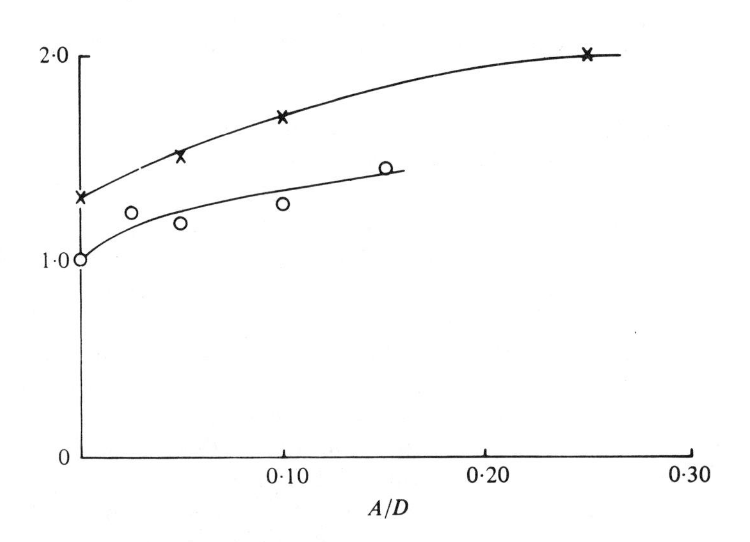

Figure 12 Square section—maximum value of the rms coefficient of fluctuating lift, C_{yrms}, measured in the lock-in range, versus oscillation amplitude. ×, estimated from pressure measurements (Bearman & Obasaju 1982); ○, force measurements (Nakamura & Mizota 1975).

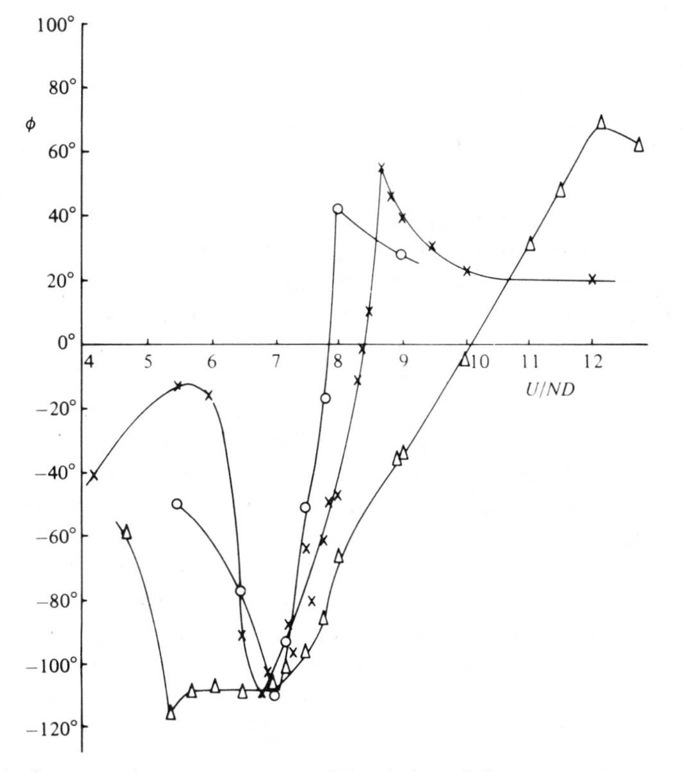

Figure 13 Square section—measurements of the phase angle between suction, measured at the center of a side face, and cylinder displacement versus reduced velocity. ○, $\bar{y}/D = 0.05$; ×, $\bar{y}/D = 0.1$; △, $\bar{y}/D = 0.25$ (Bearman & Obasaju 1982).

4. PREDICTIONS OF VORTEX-INDUCED OSCILLATIONS

Although this review is aimed primarily at providing a reappraisal of experimental results, it would not be possible to conclude without some brief discussion of prediction methods. Once again it is the circular cylinder that has received the most attention. Sarpkaya (1978) and Staubli (1981) have used their measurements of in-phase and out-of-phase fluctuating lift forces to calculate the response of freely vibrating cylinders using Equations 3.2–3.5. The results are encouraging and demonstrate that data obtained from forced-vibration experiments are relevant to flexibly mounted bluff bodies. A complicating factor in these investigations, however, is that forced-vibration data from one experimental setup have been used to predict oscillations measured in a totally different experiment where some of the many parameters that can affect circular-cylinder flow are different. Sarpkaya (1978) has predicted the maximum amplitudes measured by Griffin & Koopmann (1977) at Reynolds numbers between 300 and 1000 using forced-vibration data measured over the range from 5000 to 25,000. The good agreement achieved suggests that the Reynolds number may not be an important parameter for oscillating circular cylinders. A more extensive computation program has been carried out by Staubli (1981), and an example of his predictions of the amplitudes measured by Feng (1978) for a low mass-damping parameter are shown in Figure 14. There is some indication of a different behavior for increasing and decreasing reduced velocity, which is a feature of Feng's measurements. Sarpkaya (1979) has described the conflicting views surrounding the data of Feng, and discusses the cases for and against the hysteresis effect being due to certain structural nonlinearities. Staubli's results suggest that it is caused primarily by nonlinearities in the force coefficients, rather than a nonlinearity in the mechanical setup.

Hartlen & Currie (1970), inspired by Bishop & Hassan's (1964) observation that an oscillating cylinder/wake combination possesses the characteristics of a nonlinear oscillator, have explored the use of a van der Pol oscillator-type equation to represent the phenomenon. Their model, which incorporates a nonlinear damping term in an equation for C_y, showed considerable promise but, as pointed out by Parkinson (1974), is unable to

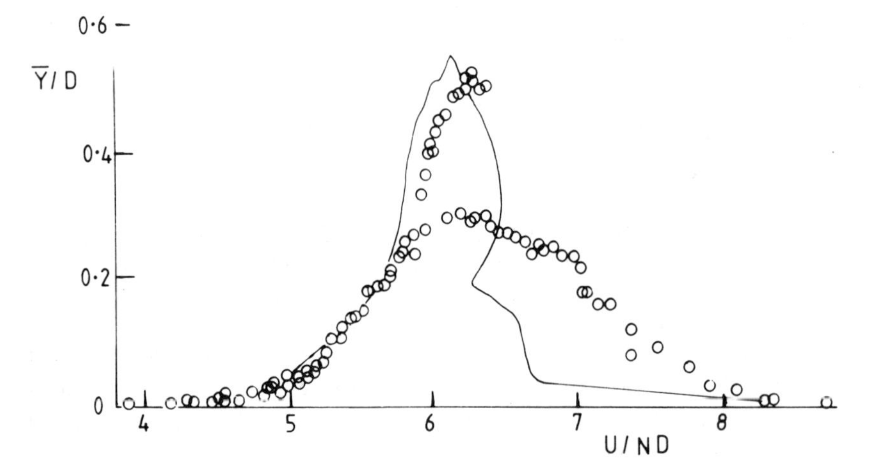

Figure 14 Response of a freely vibrating cylinder with $2M\delta_s/\rho D^2 = 0.4$. ○, experiment (Feng 1968); ———, numerical calculation (Staubli 1981).

predict the large-amplitude behavior observed in Feng's experiments. Various modifications to the basic Hartlen & Currie formulation have been proposed by Skop & Griffin (1973), Landl (1975), and E. Berger [reported by Bearman & Graham (1980)], but the solutions are unable to reproduce all of the observed effects. Despite an attempt by Iwan & Blevins (1974) to show that a van der Pol oscillator-type equation can be derived from basic fluid-mechanics principles, it appears that the method does not adequately describe the interaction between vortex shedding and body motion.

Prediction methods that accurately model the flow field around a bluff body are the ones most likely to be successful. Presumably a time will come when it will be practical to solve the Navier-Stokes equations for the flow around a bluff body at high Reynolds number using a computer. It is difficult to prophesy when this will happen, but in the meantime some researchers are experimenting with the use of the discrete-vortex method to model bluff-body flows, as mentioned in an earlier section. Sarpkaya & Schoaff (1979a) developed the model for a stationary circular-cylinder flow and then applied the same technique to a freely vibrating cylinder [see Sarpkaya & Schoaff (1979b) and Sarpkaya (1979)]. The application of this technique to oscillating bluff bodies is at an early stage, but the results are sufficiently encouraging to suggest that it could prove to be a useful method for various bluff-body shapes.

5. CONCLUDING REMARKS

During the past ten years or so there have been significant advances in the understanding of vortex shedding from oscillating bluff bodies. Nevertheless, the problem is by no means solved and a number of aspects require further study. A major uncertainty is in the mechanism that determines the phase of the vortex-induced force relative to the body motion. A related unsettled question is the role of the afterbody shape in vortex-induced oscillations of bluff bodies. Shapes other than the circle need to be studied in detail; bodies with fixed separation and a significant afterbody, such as a D-shape or a triangular section, could provide useful additional test data to prove prediction methods. The connection between free and forced oscillations needs to be pursued further, and an attempt should be made to predict free-oscillation response using data from forced oscillations carried out under dynamically similar conditions. The mathematical modeling of vortex-induced oscillations, using nonlinear oscillator theory and flow-field computation techniques, is continuing to be developed. The problem is a difficult one, but the gap between theory and experiment is being closed.

Literature Cited

Abernathy, F. H., Kronauer, R. E. 1962. The formation of vortex streets. *J. Fluid Mech.* 13:1–20

Achenbach, E. 1974. Vortex shedding from spheres. *J. Fluid Mech.* 62:209–21

Angrilli, F., Di Silvio, G., Zanardo, A. 1974. Hydroelasticity study of a circular cylinder in a water stream. In *Flow Induced Structural Vibrations*, ed. E. Naudascher, pp. 504–12. Berlin: Springer. 774 pp.

Archibald, F. S. 1975. Self-excitation of an acoustic resonance by vortex shedding. *J. Sound Vib.* 38:81–103

Bearman, P. W., Currie, I. G. 1979. Pressure-fluctuation measurements on an oscillating circular cylinder. *J. Fluid Mech.* 91:661–77

Bearman, P. W., Davies, M. E. 1977. The flow about oscillating bluff structures. *Proc. Int. Conf. Wind Eff. Build. Structures, 4th*, ed. K. J. Eaton, pp. 285–95. Cambridge Univ. Press. 845 pp.

Bearman, P. W., Graham, J. M. R. 1980. Vortex shedding from bluff bodies in oscillatory flow: A report on Euromech 119. *J. Fluid Mech.* 99:225–45

Bearman, P. W., Obasaju, E. D. 1982. An experimental study of pressure fluctuations on fixed and oscillating square-section cylinders. *J. Fluid Mech.* 119:297–321

Bearman, P. W., Trueman, D. M. 1972. An investigation of the flow around rectangular cylinders. *Aeronaut. Q.* 23:229–37

Bearman, P. W., Graham, J. M. R., Naylor, P., Obasaju, E. D. 1981. The role of vortices in oscillatory flow about bluff cylinders. *Int. Symp. Hydrodyn. Ocean Eng., Trondheim, Norway*, 1:621–43

Berger, E., Wille, R. 1972. Periodic flow phenomena. *Ann. Rev. Fluid Mech.* 4:313–40

Bishop, R. E. D., Hassan, A. Y. 1964. The lift and drag forces on a circular cylinder oscillating in a flowing fluid. *Proc. R. Soc. London Ser. A* 277:51–75

Blevins, R. D. 1977. *Flow-Induced Vibrations*. New York: Van Nostrand Reinhold. 363 pp.

Bloor, M. S., Gerrard, J. H. 1966. Measurements of turbulent vortices in a cylinder wake. *Proc. R. Soc. London Ser. A* 294:319–42

Calvert, J. R. 1967. Experiments on the low-speed flow past cones. *J. Fluid Mech.* 27:273–89

Clements, R. R. 1973. An inviscid model of two-dimensional vortex shedding. *J. Fluid Mech.* 57:321–36

Davies, M. E. 1976. A comparison of the wake structure of a stationary and oscillating bluff body, using a conditional averaging technique. *J. Fluid Mech.* 75:209–31

Deffenbaugh, F. D., Marshall, F. J. 1976. Time development of the flow about an impulsively started cylinder. *AIAA J.* 14:908–13

Den Hartog, J. P. 1934. The vibration problems in engineering. *Proc. Int. Congr. Appl. Mech., 4th, Cambridge*, pp. 36–53

Fage, A., Johansen, F. C. 1927. The flow of air behind an inclined flat plate of infinite span. *British ARC R & M 1102*

Fail, R., Lawford, J. A., Eyre, R. C. W. 1959. Low speed experiments on the wake characteristics of flat plates normal to an air stream. *British ARC R & M 3120*

Feng, C. C. 1968. *The measurement of vortex-induced effects in flow past stationary and oscillating circular and D-section cylinders*. MASc thesis. Univ. Br. Columbia, Vancouver

Gaster, M. 1969. Vortex shedding from slender cones at low Reynolds numbers. *J. Fluid Mech.* 38:565–76

Gaster, M. 1971. Some observations on vortex shedding and acoustic resonances. *British ARC CP 1141*

Gerrard, J. H. 1966. The mechanics of the formation region of vortices behind bluff bodies. *J. Fluid Mech.* 25:401–13

Gerrard, J. H. 1967. Numerical computation of the magnitude and frequency of the lift on a circular cylinder. *Philos. Trans. R. Soc. London Ser. A* 761:137–62

Griffin, O. M. 1980. OTEC cold water pipe design for problems caused by vortex-excited oscillations. *Naval Res. Lab. Memo. Rep. 4157*

Griffin, O. M., Koopmann, G. H. 1977. The vortex-excited lift and reaction forces on resonantly vibrating cylinders. *J. Sound Vib.* 54:435–48

Griffin, O. M., Ramberg, S. E. 1974. The vortex-street wakes of vibrating cylinders. *J. Fluid Mech.* 66:553–76

Griffin, O. M., Ramberg, S. E. 1982. Some recent studies of vortex shedding with application to marine tubulars and risers. *J. Energy Resour. Technol.* 104:2–13

Hardwick, J. D., Wootton, L. R. 1973. The use of model and full-scale investigations on marine structures. *Int. Symp. Vib. Probl. Ind., Keswick, U.K.*, Pap. 127

Hartlen, R. T., Currie, I. G. 1970. Lift-oscillator model of vortex-induced vibrations. *J. Eng. Mech. Div. ASCE* 96:577–91

Honji, H., Taneda, S. 1968. Vortex wakes of oscillating circular cylinders. *Res. Inst. Appl. Mech.* 16, Rep. 54

Iwan, W. D., Blevins, R. D. 1974. A model for

vortex induced oscillation of structures. *J. Appl. Mech.* 41:581–86

Jones, C. W. 1968. Unsteady lift forces generated by vortex shedding about a large, stationary and oscillating cylinder at high Reynolds number. *ASME Pap. 68-FE-36*

King, R. 1974. Vortex-excited oscillations of a circular cylinder in steady currents. *Offshore Tech. Conf. Pap. OTC 1948*

Landl, R. 1975. A mathematical model for vortex-excited vibrations of bluff bodies. *J. Sound Vib.* 42:219–34

Lee, B. E. 1975. The effect of turbulence on the surface pressure field of a square prism. *J. Fluid Mech.* 69:263–82

Mair, W. A., Maull, D. J. 1971. Bluff bodies and vortex shedding—a report on Euromech 17. *J. Fluid Mech.* 45:209–24

Maull, D. J., Young, R. A. 1973. Vortex shedding from bluff bodies in a shear flow. *J. Fluid Mech.* 60:401–9

Meier-Windhorst, A. 1939. Flatterschwingungen von Zylindern im gleichmässigen Flüssigkeitsstrom. *München Tech. Hochsch., Hydraul. Inst. Mitt.* 9:1–22

Morkovin, M. V. 1964. Flow around circular cylinders. A kaleidoscope of challenging fluid phenomena. *ASME Symp. Fully Sep. Flows*, pp. 102–18

Nakamura, Y., Mizota, T. 1975. Unsteady lifts and wakes of oscillating rectangular prisms. *J. Eng. Mech. Div. ASCE* 101:855–71

Naudascher, E., Rockwell, D. 1980. *Practical Experiences with Flow-Induced Vibrations.* Berlin: Springer. 849 pp.

Naumann, A., Morsbach, M., Kramer, C. 1966. The conditions of separation and vortex formation past cylinders. *AGARD Conf. Proc. Sep. Flows* 4(2):539–74

Otsuki, Y., Washizu, K., Tomizawa, H., Ohya, A. 1974. A note on the aeroelastic instability of a prismatic bar with square section. *J. Sound Vib.* 34:233–48

Parker, R. 1966. Resonance effects in wake shedding from parallel plates: some experimental observations. *J. Sound Vib.* 4:62–72

Parker, R. 1967. Resonance effects in wake shedding from parallel plates: calculations of resonant frequencies. *J. Sound Vib.* 5:330–43

Parkinson, G. V. 1974. Mathematical models of flow-induced vibrations. In *Flow Induced Structural Vibrations*, ed. E. Naudascher, pp. 81–127. Berlin: Springer. 774 pp.

Peltzer, R. D. 1982. Vortex shedding from a vibrating cable with attached spherical bodies in a linear shear flow. *Naval Res. Lab. Memo. Rep. 4940*

Pocha, J. J. 1971. *On unsteady flow past cylinders of square cross section.* PhD thesis. Queen Mary Coll., Univ. London

Ramberg, S. E., Griffin, O. M. 1976. Velocity correlation and vortex spacing in the wake of a vibrating cable. *J. Fluids Eng.* 98:10–18

Roshko, A. 1954. On the drag and shedding frequency of two-dimensional bluff bodies. *NACA Tech. Note No. 3169*

Roshko, A. 1961. Experiments on the flow past a circular cylinder at very high Reynolds number. *J. Fluid Mech.* 10:345–56

Sarpkaya, T. 1975. An inviscid model of two-dimensional vortex shedding for transient and asymptotically steady separated flow over an inclined plate. *J. Fluid Mech.* 68:109–28

Sarpkaya, T. 1978. Fluid forces on oscillating cylinders. *J. Waterw., Port, Coastal Ocean Div. ASCE* 104:275–90

Sarpkaya, T. 1979. Vortex-induced oscillations. *J. Appl. Mech.* 46:241–58

Sarpkaya, T., Schoaff, R. L. 1979a. Inviscid model of two-dimensional vortex shedding by a circular cylinder. *AIAA J.* 17:1193–1200

Sarpkaya, T., Schoaff, R. L. 1979b. A discrete vortex analysis of flow about stationary and transversely oscillating circular cylinders. *Tech. Rep. NPS-69SL79011*, Naval Postgrad. Sch., Monterey, Calif.

Scruton, C. 1963. On the wind-excited oscillations of stacks, towers and masts. In *Wind Effects on Buildings and Structures*, pp. 798–832. London: Her Majesty's Stationery Off. 852 pp.

Sedrak, M. 1971. Widerstandsmessungen am schwingenden Zylinder bei kleinen Reynolds-Zahlen. *DLR-FB-71-42*. Berlin: DFVLR Inst. Turbul.

Skop, R. A., Griffin, O. M. 1973. A model for the vortex-excited resonant response of bluff cylinders. *J. Sound Vib.* 27:225–33

Stansby, P. K. 1976. Base pressure of oscillating circular cylinders. *J. Eng. Mech. Div. ASCE* 102:591–600

Stansby, P. K. 1981. A numerical study of vortex shedding from one and two circular cylinders. *Aeronaut. Q.* 32:48–71

Stansby, P. K., Dixon, A. G. 1982. The importance of secondary shedding in two-dimensional wake formation at very high Reynolds number. *Aeronaut. Q.* 33:105–23

Staubli, T. 1981. Calculation of the vibration of an elastically mounted cylinder using experimental data from forced oscillation. *ASME Symp. Fluid/Struct. Interactions Turbomachinery*, pp. 19–24

Tanida, Y., Okajima, A., Watanabe, Y. 1973. Stability of a circular cylinder oscillating in uniform flow or in a wake. *J. Fluid Mech.* 61:769–84

Tanner, T. 1972. A method for reducing the base drag of wings with blunt trailing edges. *Aeronaut. Q.* 23:15–23

Vickery, B. J. 1966. Fluctuating lift and drag on a long cylinder of square cross-section in a smooth and in a turbulent stream. *J. Fluid Mech.* 25:481–94

Wawzonek, M. A., Parkinson, G. V. 1979. Combined effects of galloping instability and vortex resonance. In *Proc. Int. Conf. Wind. Eff. Build. Structures, 5th, Fort Collins, Colo.*, Pap. VI-2:1–12

Welsh, M. C., Gibson, D. C. 1979. Interaction of induced sound with flow past a square leading edged plate in a duct. *J. Sound Vib.* 67:501–11

Wilkinson, R. H. 1974. *On the vortex-induced loading on long bluff cylinders.* PhD thesis. Univ. Bristol, U.K.

Wootton, L. R. 1969. The oscillations of large circular stacks in wind. *Proc. Inst. Civ. Eng.* 43:573–98

Wootton, L. R., Warner, M. H., Sainsbury, R. N., Cooper, D. H. 1972. Oscillations of piles in marine structures. A resume of the full-scale experiments at Immingham. *CIRIA Tech. Rep. 41*

Zdravkovich, M. M. 1981. Modification of vortex shedding in the synchronization range. *ASME Pap. 81-WA/FE-25.* 8 pp.

Ann. Rev. Fluid Mech. 1984. 16: 223–44

MODERN OPTICAL TECHNIQUES IN FLUID MECHANICS

Werner Lauterborn and Alfred Vogel

Third Physical Institute, University of Göttingen, Göttingen, Federal Republic of Germany

INTRODUCTION

Optical techniques are widely used in fluid mechanics to observe and measure properties of flow fields such as velocities or densities. Many of these techniques are qualitative but of great value in guiding intuition for further research by quantitative means. Beautiful examples can be seen in the *Album of Fluid Motion* (Van Dyke 1982). Optical techniques are usually known for their largely nonintrusive properties as compared with methods like the Pitot tube or the hot-wire technique. The last few years, however, have seen some examples where light has been used not only to probe fluid flows but to generate them (Lauterborn 1980). This gives rise to a new classification of optical techniques in fluid mechanics (see Figure 1). Flow-visualization techniques use light as an information carrier where the information is impressed on the light beam by the fluid flow. Flow-generation techniques use light as an energy carrier to initiate fluid flow by radiation pressure, heating, or optical breakdown.

Flow-visualization techniques may be coarsely subdivided into two categories: those that make use of light scattered by tiny particles in the fluid and those that make use of variations in refractive index. Among the methods that rely on scattered light, laser Doppler anemometry is now a standard means of obtaining fluid velocities. This method and its various refinements are well documented (Durst et al. 1976, Durrani & Greated 1977, Drain 1980, Schulz-DuBois 1983) and are not discussed here. In laser Doppler anemometry, the fluid velocity can be measured with high accuracy as a function of time but only at a single point in the fluid at any given time. The ultimate aim, of course, is the simultaneous determination of fluid velocities in a whole volume of a fluid. First steps in this direction

0066-4189/84/0115-0223$02.00

have been taken with the invention of speckle photography, which can give the instantaneous velocity field over a complete plane of interest in the fluid. The potential applications to fluid mechanics of speckle photography have not yet been fully explored, but it is certainly a highly attractive technique.

Among the methods making use of refractive-index variations, only holographic methods are discussed here. The classical interferometric and spatial-filtering techniques are well established and are covered by several review articles and monographs (e.g. Hauf & Grigull 1970, Merzkirch 1974, 1981, Goodman 1968); however, we briefly note some interesting developments that have taken place. Smeets (1974) has given a classification of optical techniques based on differential interferometry, and Oertel (1980) has supplied a nice illustrative example from Rayleigh-Bénard convection. The state of the art is such that the optical path modulation $\phi(x, y, t)$ by a phase object and any of its first and second derivatives in space and time (except $\partial^2\phi/\partial t^2$) can be measured with high sensitivity. Further, the unsteady parts of a fluid flow can be made visible selectively. Fluid dynamicists looking for a simple system of flow visualization may try the common phase-contrast method, which, suitably applied, will give results similar to those obtainable with much more costly and vibration-sensitive interferometers (Anderson & Taylor 1982).

As is the case in almost all branches of science and technology (and not

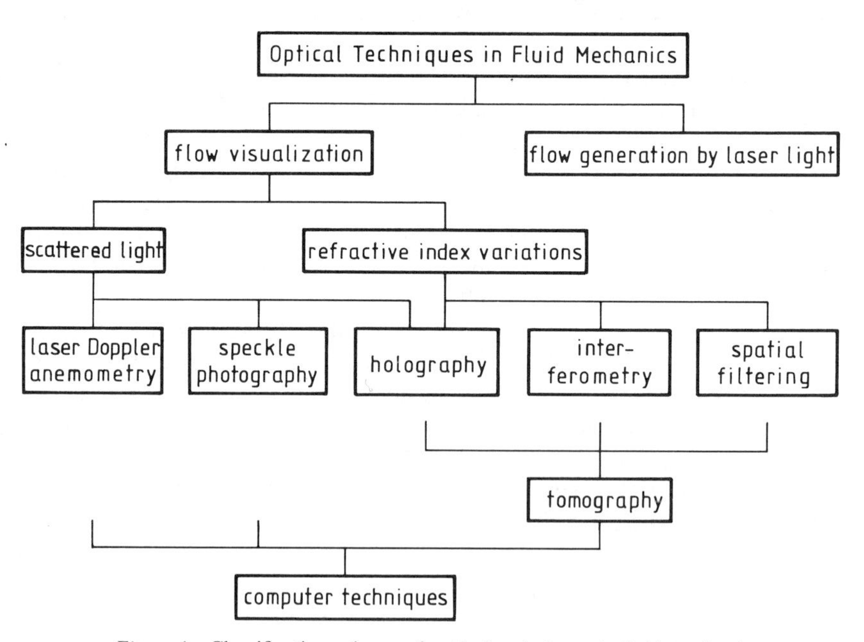

Figure 1 Classification scheme of optical techniques in fluid mechanics.

only there), the computer has slowly but steadily become more important in conducting experiments and doing the necessary data processing. The most obvious example in fluid mechanics is phase tomography, a method for obtaining information about phase objects in two or three dimensions. Phase tomography is still in its infancy but in a few years will be worthy of its own review, as will computer techniques in experimental fluid mechanics.

HOLOGRAPHY

Holography is a method for recording and reconstructing light waves. Both the amplitude and phase of a light wave, called the *object wave*, are recorded and may be reconstructed with the help of a reference wave. Thus, the full information content of a light wave can be stored and made available for a later time. This property has opened up new possibilities for investigating fluid flows:

1. Light waves from phase objects or particle distributions can be stored and analyzed later; this is done by applying various classical methods such as phase-contrast or schlieren techniques on the same reconstructed wavefront.
2. Two or more light waves originating at different instants in time can be stored and compared upon reconstruction (double-exposure interferometry).
3. The light wave belonging to one state of a flow can be recorded and later on be continuously compared with the light wave corresponding to the current state (real-time interferometry).

A reader not familiar with the basic ideas of holography may consult Collier et al. (1971), Vest (1979), or Ostrovsky et al. (1980).

In holographic flow visualization the object wave is a light wave that has passed through the fluid. In this geometry, usually only small optical path differences are introduced into the beam, and the coherence requirements are modest. The fluid may be illuminated with collimated or diffused light. Only diffuse illumination allows an investigation of the flow from different directions, but it also yields a grainy image due to the statistical superposition of many different wavelets from different directions. This so-called speckle noise is intimately connected with coherent light and cannot be avoided. Some resolution is lost in practice when speckles appear in the image. In speckle photography, however, good use can be made of the speckles for measuring flow velocities (see below).

Light sources in holography and holographic interferometry are continuous-wave lasers, such as He-Ne, argon, and dye lasers, or pulsed lasers, such as ruby, frequency-doubled Nd: YAG, or cavity-dumped argon lasers. Pulsed lasers are necessary for fast-moving flows in order to essentially freeze the interference pattern. The reconstruction is usually done with a continuous-wave laser of similar wavelength to avoid too strong a distortion in the image. In high-precision measurements the same wavelength for recording and reconstruction is necessary. For studying the dynamics of rapidly moving flow fields, series of holograms at high framing rates have to be taken (Ebeling 1977, Merboldt & Lauterborn 1982). For this purpose, and for high-speed photography of interferograms in real-time interferometry, a cavity-dumped argon laser or a multiply Q-switched ruby laser with regular spiking may be used as a light source. Solid-state pulsed-laser technology is described by Koechner (1976).

Ordinary holography has found applications in two-phase flows for particle sizing and in wavefront recording for subsequent flow visualization by shadowgraph, schlieren, or other filter techniques. Flow visualization via holographic recording has the advantage that several different visualization methods may be applied to the same wavefront and that the method applied can be optimized for maximum sensitivity without time constraints. This is of particular value for investigating high-speed flow events, which often are difficult to reproduce (Trolinger 1974). A drawback of the method is that speckles appear in the image. The technique best suited for reducing this speckle noise is differential interferometry (Smigielski & Hirth 1970).

In many two-phase flows (e.g. water with cavitation bubbles, boiling liquids, aerosols, sprays, or combustion processes), the flow essentially consists of one phase, with the other phase distributed in the form of small particles. As the depth of field in holography can be made large at high resolution throughout the volume, it is well suited to determining particle sizes as well as their spatial distribution in particle fields. Usually in-line holography is applied with opaque, small particles (Thompson 1974, Cartwright et al. 1980). For flows with semitransparent particles or with particles of largely different sizes, off-axis holography is preferable (Prikryl & Vest 1982a, Lennert et al. 1977). In this case diffuse illumination can be applied, which gives good localization in depth, albeit at the expense of resolution (Briones et al. 1978). To determine the size and location of the particles, the real image of the hologram may be scanned by a TV camera or other optical devices, with subsequent digitizing for picture processing in a computer (Haussmann & Lauterborn 1980).

Holographic Interferometry

The holographic technique most widely used in fluid mechanics is holographic interferometry, i.e. the comparison by holographic means of two light waves via their interference. Basically two methods can be distinguished: double-exposure and real-time interferometry. In double-exposure interferometry, two light waves that are present at different times are stored on one holographic plate and reconstructed simultaneously. The variations of refractive index between the exposures give rise to phase shifts and thus to an interference pattern upon superposition of the reconstructed waves. The interference pattern can be analyzed as in conventional interferometry to determine the underlying changes in the physical properties of the fluid, such as mass density, temperature, chemical species concentration, or electron number density.

One advantage of holographic interferometry over conventional interferometry is that light waves from the same object existing at different instants in time can be compared. As the light beams pass the same way, disturbances resulting from imperfections of optical elements are eliminated. For the same reason it is possible to observe flow fields through test models made of transparent materials. Thus, densities in corner-flow regimes can be measured that are not accessible to conventional optical methods.

A second advantage is that diffuse illumination can be used. A double-exposure hologram of a diffusely illuminated flow has stored projections of the optical path-length difference through the fluid from different directions (Heflinger et al. 1966). When suitable processing methods are applied (tomography; see below), three-dimensional asymmetric refractive-index distributions can be investigated. Diffuse illumination leads to strong localization of the interference fringes, not necessarily near the object under investigation. The problems connected with fringe localization are rather complex and are not discussed here (see Vest 1979, Wernicke & Osten 1982, Schumann & Dubas 1979). To give an example of double-exposure holographic interferometry with a phase object, a glass of wine set into oscillatory motion by a violin bow is shown in Figure 2.

When one works with plane object waves, reference fringes can easily be introduced by tilting the object beam or the holographic plate through a small angle between the two exposures (finite-fringe method). Reference fringes are fringes deliberately introduced into the image to decide on the sign of the phase difference and to alter the sensitivity of the method. Usually they are made parallel to the gradient of interest. If the gradient is not known beforehand, a slightly more complex setup with two reference beams may be used that offers the possibility of introducing reference fringes a posteriori (Trolinger 1979). The two superimposed holograms are each taken with a separate reference beam from a different direction. Reconstruction is done with both reference waves simultaneously. Tilting one reference beam or the holographic plate during reconstruction yields reference fringes with adjustable spacing and orientation.

Figure 2 A glass of wine set into oscillatory motion by a violin bow (*upper left corner*). Reconstruction from a double-exposure hologram taken with a Q-switched ruby laser. Pulse separation is 300 μs. (Hologram taken by Lauterborn, Hinsch & Bader on occasion of an experimental course on "Laser Physics and Holography" at the Third Physical Institute, University of Göttingen in 1972.)

Holographic real-time interferometry allows continuous comparison of the development in time of unsteady flows with a previously taken state of rest or some other initial state. To this end the initial state is holographically recorded, and the reconstructed wave is brought to interference with the current state. The method is especially suited for the investigation of fast and nonreproducible events. In connection with high-speed cinematography, the dynamics of the fluid flow can be investigated. Light-intensity problems, which arise in high-speed cinematography of real-time interferograms, may be overcome by using phase holograms. These holograms have a high diffraction efficiency and can be obtained with silver-halide emulsions when appropriate bleaching techniques are applied (Phillips & Porter 1976, Benton 1979). A special problem in real-time interferometry is the superposition of the reconstructed and the current object wave. The hologram has to be developed in situ or brought exactly back into its original position. Usually this takes some time, and sensitive experiments may suffer from the time delay. The optimum solution in such cases (and also very convenient otherwise) is the use of a holographic camera working with photothermoplastic film (Urbach 1977). Such films allow for almost instant processing and additionally give phase holograms. The spatial-frequency passband characteristic of thermoplastic materials should be taken into account by choosing the appropriate angle between object and reference wave.

Heterodyne Techniques

The crucial problem in interferometry does not seem to be the generation of fringes but the accuracy and processing speed of evaluating the interferometric images. When measurements are made on photographs, the phase differences can be determined with sufficient accuracy only along the fringe maxima and minima. But even the centerlines of the fringes can be located to no better than approximately one tenth of their spacing. The situation becomes worse when fractional fringe orders are to be determined, because nonlinear interpolation is required. Higher accuracy is obtained by using TV techniques and methods of digital picture processing to store and process the fringe pattern (Schlüter 1980, Kreis & Kreitlow 1980, Takeda et al. 1982).

An even better way to overcome the accuracy and processing-speed problem is the use of techniques that permit direct measurements of the optical phase differences in the image of the object. An elegant method that is both fast and accurate is heterodyne holographic interferometry, which has been reviewed by Dändliker (1980). The basic idea is the introduction of a small frequency shift $\Delta\omega$ between the optical frequencies ω_1 and ω_2 of the two interfering light waves. This leads to moving fringes. The optical phase difference of the two light fields at a given location in the image appears as the phase of the intensity modulation at the beat frequency $\Delta\omega = |\omega_1 - \omega_2|$. This phase can be measured electronically with respect to a reference signal at the same frequency. The reference signal may be obtained by a detector that is located somewhere in the interfering light field outside the phase object. Phase differences always contain ambiguities due to their modulo 2π behavior. These are easily avoided, as long as no jumps occur in a fringe pattern, by starting the phase measurements near the reference detector and counting the fringes. Heterodyne holographic interferometry can be considered a real breakthrough in interferometry, as it is characterized by a set of outstanding properties. The fringe interpolation capability is better than 10^{-3} of a fringe spacing ($\delta\phi = \pm 0.3°$). Any desired position in the image can be chosen for measurement of the phase, yielding the same accuracy. Therefore, high spatial resolution is attainable. Since phase shifts are measured, the method is independent of brightness variations across the image. Furthermore, no problems appear with the direction of the fringe shifts, and on-line data processing becomes feasible.

The heterodyne technique is not limited to holographic interferometry. Massie et al. (1981) constructed an apparatus on the basis of a Twyman-Green interferometer with a fast data-acquisition system and demonstrated a phase resolution of $\lambda/300$ with 10-μs time resolution.

The heterodyne technique can be combined with double-exposure holographic interferometry (Farrell et al. 1982). It may also be extended to include real-time holographic interferometry. Phase measurements in this case, however, are possible only for a limited number of points due to the complex instrumentation per point necessary for exact phase determination.

This drawback can be overcome by another technique, which is called *quasi-heterodyne interferometry* (see, for example, Dändliker et al. 1982, Hariharan et al. 1983). In quasi-heterodyne interferometry the relative phase between the interfering beams is changed stepwise, using at least three different values, e.g. 0° and $\pm 120°$. Measurements of the optical phase difference are based on measurements of the three corresponding distributions of the intensity I_1, I_2, and I_3 in the interference pattern. The great advantage of this technique is that whole interferograms can be evaluated at the same time by using a TV camera, with subsequent processing of the data stored [which are the three images $I_1(x, y)$, $I_2(x, y)$, and $I_3(x, y)$].

Quasi-heterodyne interferometry can also be combined with double-exposure and real-time holography. Real-time applications are, however, limited by the condition that the object may not change essentially during one cycle of three intensity measurements. Hariharan et al. (1983) report a measurement cycle length of 150 ms and a duration of 10 s for the calculation of the phases for a 100 × 100 array of points with an accuracy of $\lambda/100$.

TOMOGRAPHY

Tomography is a method of image reconstruction from projections. It has numerous applications in various scientific disciplines from medicine to space research. Tomography has only rarely been applied in fluid mechanics, despite its promising possibilities for determining arbitrary three-dimensional distributions of refractive-index changes via optical phase differences.

When the rays probing a transparent medium can be assumed to be straight lines, i.e. in the "refractionless limit," the optical path difference $\Delta\phi$, e.g. in the z-direction, is given by the "projection"

$$\Delta\phi(x, y) = \int [n(x, y, z) - n_0] \, dz, \tag{1}$$

where n_0 is the refractive index of the undisturbed medium in which the phase object to be investigated is generated. In holographic interferometry, n_0 is the refractive index of the phase object at the time of the initial exposure, and $n(x, y, z)$ is the refractive-index distribution at the time of the second exposure or at the instant a real-time interferogram is photographed.

The evaluation of an interferogram consists in determining $\Delta\phi$ and then inverting Equation (1) to get $n(x, y, z) - n_0$. In a third step, $n(x, y, z)$ is related to the underlying physical properties that are the original subject of the investigation. It is well known how to invert Equation (1) for two-dimensional phase objects or for the case of radial symmetry, where the Abel transform applies (see Vest 1979, Merzkirch 1981).

Arbitrary asymmetric phase objects require a large number of projections, each recorded from a different viewing direction. The projection of a phase object in a particular plane $z = z_0$ for one viewing angle θ is given by the Radon transform

$$\Delta\phi(r, \theta) = \int_{-\infty}^{+\infty} [n(x, y, z_0) - n_0] \, ds, \tag{2}$$

where

$$\begin{aligned} x &= r \cos\theta - r \sin\theta, \\ y &= r \sin\theta - r \cos\theta. \end{aligned} \tag{3}$$

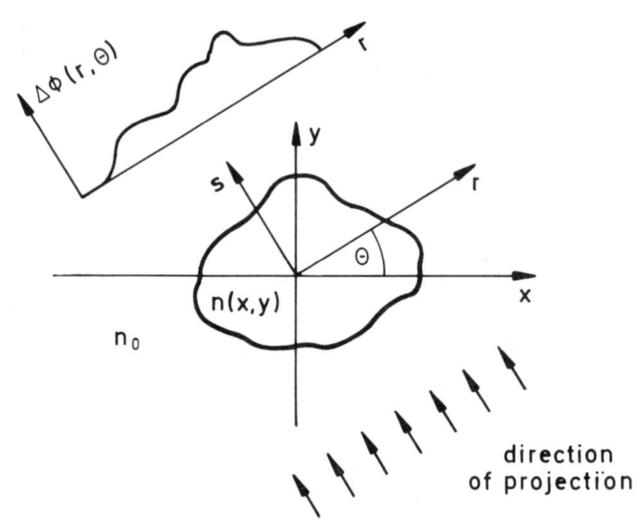

Figure 3 Diagram of a projection as used in tomography, with notation of Equations (2) and (3).

The notation of Equations (2) and (3) is illustrated in Figure 3. Tomographic methods allow $n(x, y, z_0)$ to be calculated numerically with a resolution that depends both on the number of projections $\Delta\phi(r, \theta)$ and the number of samples taken per projection. Variation of the plane $z = z_0$ yields the whole three-dimensional refractive-index distribution.

Several algorithms already exist for the inversion of the Radon transform and have been reviewed by Herman (1980). A discussion and comparison of six inversion techniques is given by Sweeney & Vest [1973; see also Herman (1979) and Crawford & Kak (1979)]. The algorithm most frequently used is the "filtered backprojection." Best suited for the determination of optical path differences for computerized tomography are visualization methods with a data output already in computer-ready form. Promising examples are the heterodyne and quasi-heterodyne techniques as well as the spatial filtering techniques, where the intensity distribution in the filtered image may be scanned by a photodetector (Häßler & Ebeling 1982).

For a complete reconstruction of a fully asymmetric phase object, data from all viewing angles between 0° and 180° are generally required. This can be accomplished in classical interferometry or spatial filtering only by successively altering the viewing angle. These techniques are, therefore, limited to steady or periodic flow fields. Holographic interferometry with diffuse object illumination, however, is ideally suited for instantaneous recording of multidirectional data. Thus, it makes possible the quantitative investigation of turbulent flows (Matulka & Collins 1971, Sweeney & Vest

1974). The problem of opaque bodies in the phase object is investigated by Prikryl & Vest (1982b).

SPECKLE PHOTOGRAPHY

Speckle photography is an ingenious technique originally invented to measure in-plane translations and deformations of solids with diffusely scattering surfaces. The method is excellently described in Vest (1979). After a long period of incubation, it has almost simultaneously been transferred to fluid mechanics by three research groups in order to measure the flow pattern in a planar section inside a liquid at a single instant (Barker & Fourney 1977, Grousson & Mallick 1977, Dudderar & Simpkins 1977).

The experimental setup is rather simple (Figure 4). A plane of interest in the liquid, seeded with microparticles as scatterers, is illuminated by a suitable coherent light source, e.g. a pulsed ruby laser or an argon laser. When the plane is imaged through a lens onto a photographic film or plate, the statistically scattered light gives rise to a speckle pattern with speckles of size

$$d_s = 1.2\lambda(m+1)f/D, \tag{4}$$

where d_s is an appropriately defined diameter of a speckle grain, λ the wavelength of the laser used, m the magnification of the imaging system, f the focal length of the lens, and D its (effective) aperture; f/D is the f-number of the camera setting. Typical speckle sizes with visible light are about 2–5 μm at $m = 1$. In usual speckle photography, two such speckle

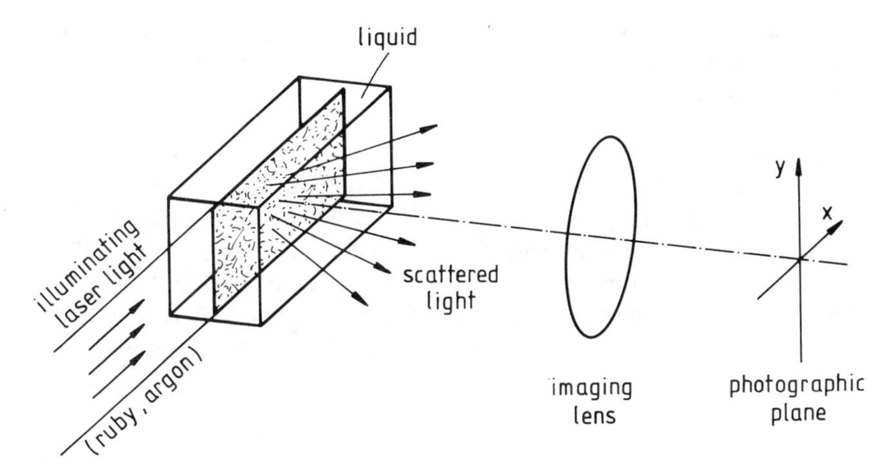

Figure 4 Experimental setup to obtain and photograph the speckle pattern generated by illuminating a liquid with a thin sheet of light.

patterns are taken in succession and superimposed on the same plate, which is often called a *specklegram.* When the time interval between the exposures is appropriately chosen and the fluid flow is not too turbulent, the speckles will have moved just a few diameters without losing their identity. This motion may be different in different areas of the image. Thus a random pattern of speckle pairs with variable distance and orientation is recorded on the plate. In this way information on particle velocities is stored and may be translated to a fluid-flow pattern by microscopic inspection. There exist, however, elegant and simple coherent optical methods to extract this information.

The first method consists in illuminating a small part of the plate (with an approximately homogeneous speckle shift) by a coherent plane wave, typically an unexpanded He-Ne laser beam. Then in the back focal plane of a converging lens (placed immediately before or behind the transparency), Young's fringes modulating a speckle pattern are observed with a spacing and an orientation directly related to the motion of the fluid. When d is the distance a scatterer moves during the time between the two exposures, the relation

$$d = \lambda_f f_T/(md_f) \tag{5}$$

holds, where λ_f is the wavelength of the laser used to form the fringes, f_T is the focal length of the lens used to get the fringes in its back focal plane (Fourier plane), m is the magnification of the imaging system used to take the speckle photograph, and d_f is the fringe spacing. The orientation of the fringes is normal to the translation of the fluid. Multiple exposure at equal time intervals can be used to increase the fringe contrast for sufficiently stationary flows. By successively scanning the photograph and measuring fringe spacing and orientation, the fluid-velocity vector field can be constructed. Simpkins & Dudderar (1978) have given an example from transient Bénard convection.

The second method of analyzing speckle photographs directly provides a picture of the spatial distribution of a selected fluid-velocity component in the illuminated plane. This is done by a spatial-filtering technique, the filter being a small hole in the Fourier plane (see Vest 1979, Grousson & Mallick 1977). Beautiful examples from Rayleigh-Bénard flow have been given by Meynart (1982). Figure 5 reproduces one of his results. A tenfold short exposure every three seconds has been used to get the specklegram of one

Figure 5 A specklegram (*top*) of a Rayleigh-Bénard convection flow in a cylindrical cell and two filtered images: (*middle*) vertical velocity component, $\Delta v_z = 15.5$ μm s^{-1}; (*bottom*) horizontal velocity component, $\Delta v_x = 18.8$ μm s^{-1} [from Meynart (1982), with permission of the author].

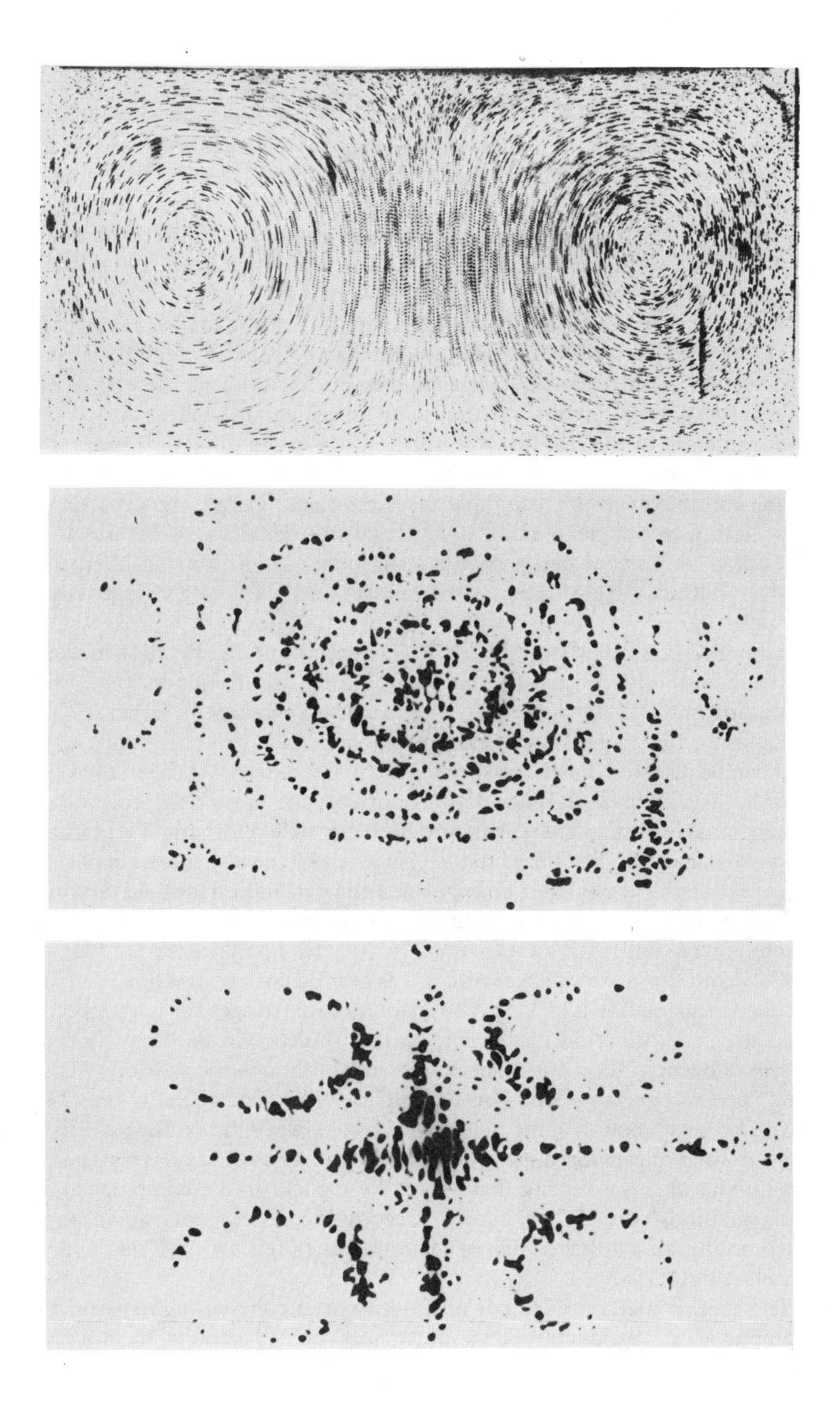

roll of stationary convection in a cylindrical cell. The upper image of Figure 5 shows a cut through the toroidal roll, with the liquid rapidly flowing up at the center of the container and slowly moving down at the wall. From one such specklegram, filtered images with isovelocity fringes can be obtained for selected velocity components. Two examples of such filtered images, each showing a different velocity component, are given in the two lower images of Figure 5.

This second method of producing velocity contour maps looks very appealing, but has some shortcomings that may prevent its widespread use. The filtered images are difficult to interpret, and only a succession of differently filtered images of the same specklegram will lead to an understanding of the real flow by composition in the observer's brain. It thus will be limited to almost stationary simple flows, a fact also suggested by the multiple-exposure technique necessary to get sharp fringes. Further, the filtering technique, a small hole in the Fourier plane, decreases the resolution and makes necessary a strong illumination to avoid intensity problems. Thus, the evaluation of the velocity field via Young's fringes is to be preferred.

Young's fringes may be analyzed for their spacing and orientation as well as their visibility. From $d \sim d_f^{-1}$ [Equation (5)] it follows that the resolution in the shift d of the scattering particles increases for larger fringe spacings d_f. But there is a natural limitation to the smallest distance d_{min} that can be resolved due to the speckle size d_s [Equation (4)]. The shift d_{min} must be larger than at least one diameter of a speckle grain, i.e. d_s, to lead to a recognizable double-point structure of the speckle photograph and thus resolvable fringes. The upper limit d_{max} is given by the decorrelation of the speckle pattern between the two exposures and depends on the fluid motion in conjunction with the time separation between the two exposures. It is roughly estimated to be approximately $20d_s$ by Simpkins & Dudderar (1978) from their experience with transient Bénard convection, but a detailed investigation is lacking. The visibility of the fringes has not yet been used to get information on the fine-scale turbulence in the flow via the degree of decorrelation of the speckle pattern, although the possibility has long been recognized (Grousson & Mallick 1977). The visibility may be measured as a function of the time separation between the exposures. First steps toward automatic data processing of Young's fringes have already been undertaken by feeding them via a TV camera into a computer and using methods of digital picture processing. So far, fringe spacing and orientation can automatically be determined (Kreitlow & Kreis 1980, Meynart 1982, 1983).

The speckle method does not necessarily need coherent light. What is required is the production of a pattern of pointlike pairs by double exposure. This can also be achieved with white light and scatterers in the liquid that are imaged onto a film. This, then, is nothing more than the usual multiple-exposure technique for flow visualization (Merzkirch 1974). The main difference is the use of Young's fringes, i.e. Fourier space, to get the displacement. This eliminates the difficulties in finding the corresponding scatter particle or speckle pairs on the exposed film. All pairs of the area of the object illuminated for shift measurement automatically combine to give Young's fringes. The speckle method can thus be viewed as an ingenious extension of the usual multiple-exposure technique for flow visualization by making use of Fourier space. An extra bonus is the easy extension to smaller displacements through the inverse resolution properties of space and Fourier space [Equation (5)].

Theoretically, the speckle technique is now well understood, but its possibilities have hardly been assessed in practice. The most advanced work is that of Meynart (1983) on instantaneous velocity-field measurements in unsteady gas flow. Some directions of possible future development are indicated in what follows. By taking specklegrams of a plane of interest in the fluid in rapid succession, the development in time of the velocity field may be obtained. By simultaneously illuminating different planes with different colors, an extension to three dimensions seems possible, e.g. by making use of particular elements of the setup of Murakami & Ishikawa (1979). These planes may be parallel or intersecting, depending on the problem. It is even conceivable that, by a combination of both methods, the three-dimensional dynamics of arbitrary fluid flows can be measured. To achieve this aim, automatic data processing will become more and more important.

Some outstanding problems in speckle photography are the selection of the appropriate light sources, scattering particles, and recording materials. Extremely bright light sources are needed because of the low scattering intensity at 90° of tiny particles and the short exposure time necessary to stop the motion. The light source should be capable of delivering two short pulses, with a small, adjustable time interval between them. This determines the range of fluid velocities that can be measured. Specklegram cinematography would need a series of double pulses, such as those used by Decker (1982) in holographic interferometric cinematography. The scattering particles should be optimized in size and reflectivity with respect to the specific problem. Closely related to the intensity problem is that of choosing the appropriate recording material. It must be able to resolve the speckle grains according to the camera setting [see Equation (4)], i.e. have a resolution of more than 300 lines per millimeter in usual cases, and should be as sensitive as possible at the wavelength used. Holographic recording materials will do, but there exist more sensitive films with sufficient resolution.

One remaining problem is inherent in speckle photography. Since no distinction can be made from Young's fringes as to which exposure of the usual two is the first, the direction of the flow cannot be determined. One solution may be given by simply moving the photographic material a certain amount between the exposures. This would lead to Young's fringes at zero motion of the object, but the direction of the object motion could be determined through an increase or decrease of the fringe spacing, together with a rotation of the fringes. Once these technical problems are overcome, speckle photography will be best suited for the investigation of multidirectional unsteady flows.

FLOW GENERATION BY LASER LIGHT

The development of high-power laser light sources has made possible the generation of fluid flow by light alone. The general kind of flow produced is determined by whether the fluid into which the laser-produced plasma expands is a gas or a liquid. When a target like a deuterium pellet is irradiated in a vacuum or a near-vacuum environment, the plasma generated can freely expand, and the initial thermal energy is rapidly converted to kinetic energy of an ordered flow. In a liquid the expansion is dominated by the inertial forces of the surrounding liquid soon after the laser pulse has stopped. In this way, special kinds of liquid flow dynamics may be studied.

The large area of laser-induced rarefied plasma flows, and its use in connection with controlled thermonuclear fusion, is outside the scope of this paper. Only the more modest laboratory experiments on laser-induced flows in liquids are reviewed here. Fluid flow is involved in the deformation of a liquid surface by radiation pressure (Ashkin & Dziedzic 1973) or absorption (e.g. Da Costa & Calatroni 1979, Emmony et al. 1976). Through the work of Ashkin & Dziedzic (1973), fluid mechanics has helped to decide on the value of the momentum of light in a dielectric medium.

When laser light is focused into the bulk of a nominally transparent liquid, dielectric or optical breakdown with bubble and shock-wave formation may occur (Askar'yan et al. 1963, Brewer & Rieckhoff 1964). Usually the optical breakdown seems to be triggered by heating of some impurity present in the liquid by absorption of the light. In this way a tiny ball of plasma of high temperature and pressure is generated, which rapidly expands to form a cavity in the liquid. Figure 6 shows the early stages of the process in two examples, one taken at 150 ns and the other at 1.75 μs after breakdown (Lauterborn & Ebeling 1977). The pictures have been taken holographically to suppress the bright white light emitted by the plasma

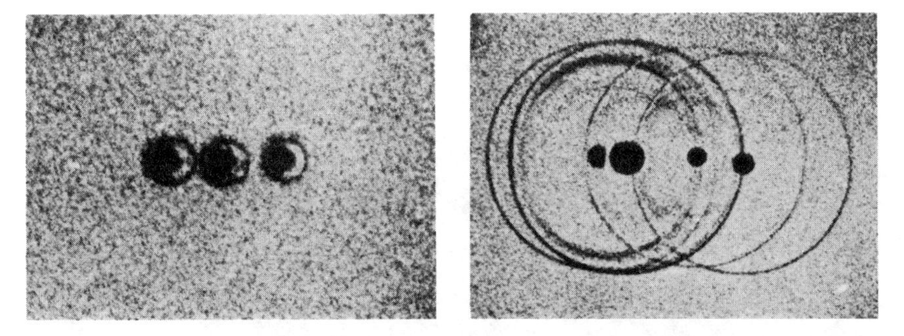

Figure 6 Bubble and shock-wave formation after Q-switched ruby-laser-induced breakdown in water. The pictures are taken via high-speed holography 150 ns (*left*) and 1.75 μs (*right*) after breakdown in two different experiments (after Lauterborn & Ebeling 1977).

inside the cavity during and immediately after breakdown. This light constitutes an incoherent background on the holographic plate and does not reproduce upon reconstruction. Bubble formation and shock-wave emission can thus easily be observed simultaneously.

Cavitation is an area in applied physics dealing with bubbles in liquids. It belongs both to fluid mechanics, through the liquid flow involved in the motion of the bubbles, and to acoustics, through the sound and shock-wave emission connected with the special kind of dynamics involved. It now also belongs to optics, because the phenomenon of laser-induced breakdown in liquids has been applied here too (Lauterborn 1974). Laser-induced bubbles can be produced free of mechanical disturbances at a given location in the liquid at a given instant of time and are thus extremely well suited for systematically studying bubble dynamics. This use of light has been called *optic cavitation* (Lauterborn & Bolle 1975, Lauterborn 1980). An example of the dynamics of a single, almost spherical, laser-produced cavitation bubble in the neighborhood of a plane solid boundary is given in Figure 7. Bubble growth, collapse, and rebound have been photographed with a rotating mirror camera at 75,000 frames s^{-1}. The most remarkable phenomenon observed is the high-speed liquid jet toward the boundary that is formed by involution of the top of the bubble in the final stage of its first collapse. The jet subsequently penetrates the lower bubble wall on its way to the boundary. A film is available that illustrates this phenomenon (Lauterborn et al. 1977).

Recently, vortex-ring formation and oscillation have been detected as part of cavitation bubble dynamics with the help of optic cavitation (see Lauterborn 1982). As vortex rings are of great interest to fluid dynamicists, an example is given in Figure 8.

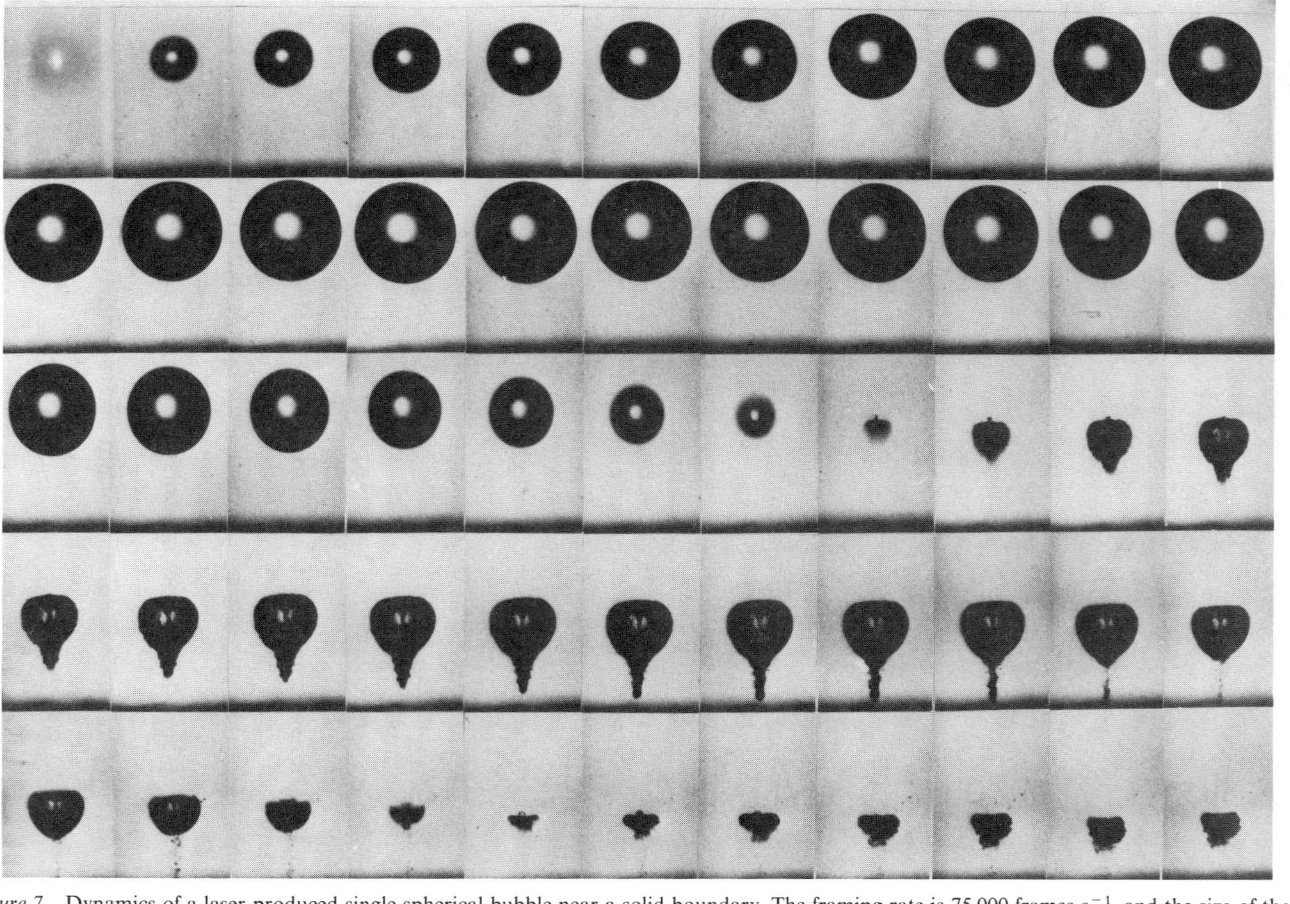

Figure 7 Dynamics of a laser-produced single spherical bubble near a solid boundary. The framing rate is 75,000 frames s^{-1}, and the size of the individual frames is 7.2×4.6 mm^2 (after Lauterborn & Bolle 1975).

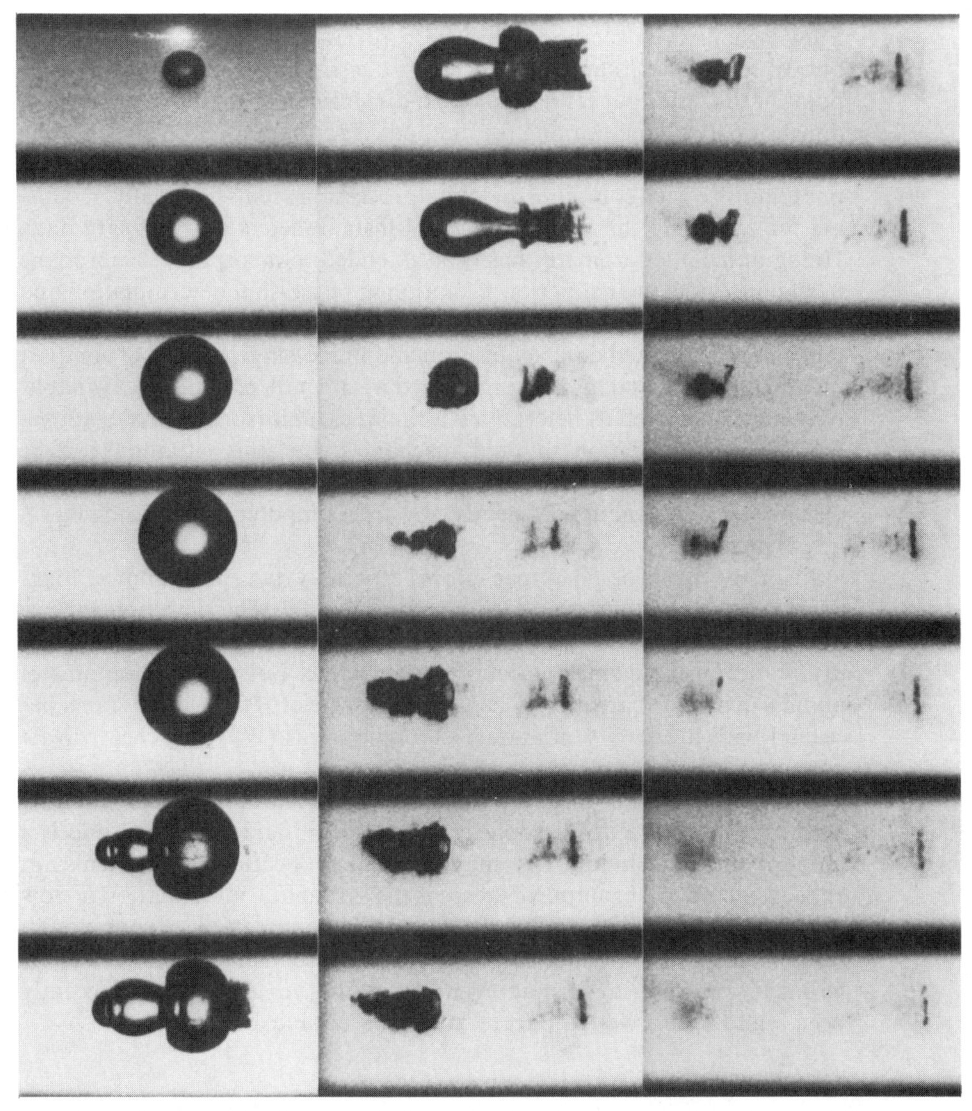

Figure 8 Vortex-ring formation during interaction of two successively produced bubbles in water. The second breakdown takes place between the fifth and sixth frame. (Frame numbering is from top to bottom and from left to right.) The framing rate is 20,000 frames s^{-1}, and the size of the individual frames is $6 \times 15\ mm^2$ (courtesy of W. Hentschel).

CONCLUSION

Flow-visualization methods have been reviewed here with emphasis on optical techniques that give velocity or density data not only at a single point in the fluid, but for a whole two-dimensional plane or even a three-dimensional volume. Among the methods described, speckle photography is the simplest to implement but with capabilities yet to be explored. Speckle photography with computerized data processing seems especially promising for obtaining the two-dimensional instantaneous flow-velocity data. Holographic interferometry has been extended to incorporate heterodyne techniques, with their superior resolution in phase-shift determination and their computer-matched properties. Tomographic methods also heavily rely on computerized data acquisition and processing. Because of complex instrumentation and processing aids that are not commonly available, tomography, as well as heterodyne techniques, unfortunately have not yet found wide application in fluid mechanics. But this situation can be expected to change in the near future. The trend of all modern flow-visualization techniques is clearly toward computerization and digital image processing.

In many areas of fluid mechanics, especially two-phase flows, high-speed holocinematographic methods are required. The development has not yet come to an end, as the holographic equivalents of the rotating drum and mirror cameras are not yet available. But small holograms of about $4 \times 4 \text{ mm}^2$ can already be taken at a rate of 200,000 holograms per second with a capacity of about 4000 holograms (W. Hentschel, private communication).

A special section of this review has been devoted to flow generation by laser light. This is a novel application of light in fluid mechanics and is a modern optical technique that may also find applications in problems not yet considered. An example of its use is the systematic investigation of flow phenomena induced by cavities in a liquid, in conjunction, of course, with flow-visualization methods like high-speed holocinematography or interferometry. Vortex-ring formation and oscillating vortex rings have recently been found in this way as part of cavitation bubble dynamics.

Literature Cited

Anderson, R. C., Taylor, M. W. 1982. Phase contrast flow visualization. *Appl. Opt.* 21:528–36

Ashkin, A., Dziedzic, J. M. 1973. Radiation pressure on a free liquid surface. *Phys. Rev. Lett.* 30:139–42

Askar'yan, G. A., Prokhorov, A. M., Chanturiya, G. F., Shipulo, G. P. 1963. The effects of a laser beam in a liquid. *Sov. Phys.-JETP* 17:1463–65

Barker, D. B., Fourney, M. E. 1977. Measuring fluid velocities with speckle patterns. *Opt. Lett.* 4:135–37

Benton, S. A. 1979. Photographic materials and their handling. In *Handbook of Optical Holography*, ed. H. J. Caulfield, pp. 349–66. New York: Academic. 638 pp.

Brewer, R. G., Rieckhoff, K. E. 1964. Stimulated Brillouin scattering in liquids. *Phys. Rev. Lett.* 13:334a–36a

Briones, R. A., Heflinger, L. O., Wuerker, R. F. 1978. Holographic microscopy. *Appl. Opt.* 17:944–50
Cartwright, S. L., Dunn, P., Thompson, B. J. 1980. Particle sizing using far-field holography: new developments. *Opt. Eng.* 19:727–33
Collier, R. J., Burckhardt, C. B., Lin, L. H. 1971. *Optical Holography.* New York/San Francisco/London: Academic. 605 pp.
Crawford, C. R., Kak, A. C. 1979. Aliasing artifacts in computerized tomography. *Appl. Opt.* 18:3704–11
Da Costa, G., Calatroni, J. 1979. Transient deformation of liquid surfaces by laser-induced thermocapillarity. *Appl. Opt.* 18:233–35
Dändliker, R. 1980. Heterodyne holographic interferometry. In *Progress in Optics*, ed. E. Wolf, pp. 1–84. Amsterdam/New York: North-Holland. 362 pp.
Dändliker, R., Thalmann, R., Willemin, J. F. 1982. Fringe interpolation by two-reference-beam holographic interferometry: reducing sensitivity to hologram misalignment. *Opt. Commun.* 42:301–6
Decker, A. J. 1982. Holographic cinematography of time-varying reflecting and time-varying phase objects using a Nd:YAG Laser. *Opt. Lett.* 7:122–23
Drain, L. E. 1980. *The Laser Doppler Technique.* Chichester/New York: Wiley. 241 pp.
Dudderar, T. D., Simpkins, P. G. 1977. Laser speckle photography in a fluid medium. *Nature* 270:45–47
Durrani, T. S., Greated, C. A. 1977. *Laser Systems and Flow Measurements.* New York: Plenum. 289 pp.
Durst, F., Melling, A., Whitelaw, J. H. 1976. *Principles and Practice of Laser-Doppler-Anemometry.* London: Academic. 412 pp.
Ebeling, K. J. 1977. Hochfrequenzholografie mit dem Rubinlaser. *Optik* 48:383–97, 481–90 (In German)
Emmony, D. C., Geerken, B. M., Straaijer, A. 1976. The interaction of 10.6 μm laser radiation with liquids. *Infrared Phys.* 16:87–92
Farrell, P. V., Springer, G. S., Vest, C. M. 1982. Heterodyne holographic interferometry: concentration and temperature measurements in gas mixtures. *Appl. Opt.* 21:1624–27
Goodman, J. W. 1968. *Introduction to Fourier Optics.* San Francisco: McGraw-Hill. 287 pp.
Grousson, R., Mallick, S. 1977. Study of flow pattern in a fluid by scattered laser light. *Appl. Opt.* 16:2334–36
Häßler, W., Ebeling, K. J. 1982. Bestimmung dreidimensionaler Ultraschallfelder mit einem tomographisch-optischen Verfahren. In *Fortschritte der Akustik FASE/DAGA 1982*, pp. 635–38. Göttingen: Drittes Phys. Inst. 1298 pp. (In German)
Hariharan, P., Oreb, B. F., Brown, N. 1983. Real-time holographic interferometry: a microcomputer system for the measurement of vector displacements. *Appl. Opt.* 22:876–80
Hauf, W., Grigull, U. 1970. Optical methods in heat transfer. *Adv. Heat Transfer* 6:131–366
Haussmann, G., Lauterborn, W. 1980. Determination of size and position of fast moving gas bubbles in liquids by digital 3-D image processing of hologram reconstructions. *Appl. Opt.* 19:3529–35
Heflinger, L. O., Wuerker, R. F., Brooks, R. E. 1966. Holographic interferometry. *J. Appl. Phys.* 37:642–49
Herman, G. T., ed. 1979. *Image Reconstruction from Projections. Implementation and Applications.* Berlin/Heidelberg/New York: Springer. 284 pp.
Herman, G. T. 1980. *Image Reconstructions from Projections. The Fundamentals of Computerized Tomography.* New York: Academic. 316 pp.
Koechner, W. 1976. *Solid-State Laser Engineering.* New York/Heidelberg/Berlin: Springer. 620 pp.
Kreis, T. M., Kreitlow, H. 1980. Quantitative evaluation of holographic interference patterns under image processing aspects. *Eur. Congr. Opt. Appl. Metrol. (METROP), 2nd*, ed. M. Grosmann, P. Meyrueis, 210:196–202. Bellingham, Wash: Soc. Photo-Opt. Instrum. Eng. 228 pp.
Kreitlow, H., Kreis, T. M. 1980. Automatic evaluation of Young's fringes related to the study of in-plane-deformation by speckle techniques. *Eur. Congr. Opt. Appl. Metrol. (METROP), 2nd*, ed. M. Grosmann, P. Meyrueis, 210:18–24. Bellingham, Wash: Soc. Photo-Opt. Instrum. Eng. 228 pp.
Lauterborn, W. 1974. Laser-induced cavitation. *Acustica* 31:51–78 (In German)
Lauterborn, W., Bolle, H. 1975. Experimental investigations of cavitation-bubble collapse in the neighbourhood of a solid boundary. *J. Fluid Mech.* 72:391–99
Lauterborn, W., Bolle, H., Inst. Wiss. Film. 1977. Cavitation. Dynamics of laser produced cavitation bubbles in water and silicone oil. *Film E 2353, 16 mm (75 m, 7 min).* Available from: Institut für den Wissenschaftlichen Film, Nonnensteig 72, D-3400 Göttingen, Fed. Repub. Germany
Lauterborn, W., Ebeling, K. J. 1977. High-speed holography of laser-induced breakdown in liquids. *Appl. Phys. Lett.* 31:663–64
Lauterborn, W., ed. 1980. *Cavitation and Inhomogeneities in Underwater Acoustics.*

Berlin/Heidelberg/New York: Springer. 319 pp.

Lauterborn, W. 1982. Cavitation bubble dynamics—new tools for an intricate problem. *Appl. Sci. Res.* 38:165–78

Lennert, A. E., Sowls, R. E., Belz, R. A., Goethert, W. H., Bentley, H. T. 1977. Electro-optical techniques for diesel engine research. In *Experimental Diagnostics in Gas Phase Combustion Systems*, ed. B. T. Zinn, pp. 629–56. New York: Am. Inst. Aeronaut. Astronaut. 657 pp.

Massie, N. A., Hartlove, J., Jungwirth, D., Morris, J. 1981. High accuracy interferometric measurements of electron-beam pumped transverse-flow laser media with 10-μsec time resolution. *Appl. Opt.* 20: 2372–78

Matulka, R. D., Collins, D. J. 1971. Determination of three-dimensional density-fields from holographic interferograms. *J. Appl. Phys.* 42:1109–19

Merboldt, K. D., Lauterborn, W. 1982. High-speed holocinematography with acousto-optic light deflection. *Opt. Commun.* 41: 233–38

Merzkirch, W. 1974. *Flow Visualization*. New York: Academic. 250 pp.

Merzkirch, W. 1981. Density sensitive flow visualization. In *Methods of Experimental Physics: Fluid Dynamics, Part A*, ed. R. I. Emrich, 18:345–403. New York: Academic. 403 pp.

Meynart, R. 1982. Convective flow field measurement by speckle velocimetry. *Rev. Phys. Appl.* 17:301–5

Meynart, R. 1983. Instantaneous velocity field measurements in unsteady gas flow by speckle velocimetry. *Appl. Opt.* 22:535–40

Murakami, T., Ishikawa, M. 1979. Holographic measurements of velocity distribution of particles accelerated by a shock wave. *Proc. Int. Congr. High Speed Photogr. Photonics, 13th*, ed. S.-I. Hyodo, pp. 326–29. Tokyo: Jpn. Soc. Precision Eng. 851 pp.

Oertel, H. Jr. 1980. Visualization of thermal convection. In *Flow Visualization: Proc. Int. Symp. Flow Visualization, 2nd*, ed. W. Merzkirch, pp. 71–76. Washington DC: Hemisphere. 803 pp.

Ostrovsky, Y. I., Butusov, M. M., Ostrovskaya, G. V. 1980. *Interferometry by Holography*. Berlin/Heidelberg/New York: Springer. 330 pp.

Phillips, N. J., Porter, D. 1976. An advance in the processing of holograms. *J. Phys. E* 9:631–34

Prikryl, I., Vest, C. M. 1982a. Holographic imaging of semitransparent droplets or particles. *Appl. Opt.* 21:2541–47

Prikryl, I., Vest, C. M. 1982b. Holographic interferometry of transparent media using light scattered by embedded test objects. *Appl. Opt.* 21:2554–57

Schlüter, M. 1980. Analysis of holographic interferograms with a TV picture system. *Opt. Laser Technol.* 12:93–95

Schulz-DuBois, E. O., ed. 1983. *Photon Correlation Techniques in Fluid Mechanics*. Berlin: Springer. 399 pp.

Schumann, W., Dubas, M. 1979. *Holographic Interferometry from the Scope of Deformation Analysis of Opaque Bodies*. Berlin/Heidelberg/New York: Springer. 194 pp.

Simpkins, P. G., Dudderar, T. D. 1978. Laser speckle measurements of transient Bénard convection. *J. Fluid Mech.* 89:665–71

Smeets, G. 1974. Observational techniques related to differential interferometry. *Int. Congr. High Speed Photogr., 11th*, pp. 283–88. London: Chapman & Hall. 615 pp.

Smigielski, P., Hirth, A. 1970. New holographic studies of high-speed phenomena. *Proc. Int. Congr. High-Speed Photography, 9th*, ed. W. G. Hyzer, W. G. Chace, pp. 321–26. New York: Soc. Motion Pict. Telev. Eng. 604 pp.

Sweeney, D. W., Vest, C. M. 1973. Reconstruction of three-dimensional refractive index fields from multidirectional interferometric data. *Appl. Opt.* 12:2649–64

Sweeney, D. W., Vest, C. M. 1974. Measurements of three-dimensional temperature fields above heated surfaces by holographic interferometry. *Int. J. Heat Mass Transfer* 17:1443–54

Takeda, M., Ina, H., Kobayashi, S. 1982. Fourier-transform method of fringe-pattern analysis for computer-based topography and interferometry. *J. Opt. Soc. Am.* 72:156–60

Thompson, B. J. 1974. Holographic particle sizing techniques. *J. Phys. E* 7:781–88

Trolinger, J. D. 1974. *Laser Instrumentation for Flow Field Diagnostics*, ed. S. M. Bogdonoff, J. A. Smith, *AGARD-AG-186*. London: NATO Advis. Group Aerosp. Res. Dev. 121 pp.

Trolinger, J. D. 1979. Application of generalized phase control during reconstruction to flow visualization holography. *Appl. Opt.* 18:766–74

Urbach, J. C. 1977. Thermoplastic hologram recording. In *Holographic Recording Materials*, ed. H. M. Smith, pp. 161–207. Berlin/Heidelberg/New York: Springer. 252 pp.

Van Dyke, M. 1982. *An Album of Fluid Motion*. Stanford, Calif: Parabolic. 176 pp.

Vest, C. M. 1979. *Holographic Interferometry*. New York: Wiley. 465 pp.

Wernicke, G., Osten, W. 1982. *Holografische Interferometrie*. Weinheim: Physik-Verlag. 272 pp. (In German)

Ann. Rev. Fluid Mech. 1984. 16 : 245–61

STABILITY AND COAGULATION OF COLLOIDS IN SHEAR FIELDS

W. R. Schowalter

Department of Chemical Engineering, Princeton University, Princeton, New Jersey 08544

INTRODUCTION

During earlier generations the field of colloid science was fostered in academic institutions mainly by chemistry departments and as a subfield of physical chemistry. This choice for an academic home was not inconsistent with the set of fundamental tools thought to be relevant to study of particulate matter in the size range of approximately 0.1 to 10 μm, the range of length scale generally considered to encompass the colloidal domain. It is an awkward regime because particles of this size are often subject to chaotic fluctuations known as Brownian motion (Russel 1981), but the particles are clearly too large to be usefully described by some analog of molecular kinetic theory. Thus one must deal with a region of mechanics that is neither truly continuum nor molecular in nature. Because the center of interest in colloids was among chemists, the properties studied were often described in terms of thermodynamics and hence were restricted to systems at equilibrium. Many examples of equilibrium properties are found in the two-volume treatise of Alexander & Johnson (1949). A useful modern introduction to the subject has been furnished by Hiemenz (1977).

Although much of the early work dealt with equilibrium properties of colloids, dynamic processes were not totally ignored. For example, Smoluchowski (1917) presented remarkably robust analyses of the dynamics of colloidal coagulation. The impetus for recent interest by fluid mechanists in the dynamics of colloids is associated with corresponding advances in fluid mechanics at low Reynolds numbers. Nonequilibrium processes of multiphase systems where the continuous phase is a liquid and the discontinuous phase is in the approximate range of 0.1 to 10 μm have been termed *microhydrodynamic* processes (Batchelor 1976a).

0066-4189/84/0115-0245$02.00

Microhydrodynamics, then, refers to the description of particle motions where inertial and gravitational forces usually can be neglected, but the effects of Brownian motion usually cannot.

In this review we are concerned with the stability of colloidal suspensions. Given a suspension of material in the colloidal size range, we ask whether that suspension will remain dispersed (i.e. will be stable to the actions of an imposed bulk flow and to Brownian motion) or will undergo agglomeration (i.e. will be unstable). In the latter situation, of course, we also wish to know the *kinetics* of the coagulation process. This review does not explicitly include two-phase systems where the continuous phase is a gas (Friedlander 1977) and Brownian effects often predominate. However, with a liquid continuous phase there is ample opportunity for the application of fluid mechanics, and it is the stability of these systems that we address.

PRACTICAL MOTIVATIONS

Colloid stability, in the sense defined above, is a topic pertinent to diverse segments of technology. A central issue in the formulation of paint recipes is the necessity of obtaining a product with a long shelf life. Insecticides and pharmaceuticals are often applied in the form of a particulate suspension, and one wishes to know those conditions under which the suspension will remain stable when it is subjected to shearing forces..Most of the pulp and paper industry is based on two-phase flow processes. Dynamics of colloidal systems is inherent in the formulation and application of coatings on paper or board products. A special case of this application is the whole printing industry, since printing ink is an archetypal colloidal suspension.

In the examples mentioned above one generally wishes to insure that a colloidal suspension is stable. There are many where *instability* is the desired goal, and many separation processes rely on the action of gravity upon a density difference to effect a separation. When the discontinuous phase, be it a liquid (emulsion) or solid (suspension), is in the colloidal size range, we have already noted that gravity separation may be ineffectual. One then wishes to use mechanisms such as shear-induced coagulation to promote the separation. An understanding of the principles governing this type of coagulation is now emerging.

FORCES ACTING AT A DISTANCE

We consider two spherical particles suspended in a liquid and ask what forces are exerted by one particle on the other. Because the particles are colloids, inertia and gravity are assumed to be inconsequential.

DLVO Theory

The classical picture of forces between identical colloidal particles is a linear superposition of (*a*) attraction due to London dispersion forces and (*b*) repulsion caused by electrical double layers surrounding each particle. This approach, developed by Derjaguin and Landau in the USSR and Verwey and Overbeek in The Netherlands, is known as the DLVO theory (Hiemenz 1977) or the DOV theory (Sonntag & Strenge 1972).

London dispersion forces are due to interactions of induced dipoles in the particles. Quantitative values for the force are difficult to calculate from first principles (Hough & White 1980). A common form, first derived by Hamaker (1937), is given by

$$V_{\mathrm{A}} = -\frac{A}{6}\left\{\frac{2}{r^2-4}+\frac{2}{r^2}+\ln\left(\frac{r^2-4}{r^2}\right)\right\}, \tag{1}$$

where V_{A} is the potential between two identical spherical particles separated by a dimensionless distance r between particle centers. The separation distance is scaled on the particle radius a, and the constant A is known as the Hamaker constant. Equation (1) is the result of an integration of atom-atom pair interaction potentials and the assumption that electromagnetic waves are propagated at infinite velocity. The validity of this assumption of no "retardation" of the wave velocity depends on the magnitude of the particle separation relative to λ, the London wavelength. Typically, $\lambda = O(10^{-1}\ \mu\mathrm{m})$. The influence of retardation has been given quantitative form by Schenkel & Kitchener (1960), who suggest that one use

$$V_{\mathrm{A}} = \frac{-A}{12(r-2)}\left[\frac{1}{1+1.77p}\right], \qquad (r-2) \ll 1 \quad \text{and} \quad p \leq 0.57, \tag{2}$$

$$V_{\mathrm{A}} = \frac{-A}{(r-2)}\left[\frac{2.45}{60p}-\frac{2.17}{180p^2}+\frac{0.59}{420p^3}\right], \qquad (r-2) \ll 1 \quad \text{and} \quad p > 0.57, \tag{3}$$

instead of Equation (1) at small separations. At large separations, one can use the approximate expression (Overbeek 1949)

$$V_{\mathrm{A}} = -\frac{16A}{9r^6}\left[\frac{2.45}{p}-\frac{2.04}{p^2}\right], \qquad r \gg 1, \tag{4}$$

where $p = 2\pi(r-2)a/\lambda$.

DLVO theory amounts to linear combination of one of the above forms for V_{A} with an equation describing the energy of repulsion due to the presence of a charged species on the particle surfaces. This leads to a distribution of countercharge from ionic species in the aqueous phase in which the particles are suspended. Thus, there is a diffuse ionic double layer surrounding each particle. The double layer is characterized by a solution of the Poisson-Boltzmann equation. Under restrictions discussed in some detail by Sonntag & Strenge (1972), the interparticle potential is given by

$$V_R = \frac{\pm \varepsilon a \psi_0^2}{2} \ln\{1 \pm \exp(-a\kappa(r-2))\}. \tag{5}$$

The positive signs correspond to the case of a constant surface potential ψ_0 on each particle. The suspending medium has dielectric constant ε, and κ^{-1} is a measure of the double-layer thickness. Equation (5) can be used with the minus signs for a boundary condition of constant surface charge σ_0, where

$$\sigma_0 = \frac{\psi_0 \varepsilon (1 + a\kappa)}{4\pi a}. \tag{6}$$

In either case, Equation (5) applies only for thin double layers ($a\kappa \gg 1$).

The potential curve $V(r) = V_A + V_R$ is especially interesting because its shape is sensitive to the governing parameters. London attraction dominates at large and at very small separations, while the shape of the potential curve at intermediate separations reflects the influence of double-layer repulsion. As shown in Figure 1, the potential can be entirely attractive or both attractive and repulsive, depending on the thickness of the double layer. Addition of salt to a system increases the ionic strength and decreases κ^{-1}, thus decreasing the range of double-layer repulsion. Coagulation under conditions of no repulsion ($a\kappa = 10^2$ in Figure 1) is known as rapid flocculation. On the other hand, in the case of $a\kappa = 10$ it is possible for coagulation to take place in either a relatively weak secondary minimum or in the much stronger primary minimum. In the latter case a minimum occurs because at separations approaching zero interparticle distance there is a short-range but strong Born repulsion, which is not shown in Figure 1.

From the above remarks it is clear that, for the systems being discussed here, a stable colloidal dispersion can be destabilized (i.e. made to flocculate) by addition of electrolyte. The critical coagulation concentration may be defined as the concentration at which the potential curve simultaneously displays $V(r) = dV/dr = 0$ at a separation other than $r \to \infty$.

An assessment of the current strengths and weaknesses of DLVO theory as it applies to colloid stability has been furnished by one of the architects of the theory (Overbeek 1982).

Non-DLVO Effects

It is not the purpose of this review to dwell on the physicochemical forces between colloidal particles. Nevertheless, these forces are an important factor affecting colloid stability, and one must be aware of the shortcomings of DLVO theory. From the book by Mahanty & Ninham (1976), one can learn how the attractive London force has been described from a field point of view that includes the effect of the medium in which two interacting particles are placed. In this way, effects of electromagnetic retardation can be incorporated naturally into a basic model. A case in point is the recent calculation by Pailthorpe & Russel (1982), who used spectral absorption data to compute the dispersion forces for polystyrene spheres in water. Casting their results into the DLVO framework, they found significant

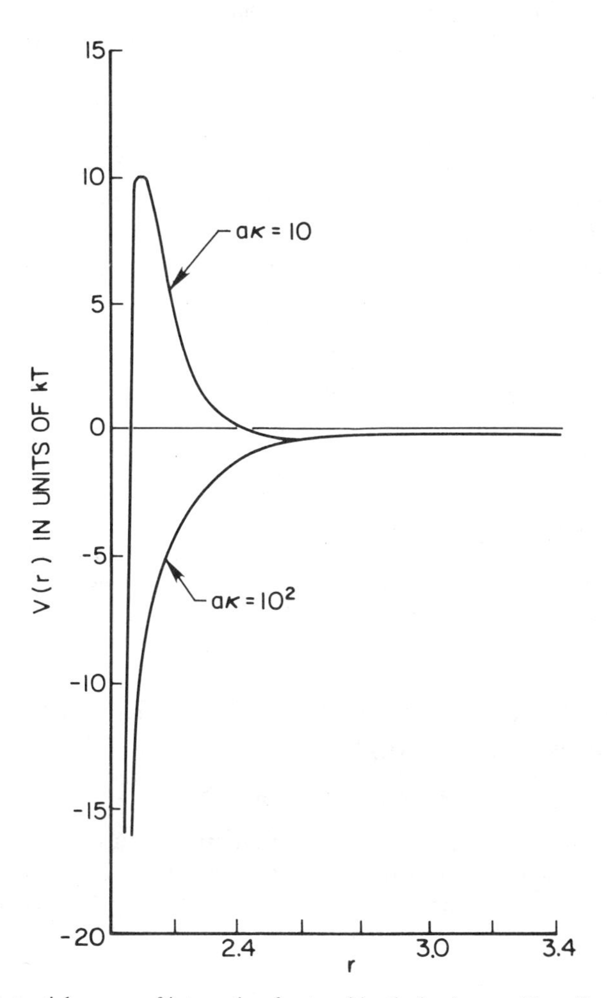

Figure 1 Potential energy of interaction for two identical spheres with radius $a = 10\ \mu m$. $A = 10^{-19}$ J and $\psi_0 = 25.6$ mV (taken from Overbeek 1949).

effects due to retardation for $r > 5$ nm, and variations in the Hamaker "constant" with material and with separation distance.

Approximations commonly invoked to determine the double-layer force have also been evaluated in the light of more-exact solutions, such as those in which finite-element methods have been employed. An example is the recent work of Chan & Chan (1983), who have obtained an exact solution for the potential between two spheres of the same diameter when the spheres have the same or opposite (but constant) surface potentials.

A fundamental development of recent years is the measurement of what are called *solvation*, *hydration*, or *structural* forces. These are repulsive forces manifest at small separations and evidently due to physics entirely outside of the DLVO picture described above (Ninham 1982). These repulsive forces, which appear to be due to a structural effect of the suspending fluid near particle surfaces, can become important at separations below approximately 5 nm. When dissolved ions are present, the nature of these ions can have an important bearing on close-range forces. Pashley (1981) has interpreted these results in terms of adsorbed layers of hydrated ions at the particle surface. Recent measurements have indicated that the interparticle force can be oscillatory at small separations. For example, Christenson et al. (1982) find distance-dependent oscillations in magnitude and sign of the force between macroscopic surfaces in cyclohexane at separations below 5 nm. Israelachvili & Pashley (1983) report similar oscillations in 10^{-3} M KCl solution at pH $= 5.5$ when the particle separation is below 1.5 nm.

Non-DLVO effects that are significant only at small distances from a particle surface have been ignored in stability calculations and interpretation of measurements. Nevertheless, situations exist for which these forces have a profound effect on stability. It is clear that as we learn more about the nature of short-range forces, they will be used to design desired stability characteristics for colloidal dispersions.

HYDRODYNAMIC INTERACTIONS BETWEEN IDENTICAL SPHERES

Influence on Brownian Coagulation

It was mentioned in the Introduction that Brownian motion is usually a contributing factor to the microhydrodynamics of colloids. Hence, although this review is primarily concerned with stability to shearing, it would be inappropriate not to recount the role of hydrodynamics on coagulation that is driven by Brownian motion. In Smoluchowski's (1917) analysis of Brownian coagulation, he considered the frequency with which a sphere of radius a_i would experience collision with spheres of radius a_j. The movement of the particles was assumed to be governed by the Stokes-Einstein expression for the Brownian diffusion coefficient. From this model of the coagulation process, one can write a kinetic expression for the formation and destruction of aggregates of any size, each aggregate being taken as spherical in shape. One obtains, after certain simplifications (Schowalter 1982), a second-order rate expression, so that the number concentration of aggregates of all sizes, N_{tot}, at time t is related to the initial number of particles N_0 by

$$\frac{N_0}{N_{tot}} = 1 + \frac{t}{t^S_{1/2}}, \tag{7}$$

where the half-life for the particles is

$$t^S_{1/2} = \frac{3\mu}{4kTN_0}. \tag{8}$$

Here μ is the viscosity of the suspending medium, k the Boltzmann constant, and T the absolute temperature. It is clear that two important effects have been neglected in the development of Equations (7) and (8): the interparticle forces described in the preceding sections, and the effect of particle proximity on hydrodynamic resistance to Brownian motion. Viscous drag on a sphere in the vicinity of a second sphere is well understood. If an interparticle potential $V(r)$ accounts for forces acting at a distance, and if the effect of particle proximity on hydrodynamic drag is given by a parameter $C(r)$, then Equation (7) can be used to describe coagulation kinetics. For the case where two interacting spheres are identical, one replaces $t^S_{1/2}$ in that equation by

$$t_{1/2} = 4t^S_{1/2} \int_2^\infty \frac{\exp[V(r)/kT]}{C(r)r^2}\, dr. \tag{9}$$

Values of $C(r)$ are available, for example, in Spielman (1970).

For sufficiently dilute suspensions, the linearity of N_0/N_{tot} with time, as predicted by Equation (7), has been confirmed in experiments (Zeichner & Schowalter 1979).

Influence of Shearing on Coagulation

Curtis & Hocking (1970) were the first to consider binary interactions between spheres in a laminar shear field ($v_x = Gy$, $v_y = v_z = 0$) when, in addition to hydrodynamic forces, the spheres experience induced-dipole attraction and thus undergo "rapid" flocculation (as opposed to the "slow" flocculation possible when there is also a double-layer repulsive force). They applied their calculations to experiments of coagulation in laminar shear and obtained reasonable estimates of the Hamaker constant, although the bispherical coordinate system used by Curtis & Hocking to describe the hydrodynamic portion of the problem is not appropriate at small interparticle separations. Batchelor & Green (1972) improved on the bispherical coordinate solution used by Lin et al. (1970) to describe the purely hydrodynamic problem of two spheres placed in a velocity field that, at large distances from the spheres, is characterized by a velocity $\mathbf{u}^0$ varying linearly with position. Thus,

$$\mathbf{u}^0 = \omega \times \mathbf{r} + \mathbf{E} \cdot \mathbf{r}, \tag{10}$$

where $\omega = \frac{1}{2}(\nabla \times \mathbf{u}^0)$ and $\mathbf{E} = \frac{1}{2}[\nabla \mathbf{u}^0 + (\nabla \mathbf{u}^0)^{\mathrm{T}}]$ are the constant ambient spin vector and rate-of-deformation tensor, respectively. Batchelor & Green (1972) showed that, if the spheres have equal radii, the velocity of a second sphere relative to a test sphere at the origin is

$$\mathbf{v}_{\mathrm{f}} = \mathbf{u}^0 - \left\{ A(r)\frac{\mathbf{rr}}{r^2} + B(r)\left[\mathbf{I} - \frac{\mathbf{rr}}{r^2}\right]\right\} \cdot \mathbf{E} \cdot \mathbf{r}. \tag{11}$$

Here $A(r)$ and $B(r)$ are known functions.

Because of the linearity of the governing equations of motion, it is possible to add hydrodynamic and other effects so that the relative particle velocity $\mathbf{v}$ is

$$\mathbf{v} = \mathbf{v}_{\mathrm{c}} + \mathbf{v}_{\mathrm{f}}. \tag{12}$$

In Equation (12) $\mathbf{v}_{\mathrm{c}}$ corresponds to the relative velocity induced by physicochemical forces, such as are described by DLVO theory. Then

$$\mathbf{F}_{\mathrm{c}} = -\nabla V(r), \tag{13}$$

which gives rise to the velocity

$$\mathbf{v}_{\mathrm{c}} = \frac{C(r)}{6\pi\mu a}\mathbf{F}_{\mathrm{c}}, \tag{14}$$

where $C(r)$ is the known hydrodynamic function used in Equation (9).

In addition to its applicability at small separations, Equation (11), which is valid for any constant $\nabla \mathbf{u}^0$, allows the study of binary interactions in a general linear flow field. Hence, one is not restricted to laminar shearing.

These equations are applied to coagulation kinetics by computing, as with Brownian coagulation, the collision rate j of spherical particles with an identical test sphere and comparing the results with j_{S}^0, Smoluchowski's (1917) computation under the assumption that all particles move along streamlines corresponding to $\mathbf{E}$. For laminar shearing, one obtains

$$j_{\mathrm{S}}^0 = \frac{32}{3}a^3 G n_{\infty}, \tag{15}$$

where n_∞ is the number density of particles far from the test sphere. We define a capture efficiency α_0 by

$$\alpha_0 = \frac{j}{j_S^0}. \tag{16}$$

It is of interest to compare capture efficiencies for laminar shearing with those for uniaxial extension flow ($v_x = -Gx$, $v_y = -Gy$, $v_z = 2Gz$). In this case the collision frequency analogous to Equation (15) is

$$j_E^0 = \frac{64\pi}{3\sqrt{3}} a^3 G n_\infty, \tag{17}$$

and one can define a capture efficiency analogous to Equation (16).

Van de Ven & Mason (1976a) were the first to combine Equation (11) with interparticle forces predicted from DLVO theory. They restricted their calculations to laminar shear flow ($v_x = Gy$) and, evidently, to trajectories in the xy-plane. Similar calculations were performed independently by Zeichner & Schowalter (1977), who included nonequatorial trajectories and considered uniaxial extensional as well as laminar-shearing flow. From the work of Kao & Mason (1975), one would expect vorticity to have a significant effect on capture efficiency, and indeed it does. If one assumes that, far from the test sphere, particles are uniformly distributed and that one need only consider binary encounters with the test sphere, then values of the collision rate j and the capture efficiency can be computed from the equations presented above. The effect of vorticity is shown in Figure 2, as well as the differences introduced by retardation of electromagnetic wave propagation.

The results in Figure 2 do not include the possibility that some trajectories will undergo one or more orbits about the test sphere; the effect of such trajectories has recently been addressed by Feke & Schowalter (1983). Note that when the physicochemical force is entirely attractive, permanent orbits related to the closed streamlines present in laminar shear flow (see, for example, Arp & Mason 1977) cannot occur.

Returning to the role of vorticity, one learns from Figure 2 that, based on comparisons at the same flow number, $\mathrm{Fl} = 6\pi\mu a^3 G/A$, it is not possible to draw a simple comparative conclusion about capture efficiency for laminar shearing relative to uniaxial extension. At sufficiently high values of G, capture efficiency is higher for extensional flow. This conclusion must, however, be tempered with the recognition that there are other bases, such as conditions for equal energy dissipation (see Zeichner & Schowalter 1977), on which to make the comparison. Also, no allowance has been made for the effect of hydrodynamic forces on rupture of particle pairs (known in colloid science as *repeptization*).

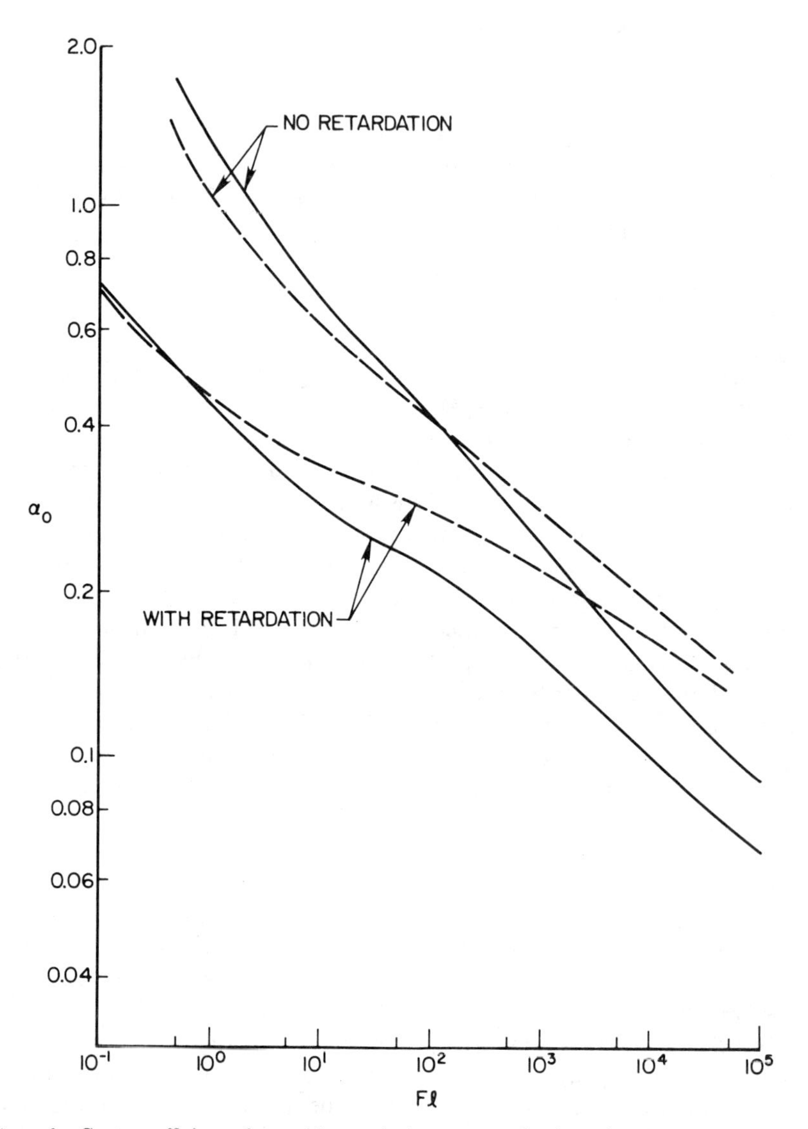

Figure 2 Capture efficiency for rapid coagulation. ———: laminar shear; — — —: uniaxial extension. For retardation, $\lambda/a = 10^{-2}$. Fl $= 6\pi\mu a^3G/A$ (taken from Zeichner & Schowalter 1977).

The results described thus far have included the London attractive force but no electrostatic repulsion. Because of the shape of the potential curve when repulsion and attraction are present, it is possible to have loose aggregation in the secondary minimum shown in Figure 1 or strong flocculation in the primary minimum (van de Ven & Mason 1976a, Zeichner & Schowalter 1977, van de Ven 1982). The importance of electrostatic repulsion is gauged by the value of a repulsion number, $\mathrm{Rp} = \varepsilon\psi_0^2 a/A$.

In spite of the limitations of binary hydrodynamics and DLVO theory, coagulation kinetics based on these principles is in reasonable agreement with experiments for which the physicochemical force is attractive. When electrostatic repulsion is appreciable, this agreement is less satisfying, and the reasons for disparity are not clear because of the difficulty of performing unambiguous experiments. Representative experimental results for both cases are available in Zeichner & Schowalter (1977). An excellent experimental study of trajectories of latex spheres with $a = O(1\ \mu\mathrm{m})$ in liquids with varying ionic strengths is that of Takamura et al. (1981). In highly viscous media (93% glycerol, 7% water), unexpected short-range repulsion was observed. The authors speculated that non-DLVO solvation forces could be present. A fluorescence technique has recently been adapted to the study of aggregating colloids (Cummins et al. 1983), and this method appears promising enough to warrant further application.

The Combination of Brownian Motion and Bulk Shearing

Having achieved some understanding of these two cases taken separately, one next seeks an analysis valid when both are present. Again, we are dealing with a linear problem in the sense that Equation (12) can be modified by inclusion of an additional velocity

$$\mathbf{v}_\mathrm{B} = -\mathbf{D}(\mathbf{r})\cdot\nabla \ln P(\mathbf{r}), \tag{18}$$

where $\mathbf{D}(\mathbf{r})$ is the Brownian diffusion coefficient for equal spheres given by (Batchelor 1976b)

$$\mathbf{D} = \frac{kT}{3\pi\mu a}\left\{\frac{C(r)}{2}\frac{\mathbf{rr}}{r^2} + H(r)\left[\mathbf{I} - \frac{\mathbf{rr}}{r^2}\right]\right\}, \tag{19}$$

and $H(r)$ is known. In Equation (18), $P(\mathbf{r})$ is the conditional pair distribution function, which must now be calculated from the dimensionless equation (Feke & Schowalter 1983)

$$\nabla\cdot\left[\left(\mathbf{v}_\mathrm{f} + \frac{1}{\mathrm{Fl}}\,C(r)\mathbf{F}_\mathrm{c}\right)P(\mathbf{r})\right] = \frac{1}{\mathrm{Pe}}\nabla\cdot[\mathbf{D}\cdot\nabla P(\mathbf{r})], \qquad r \geq 2. \tag{20}$$

The Péclet number, $Pe = 3\pi\mu a^3 G/(kT)$, is, of course, a measure of the importance of convection relative to Brownian diffusion.

The solution of this equation and the subsequent calculation of collection efficiencies have required further simplifying assumptions. It is tempting to hope that the processes of Brownian- and shear-induced coagulation are linearly independent, even though Equation (20) provides little encouragement in that regard. Many years ago, Swift & Friedlander (1964) made this assumption and reported results based upon it. More recently, Equation (20) has been solved for cases of low, but nonzero, and high, but finite, Péclet numbers. Van de Ven & Mason (1977) (see also van de Ven 1982) report that the increase in capture efficiency over unity [the capture efficiency being Equation (16) adapted to Smoluchowski's analysis for Brownian coagulation] is proportional to $(Pe)^{0.5}$ for small values of Pe. At the other end of the scale, Feke & Schowalter (1983) have studied the effect of Brownian motion for $Pe \gg 1$ in both uniaxial extension and laminar shearing. The computations in the latter case can be difficult because of the effect of orbiting trajectories on the spatial dependence of the probability distribution function. A comparison of theory and experiment is shown in Figure 3. Under experimental conditions where one expects the theory to apply, the degree of agreement is encouraging. A comparison of the assumption of linear independence of shear and Brownian effects against the more rigorous use of Equation (20) is shown in Figure 4.

HETEROCOAGULATION

The dynamics in the illustrations thus far have been limited to binary encounters between identical spheres, although these results have been freely applied to population balance equations in which multiplets are allowed. It is not difficult to rewrite the equations so that they describe hydrodynamics of relative motion between two spheres of arbitrary size. In a series of papers, Adler (1981a,b) has looked in detail at the consequences for coagulation of binary encounters between unequal spheres in a shear field. Except for this difference his approach is similar to those already described. In light of the complexities present for homocoagulation, one is not surprised to read that, "... it was not possible to represent the results by a single formula when colloidal forces were present." From Adler's results it appears that homocoagulation is generally, but not always, favored. The reader is referred to his papers for those specific cases that have been calculated.

Floc Dynamics and Structure

When one contemplates the difficulties inherent in an exact hydrodynamic and probabilistic treatment of coagulation beyond the binary level, it is easy to understand why such analyses are absent. We are given some insight into the subject by van de Ven & Mason (1976b), who considered the purely hydrodynamic issues associated with chains of identical spheres in laminar shear flow. An important feature is the distinction between touching and nontouching multiplets; the dynamic differences are appreciable, and can correspond to differences between capture in strong primary or weak secondary minima. Sonntag et al. (1982) have discussed the kinetics of chain formation and the equilibrium between chain formation and chain disaggregation. However, they neglected the effect of shear and considered only the action of Brownian diffusion. Adler & Mills (1979) used a very different approach to the problem of motion and resistance to rupture of an aggregate in a shear field. A floc was modeled as a porous sphere, and flow through interstices of the sphere was described by the Brinkman equation. A good review of the assumptions associated with the Brinkman equation and the limits to its validity is available in Higdon & Kojima (1981).

The goal, of course, is to apply fluid-dynamic principles that will help us

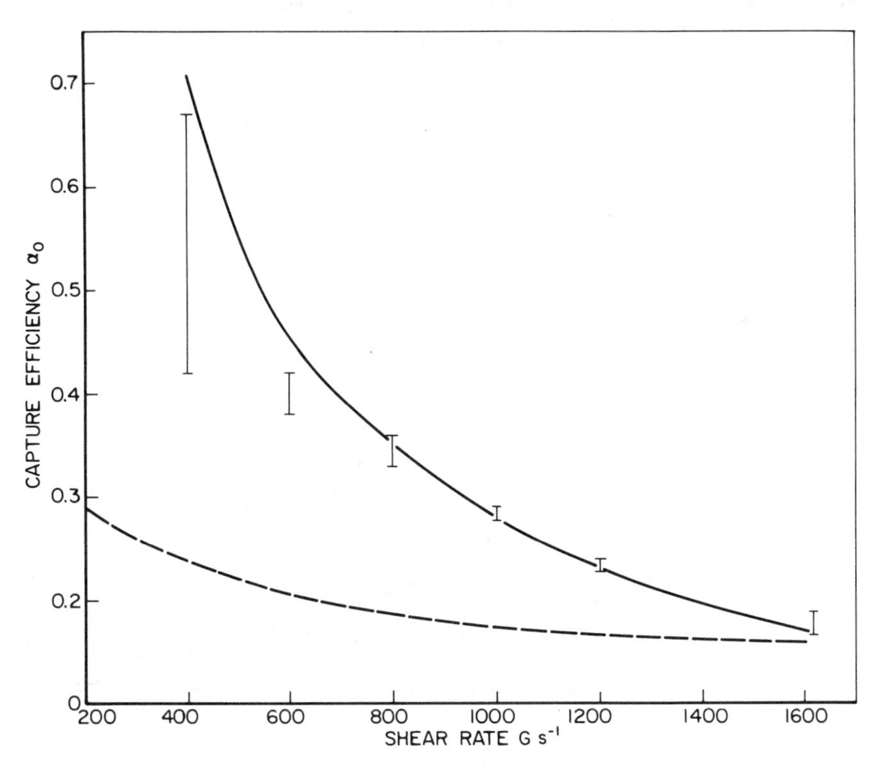

Figure 3 Comparison of theory and experiment (Feke 1981). ——— : prediction for $Pe = \infty$ $(D \equiv 0)$; ———: prediction for $Pe \gg 1$. Error bars indicate data for coagulation of polystyrene spheres ($a \cong 0.33\ \mu m$, $19 < Pe < 150$) in a laminar shear field.

to understand details of coagulation at the multiplet level. That goal remains unattained. Some progress has been made in attempting to rationalize the "elastic floc model" for clusters of "flocculi," which in turn are composed of large numbers of primary particles. Floc size and structure are determined by the magnitude and history of shearing to which a suspension has been subjected. The elastic floc model is due to R. J. Hunter and his collaborators. (See, in particular, Hunter 1982.) Through measurements of the rheology of a flocculated suspension, one attempts to infer floc structure and strength. Fluid-mechanical consequences of the elastic floc model have been reported by van de Ven & Hunter (1977). They considered the various mechanisms that are present for energy dissipation during laminar shearing of a suspension of elastic flocs. It is surprising to learn that the dominant source of dissipation is from the movement of suspending fluid in interstices in a floc as it is compressed, albeit by fractions of a nanometer, during collision.

Thus far we have alluded only briefly to disaggregation. Because the process is most relevant to multiplets, any description suffers from the lack of a firm fluid-mechanical basis. An example of the current level of

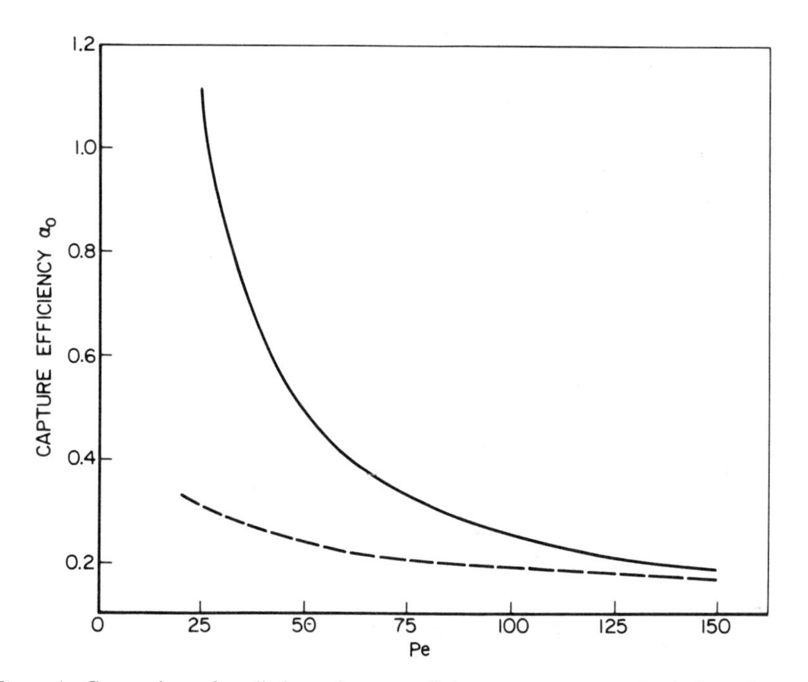

Figure 4 Comparison of predictions of capture efficiency. – – – : assuming independence of coagulation due to Brownian motion and laminar shearing; ——— : allowing for interdependence of the two coagulation mechanisms (Feke 1981).

knowledge is available in a recent work by Pandya & Spielman (1982), who studied floc splitting and erosion in a turbulent field.

PERTINENT RESULTS FROM SEDIMENTATION

It was stated at the outset that a distinguishing feature of microhydrodynamics is the unimportance of gravitational effects. We wish to end with a qualification to that claim. One of the oldest techniques for separation of suspensions is centrifugation, which is of course equivalent to adding a gravitational body force. Thus, research primarily concerned with sedimentation can be relevant to coagulation of colloids. We mention just two recent examples. Buscall et al. (1982) followed sedimentation of 1.55-μm spheres under conditions where both physicochemical and hydrodynamic effects are important; and Batchelor & Wen (1982) have carried out a few calculations for sedimentation in polydisperse systems, where one simultaneously accounts for hydrodynamic interactions, Brownian diffusion, and interparticle attraction or repulsion. The objective of these researches is to provide a sedimentation coefficient that reflects departures from sedimentation predicted from Stokes law for an isolated sphere. This coefficient clearly depends on the same fundamental interactions as those discussed in this review. Hence, the work is also relevant to coagulation of colloids in shear fields.

NEXT STEPS

This summary has shown how the principles of fluid mechanics can be productively applied to advance our understanding of interesting and important problems in colloid science. We now close with a list of some remaining tasks.

1. Studies of interactions between coagulation in the primary and secondary minimum during shearing. Theoretical results for coagulation when the interparticle potential curve contains a secondary minimum (Figure 1) are not complete. In the presence of Brownian motion, the problem is formidable.
2. Careful experiments of slow coagulation during shearing.
3. An understanding of stability in the presence of polymer additives. We have said nothing about the role played by macromolecules dissolved in the suspending medium during coagulation. Yet it is clear that these can have a large effect on colloid stability (Sperry et al. 1981). Feigen & Napper (1980) have attempted to explain the role of dissolved polymer using equilibrium thermodynamics, but a dynamic treatment that applies to the kinetics of coagulation is needed. It is also known that

polymer adsorption on particles and consequent bridging between polymers can affect coagulation (van de Ven 1981). Constructive combinations of fluid mechanics and polymer science are required.

4. A more detailed picture of floc aggregation and rupture. In order to refine the elastic floc model, further observational efforts with controlled shearing [along the lines of Kao & Mason (1975), for example] are needed.

ACKNOWLEDGMENT

This review was made possible through the continuing support of the Xerox Corporation and a grant from the Exxon Education Foundation. Helpful comments were supplied by D. L. Feke, R. L. Hoffman, R. T. Mifflin, R. M. Pashley, W. B. Russel, and D. A. Saville.

Literature Cited

Adler, P. M. 1981a. Heterocoagulation in shear flow. *J. Colloid Interface Sci.* 83:106–15

Adler, P. M. 1981b. Interaction of unequal spheres. I. Hydrodynamic interaction: colloidal forces. *J. Colloid Interface Sci.* 84:461–74

Adler, P. M., Mills, P. M. 1979. Motion and rupture of a porous sphere in a linear flow field. *J. Rheol.* 23:25–37

Alexander, A. E., Johnson, P. 1949. *Colloid Science*, Vols. 1, 2. London: Oxford. 837 pp.

Arp, P. A., Mason, S. G. 1977. The kinetics of flowing dispersions. VIII. Doublets of rigid spheres (theoretical). *J. Colloid Interface Sci.* 61:21–43

Batchelor, G. K. 1976a. Developments in microhydrodynamics. In *Theoretical and Applied Mechanics*, ed. W. T. Koiter, pp. 33–55. Amsterdam: North-Holland. 260 pp.

Batchelor, G. K. 1976b. Brownian diffusion of particles with hydrodynamic interaction. *J. Fluid Mech.* 74:1–29

Batchelor, G. K., Green, J. T. 1972. The hydrodynamic interaction of two small freely-moving spheres in a linear flow field. *J. Fluid Mech.* 56:375–400

Batchelor, G. K., Wen, C. S. 1982. Sedimentation in a dilute polydisperse system of interacting spheres. Part 2. Numerical results. *J. Fluid Mech.* 124:495–528

Buscall, R., Goodwin, J. W., Ottewill, R. H., Tadros, Th. F. 1982. The settling of particles through Newtonian and non-Newtonian media. *J. Colloid Interface Sci.* 85:78–86

Chan, B. K. C., Chan, D. Y. 1983. Electrical double-layer interaction between spherical colloidal particles: an exact solution. *J. Colloid Interface Sci.* 92:281–83

Christenson, H. K., Horn, R. G., Israelachvili, J. N. 1982. Measurement of forces due to structure in hydrocarbon liquids. *J. Colloid Interface Sci.* 88:79–88

Cummins, P. G., Staples, E. J., Thompson, L. G., Smith, A. L., Pope, L. 1983. Size distribution measurements of nonaggregating and aggregating dispersions using a modified flow ultramicroscope. *J. Colloid Interface Sci.* 92:189–97

Curtis, A. S. G., Hocking, L. M. 1970. Collision efficiency of equal spherical particles in a shear flow. *Trans. Faraday Soc.* 66:1381–90

Feigen, R. I., Napper, D. H. 1980. Depletion stabilization and depletion flocculation. *J. Colloid Interface Sci.* 75:525–41

Feke, D. L. 1981. *Kinetics of flow-induced coagulation with weak Brownian diffusion.* PhD thesis. Princeton Univ., N.J. 150 pp.

Feke, D. L., Schowalter, W. R. 1983. The effect of Brownian diffusion on shear-induced coagulation of colloidal dispersions. *J. Fluid Mech.* 133:17–35

Friedlander, S. K. 1977. *Smoke, Dust and Haze.* New York: Wiley. 317 pp.

Hamaker, H. C. 1937. The London-van der Waals attraction between spherical particles. *Physica* 4:1058–72

Hiemenz, P. C. 1977. *Principles of Colloid and Surface Chemistry.* New York: Marcel Dekker. 516 pp.

Higdon, J. J. L., Kojima, M. 1981. On the calculation of Stokes' flow past porous particles. *Int. J. Multiphase Flow* 7:719–27

Hough, D. B., White, L. R. 1980. The calculation of Hamaker constants from Lifshitz theory with applications to wetting phenomena. *Adv. Colloid Interface Sci.* 14:3–41

Hunter, R. J. 1982. The flow behavior of coagulated colloidal dispersions. *Adv. Colloid Interface Sci.* 17:197–211

Israelachvili, J. N., Pashley, R. M. 1983. Molecular layering of water at surfaces and the origin of repulsive hydration forces. *Nature*. In press

Kao, S. V., Mason, S. G. 1975. Dispersion of particles by shear. *Nature* 253:619–21

Lin, C. J., Lee, K. J., Sàther, N. F. 1970. Slow motion of two spheres in a shear field. *J. Fluid Mech.* 43:35–47

Mahanty, J., Ninham, B. W. 1976. *Dispersion Forces*. New York: Academic. 236 pp.

Ninham, B. W. 1982. Hierarchies of forces: the last 150 years. *Adv. Colloid Interface Sci.* 16:3–15

Overbeek, J. Th. G. 1949. The interaction between colloidal particles. In *Colloid Science*, ed. H. R. Kruyt, 1:276. Amsterdam: Elsevier. 389 pp.

Overbeek, J. Th. G. 1982. Strong and weak points in the interpretation of colloid stability. *Adv. Colloid Interface Sci.* 16: 17–30

Pailthorpe, B. A., Russel, W. B. 1982. The retarded van der Waals interaction between spheres. *J. Colloid Interface Sci.* 89:563–66

Pandya, J. D., Spielman, L. A. 1982. Floc breakage in agitated suspensions: theory and data processing strategy. *J. Colloid Interface Sci.* 90:517–31

Pashley, R. M. 1981. DLVO and hydration forces between mica surfaces in Li^+, Na^+, K^+, and Cs^+ electrolyte solutions: a correlation of double-layer and hydration forces with surface cation exchange properties. *J. Colloid Interface Sci.* 83:531–46

Russel, W. B. 1981. Brownian motion of small particles suspended in liquids. *Ann. Rev. Fluid Mech.* 13:425–55

Schenkel, J. H., Kitchener, J. A. 1960. A test of the Derjaguin-Verwey-Overbeek theory with a colloidal suspension. *Trans. Faraday Soc.* 56:161–73

Schowalter, W. R. 1982. The effect of bulk motion on coagulation rates of colloidal dispersions. *Adv. Colloid Interface Sci.* 17:129–47

Smoluchowski, M. 1917. Versuch einer mathematischen Theorie der Koagulationskinetik kolloider Lösungen. *Z. Phys. Chem.* 92:129–68

Sonntag, H., Florek, Th., Schilov, V. 1982. Linear coagulation and branching under condition of slow coagulation. *Adv. Colloid Interface Sci.* 16:337–40

Sonntag, H., Strenge, K. 1972. *Coagulation and Stability of Disperse Systems*. New York: Halsted. 139 pp.

Sperry, P. R., Hopfenberg, H. R., Thomas, N. L. 1981. Flocculation of latex by water-soluble polymers: experimental confirmation of a nonbridging, nonadsorptive, volume-restriction mechanism. *J. Colloid Interface Sci.* 82:62–76

Spielman, L. A. 1970. Viscous interactions in Brownian coagulation. *J. Colloid Interface Sci.* 33:562–71

Swift, D. L., Friedlander, S. K. 1964. The coagulation of hydrosols by Brownian motion and laminar shear flow. *J. Colloid Sci.* 19:621–47

Takamura, K., Goldsmith, H. L., Mason, S. G. 1981. The microrheology of colloidal dispersions. XII. Trajectories of orthokinetic pair-collisions of latex spheres in a simple electrolyte. *J. Colloid Interface Sci.* 82:175–89

van de Ven, T. G. M. 1981. Effects of polymer bridging on selective shear flocculation. *J. Colloid Interface Sci.* 81:290–91

van de Ven, T. G. M. 1982. Interactions between colloidal particles in simple shear flow. *Adv. Colloid Interface Sci.* 17:105–27

van de Ven, T. G. M., Hunter, R. J. 1977. The energy dissipation in sheared coagulated sols. *Rheol. Acta* 16:534–43

van de Ven, T. G. M., Mason, S. G. 1976a. The microrheology of colloidal dispersions. IV. Pairs of interacting spheres in shear flow. *J. Colloid Interface Sci.* 57:505–16

van de Ven, T. G. M., Mason, S. G. 1976b. The microrheology of colloidal dispersions. VI. Chains of spheres in shear flow. *J. Colloid Interface Sci.* 57:535–46

van de Ven, T. G. M., Mason, S. G. 1977. The microrheology of colloidal dispersions. VIII. Effect of shear on perikinetic doublet formation. *Colloid Polym. Sci.* 255:794–804

Zeichner, G. R., Schowalter, W. R. 1977. Use of trajectory analysis to study stability of colloidal dispersions in flow fields. *AIChE J.* 23:243–54

Zeichner, G. R., Schowalter, W. R. 1979. Effects of hydrodynamic and colloidal forces on the coagulation of dispersions. *J. Colloid Interface Sci.* 71:237–53

Ann. Rev. Fluid Mech. 1984. 16: 263–85

AEROACOUSTICS OF TURBULENT SHEAR FLOWS[1]

M. E. Goldstein

National Aeronautics and Space Administration, Lewis Research Center, Cleveland, Ohio 44135

1. *Introduction*

This article is primarily concerned with the generation of sound by high-Reynolds-number turbulent shear flows. Much of our understanding of this phenomenon is based on the acoustic analogy introduced by Lighthill (1952, 1954) to deal with the problem of jet noise. In this approach the flow is assumed to be known and the sound field is calculated as a small by-product of that flow. Lighthill achieved considerable success in explaining some of the most prominent features of the experimentally observed jet sound field, but when more-detailed experiments were conducted by Lush (1971), Ahuja & Bushell (1973), and others, it became clear that there were other more subtle features that could not be explained by Lighthill's analogy. Dowling et al. (1978) developed an extension that can account for such effects, but, in the main, attempts to explain the new observations were based on more-complex analogies such as the ones developed by Phillips (1960), Lilley (1974), Howe (1975), and Yates & Sandri (1975). All of these analogies involve, in one form or another, a nonlinear wave operator that eventually must be linearized before meaningful calculations can be carried out.

There have been a number of excellent reviews of the acoustic-analogy approach (e.g. Ribner 1981), including the last article on aerodynamic noise to appear in this series (Ffowcs Williams 1977), and there is probably very little that I can add to what has already been said. I therefore emphasize an alternative approach, which has not been reviewed so extensively, but which usually leads to the same results as the acoustic analogy. It amounts to nothing more than calculating the unsteady flow that produces the sound along with its resulting sound field, starting from some prescribed upstream state that, ideally, is specified just ahead of the region where the sound generation takes place.

In order to make progress with this approach it is necessary to linearize the governing equations about an appropriate known mean flow, but, as previously indicated, this is also necessary in the acoustic-analogy approach, and it seems better to introduce this linearization at the outset and carry through its consequences in a systematic fashion.

The use of linearized theory to calculate turbulent flows or, better yet, changes in turbulent flows is a branch of turbulence theory now known as "rapid-distortion theory" [see Moffatt (1981), p. 40]. Its validity depends upon the following conditions being satisfied (Hunt 1973): (*a*) that $u'/U \ll 1$, where u' is the rms turbulence velocity and U is the local mean-flow velocity; and (*b*) that the interaction or change being calculated be completed in a time τ_{I} that is short compared with τ_{decay}, where τ_{decay} is the decay time or lifetime of a typical turbulent eddy of order l/u'. Here, l denotes the characteristic size of the turbulent eddies.

In practice most rapid-distortion calculations are based on the inviscid equations, an application that is justified when both the mean flow and turbulence Reynolds numbers are large. *The radiated sound field can be determined as a by-product of any such rapid-distortion-theory calculation, provided compressibility effects are retained.*

2. *Solid-Surface Interactions: The Linear Theory*

In a high-Reynolds-number turbulent airjet such as shown schematically in Figure 1, the maximum turbulence level occurs along the centerline of the initial mixing layer, indicated by the dashed line in the figure. Here the ratio of the rms turbulence velocity to the local mean-flow velocity is roughly equal to 0.24 (Bradshaw et al. 1964, Hussain & Husain 1980), which, although not all that small, still constitutes an acceptable "small para-

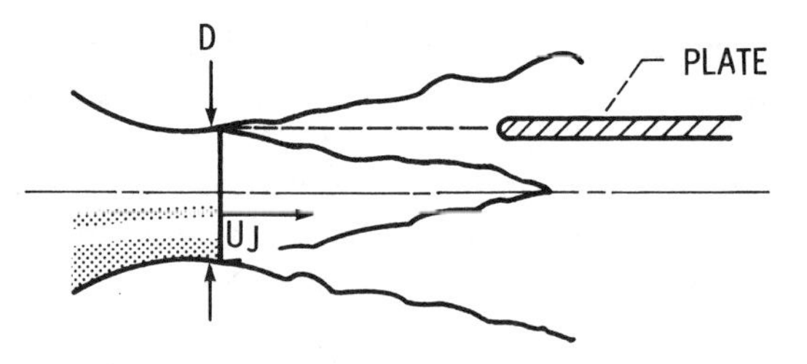

Figure 1 Plate embedded in turbulent airjet. Note that the leading edge of the plate is positioned in the region of maximum turbulence intensity so that its sound field will exceed the background jet noise by the maximum amount.

meter" to many applied mathematicians. Condition (*a*) in Section 1 is therefore reasonably well satisfied.

Now suppose that a large but infinitesimally thin flat plate is inserted into the flow in the manner indicated in the figure. Then the interaction between the turbulence and the leading edge will be completed in a time $\tau_I = O(l/U)$, which, in view of the smallness of the turbulence intensity, is fairly small compared with $\tau_{\text{decay}} = O(l/u')$. Thus, inviscid rapid-distortion theory applies, and the interaction between the turbulence and the edge can be calculated by linearizing the inviscid equations (the Euler equations) about the mean flow.

Since the ratio of the cross-stream to streamwise components of the mean-flow velocity is of the order of $(u'/U)^2$ (Tennekes & Lumley 1972, pp. 129–30), there will certainly be no loss in accuracy if this flow is taken to be to be a unidirectional transversely sheared flow. And since such a flow is itself a solution of the inviscid equations (for any velocity profile), this leads to a rational perturbation procedure that can, in principle, be carried to arbitrary order without internal inconsistency. The governing equations are now the same as those used in inviscid stability theory, i.e. the Rayleigh equations (see Betchov & Criminale 1967, pp. 175–79), and as already indicated, the radiated sound field can be determined as part of the solution to these equations when compressibility effects are retained.

2.1 REPRESENTATION OF THE TURBULENCE However, before calculating this interaction from these equations, one must decide which solutions should be used to represent the incident turbulence. For the very simple special case of a completely uniform mean flow, any solution for the unsteady velocity/pressure fluctuations can be decomposed into the sum of an "acoustic solution" that carries no vorticity and a vortical solution, which produces no pressure fluctuations and which is often referred to as the "gust" or "hydrodynamic" solution; the latter solution is used to represent the incident turbulence in most problems that involve the interaction of turbulence with solid surfaces (Goldstein 1976b, pp. 220–21). Its suitability for this purpose is largely due to the following reasons:

1. It is defined everywhere in space in the absence of solid surfaces so that it will represent the turbulence that would exist if the surfaces were not present.
2. It involves two arbitrary "convected" quantities that can be specified as upstream boundary conditions to describe the turbulence entering the interaction zone in any given problem.[2]
3. It has no acoustic field at subsonic speeds.

[2] This turns out to be just the appropriate degree of generality for this purpose. Note that the vorticity is a convected quantity when the mean flow is uniform, and that it has only two independent components, since its divergence must vanish.

The decomposition of the solution into acoustic and vortical parts is no longer possible when the mean flow is nonuniform, but the compressible Rayleigh equations still possess a solution that has the three properties listed above and, in fact, approaches the vortical solution on a uniform mean flow in the limit as the mean flow approaches a uniform flow (Goldstein 1978, Möhring 1976). Then it is the natural generalization of the latter to nonuniform flows, and it is therefore appropriate to refer to it as the gust or hydrodynamic solution and to use it to represent the incident turbulence.

In the general case this solution can be written as (Goldstein 1978, 1979, Case 1960)

$$u_\sigma = A_\sigma\left(\frac{x_1}{U(\mathbf{x}_t)} - t, \mathbf{x}_t\right) + \int_{-\infty}^{\infty}\int_{-\infty}^{\infty} G_\sigma(\mathbf{x}, t|\mathbf{y}, \tau)\omega_c\left(\frac{y_1}{U(\mathbf{y}_t)} - \tau, \mathbf{y}_t\right) d\mathbf{y}\, d\tau, \quad (\sigma = 1, 2, 3, 4) \quad (1)$$

where u_σ denotes one of the perturbation velocity components when $\sigma = 1$, 2, 3, and u_4 denotes the associated normalized pressure fluctuation. Here, t is the time; x_1, x_2, x_3 are Cartesian coordinates with x_1 in the mean-flow direction and $\mathbf{x}_t = \{x_2, x_3\}$ in the transverse direction; G_σ is a free space vector Green's function for the compressible Rayleigh equations, which is slightly unusual in that it is defined by placing the convective derivative (based on the mean-flow velocity) of the delta function on the right side of the Rayleigh equation, rather than the delta function itself; ω_c is a convected quantity that can be arbitrarily specified as an upstream boundary condition; and A_σ is another convected quantity. The fourth component of A_σ is identically zero, and the remaining three components form a three-dimensional vector that has zero divergence and is perpendicular to the gradient of the mean-flow velocity. Therefore, A_σ has one independent component, and this component can be an arbitrary function of its argument.

Equation (1) thus involves two arbitrary convected quantities. It is certainly defined over all space, since G_σ is the "free space" Green's function. That it is a homogeneous solution of the linearized Rayleigh equation can be seen by inspection. For example, substituting the second term into the Rayleigh equations will, in view of the definition of G_σ, transform the integrand into the convective derivative of $\delta(\mathbf{x}-\mathbf{y})\,\delta(t-\tau)$ times ω_c. Integration by parts produces a convective derivative of ω_c, which by construction is identically zero. This shows that the second member of Equation (1) is indeed a homogeneous solution of the Rayleigh equations. It is easy to show that the first member also has this property.

2.2 SOUND GENERATION AND ROLE OF INSTABILITY WAVES Returning now to the problem of a large flat plate embedded in a turbulent shear flow, we can, as argued above, use the gust or hydrodynamic solution to represent the incident turbulence. But this solution does not satisfy the zero normal velocity boundary condition at the plate, and it is necessary to add another solution to cancel this component of velocity. However, unlike the gust this latter solution does not vanish exponentially fast at infinity but instead behaves like a propagating acoustic wave there (Goldstein 1979). This shows that the plate is able to "scatter" the nonpropagating motion associated with the gust into a propagating acoustic wave (Ffowcs Williams & Hall 1970).

However, the problem also possesses an eigensolution associated with the spatially growing instability wave that can propagate downstream from the edge on the inflectional mean-velocity profile (Crighton & Leppington 1974, Goldstein 1981). The solution is therefore not unique! It can be made unique by requiring that it remain bounded at infinity, since this would eliminate the eigensolution, which grows without bound. But since the linearization is only valid in the vicinity of the leading edge, it is probably not appropriate to impose a "boundary"-type condition far downstream in the flow because all sorts of nonlinear effects will have had a chance to intervene (Rienstra 1979).

One, therefore, ought to look for an alternative way to make the solution unique. This can be done by treating the steady-state solution, which is, of course, the one of interest here, as the long time limit of the solution to an initial-value problem and then imposing a "causality condition" in the sense that the solution is required to be identically zero before the initial time when the incident disturbance is "turned on" (Crighton & Leppington 1974). But it is not clear that an initial condition imposed in the distant past should be relevant to the steady-state solution (Rienstra 1979). One might therefore consider a third way of making the solution unique. This can be done by using the eigensolution to eliminate the leading-edge singularity that appears in both the bounded and causal solutions, i.e. by satisfying a leading-edge "Kutta condition" (Goldstein 1981).

Physically, this may be rationalized by noting that the instability wave represents vortex shedding downstream of the edge (Rockwell 1983). It would then be reasonable to expect that such shedding would take place in a way that eliminates the singularity in the inviscid solution and thereby prevents any flow separation that would otherwise occur at a very sharp edge.

2.3 COMPARISON WITH DATA It is not clear which of these three solutions is correct, but Goldstein (1979) compared the theory with the data of Olsen (1976), who measured the sound radiated in one-third-octave frequency bands as a function of the angle from the jet axis in a plane perpendicular to that of the plate. Comparison of experiment and theory is shown in Figure 2.

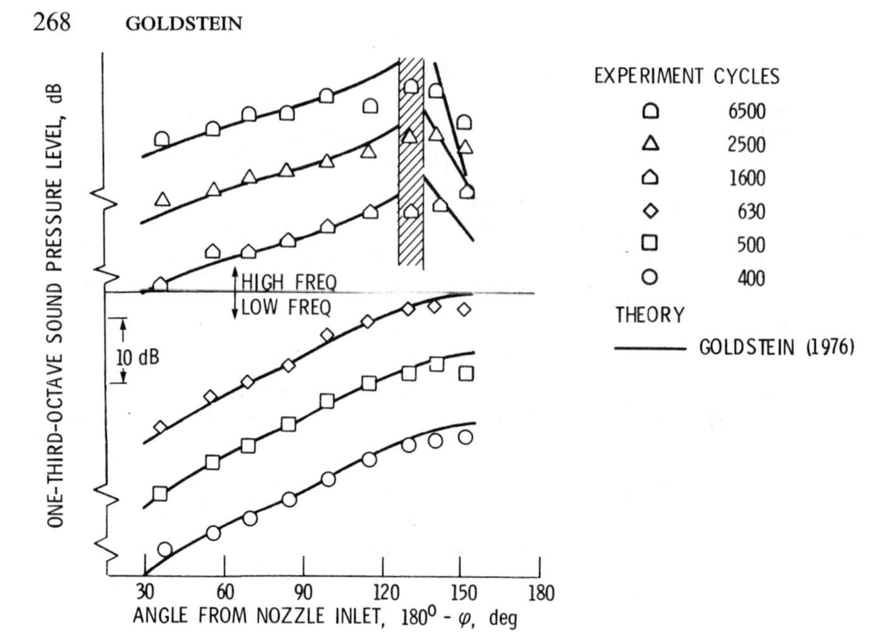

Figure 2 Comparison of causal or leading-edge Kutta condition solution with data of Olsen (1976) for $U_J = 700$ ft s^{-1}.

The top of the figure corresponds to the high-frequency limit where the instability waves are "cut off" and the issues of causality and Kutta conditions are irrelevant. However, the low-frequency *causal* solution, which is shown at the bottom, is strongly affected by the instability wave. The agreement between experiment and theory is good, but the causal and leading-edge Kutta conditions have the same low-frequency limit and one cannot conclude from this comparison which is correct. However, the bounded solution behaves quite differently in this limit and consequently does not agree with the data.

3. *Sound Generated by Turbulence Interacting with Itself: The Jet-Noise Problem*

Having achieved some success in using linear theory to calculate the sound generated by turbulence interacting with an edge, it is natural to consider using it to calculate the sound generated by turbulence interacting with itself, i.e. to deal with the problem of jet noise. We have seen that the ratio rms turbulence velocity to local mean-flow velocity is reasonably small in the region of maximum turbulence level, and the first requirement for the validity of rapid-distortion theory is therefore satisfied. However, the interaction time τ_I, which in the present context should be taken as the time for the sound generation to occur, is now equal to the decay time τ_{decay} of the turbulence, and the second requirement is no longer satisfied.

One might still attempt to introduce the same nondimensionalization as before and carry the linearization (in terms of the same small parameter) to its logical conclusion. Like the more ad hoc acoustic-analogy approach, this systematic procedure assures that all appropriate conservation laws will be satisfied and that the acoustic sources will be of the appropriate multipole order. However, it has certain advantages over the acoustic analogy in that it may apply to some physically realizable flow, which is hopefully not too different from the real turbulent flow of interest. In addition, as shown below, it provides a method for identifying acoustic sources and distinguishing acoustic and nonacoustic components of the unsteady motion. More importantly, it provides a useful tool for assessing the internal consistency of the various analyses described below.

3.1 THE BASIC EQUATION The lowest-order equations are of course the same as before, i.e. they are the compressible Rayleigh's equations. It is well known (Betchov & Criminale 1967, pp. 175–79) that the velocity components can be eliminated between these equations to obtain a single equation for the normalized first-order pressure fluctuation $\Pi_1 \equiv p_1/\rho_0 C_0^2$, where p_1 is the actual first-order pressure fluctuation, $\rho_0(\mathbf{x}_t)$ is the mean-flow density, and $C_0(\mathbf{x}_t)$ is the mean-flow sound speed. It can be written symbolically as

$$L\Pi_1 = 0, \tag{2}$$

where L denotes the third-order linear-wave operator

$$L \equiv \frac{D}{Dt}\left(\frac{D^2}{Dt^2} - \nabla \cdot C_0^2 \nabla\right) + 2C_0^2(\nabla U)\cdot\nabla\frac{\partial}{\partial x_1}, \tag{3}$$

and $D/Dt \equiv \partial/\partial t + U\ \partial/\partial x_1$ is the convective derivative based on the mean-flow velocity $U(\mathbf{x}_t)$.

Since solid boundaries are acoustically irrelevant for the turbulence self-noise problem, it is appropriate to suppose that the flow is defined over all space. Then we have seen that Equation (2) possesses a homogeneous solution, which is given by Equation (1) (actually by the fourth component of that equation). But one must also add to this gust or hydrodynamic solution any (spatially growing) instability-wave solutions, which can exist whenever the mean flow is inflectional (Betchov & Criminale 1967, pp. 104–6). It has often been argued (Crighton 1979, 1981, Liu 1974, Tam & Chen 1979, Haertig 1979, and others) that these latter solutions represent the large-scale turbulent structures, in which case it would not be unreasonable to identify the gust solution with the "fine-grained" (or relatively "fine-grained") turbulent motions.

However, there are experimentally observed large-scale motions that, on a global basis, seem to bear little resemblance to any motion that can be represented by either the gust or linear instability-wave solutions. This should come as no surprise, since we have already noted that the linearized solution can only be valid over relatively small streamwise distances.

We have seen that the gust solution produces no acoustic radiation at subsonic speeds, and the same is true for the instability waves. The analysis must therefore be carried to the next order if it is to be used to calculate radiated sound. The second-order normalized pressure fluctuation Π_2 again satisfies a third-order wave equation, but it is more convenient to work with the isentropic density fluctuation

$$\Pi \equiv \Pi_2 - \frac{k-1}{2}\Pi_1^2, \tag{4}$$

where k is the specific heat ratio. Then Π satisfies

$$L\Pi = \gamma, \tag{5}$$

which, except for the inhomogeneous source term

$$\gamma \equiv \frac{D}{Dt}\nabla\cdot\mathbf{f} - 2\frac{\partial\mathbf{f}}{\partial x_1}\cdot\nabla U, \tag{6}$$

is the same as Equation (2) for the first-order normalized pressure fluctuation.

Equation (6) is identical to the source term that would be produced by an externally applied fluctuating force per unit mass $\mathbf{f} \equiv \{f_1, f_2, f_3\}$ and might therefore be thought of as a dipole-type source, since a fluctuating force is known to produce such a source in the absence of mean flow. However, the force $\mathbf{f}$ cannot be arbitrarily specified, but is now given as a quadratic function of the first-order solutions, viz.

$$f_i \equiv \frac{\partial}{\partial x_j} u_i^{(1)} u_j^{(1)} + C_1^2 \frac{\partial}{\partial x_i}\Pi_1, \qquad (i, j = 1, 2, 3) \tag{7}$$

where $u_i^{(1)}$ are the first-order velocity fluctuations and $C_1^2 = k\mathscr{R}T_1$ is the first-order square sound speed fluctuation ($\mathscr{R}$ being the gas constant and T_1 being the first-order temperature fluctuation).

Equation (5), with some relatively minor differences, was first derived by Lilley (1974), who used the acoustic-analogy approach; it is now commonly referred to as Lilley's equation. In the present approach, it arises as the equation for the composite second-order pressure fluctuation Π with a source term γ that involves only first-order solutions. Since these solutions, which satisfy the homogeneous equation (2), have no acoustic fields at subsonic speeds (except as noted in footnote 4 below), while the second-order solution does, this provides a mechanism for identifying acoustic and nonacoustic parts of the unsteady motion.

3.2 THE SOURCES OF SOUND The second term in Equation (7) represents a dipole-type source due to the temperature fluctuations in the flow (Tester & Morfey 1976). We do not discuss it further. The first term, which is the divergence of the fluctuating (first-order) Reynolds stress $u_i^{(1)}u_j^{(1)}$, is the same source that would be produced by an externally applied fluctuating stress field and might therefore be interpreted as a quadrupole-type source.

It can be further decomposed into a number of subsources by separating the first-order solution $u_i^{(1)}$ into its gust and linear (spatially growing) instability-wave components and, as before, identifying the gust with the fine-grained turbulent motion. It is, however, important to notice that the linear instability waves, which grow without bound on a parallel mean flow, will ultimately produce an unbounded source term in Equation (5) when these waves are included in the first-order solution. It would then be inappropriate to use Equation (5) to calculate the acoustic field, since it is its global, and not its local, solutions that must be used in such a calculation.[3] However, the real flow is only locally parallel, and the slowly varying (rather than the parallel-flow) approximation should be used to represent the instability waves,[4] as was done by Crighton & Gaster (1976), Tam & Morris (1980), and others. Then, since the local growth rate of the instability wave varies with the thickness of the jet or shear layer (it first increases, reaches a maximum, and then becomes negative as the thickness increases), the source term in Equation (5) will remain bounded, and the first term in **f** should then describe the sound generation due to the following types of interactions:

1. Linear instability wave–fine-grained turbulence.
2. Linear instability wave–linear instability wave.
3. Fine-grained turbulence–fine-grained turbulence.

However, this list may be incomplete or even inappropriate, since there are other types of large-scale motions in the jet that are not globally representable by either the gust or instability wave. In fact, Hussain &

[3] A more subtle, but not entirely unrelated, difficulty occurs as a result of the secular terms that can appear in the source function γ because the formal expansion becomes nonuniformly valid in the streamwise direction. They can be eliminated by using singular perturbation techniques such as the method of multiple scales, which is an extension of the slowly varying approximation (Nayfeh 1973, pp. 288 ff.).

[4] It is then possible, as argued by Tam & Morris (1980), that the linear instability waves (when corrected for nonuniformities at transverse infinity) will be able to radiate sound directly even at subsonic speeds, but this point still needs further clarification.

Zaman (1981) argue that subsonic jet noise is primarily the result of a breakdown of the toroidal vortical structures (which can be viewed as finite-amplitude instability waves on the initial shear layer) through a secondary instability that leads to the development of smaller-scale azimuthal lobes on the primary structure (see Figure 3).

They point out that this breakdown occurs at the end of the potential core, which is thought to be the region of maximum sound generation (Bishop et al. 1971). However, the large-scale structures undergo considerable further breakdown and evolution in this region, especially at large Reynolds numbers, and these latter events will probably generate most of the sound. The process will then be covered by item number 3 in the preceding list, which we discuss in considerable detail below.

In any case, it is clear that this list should only be taken as an indication of the types of interactions that can occur and should not be considered to be the result of a rigorous analysis. One might then choose to ignore it entirely and argue, as was done by Balsa (1982), that one should use the experimentally observed turbulent motions to replace the first-order solutions that appear in the source term (6), which is, in effect, what is done in the acoustic-analogy approach.

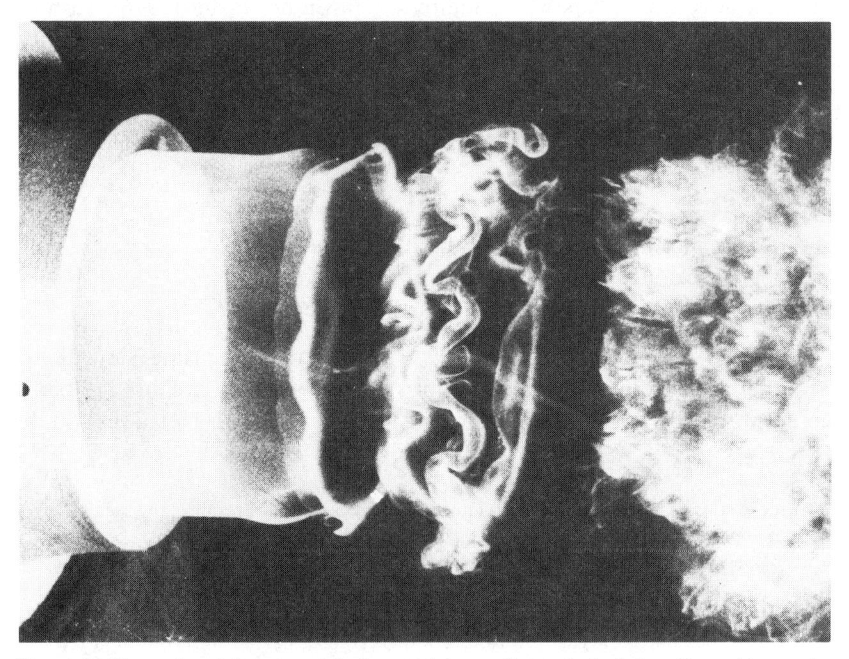

Figure 3 Vortex breakdown at end of potential core (in a relatively low-Reynolds-number jet). Courtesy of Profs. M. Van Dyke and A. Michalke.

However, it seems to me that this list cannot be dismissed that easily. Its first item, i.e. the instability wave–fine-grained turbulence interaction, has only been considered very briefly in the literature (Ffowcs Williams & Kempton 1978, Berman 1979, Liu 1974) and, to my knowledge, only limited quantitative results have been obtained. It represents the sound generation due to the shaking of the instability waves by the fine-grained turbulence and is likely to emerge as an important source mechanism in relatively low-Reynolds-number flows.

The instability wave–instability wave interaction may be related to the vortex-pairing events that occur in the initial mixing region of a high-speed jet.[5] These events are best studied experimentally by exciting the jet with an external acoustic source tuned to the most unstable frequency of the shear layer at the nozzle lip, as was done by Kibens (1980). He found that this caused the natural broadband noise of the jet to be suppressed and that most of the sound was then generated at subharmonics of the excitation frequency. By taking measurements in the near and far fields, Kibens (1980) showed that there was no Doppler shift in frequency; this indicates that the sound was generated by nonconvecting sources within the jet, whose locations he subsequently identified with the vortex-pairing locations.

3.3 THE LIGHTHILL RESULT The fine-grained turbulence–fine-grained turbulence interaction is essentially the mechanism originally considered by Lighthill (1952, 1954). In this case it is appropriate to neglect variations in retarded time across the turbulent eddies, since the time $l/C_0(1-M_c\cos\theta)$ for a sound wave to cross a turbulent eddy will be small compared with the characteristic time l/u' of the sound source at subsonic jet velocities (Goldstein 1976b, pp. 82–83). Here M_c is the "convection Mach number" of the turbulence, and θ is the angle between the downstream jet axis and the line connecting the source point and the observation point. The sound radiated by any given turbulent eddy will then be independent of that radiated by any other eddy, and each eddy will behave like a random point quadrupole source moving downstream with the "convection velocity" of the turbulence. The entire sound field of the jet can then be estimated by calculating the sound radiated by a "typical"

[5] Quadratic interactions between two-dimensional (or between axisymmetric) instability waves produce only subsonically traveling waves on a subsonic parallel flow. These waves do not radiate sound, but the quadratic interactions may produce supersonically traveling waves when the slowly varying approximation is used to represent the instability waves (Tam & Morris 1980). However, the relevant calculation is not entirely free from internal inconsistency, since one must ultimately abandon the slowly varying approximation in order to calculate the acoustic field, i.e. it is necessary to treat the slowly varying part of the first-order solution as a rapidly varying quantity in order to calculate the acoustic field from Equations (5), (6), and (7).

turbulent eddy. This was done, in fact, by Lighthill (and later corrected by Ffowcs Williams 1963) for the case of zero mean flow.

The Lighthill (1952)/Ffowcs Williams (1963) result implies that the mean square pressure $\overline{p^2}$ radiated in any proportional frequency band at a fixed source frequency Ω, where $\omega = \Omega/(1-M_c \cos \theta)$ is the actual frequency of the sound, behaves like

$$\overline{p^2} \sim \frac{f(\Omega)}{(1-M_c \cos \theta)^5}, \tag{8}$$

so that its "directivity pattern" is primarily determined by the Doppler factor $(1-M_c \cos \theta)$ raised to the -5 power. These inverse 5 Doppler factors produce a highly directional radiation pattern at high subsonic Mach numbers.

3.4 SOLUTIONS OF LILLEY'S EQUATION Solutions of Equation (5) with γ treated as a moving point source can be interpreted as corrections to the Lighthill (1952)/Ffowcs Williams (1963) result (8) that account for the effects of the nonuniform surrounding mean flow. A number of investigators (Mani 1976, Balsa 1976, 1977, Berman 1974, Goldstein 1975, 1976a, 1982, Tester & Morfey 1976, Scott 1979, Lilley 1974, etc.) have therefore calculated the acoustic radiation from point quadrupole sources moving through transversely sheared mean flows. Solutions must, in general, be obtained numerically, but relatively simple closed-form (or nearly closed-form) solutions can be obtained in the low- and high-frequency limits $\omega D/U_J \ll 1$ and $\omega D/U_J \gg 1$, respectively, where D denotes the jet diameter (see Figure 1) and U_J denotes the jet velocity. Low-frequency solutions were obtained for a round jet with arbitrary mean-velocity profile by Goldstein (1975, 1976a) and Balsa (1977).

All components of an idealized convecting quadrupole exhibit directivity patterns given by inverse Doppler factors times sines and cosines of the observation angle when there is no mean flow. The low-frequency analyses show that only the x_1-x_1 and x_1-r quadrupole components (where r is the radial coordinate) retain this property when embedded in a nonuniform mean flow. The remaining quadrupole components exhibit directivity patterns that are given by more-complex formulas, which depend on the complete mean-velocity profile and the location of the sources within the jet (Goldstein 1976a).

However, the low-frequency analyses uncovered a very surprising result—namely, that the mean flow causes certain quadrupole components to emit sound much more efficiently than they would in its absence; the mean square pressures in the absence and presence of the mean flow are, respectively, $O(\Omega^4)$ and $O(\Omega^2)$ as $\Omega \to 0$. The acoustic field of the x_1-r

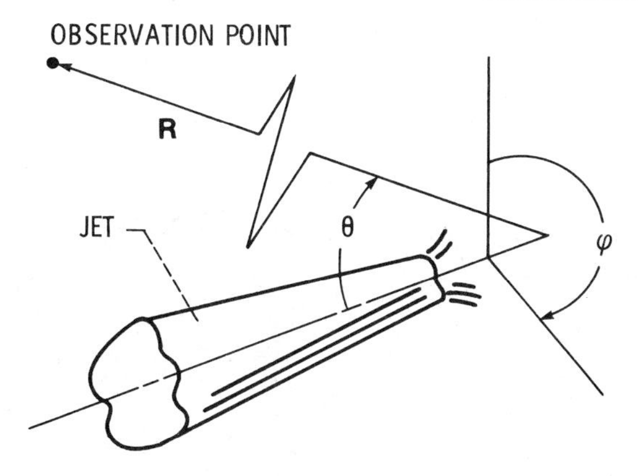

Figure 4 Coordinates for observation point.

quadrupole, which is the only one of these more efficient sources that can be expressed in simple Doppler-factor form, is proportional to the local mean-velocity gradient. It is worth noting that this source arises as much from the first member of the source term (6) as from the second, even though the former does not explicitly involve the mean-velocity gradient (Balsa 1977).

Observed low-frequency jet-noise directivity patterns will therefore depend on complex properties of the jet turbulence and mean flow that are difficult to estimate. But the mathematical results do show that they will always be more directional than Lighthill's inverse 5 Doppler factors would indicate. In fact, they show that the low-frequency sound will be more concentrated on the downstream axis than Lighthill's result implies, with the on-axis sound being produced by the quadrupoles with one axis in the streamwise direction.

The high-frequency solution, which was studied by Lilley (1974), Tester & Burrin (1974), Balsa (1976, 1977), Berman (1974), and Goldstein (1976b, 1982),[6] exhibits a "zone of silence" on the downstream jet axis. The acoustic field is exponentially small in this region, which is circumferentially asymmetric when the jet is nonaxisymmetric and/or the sound source is located off-axis. It will fill the entire range of circumferential angles when θ is sufficiently close to the downstream jet axis (see Figure 4), but will only occupy a limited range of angles (say $\varphi_{\min} < \varphi < \varphi_{\max}$) at larger values[7] of θ (say $\theta_{c_{\min}} < \theta < \theta_{c_{\max}}$) and finally it will disappear completely when $\theta > \theta_{c_{\max}}$ (see Figure 5).

[6] It was previously studied by Pao (1973) for a related equation obtained by Phillips (1960).

[7] These remarks only apply to subsonic isothermal jets with monotonic or nearly monotonic mean-velocity profiles. A host of complex interference effects can occur when these restrictions are relaxed.

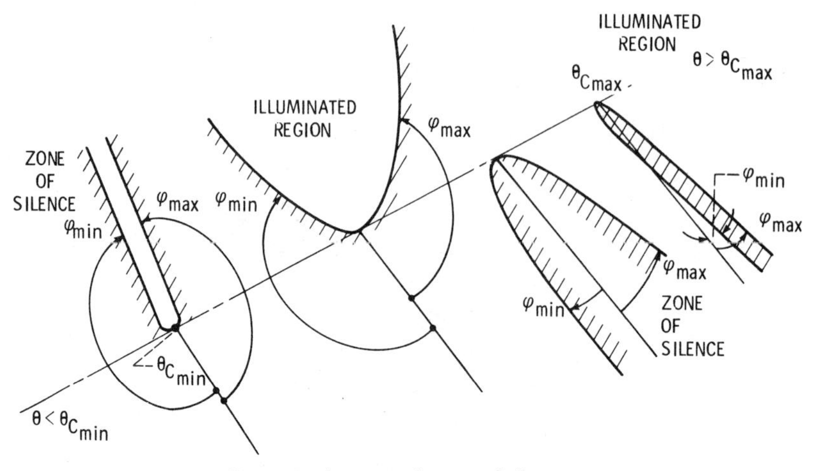

Figure 5 Asymmetric zone of silence.

As in geometric optics, the sound propagates along distinct rays in the high-frequency approximation. Only one ray can reach the observer when $\theta > \theta_{c_{max}}$, but there will be at least two rays reaching the observer when $\theta_{c_{min}} < \theta < \theta_{c_{max}}$—a direct ray and a ray reflected from the boundary of the zone of silence (Goldstein 1982). The corresponding sound waves can then interfere, but the interference term will be a rapidly oscillating function of angle and, since all acoustic measurements involve some form of spatial averaging, will probably not be experimentally observable.

The measured mean square pressure in any proportional frequency band of fixed source frequency Ω will then be the sum of the mean square pressures for each ray reaching the observer. For a convecting quadrupole source, corrected for the Ffowcs Williams (1963) effect, this is given by (Goldstein 1982)

$$\overline{p^2} \propto \frac{\rho_\infty^2 \Omega^5 \left| \sum_{i,j=1}^{3} Q_{ij} v_i v_j \right|^2}{(4\pi R C_\infty C_0)^2 (1 - M \cos\theta)^2 (1 - M_c \cos\theta)^5} \Delta(\varphi), \tag{9}$$

where R is the distance between the source point and the observation point shown in Figure 4, ρ_∞ and C_∞ are the density and sound speeds at infinity, and M is the Mach number based on the mean flow at the source location and the speed of sound at infinity. In addition, Q_{ij} denote the relative quadrupole strengths,

$$v_1 = q_0 \cos\lambda, \qquad v_2 = q_0 \sin\lambda, \qquad v_3 = \cos\theta, \tag{10}$$

$$q_0 = \sqrt{\left(\frac{1 - M\cos\theta}{C_0/C_\infty}\right)^2 - \cos^2\theta}, \tag{11}$$

λ is the initial circumferential angle made by the acoustic ray associated with Equation (9), and Δ denotes a "circumferential directivity factor" that depends on the circumferential observation angle φ, the location of the sound source within the jet, and the mean-velocity and temperature profiles of the jet.

Equation (9) is an *exact high-frequency result* that applies to jets of any cross section and with any transverse mean-velocity and temperature profiles, but λ and Δ must be calculated by solving a second-order ordinary differential equation in the general case. However, they are given by relatively simple analytic formulas (Goldstein 1982) for off-axis sources at arbitrary locations in a circular jet with arbitrary velocity and temperature profiles.

The circumferential directivity factor can be used to study the effects of nonaxisymmetric jet velocity and temperature profiles in reducing jet noise below the flight path of a jet aircraft, which is of considerable technological importance (von Glahn & Goodykoontz 1980). However, our primary interest here is in the azimuthal directivity pattern, which is unaffected by this factor.

Since the local mean-flow Mach number and the turbulence-convection Mach number are usually not very different in the regions of peak turbulence, Equation (9) shows that the inverse-Doppler-factor exponent is increased from 5 to 7 in the high-frequency limit, which by itself would cause the high-frequency sound (like the low-frequency sound) to be more directional than Lighthill's equation (8) would predict. If, however, following Balsa (1977), the quadrupole is assumed to be isotropic so that

$$Q_{ij} = \delta_{ij} Q_0, \qquad (i, j = 1, 2, 3)$$

where δ_{ij} is the Kronecker delta, it follows from Equations (10) and (11) that

$$\left| \sum_{i,j=1}^{3} Q_{ij} v_i v_j \right|^2 = \left(\frac{1 - M \cos \theta}{C_0 / C_\infty} \right)^4 Q_0^2,$$

which more than compensates for the additional two Doppler factors in the denominator of Equation (9) and produces a net azimuthal directivity pattern that is given by three inverse Doppler factors.

3.5 SOUND GENERATION DUE TO STREAMWISE VARIATIONS IN MEAN FLOW The expansion of Section 3.1 can be formally continued to the third order. At this stage, interactions between the first-order perturbation solution and the streamwise variations in the mean flow will appear in the source term. [Recall that the ratio of the cross-stream to streamwise components of the mean-flow velocity is $O((u'/U)^2)$, while the first-order solution is $O(u'/U)$.] Then, by decomposing the first-order solution into its gust and instability-wave components and making the connection between the gust and fine-grained turbulence that we discussed above, we infer that the source term now describes sound generation due to (*a*) the fine-grained turbulence interacting with streamwise variations (i.e. spreading) of the mean flow and (*b*) the instability waves interacting with the streamwise mean-flow variations. The first mechanism has, to my knowledge, not yet been considered in the literature. The second mechanism has (to some degree) already been incorporated into the expansion by using the slowly varying approximation to describe the instability waves (see Section 3.2 above). It has been analyzed by Crow & Champagne (1971), Ffowcs Williams & Kempton (1978), Huerre & Crighton (1983), Liu (1974) and Tam & Morris (1980). All of these analyses are subject to the inconsistency noted in footnote 5 above. Moreover, Tam & Morris (1980) ultimately conclude that this source is not important at subsonic speeds, consistent with the findings of Moore (1977), who studied the phenomenon experimentally by artificially exciting a jet under conditions that tended to minimize vortex pairing. Unlike Kibens (1980), Moore found that the broadband noise was usually increased rather than suppressed by the external excitation. He concluded that the instability wave, while not radiating noise directly, acted as a conduit through which energy could be transferred to the small-scale turbulent motion.

3.6 COMPARISON OF EXPERIMENT AND THEORY Figure 6 is a plot of the sound radiated by a high-Reynolds-number turbulent airjet in one-third-octave frequency bands at fixed source frequencies as a function of azimuthal angle measured from the downstream jet axis (Olsen & Friedmann 1974). The jet velocity was 994 ft s^{-1}. The data indicated by the

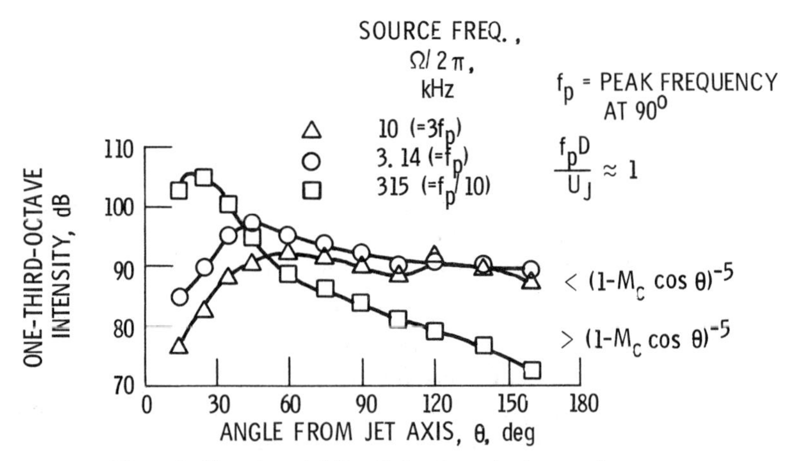

Figure 6 Experimental directivity at constant source frequency.

open circles coincide with the peak frequency of the jet noise at 90° from the jet axis, which corresponds to a Strouhal number $\Omega D/2\pi U_J$ of unity. The squares represent low-frequency data corresponding to one tenth of this value, and the triangles represent relatively high-frequency data corresponding to about three times the peak frequency.

The low-frequency data are more directional than Lighthill's equation (8) would indicate. This may be due to the effect of the surrounding mean flow, which, as we have seen, causes the low-frequency sound to be more concentrated on the downstream jet axis, but (as already noted) it is necessary to make specific assumptions about source locations, relative quadrupole strengths, and mean profile shapes before explicit calculations can be made. This was, in fact, done by Mani[8] (1976), who assumed the sound source to be located on-axis in a slug flow velocity profile to simplify his calculation. However, this model effectively precludes the exceptionally efficient low-frequency sources discussed above. Mani (1976) also introduced a specific assumption about the relative quadrupole strengths, which he attempted to justify at least partially by evoking an analysis due to Ribner (1969).

His calculations are compared with data taken by Lush (1971) in Figure 7, which is a plot of the sound radiated in a one-third-octave frequency band at a very low constant source Strouhal number of 0.03 and at three different jet velocities. The theoretical curves are adjusted to pass through the data at 60° from the downstream jet axis, rather than at 90° as is usually done.[9] The agreement appears to be fairly good, but it is probably necessary to test the sensitivity of the analysis to its numerous assumptions before definite conclusions can be drawn.

The remaining curves in Figure 6 (f_p and $3f_p$) exhibit a "zone of silence" on the downstream jet axis, as predicted by the high-frequency solution, and are less directional outside this zone than Lighthill's five inverse Doppler factors [see Equation (8)] would indicate. The latter effect could be due to the cancellation between the various components of the quadrupole source that occurs in the high-frequency solution (9) when this source is assumed to be isotropic. Then, as we have seen, the one-third-octave band pressure fluctuations will vary like three inverse Doppler factors.

Figure 8 is a plot of the jet noise radiated in one-third-octave frequency bands at constant source frequency as a function of angle measured from the downstream jet axis. The measurements, which are indicated by the

[8] Mani's analysis is not restricted to low frequencies, and he did not explicitly take the low-frequency limit of his result. But he did carry out numerical calculations at a very low frequency, and we consider only these results.

[9] The usual argument is that the sound is unaffected by both source-convection and mean-flow effects at 90°.

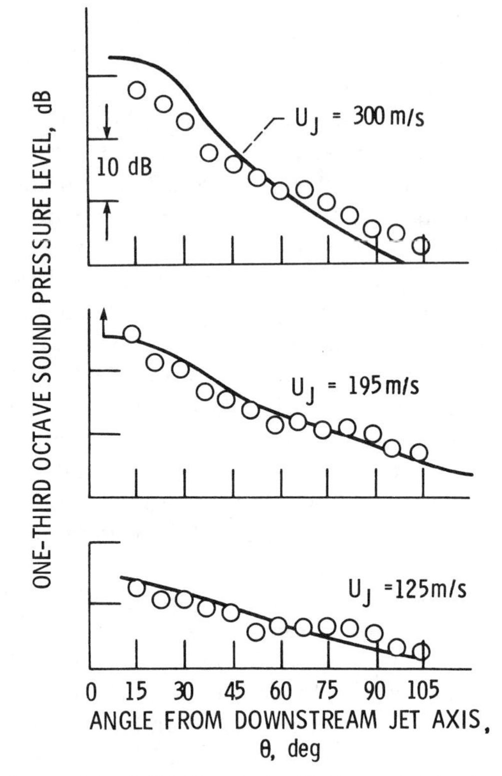

Figure 7 Comparisons of Mani's (1976) analysis with the one-third-octave directivity data of Lush (1971) for source Strouhal number $\Omega D/2\pi U_J = 0.03$.

open symbols, are due to Olsen & Friedmann (1974). They correspond to the peak frequency of the sound radiated at 90° to the jet axis (i.e. to a source Strouhal number of unity) but are taken at three different subsonic jet velocities. The Reynolds number is quite high in these experiments.

The solid curves are obtained by putting three inverse Doppler factors through the data at 90° to the jet axis. The turbulence-convection Mach number M_c is taken to be 0.62 times the jet exit Mach number based on the speed of sound at infinity, which is the value usually recommended by experimentalists. [The dashed curve is the result of using the five inverse Doppler factors implied by Equation (8).]

Although the agreement between experiment and theory is remarkably good, one might feel somewhat uncomfortable about extending the high-frequency solution to such low Strouhal numbers. However, Scott (1979) compared the exact and high-frequency solutions for fixed quadrupoles in a linear shear flow and found the results to be in close agreement, even at a

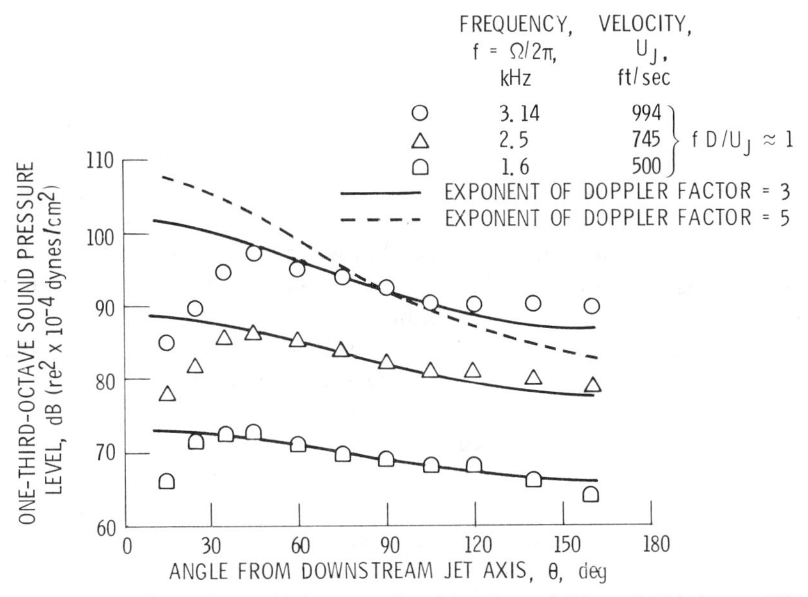

Figure 8 Experimental one-third-octave directivity data of Olsen & Friedmann (1974) plotted at constant source frequency Ω, where $\Omega/2\pi$ is equal to the peak frequency of the spectrum at 90° from the jet axis.

Strouhal number of unity. Moreover, the peak-frequency sound is believed to be generated at the end of the potential core, where the relevant length scale for computing the Strouhal number is $2D$ rather than D (see Figure 1).

There are, of course, other possible explanations for the disagreement between Lighthill's theory and the high-Reynolds-number experiments. First, the decreased directivity at the higher frequencies could be due to the scattering of the sound by the turbulence, an effect that will certainly be more important at the higher frequencies and that will tend to make the sound field less directional. It could also be due to variations in retarded time across the source (i.e. to source-coherence effects), which were neglected in the point-source models described above and which, for a fixed source size, would also be more important at the higher frequencies. Ribner (1962) and Ffowcs Williams (1963) analyzed this phenomenon and showed that it tends to diminish the increased directionality that results from source convection. However, it is important to note that their analyses are based on an assumed source model.

4. *Extensions to More-Complex Turbulent Flows*

The results of the preceding section suggest that the high-frequency solution may remain valid at frequencies that are low enough to include the most energetic portion of the jet-noise spectrum. This might remain true for other more complex turbulent flows. We may therefore be able to calculate the sound generated in such flows by finding the high-frequency solution for a point quadrupole source moving through the appropriate mean flow. Durbin (1983a) recently obtained such a solution for a completely general mean flow. His final equation is somewhat formal in that (*a*) it involves quantities that depend on the solution of a system of ordinary differential equations (which must, in general, be obtained numerically), and (*b*) it does not explicitly account for the presence of the caustics that, as we have seen, can occur in the high-frequency limit. However, the latter can easily be incorporated into the theory (Goldstein 1982), which is, in any case, explicit enough to provide considerable information about the high-frequency sound generation in these more complex flows.

Durbin (1983b) used his analysis to study the effect of mean-flow divergence on subsonic jet noise (see Figure 4). He showed that even small jet spreading will eliminate the "zone of silence" that occurs in the axisymmetric parallel-flow model, in the sense that the radiated sound is no longer exponentially small in that region but is merely greatly reduced due to a strong divergence of the acoustic rays. Unlike the parallel-flow calculation, his results indicate that the shallow-angle acoustic field is nonzero and exhibits a directivity that is independent of frequency in the limit as $\omega \rightarrow \infty$. It is worth noting that the experimental "zone of silence" directivity pattern only becomes frequency independent at very high frequencies.

5. *Supersonic Flows*

Up to now our remarks have been confined to subsonic flows, though much of what has been said also applies to supersonic flows. However, some new phenomena also come into play with supersonic flows. First, the linear gust solution (1) will no longer decay exponentially at large distances from a jet but will involve propagating wave components. These waves probably correspond to the so-called Mach wave radiation observed in the initial mixing region of supersonic jets by Lowson & Ollerhead (1968), Dosanjh & Yu (1968), and many others. They are the leading-edge shock waves (or bow waves) of the supersonically moving eddies. It is generally agreed that this phenomenon is not very important at moderate supersonic Mach numbers.

Second, the instability waves can achieve supersonic phase speed and generate sound directly at very high Mach numbers. But Tam & Morris (1980) argue that the instability waves in a slowly varying mean flow involve supersonically traveling components that can radiate significant sound even at moderate supersonic Mach numbers. This is consistent with the low-Reynolds-number supersonic jet-noise experiments of McLaughlin et al. (1975) and Troutt & McLaughlin (1982), which imply that the majority of the noise is generically related to the relatively slow growth and decay of organized wavelike structures. Whether this behavior persists at high Reynolds numbers is still an open question.

Finally, both the fine-grained turbulence and the instability waves can interact with the shock waves, which can now appear in the flow, to generate noise in a very efficient manner. This is usually referred to as the "shock-associated noise" (or shock "screetch" if there is feedback to the nozzle lip). Harper-Bourne & Fisher (1973) suggested that turbulent eddies can retain their identities long enough to pass through several shock waves and that the resulting coherence between the noise sources will have a dominant influence on the radiated sound. This will be even more true if the sound is generated by interactions between instability waves and the shock structure (Tam & Tanna 1982). Shock-associated noise is still being actively studied by a number of investigators.

ACKNOWLEDGEMENT

The author would like to thank Dr. P. A. Durbin of NASA Lewis and Profs. T. F. Balsa of the University of Arizona, C. K. W. Tam of Florida State University, and A. K. M. F. Hussain of the University of Houston for their helpful comments on this review.

Literature Cited

Ahuja, K. K., Bushell, K. W. 1973. An experimental study of subsonic jet noise and comparison with theory. *J. Sound Vib.* 30: 317–41

Balsa, T. F. 1976. The far field of high frequency convected singularities in sheared flows, with applications to jet-noise prediction. *J. Fluid Mech.* 74: 193–208

Balsa, T. F. 1977. The acoustic field of sources in shear flow with application to jet noise: convective amplification. *J. Fluid Mech.* 79: 33–47

Balsa, T. F. 1982. *A review of recent developments in the theory of jet noise.* Presented at Ann. Meet. Soc. Eng. Sci., 19th, Rolla, Mo.

Berman, C. H. 1974. Noise from nonuniform turbulent flows. *AIAA Pap. 72-2, Aerosp. Sci. Meet., 12th, Washington DC*

Berman, C. H. 1979. Some analytical considerations in jet noise prediction. In *Mechanics of Sound Generation in Flows,* ed. E.-A. Müller, pp. 160–66. Berlin: Springer. 300 pp.

Betchov, R., Criminale, W. O. 1967. *Stability of Parallel Flows.* New York: Academic. 330 pp.

Bishop, K. A., Ffowcs Williams, J. E., Smith, W. 1971. On the noise sources of the unsuppressed high-speed jet. *J. Fluid Mech.* 50: 21–32

Bradshaw, P., Ferriss, D. H., Johnson, R. F. 1964. Turbulence in the noise-producing region of a circular jet. *J. Fluid Mech.* 19: 591–624

Case, K. M. 1960. Stability of an idealized atmosphere. I. Discussion of results. *Phys. Fluids* 3: 149–54

Crighton, D. G. 1979. Why do the acoustics and the dynamics of a hypothetical mean flow bear on the issue of sound generation by turbulence? In *Mechanics of Sound Generation in Flows,* ed. E.-A. Müller, pp. 1–11. Berlin: Springer. 300 pp.

Crighton, D. G. 1981. Acoustics as a branch of fluid mechanics. *J. Fluid Mech.* 106: 261–98

Crighton, D. G., Gaster, M. 1976. Stability of slowly diverging jet flow. *J. Fluid Mech.* 77: 397–413

Crighton, D. G., Leppington, F. G. 1974. Radiation properties of the semi-infinite vortex sheet: the initial value problem. *J. Fluid Mech.* 64: 393–414

Crow, S. C., Champagne, F. H. 1971. Orderly structure in jet turbulence. *J. Fluid Mech.* 48:547–92

Dosanjh, D. S., Yu, J. C. 1968. Noise from underexpanded axisymmetric jet flows using radial jet flow impingement. In *Aerodynamic Noise*, ed. H. S. Ribner, pp. 169–88. Univ. Toronto Press. 438 pp.

Dowling, A. P., Ffowcs Williams, J. E., Goldstein, M. E. 1978. Sound production in a moving stream. *Philos. Trans. R. Soc. London Ser. B* 288:321–49

Durbin, P. A. 1983a. *J. Sound Vib.* 91: In press

Durbin, P. A. 1983b. *J. Sound Vib.* 91: In press

Ffowcs Williams, J. E. 1963. The noise from turbulence convected at high speeds. *Philos. Trans. R. Soc. London Ser. B* 255:469–503

Ffowcs Williams, J. E. 1977. Aeroacoustics. *Ann. Rev. Fluid Mech.* 9:447–68

Ffowcs Williams, J. E., Hall, L. H. 1970. Aerodynamic sound generation by turbulent flow in the vicinity of a scattering half-plane. *J. Fluid Mech.* 40:657–70

Ffowcs Williams, J. E., Kempton, A. J. 1978. The noise from large-scale structures of a jet. *J. Fluid Mech.* 84:673–94

Goldstein, M. E. 1975. The low frequency sound from multipole sources in axisymmetric shear flows, with applications to jet noise. *J. Fluid Mech.* 70:595–604

Goldstein, M. E. 1976a. The low frequency sound from multipole sources in axisymmetric shear flows. Part 2. *J. Fluid Mech.* 75:17–28

Goldstein, M. E. 1976b. *Aeroacoustics*. New York: McGraw-Hill. 293 pp.

Goldstein, M. E. 1978. Characteristics of the unsteady motion on transversely sheared mean flows. *J. Fluid Mech.* 84:305–29

Goldstein, M. E. 1979. Scattering and distortion of the unsteady motion on transversely sheared mean flows. *J. Fluid Mech.* 91:601–32

Goldstein, M. E. 1981. The coupling between flow instabilities and incident disturbances at a leading edge. *J. Fluid Mech.* 104:217–46

Goldstein, M. E. 1982. High frequency sound emission from moving point multipole sources embedded in arbitrary transversely sheared mean flows. *J. Sound Vib.* 80:499–521

Haertig, J. 1979. Theoretical and experimental study of wavelike disturbances in round free jet with emphasis being placed on orderly structure. In *Mechanics of Sound Generation in Flows*, ed. E.-A. Müller, pp. 167–73. Berlin: Springer. 300 pp.

Harper-Bourne, M., Fisher, J. J. 1973. The noise from shock waves in supersonic jets. In *Noise Mechanisms*, pp. 11.1–11.13. *AGARD CP-131*. 345 pp.

Howe, M. S. 1975. Contributions to the theory of aerodynamic sound with application to excess jet noise and the theory of the flute. *J. Fluid Mech.* 71:625–73

Huerre, P., Crighton, D. G. 1983. Sound generation by instability waves in a low Mach number flow. *AIAA Pap. 83-0661, Aeroacoustics Conf., 8th, Atlanta, Ga.*

Hunt, J. C. R. 1973. A theory of turbulent flow round two-dimensional bluff bodies. *J. Fluid Mech.* 61:625–706

Hussain, A. K. M. F., Husain, Z. D. 1980. Turbulence structure in the axisymmetric free mixing layer. *AIAA J.* 18:1462–69

Hussain, A. K. M. F., Zaman, K. B. M. Q. 1981. The preferred mode of the axisymmetric jet. *J. Fluid Mech.* 110:39–71

Kibens, V. 1980. Discrete noise spectrum generated by an acoustically excited jet. *AIAA J.* 18:434–41

Lighthill, M. J. 1952. On sound generated aerodynamically: I.—General theory. *Proc. R. Soc. London Ser. A* 211:564–87

Lighthill, M. J. 1954. On sound generated aerodynamically: II.—Turbulence as a source of sound. *Proc. R. Soc. London Ser. A* 222:1–32

Lilley, G. M. 1974. On the noise from jets. In *Noise Mechanisms*, pp. 13.1–13.12. *AGARD CP-131*. 345 pp.

Liu, J. T. C. 1974. Developing large-scale wavelike eddites and near-jet noise field. *J. Fluid Mech.* 62:437–64

Lowson, M. V., Ollerhead, J. B. 1968. Visualization of noise from cold supersonic jets. *J. Acoust. Soc. Am.* 44:624–30

Lush, P. A. 1971. Measurement of subsonic jet noise and comparison with theory. *J. Fluid Mech.* 46:477–500

Mani, R. 1976. Influence of jet flow on jet noise. *J. Fluid Mech.* 73:753–93

McLaughlin, D. K., Morrison, G. L., Troutt, T. R. 1975. Experiments on the instability waves in a supersonic jet and their acoustic radiation. *J. Fluid Mech.* 69:73–95

Moffatt, H. K. 1981. Some developments in the theory of turbulence. *J. Fluid Mech.* 106:27–47

Möhring, W. 1976. Über Schallwellen in Scherströmungen. *Fortshritte der Akustik*, pp. 543–46. *DAGA 76*. VDI-Verlag.

Moore, C. J. 1977. The role of shear layer instability waves in jet exhaust noise. *J. Fluid Mech.* 80:321–67

Nayfeh, A. 1973. *Perturbation Methods*. New York: Wiley. 425 pp.

Olsen, W. A. 1976. Noise generated by impingement of turbulent flow on airfoils of varied chord, cylinders and other flow obstructions. *NASA Tech. Memo. X-*

73464, Lewis Res. Cent., Cleveland, Ohio

Olsen, W., Friedmann, R. 1974. Jet noise from co-axial nozzles over a wide range of geometric and flow parameters. *AIAA Pap. 74-43, Aerosp. Sci. Meet., 12th, Washington DC*

Pao, S. P. 1973. Aerodynamic noise emission from turbulent shear layers. *J. Fluid Mech.* 59:451–80

Phillips, O. M. 1960. On the generation of sound by supersonic turbulent shear layers. *J. Fluid Mech.* 9:1–28

Ribner, H. S. 1962. Aerodynamic sound from field dilations. *UTIA-86*, Univ. Toronto, Ontario

Ribner, H. S. 1969. Quadrupole correlations governing the pattern of jet noise. *J. Fluid Mech.* 38:1–24

Ribner, H. S. 1981. Perspectives on jet noise. *AIAA Pap. 81-0428, Aerosp. Sci. Meet., 19th, St. Louis*

Rienstra, S. W. 1979. *Edge influence on the response of shear layers to acoustic forcing.* PhD thesis. Tech. Hogeschool, Eindhoven. 116 pp.

Rockwell, D. 1983. Oscillations of impinging shear layers. *AIAA J.* 21:645–64

Scott, J. N. 1979. Propagation of sound waves through linear shear layers. *AIAA J.* 17:237–44

Tam, C. K. W., Chen, K. C. 1979. A statistical model of turbulence in two-dimensional mixing layers. *J. Fluid Mech.* 92:303–26

Tam, C. K. W., Morris, P. J. 1980. The radiation of sound by the instability waves of a compressible plane turbulent shear layer. *J. Fluid Mech.* 98:349–81

Tam, C. K. W., Tanna, H. K. 1982. Shock-associated noise of supersonic jets from convergent-divergent nozzles. *J. Sound Vib.* 81:337–58

Tennekes, H., Lumley, J. L. 1972. *A First Course in Turbulence.* Cambridge, Mass: MIT Press. 300 pp.

Tester, B. J., Burrin, R. H. 1974. On sound radiation from sources in parallel sheared jet flows. In *The Generation and Radiation of Supersonic Jet Exhaust Noise: Studies of Jet Noise Generation and Radiation, Turbulent Structure and Laser Velocimetry*, ed. H. E. Plumblee Jr., p. 59–87. *AFAPL-TR-74-24*, Lockheed-Georgia Co., Marietta, Ga. 190 pp.

Tester, B. J., Morfey, C. L. 1976. Developments in jet noise modelling—theoretical predictions and comparisons with measured data. *J. Sound Vib.* 46:79–103

Troutt, T. R., McLaughlin, D. K. 1982. Experiments on the flow and acoustic properties of a moderate-Reynolds-number supersonic jet. *J. Fluid Mech.* 116:123–56

von Glahn, U., Goodykoontz, J. 1980. Noise suppression due to annulus shaping as a conventional coaxial nozzle. *NASA Tech. Memo. 81461*, Lewis Res. Cent., Cleveland, Ohio

Yates, J. E., Sandri, G. 1975. Bernoulli enthalpy: A fundamental concept in the theory of sound. *AIAA Pap. 75-439, Aeroacoust. Conf., 2nd, Hampton, Va.*

Ann. Rev. Fluid Mech. 1984. 16: 287–309

COMPUTER-EXTENDED SERIES

Milton Van Dyke

Departments of Mechanical Engineering and of Aeronautics and Astronautics, Stanford University, Stanford, California 94305

1. INTRODUCTION

Fluid mechanicians around the world are increasingly making use of the technique of extending a perturbation series to high order by delegating the rapidly mounting labor to a computer. It is appropriate now to review this subject, because it is sufficiently young that the literature, though substantial, is still of manageable proportions. Almost every paper is of interest, if only for providing a salutary warning for the further development of the method.

Computer-extended series cannot compete with purely numerical methods—finite differences and finite elements—for complicated shapes. For simple geometries, however, such as spheres, paraboloids, and sinusoidal walls, they provide results that cannot yet be achieved in any other way. Another advantage is that a single computer run yields the solution for a range of the expansion quantity. The method is also economical of computer time: on a contemporary machine a few minutes will produce hundreds of terms for a linear problem governed, for example, by the Laplace equation, and dozens of terms for the Navier-Stokes equations. (To be sure, this may be preceded by days of programming and debugging, and followed by weeks of analyzing the results!)

We who study fluid mechanics are the beneficiaries of earlier work by two groups of colleagues. From those who study celestial mechanics we inherit a tradition of extending perturbation series to high order, formerly by hand, as in the French astronomer Delaunay's heroic twenty-year calculation of the theory of the motion of the moon (Delaunay 1867), and more recently by computer, as in the twenty-hour extension and correction of Delaunay's lifework by Deprit et al. (1970). Then from physicists who study critical phenomena in crystal lattices we borrow a number of ingenious devices for analyzing the structure of a function that is known only through a finite

0066-4189/84/0115-0287$02.00

number of terms of its Taylor series (Baker 1965, Gaunt & Guttmann 1974, Pearce 1978). However, we must to some degree modify and extend their methods, because they are invariably seeking a singularity on the positive axis, whereas we often find no singularities there short of infinity.

The techniques of computer extension are of as much interest as the various specific applications that have been made to fluid flow. We therefore begin with a brief survey, in Sections 2 and 3, of its three facets: calculating higher coefficients by machine, analyzing them to understand the structure of the solution, and using that information to improve the utility of the series. After that, we review in Sections 4–6 the applications that have been made to compressible, viscous, and potential flows.

2. MACHINE COMPUTATION OF COEFFICIENTS

The author had thought (Van Dyke 1976) that in fluid mechanics a computer was programmed to extend a series for the first time in 1957. However, Andrew Van Tuyl has pointed out that a full five years earlier Munk & Rawling (1952) wired the IBM 604 Calculating Punch to compute the coefficients in the Janzen-Rayleigh expansion for subsonic flow past a circle. The process was started by hand early in this century by Janzen in Germany and Rayleigh in England, who independently considered the first effects of compressibility by adding to the classical incompressible velocity potential a correction proportional to the square of the free-stream Mach number M:

$$\phi = \left(r+\frac{1}{r}\right)\cos\theta + M^2\left[\left(\frac{13}{12}\frac{1}{r}-\frac{1}{2}\frac{1}{r^3}+\frac{1}{12}\frac{1}{r^5}\right)\cos\theta + \left(-\frac{1}{4}\frac{1}{r}+\frac{1}{12}\frac{1}{r^3}\right)\cos 3\theta\right]. \tag{1}$$

Imai added the term in M^4 in 1938, and the term in M^6 in 1941. In 1952 Munk & Rawling needed eight hours handling and machine time to compute the next term. (In 1978 the same calculation took 0.3 seconds on an IBM 3032 computer.) For the maximum speed on the circle, their solution for an adiabatic exponent $\gamma = 7/5$ gives

$$q_{\max} \doteq 2+1.16667M^2+2.57833M^4+7.51464M^6+25.61M^8. \tag{2}$$

Series Based on Closed-Form Solutions

This early result is in many respects typical of the largest class of computer-extended series. It is based on a closed-form solution, so that each subsequent term is also found in closed form. The coefficients are rational fractions, but are often computed in decimal form; here the first three can, with decreasing confidence, be recognized as 2, 7/6, and 1547/600. They would, however, be irrational if γ were taken as irrational. The perturbation is a regular, not a singular one, so that the series involves no logarithms or nonintegral powers of the perturbation parameter M^2. This is by construction merely a formal power series, but we anticipate that it is in fact the Taylor-series expansion of the true solution, having a nonzero radius of convergence; and this has exceptionally in this case been proved (Frankl & Keldysh 1934). The signs approach a simple pattern—here all positive from the outset. The coefficients vary smoothly, in this case increasing to suggest a radius of convergence less than unity.

To put a problem of this class on the computer, it is imperative first to calculate several terms of the series by hand, from which to deduce the form of the general term and to use later as a check in debugging. With the general term identified, substituting it into the governing equations leads to recursion relations for the coefficients. Then a computer program can be written in a language such as FORTRAN to handle the arithmetic of calculating them in succession. It will consist primarily of nested DO-loops. How deeply they are nested depends on the number of summations in the general term and the degree of nonlinearity of the equations. The time required to compute the Nth term grows asymptotically as N^D, where D is the number of nested DO-loops in each cycle of the program.

When the coefficients are rational, one might suppose it worthwhile to compute them exactly. This has been done in several problems (Aoki 1980, Lucas 1972, Meiron et al. 1982, etc.), but at the cost of fewer terms and increased computing time. In any case, the integers involved quickly become so huge that decimal equivalents are needed for comprehension. For example, the fifth coefficient in Equation (2) is already 1,015,683,271/39,690,000. Of course, significant figures are lost when floating-point arithmetic is used, typically one figure every other term, as is the case in Equation (2). Most investigators estimate the accuracy by repeating the calculations in single- and double-precision arithmetic. Alternatively, H. Takagi (private communication) suggests running the program again with the perturbation quantity altered by a scale factor (not a power of two). Such a factor must in any case often be supplied to prevent under- or overflow.

Only seldom, in very simple or linear problems, can one compute as many terms as he would like. Extension of a series is limited by either storage, accuracy, or cost. Each of these limitations can be mitigated; but in a well-written program all three appear at about the same point, so that it is clear where to stop.

Singular Perturbations

Nearly all computer extensions are regular perturbations. No one has yet programmed a computer to apply the method of matched asymptotic expansions to a boundary layer, or multiple scales to a slowly modulated oscillation. However, Aoki (1980) points out that in extending to eighth order the expansion for plane periodic standing water waves he has treated a singular perturbation. It is, however, a "mild" singular perturbation, where one need only recognize that the period of oscillation is unknown, and expand it also in series, in order to avoid secular terms. This process was applied earlier to traveling waves by Fenton (1972) and Schwartz (1972), and then to hydrodynamic stability by Herbert (1980) and von Kerczek (1982).

Computer Algebra

In some problems it is the algebra involved in forming recursion relations that soon becomes intractable, rather than the subsequent arithmetic. Then one can resort to symbol-manipulation languages such as FORMAC, REDUCE, and MACSYMA. Fitch (1979) surveys several applications in fluid mechanics.

Unfortunately, this process suffers from so-called *intermediate expression swell*: what goes into the computer and what comes out are of manageable size, but inside it balloons out of all proportion. Consequently, whereas a FORTRAN program will typically extend a few hand-computed terms to 20 or 30, Atherton & Homsy (1973) expanded long waves at an interface to 2 terms by hand and were then able to compute only two more using REDUCE 2. For standing waves, Aoki (1980) used 70 minutes of computer time to calculate 8 terms using REDUCE 2; by contrast, for the same problem Schwartz & Whitney (1981) found 25 terms in 4 minutes using a FORTRAN program. Likewise, Hui & Tenti (1982) computed just 11 terms of the Stokes expansion for progressive waves in deep water using MACSYMA, whereas Schwartz (1972, 1974a) had earlier found 117 terms using FORTRAN.

Series Based on Numerical Solutions

In another class of computer-extended series the unperturbed problem cannot be solved in closed form, but must be integrated numerically. Then each higher term must be treated similarly. In the simplest cases there are only two coordinates and the basic flow is self-similar, being governed by a nonlinear ordinary differential equation. A classical example is the expansion of Howarth (1938) for a laminar boundary layer with external speed $1-x/8$, which is based on the Prandtl-Blasius self-similar solution for uniform speed. The skin friction is found to be proportional to

$$T = 0.4696000 - 0.3608156x - 0.0244847x^2 - 0.0197848x^3 - 0.0131465x^4 - 0.0095650x^5 - 0.0074389x^6 - 0.0060627x^7 - 0.0051128x^8 - \cdots. \quad (3)$$

Howarth's hand integration has stood the test of time; but we have given here the more precise coefficients that Curle (1979) found by computer.

Only slightly more complicated, although it involves three coordinates, is the expansion in powers of (dimensionless) time for the boundary layer on an impulsively translated cylinder. On a circle, the skin friction is proportional to

$$T = 2.2568 \sin\theta - (3.2146 \sin 2\theta)t - (0.0917 \sin\theta + 0.6310 \sin 3\theta)t^2 + (0.0587 \sin 2\theta + 0.0278 \sin 4\theta)t^3 + (-0.0746 \sin\theta + 0.0627 \sin 3\theta + 0.1610 \sin 5\theta)t^4 + \cdots. \quad (4)$$

The ordinary differential equations for at least the first three terms here have closed-form solutions (the first coefficient is $4\pi^{-1/2}$), but thereafter it has been found simpler to solve them numerically. Cowley (1983) has thus extended this series to t^{50} by computer.

3. ANALYSIS AND IMPROVEMENT OF SERIES

We are seldom satisfied with a computer-extended series as it stands. For example, we often need to apply it beyond its radius of convergence. A few techniques for improving the utility of a power series can be applied blindly, but most benefit from some knowledge of the analytic structure of the function. For that purpose it is essential to consider the perturbation quantity in its complex plane, even though only real (and often only positive) values are of physical interest.

The Pattern of Signs

It is surprising that almost all regular perturbation series in fluid mechanics prove to have a finite radius of convergence. An infinite radius is rare, but two examples are the expansion of Lam & Rott (1960) for the skin friction on a plate in an oscillating stream, and that of Curle (1981) for the steady laminar boundary layer with exponentially increasing pressure gradient. A zero radius of convergence is also infrequent (though common in singular perturbations). Lucas (1972) encountered it in laminar flow through slowly varying channels; Witting (1975) pointed out that the expansion of Fenton (1972) for the solitary wave is only asymptotic; and Vanden Broeck et al.

(1978) have shown that expansions in powers of Froude number for two-dimensional waves are divergent.

A power series eventually displays a pattern of signs determined by the singularities on its circle of convergence. If there is only one such singularity—and this is the most common situation—it must lie on the real axis for the coefficients to be real. Then the signs are ultimately fixed if the singularity is on the positive axis and alternating if it is on the negative axis, as illustrated by

$$\frac{1}{1-\varepsilon} = 1+\varepsilon+\varepsilon^2+\varepsilon^3+\cdots, \qquad \frac{1}{1+\varepsilon} = 1-\varepsilon+\varepsilon^2-\varepsilon^3+\cdots. \tag{5}$$

Of course, the final pattern may appear only after a number of terms, as for

$$(1-\varepsilon)^{5/2} = 1-\frac{5}{2}\varepsilon+\frac{15}{8}\varepsilon^2-\frac{5}{16}\varepsilon^3-\cdots. \tag{6}$$

Any other final pattern corresponds to pairs of nearest singularities at complex-conjugate points. If there is only one pair, its angles $\pm\theta$ from the positive axis can be determined by comparing with the sign pattern of the model function

$$\begin{aligned} F(\varepsilon) &= \frac{1}{2}\left(\frac{1}{e^{i\theta}-\varepsilon}+\frac{1}{e^{-i\theta}-\varepsilon}\right) = \frac{\cos\theta-\varepsilon}{1-2\varepsilon\cos\theta+\varepsilon^2} \\ &= \cos\theta+\varepsilon\cos 2\theta+\cdots+\varepsilon^n\cos(n+1)\theta+\cdots. \end{aligned} \tag{7}$$

Evidently the sign pattern repeats with period N if θ is a rational fraction of 2π: $\theta = (M/N)\cdot 2\pi$, where M and N are integers. Otherwise the pattern is random. Seemingly random patterns have been found by Baumel et al. (1982), Brady & Acrivos (1981), and Sangani & Acrivos (1982).

It is remarkable, however, that more often than not the signs show a short pattern repeated, either exactly or with occasional slips that indicate that the angle is not precisely a simple fraction of 2π. Thus Andersen & Geer (1982) find for the van der Pol oscillator a period of seven repeated six times; and for the impulsively started circle, Cowley (1983) calculates a period of six that slips once in eight cycles. In his series for the shock wave produced by an impulsively moved piston, Reddall (1978) finds a pattern of period eight for $\gamma = 7/5$, five for $\gamma = 5/3$, and two (with slips every eleven cycles) for $\gamma = \infty$.

The Ratio Test

The distance to the nearest singularity—the radius of convergence—can be estimated using any of the standard tests. When the signs are fixed or alternating, Cauchy's ratio test is the most useful, in a graphical form devised by Domb & Sykes (1957). They observe that for the simple singularities

$$f(\varepsilon) = \sum c_n \varepsilon^n = \text{const} \begin{cases} (\varepsilon_0 \pm \varepsilon)^\alpha & \alpha \neq 0, 1, 2, \ldots \\ (\varepsilon_0 \pm \varepsilon)^\alpha \log(\varepsilon_0 \pm \varepsilon), & \alpha = 0, 1, 2, \ldots \end{cases} \tag{8}$$

the inverse ratio of the Taylor coefficients c_n is exactly linear in $1/n$:

$$\frac{c_n}{c_{n-1}} = \mp \frac{1}{\varepsilon_0} \left(1 - \frac{1+\alpha}{n}\right). \tag{9}$$

Happily, most solutions of physical problems have nearest singularities of this kind. Hence the radius of convergence ε_0 can be estimated by linear extrapolation of a plot of c_n/c_{n-1} versus $1/n$, now usually called a *Domb-Sykes plot.* At the same time, the slope indicates the exponent α. The extrapolation can be refined by fitting higher-order polynomials in $1/n$, which is conveniently carried out numerically by forming a *Neville table* (Gaunt & Guttmann 1974).

Padé Approximants

When the sign pattern is more complicated, the nearest singularities can be located by forming Padé approximants (Baker 1965). The $[M/N]$ Padé approximant is a rational fraction with numerator of degree M, and denominator of degree N and constant term 1, that when expanded agrees with the given series to degree $M+N$. As M and N increase, a Padé approximant performs an analytic continuation of the series outside its radius of convergence in ways that are still not fully understood. It is clear that it can approximate a pole by zeros of the denominator. With branch points it extracts a single-valued function by inserting branch cuts, which it simulates by lines of alternating poles and zeros.

Improvement of Series

Padé approximants are the most common method of improving a power series "blindly"—making no use of information about the structure of the function. Related techniques include Aitken's method, the transformation of Shanks (1955), continued fractions, and the method of recurrence relations of Guttmann & Joyce (1972). These are the only recourse when the series is short with a random sign pattern. For example, the eight-term expansion of Takagi (1972) for a spinning sphere converges only for Reynolds numbers less than 12, but the [3/4] and [4/3] Padé approximants are accurate past Reynolds number 60 (Van Dyke 1978a).

When enough terms are available to permit some estimate of the analytic structure, that information usually suggests more effective ways of improving the series (Van Dyke 1974). For example, in his expansion for periodic progressive water waves, Schwartz (1974a) found a simple square root on the positive real axis, indicating that the solution is double valued. He then extended the range of convergence by simply reverting the series. More often a singularity on the positive axis has physical significance, in that it signifies a breakdown of the assumptions. We then want to identify it, extract its expansion from the series multiplicatively or additively, and proceed to explore the secondary and higher confluent singularities.

When the nearest singularity is not on the positive axis, the range of convergence of the series can be enlarged by analytic continuation. The simplest situation is a singularity on the negative axis, at $\varepsilon = -\varepsilon_0$. It can be mapped away to infinity by making an Euler transformation, recasting the series in powers of $\varepsilon/(\varepsilon_0+\varepsilon)$. Similarly, a complex-conjugate pair located at distance ε_0 and angles $\pm\theta$ from the positive axis can be mapped to infinity with a generalization of the Euler transformation, recasting in powers of $\varepsilon/(\varepsilon_0^2-2\varepsilon_0\varepsilon\cos\theta+\varepsilon^2)^{1/2}$. Other transformations have been studied by Pearce (1978).

Purely asymptotic series, with zero radius of convergence, can be summed in various ways, though the significance of the result is not always clear. Fenton (1972) and Vanden Broeck et al. (1978) use the Shanks transformation, while Longuet-Higgins & Fenton (1974) find greater accuracy with Padé approximants. Other techniques are described by Buchanan (1976).

4. COMPRESSIBLE FLOWS

The Transonic Controversy

Munk & Rawling's pioneering application of the computer to subsonic flow past a circle has been extended repeatedly with the aim of resolving the *transonic controversy*: does a shock wave invariably form on a body as soon as the flow becomes locally sonic? Simasaki (1955) succeeded in carrying the Janzen-Rayleigh expansion to order M^{10} by hand; Hoffmann (1974a) extended it to M^{12} by computer; and W. C. Reynolds (unpublished) calculated two further terms. Most recently, Van Dyke & Guttmann (1983) have carried the series to 29 terms (order M^{56}), and concluded that smooth shock-free flow persists up to a free-stream Mach number 1.1 percent above the critical.

That conclusion has been cast in doubt by a complementary computation by Bollmann (1982). He solves the subsonic flow past the sinusoidal wall $y = \tau \sin x$ in the transonic small-disturbance approximation by expanding in powers of the similarity parameter $\tau(\gamma+1)(1-M^2)^{-3/2}$. Kaplan (1952) had calculated 8 terms by hand, and Bollmann wrote a FORTRAN program to compute 40 terms. Analyzing the series for maximum speed using Padé approximants, he concludes that the solution breaks down as soon as the flow becomes sonic. He attributes the failure of other investigators to detect the breakdown to the fact that it is exponentially weak.

Transonic Nozzles

Martensen & von Sengbusch (1958) treated the inverse problem of transonic flow through a plane or axisymmetric Laval nozzle with a prescribed linear variation of speed along the axis. They programmed an early computer to find the double Taylor series in powers of the axial and transverse coordinates up to order 23. Later, Van Tuyl (1973, 1976) admitted more general variations of centerline speed and carried double expansions, centered at several points along the axis, to order 48.

Although Padé approximants have been generalized to several variables, most methods of analysis and improvement require a power series in a single variable. Accordingly, Van Tuyl has formed Padé approximants along various straight lines, parabolas, and other curves in the flow field.

Supersonic Flow Past Blunt Bodies

The difficult problem of calculating the flow past a blunt body, such as a sphere, in a supersonic stream was attacked by a variety of methods in the 1940s and 1950s, including the inverse approach of taking the detached shock wave as known and expanding downstream from it in a double power series. Hand computation of that expansion was carried farthest by Cabannes (1956), who found as many as seven terms on the axis. Machine computation was introduced by Richtmyer (1957, 1960) and Lewis (1960, 1965), and carried further by Moran (1965).

Schwartz (1974b) was able to use a single power series to treat parabolic and paraboloidal shock waves at infinite Mach number in parabolic coordinates, and computed 24 terms. However, he found it expedient to revert to double-series expansions about the apex of the shock wave in treating asymmetric plane flow (Schwartz, 1975), and computed triangular arrays of coefficients of order 30 and 40.

It was soon recognized (Van Dyke 1958) that such expansions do not ordinarily converge far enough downstream to reach the surface of the body. They are restricted by a limit line in the analytic continuation upstream of the shock wave (which does not of course appear in the actual flow field). This difficulty has been overcome in various ways. Lewis (1960, 1965) rewrites the truncated double series as a new expansion about an intermediate point between the shock wave and body, which of course does not change its values anywhere, but then truncates further to a fraction of the original number of terms; he shows that this converges as the number of terms increases. Alternatively, he makes use of the differential equations again at an intermediate surface. Van Tuyl (1960), Moran (1965), and Schwartz (1974b, 1975) use Padé approximants, reducing double to single series by a variety of devices.

The nonlinear equations of shallow-water theory are analogous to those of gas dynamics for a gas with adiabatic exponent $\gamma = 2$. Accordingly, Forbes & Schwartz (1981) have adapted the methods just discussed to calculate the flow about a blunt obstacle, such as a bridge pier, penetrating a supercritical stream of water.

Shock Waves

Reddall (1972, 1978) has calculated the one-dimensional unsteady motion produced by a plane piston that moves impulsively at high speed into a quiescent perfect gas and then comes to rest at a finite distance, its displacement at dimensionless time t given by $t/(1+t)$. He expands the solution in powers of time, computing 70 terms. As mentioned already, the sign pattern varies with the adiabatic exponent γ of the gas. Reddall continues the solution analytically outside its original radius of convergence using a generalized Euler transformation, and thereby estimates the motion of the shock wave at large time.

According to the blast-wave analogy, the flow should, as time increases, approach that produced by an instantaneous intense explosion on the face of the piston, for which dimensional analysis shows the shock wave receding as the 2/3 power of time. Reddall confirms that exponent for $\gamma = \infty$, but finds it decreasing to about 0.60 at $\gamma = 7/5$, and apparently tending toward 1/2 as γ approaches unity.

In a similar way Van Dyke & Guttmann (1982) have treated a converging cylindrical or spherical shock wave. They imagine a cylindrical or spherical vessel initially filled with quiescent perfect gas that suddenly begins to contract with a very large radial speed. They expand the solution in powers of time and compute 40 terms. The signs are fixed, corresponding to singular collapse of the shock wave onto the axis. They confirm to six figures that the final collapse is governed by the self-similar solution of the second kind (as it is called in the Russian literature) that was first calculated by G. Guderley.

Moran & Van Moorhem (1969) have considered the diffraction of a plane shock wave that strikes a circle or sphere. They expand the unknown reflected shock wave in a double Taylor series, and the flow properties in triple series, in the two space coordinates and time, and compute the coefficients to a total order of 13. They enlarge the range of convergence by recasting in continued fractions, after recasting the multiple summations as single series along rays. They point out that the analysis will break down at the instant when regular reflection changes to Mach reflection. In the same way Moran (1970) has tackled the still more complicated problem of a plane shock wave striking the bow wave of a blunt body in supersonic flight.

5. VISCOUS FLOWS

Although computer-extended series were first applied to compressible flows, there is by now a considerably larger body of literature devoted to laminar viscous flows. Turbulence has so far only been approached via Taylor-Green vortices and hydrodynamic stability theory.

Creeping Motion

The Stokes equations for viscous flow at low Reynolds numbers are linear, but closed-form solutions are known for only a few simple geometries. Those can be perturbed, however, to solve more-complicated problems.

Thus Kim (1981) has treated the slow steady rotation of a circular disk about the axis of an infinite circular cylinder. He expands in powers of the ratio of disk to cylinder radius, and tabulates the first 40 coefficients in the series for the torque. The signs are fixed, corresponding to the obvious singularity when the disk touches the cylinder. A Domb-Sykes plot shows that the singularity is logarithmic. Kim estimates its coefficient, subtracts its expansion, and repeats the process to estimate the secondary singularity. He confirms these results with a separate analysis for small gap by the method of matched asymptotic expansions.

Sangani & Acrivos (1982) have calculated the creeping flow through periodic arrays of equal spheres. They expand the drag in powers of the ratio of the radius of a sphere to the maximum radius at which it would touch its neighbors, and compute 31 terms. The sign pattern becomes irregular for the simple cubic array; for the body- and face-centered cubic arrays, the signs are fixed but the values become increasingly irregular. As a consequence, although the series are found to yield accurate results up to a radius ratio of 0.85, they are useless for the touching configuration; and the authors did not succeed in improving the convergence.

Barthès-Biesel & Acrivos (1973) have discussed the motion of a liquid droplet freely suspended in a shear flow. They approximate on the basis that its shape differs only slightly from spherical, and outline how the symbolic language REDUCE has been used to handle the straightforward but complicated algebra to give the first approximation and part of the second.

Oseen Flow

The Oseen linearization about a uniform stream rather than a state of rest is known to be more accurate than the Stokes equations at low Reynolds number, although it is at best a qualitative model at higher speeds. It is of course harder to solve, so that even the sphere can be treated only approximately.

Van Dyke (1970) has expanded the drag of a sphere in powers of Reynolds number to 24 terms. A Domb-Sykes plot shows that the regularly alternating signs are the result of a simple pole on the negative axis of Reynolds number. Then an Euler transformation produces convergence (to two figures), even at infinite Reynolds number.

Hunter & Lee (1982) have extended the series to 66 terms, and using it and other techniques have shown that there is a doubly infinite array of poles in the left half-plane of Reynolds number. These have infinity as a limit point, which complicates the analysis there.

Boundary Layers

In 1908 Blasius approximated the steady laminar boundary layer on a symmetrical smooth cylinder by perturbing the self-similar solution near the stagnation point in order to obtain a series in powers of distance downstream. In the same paper, he treated the impulsive motion of a cylinder by expanding in powers of time. Over the years these two approaches have been generalized and extended, first by hand computation and then by computer.

STEADY BOUNDARY LAYERS Van Dyke (1964) examined the six-term expansion for the steady symmetrical flow past a parabola. It converges only 0.617 of the nose radius downstream from the vertex; but an Euler transformation renders it convergent even infinitely far downstream, where six terms give the skin friction just 15 percent higher than the correct flat-plate value. Combining it with an inverse expansion from downstream yields good accuracy everywhere.

Curle (1980a) has applied a similar procedure to a long cylinder, computing 20 terms in the departure downstream from the flat-plate boundary layer that forms near the front. He finds it advantageous to work with the series for the reciprocal of the skin friction because it is better behaved, and achieves five-figure accuracy everywhere. Likewise, he has treated the universal problem of the self-induced interaction in a supersonic boundary layer that is produced, for example, by an impinging shock wave (Curle 1982). He analytically continues a 15-term series solution far enough beyond separation to locate the most negative value of skin friction in the separated region.

Howarth's series (3) for the boundary layer in an adverse gradient was analyzed by Van Dyke (1974), who extracted from it the Goldstein square-root singularity at the separation point, estimated to lie at $x = 0.96$. With his refined coefficients, Curle (1979) has made a much more detailed analysis, finding separation at $x = 0.959$ and extracting the coefficient of the Goldstein singularity to three significant figures.

Curle (1981) has discussed another perturbation of the flat-plate boundary layer in an adverse gradient, with the external velocity increasing exponentially as $u_0(1-\varepsilon e^{x/c})$. He invents a device for avoiding the usual gradual erosion of accuracy in higher terms, which are dependent on their predecessors, and is thus able to compute 23 terms of an expansion in powers of x/c when ε is small. This series appears to have infinite radius of convergence, so that no recasting is necessary.

UNSTEADY BOUNDARY LAYERS All these series have been based on self-similar solutions that must be found by integrating nonlinear ordinary differential equations numerically. On the other hand, impulsive motions are based on the solution of the so-called Rayleigh problem—the abrupt translation of an infinite plane—which is linear, and is solved in closed form in terms of an exponential and its integral, the error function. As a consequence, Blasius found the second approximation also in closed form; and it is plausible that closed-form solutions exist to every order for impulsively started boundary layers. However, the algebraic computation increases so rapidly that only once—in a linear problem—has every term been found analytically. For the growth of the thermal boundary layer produced by slightly increasing the temperature of a flat plate with an established velocity boundary layer, Van Dyke (1975) identified the form of the general term, and wrote a FORTRAN program to compute 37 terms of the expansion in powers of time. Otherwise, all investigators have found it more practical to compute the higher approximations numerically.

Thus Cowley (1983) has extended Blasius's expansion (3) in powers of time for the impulsively started circle to 51 terms. He solved 676 ordinary differential equations numerically in quadruple-precision arithmetic, using 4.5 hours on an IBM 3081 computer. He finds that the flow reverses at the rear at a dimensionless time of 0.64383978, and the series for shear ceases to converge at a time of about 3.0. For later times Cowley forms Padé approximants, or alternatively continued fractions. Thus he finds that the boundary-layer equations develop a singularity at an angle of 111.5° from the front and a time of about 2.2, in good agreement with the predictions of other investigators.

By confining attention to the immediate vicinity of the rear (or front) stagnation point, Hommel (1981) has carried the expansion to 99 terms. He uses Padé approximants only to locate the nearest singularities, and then maps them away using a generalized Euler transformation. He can then compute the shear stress for an arbitrarily long time.

OTHER BOUNDARY LAYERS Closed-form solutions for the successive terms are also encountered in some steady boundary layers. Afzal (1981) has analyzed the buoyant plume from a line source of heat in a vertical stream. For weak buoyancy he expands the solution in powers of the square root of vertical distance. He gives the first two terms in closed form, but resorts to numerical integration to extend the series to 11 terms. Likewise, in treating the concentration boundary layer that forms in a solvent as it flows over a semipermeable membrane separating it from pure solute, Pedley (1981) calculates numerically an 11-term expansion in powers of $x^{1/3}$, finding it simpler to disregard the fact that each term can be expressed as a confluent hypergeometric function. In both these problems the range of convergence has been enlarged with an Euler transformation.

All the boundary-layer solutions mentioned so far are coordinate perturbations, in either streamwise distance or time. A parameter perturbation was used by Aziz & Na (1981) to obtain from the flat-plate boundary layer the Falkner-Skan family of self-similar solutions. They expand to the eleventh power of the parameter that is always called β, and then apply the Shanks transformation five times to obtain the skin friction correct to at least three figures throughout the range $-0.19884 \leq \beta \leq 2$ of practical interest. Another parameter perturbation is applied by Curle (1980b) to flow normal to a plate that slides in its own plane with speed $a \exp(\lambda u_0 t/c)$, where the velocity outside the boundary layer is $u_0 x/c$. He expands in powers of λ to 25 terms, and finds convergence limited by a simple pole at $\lambda = -2.39972378381$. An Euler transformation extends the convergence to infinity for positive λ, and then eight-figure accuracy is obtained by extracting 4 terms of the known expansion for large λ. For negative λ, successively subtracting the first four poles yields accurate results as far as $\lambda = -7$.

Navier-Stokes Equations

The Oseen model discussed above suggests solving the full Navier-Stokes equations by expanding in powers of Reynolds number. This cannot be done for flow past a sphere, however, because it is a singular perturbation. Only confined flows can be considered, or those that decay rapidly with distance.

SERIES IN REYNOLDS NUMBER Kuwahara & Imai (1969) first applied this procedure to the steady plane flow inside a circle whose boundary moves tangentially with a speed that varies around the circumference as $(1+\cos\theta)$, $\cos\theta$, or $(1+2\cos\theta)$. They computed 9 terms in powers of the Reynolds number Re, and estimated convergence up to about 30 in all three cases. For the first case, Conway (1978) has extended the series to 23 terms, and located the singularity on the negative axis of Re^2 at about -924. Mapping it away with an Euler transformation appears to yield convergence of the series for power dissipation at all Reynolds numbers.

In the same way, Hoffman (1974b) has treated the flow between infinite rotating disks. By symmetry the expansion proceeds in powers of the square of the Reynolds number $\mathrm{Re} = \omega d^2/\nu$ based on the spacing d. Eight terms suffice to show a square-root singularity on the negative axis of Re^2, which limits convergence to $\mathrm{Re} = 42.1$ for counterrotating disks and $\mathrm{Re} = 14.7$ for one disk fixed. Padé approximants, or an Euler transformation followed by a Shanks transformation, yield accurate velocity profiles at much higher speeds. A sphere spinning in quiescent fluid produces a more complicated flow. The eight terms computed by Takagi (1977) suggest convergence up to about $\mathrm{Re} = 12$ but show no clear sign pattern. Padé approximants extend the range of agreement with experiment past $\mathrm{Re} = 60$ (Van Dyke 1978a).

Laminar flow through a coiled pipe tends downstream toward a fully developed pattern with double-spiral streamlines. For loose coiling, the motion depends on only the Dean number—the product of the coiling ratio and the square of the Reynolds number. By perturbing the Poiseuille flow through a straight pipe, Van Dyke (1978b) has computed 25 terms of an expansion in powers of Dean number. He applies an Euler transformation and further refinements to deduce that the friction ratio ultimately grows as the 1/4 power of Dean number. Although this agrees with experiments on very loosely coiled pipes, it is in conflict with four separate boundary-layer analyses, which predict a 1/2-power variation and are themselves in accord with experiments on more tightly coiled pipes. Until this discrepancy is resolved, the series solution—and indeed the method itself—lie under a cloud.

In these examples with circular or spherical boundaries, each term in the expansion is found in closed form. By contrast, Walker & Homsy (1978) have treated convection in a heated porous cavity by expanding in powers of the Darcy-Rayleigh number R, where each term beyond the first is itself a doubly infinite Fourier series. They nevertheless compute 21 terms with enough accuracy to identify a square-root singularity on the negative axis of R^2, map it away with an Euler transformation, and then extract the boundary-layer limit to two-figure accuracy.

The Reynolds number is a parameter in all these solutions. However, it is a coordinate in the expansion of Hancock et al. (1981) for steady viscous flow in the corner where one plate scrapes across another. They expand in powers of the dimensionless radius $\rho \equiv Ur/\nu$ from the corner, finding the first two terms in closed form, but thereafter preferring to integrate numerically to obtain 25 terms for the 90° corner. The series for the wall stress shows the sign pattern $+ + - - -$ repeated four times, suggesting nearest singularities at angles of $\pm 2\pi/5$ in the plane of the Reynolds number ρ. This local expansion is not complete, however, because distant effects will add solutions of the homogeneous equation, with nonintegral exponents. These will appear after the term in ρ^3 in the case of the right-angled corner.

OTHER PARAMETER EXPANSIONS Lucas (1972) has treated laminar flow through a symmetric plane nozzle by assuming it to be slowly varying and expanding in powers of a characteristic wall slope—an approximation invented by Blasius in 1910. He computes the coefficients of the series to twelfth order for a general shape, and as high as thirty-fifth order for some special shapes. He concludes that in most cases the series are only asymptotic, with zero radius of convergence, but can be summed using Padé approximants.

Weidman & Redekopp (1976) have investigated the self-similar solution for an unbounded fluid rotating above an infinite disk that is itself rotating. With the aim of clarifying the situation for negative values of the Rossby number ε—the ratio of fluid to disk speed—they expand in powers of ε and compute nine terms by numerical integration of ordinary differential equations. From Domb-Sykes plots they concluce that there is a 4/3-power singularity at $\varepsilon = -0.1433$. However, their extrapolation was too rash, for Zandbergen & Dijkstra (1977) show by detailed numerical integration that the solution turns back onto a second branch at $\varepsilon = -0.16054$, which corresponds to a 1/2-power singularity there. (In fact, Weidman & Redekopp's analysis is self-inconsistent, for they divide out the putative 4/3-power singularity only to find that new Domb-Sykes plots indicate a canceling $-4/3$-power singularity at the same point.)

TAYLOR-GREEN VORTICES Considerable effort has been devoted to the flow in a cubical periodic array of vortices that Taylor & Green (1937) introduced as a model for the stretching of vortex lines in three dimensions and the consequent grinding down of large eddies into smaller ones. They started with the initial velocity field

$$u = \cos x \sin y \sin z, \qquad v = -\sin x \cos y \sin z, \qquad w = 0, \tag{10}$$

(though later investigators have found it convenient to shift the origin), and expanded by hand the solution of the Navier-Stokes equations in powers of time. They found the mean-square vorticity and mean rate of dissipation of energy to order t^5. Deissler (1970) suggested programming the analysis for a computer, and Van Dyke (1975) thus extended it to t^{10}. At high Reynolds number, convergence seemed to be limited to about $t = 2$ by a singularity on the negative axis.

The limit of infinite Reynolds number is of special interest, because it has from time to time been conjectured that the stretching of vortex lines will, if not offset by viscosity, lead to a singularity of the flow in a finite time. Therefore, Morf et al. (1980, 1981) extended the inviscid solution for the mean-square vorticity, or *enstrophy*, to order t^{44}, which required seven hours on a CDC 7600 computer. Using Padé approximants and an Euler transformation, they find evidence of a singularity on the positive axis at about $t = 5.2$, although it is masked by closer nonphysical singularities in the complex plane. However, Brachet et al. (1983) have greatly streamlined the computation and extended it to t^{80}, and conclude that they cannot say with certainty whether or not a singularity develops at finite time.

HYDRODYNAMIC STABILITY Only recently have computer-extended series been applied to problems of hydrodynamic stability. Von Kerczek (1982) has examined the linear stability of the oscillatory plane Poiseuille flow produced by a pressure gradient that is the sum of a steady term plus one varying sinusoidally in time. After reducing the time-dependent Orr-Sommerfeld equation to a system of ordinary differential equations using a Galerkin-like method, he expands each solution in powers of a parameter that measures the ratio of oscillatory to steady velocities. Computing up to 40 terms, he uses the Domb-Sykes plot and Neville tables to estimate the radius of convergence, and outside that range sums the series by the Shanks transformation. He finds that the imposed oscillations stabilize the flow except at very low and very high frequencies.

For nonlinear stability, Herbert (1980, 1981, 1983) has studied expansions in powers of the amplitude of the disturbance. He finds it necessary first to devote considerable attention to developing rational perturbation schemes that can be carried to arbitrarily high order. He then shows that the shortcomings of previous methods can be removed by introducing a well-defined amplitude. For plane Poiseuille flow he carries the expansion to eight terms by computer (Herbert 1980). Although the series are too short to estimate the radius of convergence, the terms alternate regularly in sign, which suggests that convergence is limited by a nonphysical singularity and could be enlarged. In applying the same procedures to circular Couette flow (Herbert 1981), he obtains results outside the range of convergence by means of Padé approximants.

6. POTENTIAL FLOWS

Incompressible irrotational flows, the simplest class of fluid motion, were among the last to be treated by computer-extended series. However, most problems have been complicated by the presence of an unknown free surface. One that is not is the model problem of potential flow past a sinusoidal wall $y = \varepsilon \cos x$. Kaplan (1954) expanded the solution in powers of ε to 9 terms by hand. Whitley (1982) has extended the series to 50 terms in 25 seconds on the IBM 3033 computer (and could have calculated many more). The repeated pattern of signs $++--$ and a Domb-Sykes plot reveal conjugate singularities at $\varepsilon = \pm i$, which are mapped away by recasting in powers of $\varepsilon/(1+\varepsilon^2)^{1/2}$; and the new series converges for arbitrarily large amplitude ε.

Water Waves

Free-surface waves have benefited from computer extension more than any other branch of fluid mechanics. Hand computation was started in the last century by Stokes, who treated a periodic train of waves propagating without change of form by expanding to as high as the fifth power of the first Fourier coefficient for the free surface. Schwartz (1972, 1974a) extended that expansion by computer and discovered that it fails to converge to the highest wave. He removed that defect by expanding instead in powers of the wave height itself, which he carried to 117 terms for deep water. At the same time, Fenton (1972) extended the series for a solitary wave to ninth order by computer.

Inspired by that work, Longuet-Higgins (1975) and Cokelet (1977) have refined the series for progressive waves, Hogan (1980) has added surface tension, and Holyer (1979) has treated interfacial waves between two unbounded fluids of different densities. Aoki (1980) and Schwartz & Whitney (1981) have solved standing waves, which are more difficult because they cannot be reduced to steady flow. The series for a solitary wave has been refined by Longuet-Higgins & Fenton (1974), and criticized by Witting (1975). Cnoidal waves have been dealt with by Fenton (1979). We have simply mentioned these papers without detailed comment because all this work has recently been reviewed in these pages by Schwartz & Fenton (1982).

Roll-Up of a Vortex Sheet

In the early days of aeronautics, Lanchester and Prandtl sketched the vortex sheet rolling up behind the trailing edge of a lifting wing. Following earlier work, Schwartz (1981) has treated the analogous unsteady two-dimensional sheet rolling up in the course of time by expanding in powers of time. For an initial lift distribution proportional to $(1-x^2)^{3/2}$, he computes 21 terms in 2.2 seconds on the IBM 3033 machine. Forming complex Domb-Sykes plots, he deduces that the sheet remains analytic only up to a dimensionless time of 0.39. Thereafter each edge rolls up as an exponential spiral.

Similarly, Meiron et al. (1982) have examined the Kelvin-Helmholtz instability of an initially flat infinite vortex sheet to a sinusoidal perturbation of its initial vorticity, of the form $1+a \sin x$. They compute 7 terms of an expansion in powers of time in closed form using MACSYMA, and then continue with floating-point arithmetic to 24 or 38 terms for several values of the amplitude a. They draw Domb-Sykes plots for the mean-square gradients, of various orders, of the interface, and conclude that the solution probably becomes singular, with weakly divergent curvature, at a finite time, which decreases as the amplitude of the perturbation increases.

Rise of a Bubble

The motion of a bubble released from rest seems a relatively simple problem, but it turns out to be one of the most recalcitrant problems yet attacked by series in fluid mechanics. Baumel et al. (1982) consider an infinite horizontal bubble, initially circular, in an unbounded, incompressible, inviscid liquid, with gravity acting but surface tension neglected. They study its motion after release by expanding the flow in powers of t^2, and compute 150 terms. They restrict attention to the top and bottom of the bubble, although earlier Gammel (1976) considered the whole contour. They find that convergence is limited to a dimensionless time t slightly less than unity by a complex-conjugate pair of branch-point singularities located less than 20° from the real axis. The standard way of summing such a series beyond its radius of convergence is by means of Padé approximants. However, Padé approximants, being single-valued, necessarily connect branch points by cuts of their own devising. In this case the cut runs across the real axis. The authors tried to overcome this limitation by a number of ingenious devices, but in the end succeeded in calculating reliable results out to only twice the radius of convergence.

7. CONCLUDING REMARKS

The use of computer-extended series, like the computer itself, is in a state of rapid development, and it is impossible to predict its final evolution. The method has achieved some unqualified successes, notably for water waves. In other cases it is limited by technical difficulties, as for the rising bubble. In a few problems, exemplified by laminar flow through a pipe and subsonic flow past a circle, it has led to controversial results.

It is not clear whether in general we need more terms, or better ways of analyzing them. Certainly faster and cheaper computers will continually make more terms available. We can always profit from the greater experience of our colleagues in statistical mechanics; and it is disquieting to learn that adding a few more terms may sometimes alter trends rather abruptly. Thus Hunter & Baker (1979), in analyzing three-dimensional lattices, observe a striking "onset of convergence" after a number of terms that varies considerably from one case to another. In fluid mechanics we have seen that, on the one hand, as few as 6 terms serve to describe the boundary layer everywhere on a parabola; on the other hand, 44 terms give erroneous results for the inviscid Taylor-Green problem and even 80 are not enough to settle the question of a blowup at finite time.

Clearly the technique of computer-extended series is not yet ready to be applied routinely. Until we understand it better, it should be regarded as a useful but dangerous tool, to be used with the utmost care and scepticism.

ACKNOWLEDGEMENT

The writing of this article, together with the research of the author and his students for the past four years, has been supported by the National Science Foundation under Grant ENG-7824412.

Literature Cited

Afzal, N. 1981. Mixed convection in a two-dimensional buoyant plume. *J. Fluid Mech.* 105:347–68

Andersen, C. M., Geer, J. F. 1982. Power series expansions for the frequency and period of the limit cycle of the van der Pol equation. *SIAM J. Appl. Math.* 42:678–93

Aoki, H. 1980. Higher order calculation of finite periodic standing gravity waves by means of the computer. *J. Phys. Soc. Jpn.* 49:1598–1606

Atherton, R. W., Homsy, G. M. 1973. Use of symbolic computation to generate evolution equations and asymptotic solutions to elliptic equations. *J. Comput. Phys.* 13:45–58

Aziz, A., Na, T. Y. 1981. New approach to the solution of Falkner-Skan equation. *AIAA J.* 19:1242–44

Baker, G. A. 1965. The theory and application of the Padé approximant method. *Advances in Theoretical Physics* 1:1–58. New York: Academic

Barthès-Biesel, D., Acrivos, A. 1973. On computer generated analytic solutions to the equations of fluid mechanics. The case of creeping flows. *J. Comput. Phys.* 12:403–11

Baumel, R. T., Burley, S. K., Freeman, D. F., Gammel, J. L., Nuttall, J. 1982. The rise of a cylindrical bubble in an inviscid liquid. *Can. J. Phys.* 60:999–1007

Bollmann, G. 1982. Potential flow along a wavy wall and transonic controversy. *J. Eng. Math.* 16:197–207

Brachet, M. E., Meiron, D. I., Orszag, S. A., Nickel, B. G., Morf, R. H., Frisch, U. 1983. Small-scale structure of the Taylor-Green vortex. *J. Fluid Mech.* 130:411–52

Brady, J. F., Acrivos, A. 1981. Steady flow in a channel or tube with an accelerating surface velocity. An exact solution to the Navier-Stokes equations with reverse flow. *J. Fluid Mech.* 112:127–50

Buchanan, D. J. 1976. Analysis and improvement of divergent series. *Q. J. Mech. Appl. Math.* 29:127–35

Cabannes, H. 1956. Tables pour la détermination des ondes de choc détachées. *Rech. Aéronaut.* 49:11–15

Cokelet, E. D. 1977. Steep gravity waves in water of arbitrary uniform depth. *Philos. Trans. R. Soc. London Ser. A* 286:183–230

Conway, B. A. 1978. Extension of Stokes series for flow in a circular boundary. *Phys. Fluids* 21:289–90

Cowley, S. J. 1983. Computer extension and analytic continuation of Blasius' expansion for impulsive flow past a circular cylinder. *J. Fluid Mech.* In press

Curle, N. 1979. Analysis of certain slowly converging series. *J. Inst. Math. Its Appl.* 23:265–75

Curle, S. N. 1980a. Calculation of the axisymmetric boundary layer on a long thin cylinder. *Proc. R. Soc. London Ser. A* 372:555–64

Curle, S. N. 1980b. An unsteady boundary-layer problem—a critical test of extended series expansions. *J. Inst. Math. Its Appl.* 25:199–209

Curle, N. 1981. Development and separation of a laminar boundary layer with an exponentially increasing pressure gradient. *Q. J. Mech. Appl. Math.* 34:383–95

Curle, N. 1982. Self-induced interactions. *Proc. R. Soc. London Ser. A* 379:217–30

Deissler, R. G. 1970. Nonlinear decay of a disturbance in an unbounded viscous fluid. *Appl. Sci. Res.* 21:393–410

Delaunay, C. E. 1867. *Théorie du Mouvement de la Lune*. Paris: Mallet-Bachelier

Deprit, A., Henrard, J., Rom, A. 1970. Lunar ephemeris: Delaunay's theory revisited. *Science* 168:1569–70

Domb, C., Sykes, M. F. 1957. On the susceptibility of a ferromagnetic above the Curie point. *Proc. R. Soc. London Ser. A* 240:214–28

Fenton, J. 1972. A ninth-order solution for the solitary wave. *J. Fluid Mech.* 53:257–71

Fenton, J. D. 1979. A high-order cnoidal wave theory. *J. Fluid Mech.* 94:129–61

Fitch, J. P. 1979. The application of symbolic algebra to physics—a case of creeping flow. *Symbolic and Algebraic Computation, Lecture Notes in Computer Science* 72:30–41. Berlin: Springer

Forbes, L. K., Schwartz, L. W. 1981. Supercritical flow past blunt bodies in shallow water. *Z. Angew. Math. Phys.* 32:314–28

Frankl, F. I., Keldysh, M. V. 1934. The exterior Neumann problem for nonlinear elliptic differential equations with application to the theory of a wing in a compressible gas. *Izv. Akad. Nauk SSSR* 12:561–601 (In Russian)

Gammel, J. L. 1976. The rise of a bubble in a fluid. *Padé Approximants Method and its Applications to Mechanics, Lecture Notes in Physics* 47:141–63. Berlin: Springer

Gaunt, D. S., Guttmann, A. J. 1974. Asymptotic analysis of coefficients. *Phase Transitions and Critical Phenomena* 3:181–243. New York: Academic

Guttmann, A. J., Joyce, G. S. 1972. On a new method of series analysis in lattice statistics. *J. Phys. A* 5:L81–84

Hancock, C., Lewis, E., Moffatt, H. K. 1981. Effects of inertia in forced corner flows. *J. Fluid Mech.* 112:315–27

Herbert, T. 1980. Nonlinear stability of parallel flows by high-order amplitude expansions. *AIAA J.* 18:243–48

Herbert, T. 1981. Numerical studies on nonlinear hydrodynamic stability by computer-extended perturbation series. *Int. Conf. Numer. Methods Fluid Dyn., 7th, Lecture Notes in Physics* 141:200–5. Berlin: Springer

Herbert, T. 1983. On perturbation methods in nonlinear stability theory. *J. Fluid Mech.* 126:167–86

Hoffman, G. H. 1974a. Extension of perturbation series by computer: Symmetric subsonic potential flow past a circle. *J. Méc.* 13:433–47

Hoffman, G. H. 1974b. Extension of perturbation series by computer: Viscous flow between two infinite rotating disks. *J. Comput. Phys.* 16:240–58

Hogan, S. J. 1980. Some effects of surface tension on steep water waves. Part 2. *J. Fluid Mech.* 96:417–45

Holyer, J. Y. 1979. Large amplitude progressive interfacial waves. *J. Fluid Mech.* 93:433–48

Hommel, M. J. 1981. *The laminar, unsteady flow of a viscous fluid at a stagnation point*. PhD dissertation. Stanford Univ., Calif.

Howarth, L. 1938. On the solution of the laminar boundary layer equations. *Proc. R. Soc. London Ser. A* 164:547–79

Hui, W. H., Tenti, G. 1982. A new approach to steady flows with free surfaces. *Z. Angew. Math. Phys.* 33:569–89

Hunter, C., Lee, S. M. 1982. The Oseen drag of a sphere. *SIAM Anniv. Meet., 30th, July 19–23, Stanford Univ.*, p. 55 (Abstr.)

Hunter, D. L., Baker, G. A. Jr. 1979. Methods of series analysis. III. Integral approximant methods. *Phys. Rev. B* 19:3808–21

Kaplan, C. 1952. On a solution of the nonlinear differential equation for transonic flow past a wave-shaped wall. *NACA Tech. Note 2383*

Kaplan, C. 1954. Incompressible flow past a sinusoidal wall of finite amplitude. *NACA Tech. Note 3069*

Kim, M.-U. 1981. Slow rotation of a disk in a fluid-filled circular cylinder. *J. Phys. Soc. Jpn.* 50:4063–67

Kuwahara, K., Imai, I. 1969. Steady, viscous flow within a circular boundary. *Phys. Fluids, Suppl.* 2:94–101

Lam, S. H., Rott, N. 1960. Theory of linearized time-dependent boundary layers. *AFORS TN-60-1100*, Cornell Univ., Ithaca, N.Y.

Lewis, G. 1960. Two methods using power series for solving analytic initial value problems. *New York Univ., Inst. Math. Sci., Rep. NYO-2881 PHYSICS*

Lewis, G. E. 1965. Analytic continuation using numerical methods. *Methods in Computational Physics* 4:45–81. New York: Academic

Longuet-Higgins, M. S. 1975. Integral properties of periodic gravity waves of finite amplitude. *Proc. R. Soc. London Ser. A* 342:157–74

Longuet-Higgins, M. S., Fenton, J. D. 1974. On the mass, momentum, energy and circulation of a solitary wave. II. *Proc. R. Soc. London Ser. A* 340:471–93

Lucas, R. D. 1972. *A perturbation solution for viscous incompressible flow in channels*. PhD dissertation. Stanford Univ., Calif.

Martensen, E., von Sengbusch, K. 1958. Numerische Darstellung von ebenen und rotationssymmetrischen transsonischen Düsenströmungen mit gekrümmtem Schalldurchgang. *Mitt. 19 der AVA und des MPI für Strömungsforschung*. Göttingen

Meiron, D. I., Baker, G. R., Orszag, S. A. 1982. Analytic structure of vortex sheet dynamics. Part 1. Kelvin-Helmholtz instability. *J. Fluid Mech.* 114:283–98

Moran, J. P. 1965. The inverse blunt-body problem. Part II of *Two problems in gas-*

dynamics. PhD dissertation. Cornell Univ., Ithaca, N.Y.

Moran, J. P. 1970. Initial stages of axisymmetric shock-on-shock interaction for blunt bodies. *Phys. Fluids* 13:237–48

Moran, J. P., Van Moorhem, W. K. 1969. Diffraction of a plane shock by an analytic blunt body. *J. Fluid Mech.* 38:127–36

Morf, R. H., Orszag, S. A., Frisch, U. 1980. Spontaneous singularity in three-dimensional, inviscid, incompressible flow. *Phys. Rev. Lett.* 44:572–75

Morf, R. H., Orszag, S. A., Meiron, D. I., Frisch, U., Meneguzzi, M. 1981. Analytic structure of high Reynolds number flows. *Int. Conf. Numer. Methods Fluid Dyn., 7th, Lecture Notes in Physics* 121:292–98. Berlin: Springer

Munk, M., Rawling, G. 1952. Calculation of compressible subsonic flow past a circular cylinder. *US Nav. Ordnance Lab. NAVORD Rep. 2477*

Pearce, C. J. 1978. Transformation methods in the analysis of series for critical properties. *Adv. Phys.* 27:89–148

Pedley, T. J. 1981. The interaction between stirring and osmosis. Part 2. *J. Fluid Mech.* 107:281–96

Reddall, W. F. III. 1972. *The asymptotic trajectory of a strong planar shock wave arising from a non-self-similar piston motion*. PhD dissertation. Stanford Univ., Calif.

Reddall, W. F. 1978. Strong-shock trajectory by series extension and improvement. *Bull. Am. Phys. Soc.* 23:966

Richtmyer, R. D. 1957. Detached-shock calculations by power series. I. *New York Univ., Inst. Math. Sci., Rep. PHYSICS NYO-7973*

Richtmyer, R. D. 1960. Power series solution, by machine, of a nonlinear problem in two-dimensional fluid flow. *Ann. NY Acad. Sci.* 86:828–43

Sangani, A. S., Acrivos, A. 1982. Slow flow through a periodic array of spheres. *Int. J. Multiphase Flow* 8:343–60

Schwartz, L. W. 1972. *Analytic continuation of Stokes's expansion for gravity waves*. PhD dissertation. Stanford Univ., Calif.

Schwartz, L. W. 1974a. Computer extension and analytic continuation of Stokes' expansion for gravity waves. *J. Fluid Mech.* 62:553–78

Schwartz, L. W. 1974b. Hypersonic flows generated by parabolic and paraboloidal shock waves. *Phys. Fluids* 17:1816–21

Schwartz, L. W. 1975. Series solution for the planar asymmetric blunt-body problem. *Phys. Fluids* 18:1630–38

Schwartz, L. W. 1981. A semi-analytic approach to the self-induced motion of vortex sheets. *J. Fluid Mech.* 111:475–90

Schwartz, L. W., Fenton, J. D. 1982. Strongly nonlinear waves. *Ann. Rev. Fluid Mech.* 14:39–60

Schwartz, L. W., Whitney, A. K. 1981. A semi-analytic solution for nonlinear standing waves in deep water. *J. Fluid Mech.* 107:147–71

Shanks, D. 1955. Non-linear transformations of divergent and slowly convergent sequences. *J. Math. & Phys.* 34:1–42

Simasaki, T. 1955. On the flow of a compressible fluid past a circular cylinder. I. *Bull. Univ. Osaka Prefect Ser. A* 3:21–30

Takagi, H. 1977. Viscous flow induced by slow rotation of a sphere. *J. Phys. Soc. Jpn.* 42:319–25

Taylor, G. I., Green, A. E. 1937. Mechanism of the production of small eddies from large ones. *Proc. R. Soc. London Ser. A* 158:499–521

Vanden Broeck, J.-M., Schwartz, L. W., Tuck, E. O. 1978. Divergent low-Froude-number series expansion of nonlinear free-surface flow problems. *Proc. R. Soc. London Ser. A* 361:207–24

Van Dyke, M. D. 1958. A model of supersonic flow past blunt axisymmetric bodies, with application to Chester's solution. *J. Fluid Mech.* 3:515–22

Van Dyke, M. 1964. Higher approximations in boundary-layer theory. Part 3. Parabola in uniform stream. *J. Fluid Mech.* 19:145–59

Van Dyke, M. 1970. Extension of Goldstein's series for the Oseen drag of a sphere. *J. Fluid Mech.* 44:365–72

Van Dyke, M. 1974. Analysis and improvement of perturbation series. *Q. J. Mech. Appl. Math.* 27:423–50

Van Dyke, M. 1975. Computer extension of perturbation series in fluid mechanics. *SIAM J. Appl. Math.* 28:720–34

Van Dyke, M. 1976. Extension, analysis, and improvement of perturbation series. *Symp. Nav. Hydrodyn., 10th*, pp. 449–57. Washington DC: Govt. Print. Off.

Van Dyke, M. 1978a. Semi-analytical applications of the computer. *Fluid Dyn. Trans.* 9:305–20

Van Dyke, M. 1978b. Extended Stokes series: laminar flow through a loosely coiled pipe. *J. Fluid Mech.* 86:129–45

Van Dyke, M., Guttmann, A. J. 1982. The converging shock wave from a spherical or cylindrical piston. *J. Fluid Mech.* 120:451–62

Van Dyke, M., Guttmann, A. J. 1983. Subsonic potential flow past a circle and the transonic controversy. *J. Austral. Math. Soc. Ser. B* 24:243–61

Van Tuyl, A. H. 1960. The use of rational approximations in the calculation of flows with detached shocks. *J. Aero/Sp. Sci.* 27:559–60

Van Tuyl, A. H. 1973. Calculation of nozzle flows using Padé fractions. *AIAA J.* 11:537–41

Van Tuyl, A. H. 1976. The use of Padé fractions in the calculation of nozzle flows. *Padé Approximants Method and its Applications to Mechanics, Lecture Notes in Physics* 47:225–42. Berlin: Springer

von Kerczek, C. H. 1982. The instability of oscillatory plane Poiseuille flow. *J. Fluid Mech.* 116:91–114

Walker, K. L., Homsy, G. M. 1978. Convection in a porous cavity. *J. Fluid Mech.* 87:449–74

Weidman, P. D., Redekopp, L. G. 1976. On the motion of a rotating fluid in the presence of an infinite rotating disk. *Arch. Mech. Stosow.* 28:1011–24

Whitley, N. L. 1982. Potential flow past a sinusoidal wall of finite height. Part 2 of *Three problems in potential flow*. PhD dissertation. Stanford Univ., Calif

Witting, J. 1975. On the highest and other solitary waves. *SIAM J. Appl. Math.* 28:344–63

Zandbergen, P. J., Dijkstra, D. 1977. Non-unique solutions of the Navier-Stokes equations for the Karman swirling flow. *J. Eng. Math.* 11:167–88

Ann. Rev. Fluid Mech. 1984. 16: 311–36

DYNAMIC PARAMETERS OF GASEOUS DETONATIONS

John H. S. Lee

Department of Mechanical Engineering, McGill University, Montreal, Quebec, Canada H3A 2K6

INTRODUCTION

In addition to gases, flammable liquids and solids in the form of fine droplets and dust particles also form explosive mixtures with air. An explosive mixture can, in general, support two modes of combustion. The slow laminar deflagration mode is at one extreme; here the flame propagates at typical velocities of the order 1 m s^{-1} relative to the unburned gases and there is negligible overpressure development when the explosion is unconfined. At the other extreme is the detonation mode, in which the detonation wave propagates at about 2000 m s^{-1} accompanied by an overpressure rise of about 20 bars across the wave. The propagation of laminar deflagrations is governed by the molecular diffusion of heat and mass from the reaction zone to the unburned mixture. The propagation of detonations depends on the adiabatic shock compression of the unburned mixtures to elevated temperatures to bring about autoignition. The very strong exponential temperature dependence of chemical reaction rates in general makes possible the rapid combustion in the detonation mode. Two-phase liquid droplets or dust-air mixtures are similar, but they require more physical processes (e.g. droplet break-up, phase change, mixing, etc.) prior to combustion. Thus, characteristic time or length scales associated with the combustion front are usually much larger than those of homogeneous gaseous fuel-air mixtures. The essential mechanisms of propagation of the combustion waves, however, are similar. In between the two extremes of laminar deflagration and detonation, there is an almost continuous spectrum of burning rates where turbulence plays the dominant role in the combustion process. Due to space limitations, only homogeneous gaseous fuel-air detonations are considered in this article.

0066-4189/84/0115-0311$02.00

The problem of fuel-air explosions has received considerable attention in recent years in connection with the safety aspects of large-scale transport and storage of liquified petroleum or natural gases. The Three Mile Island incident also raises the question of the possibility and consequences of a hydrogen-air explosion inside a nuclear-reactor containment. Two-phase liquid sprays and dust explosions are also of concern in accidental fuel release in petrochemical plants and in grain elevators and coal mines. As a result of the current worldwide interest in fuel-air explosions, active research involving large-scale field tests has been carried out in recent years. In this paper, we single out the gaseous fuel-air detonation part of fuel-air explosions, reporting on the progress made and the current state of the art.

Accidental fuel-air explosions are discussed in the recent monographs of Gugan (1978), Bartknecht (1978), and Baker et al. (1983), as well as in an excellent general review article by Strehlow (1980). In addition, the proceedings of a recent specialist meeting on fuel-air explosions contains progress reports of current activities at various laboratories around the world (Lee & Guirao 1982). The present paper, therefore, avoids duplication of material that can be found elsewhere. Efforts are made, however, to keep the paper self-contained.

THE DYNAMIC PARAMETERS OF DETONATIONS

The classical Chapman-Jouguet theory, in essence, seeks the unique solution of the one-dimensional conservation equations across the detonation front in which the flow behind the wave is sonic. It does not require a knowledge of the structure of the wave itself, and involves only an equilibrium thermodynamic calculation for the detonation states (i.e. the detonation velocity, pressure, temperature, and density ratios across the wave, and the equilibrium composition of the product gases). The equilibrium, or "static," detonation states obtained from the classical Chapman-Jouguet theory agree surprisingly well with experimental observations, even in near-limit conditions when the flow structure near the front is highly three-dimensional.

The parameters requiring a knowledge of the structure, and hence the chemical reaction rates, are the detonability limits, the initiation energy, the critical tube diameter, and the thickness of the reaction zone. We henceforth refer to these parameters as the dynamic detonation parameters to distinguish them from the equilibrium "static" detonation states obtained from the Chapman-Jouguet theory. A one-dimensional model for the detonation structure was proposed in the early 1940s by Ya. B. Zel'dovich, W. Döring, and J. von Neumann (i.e. the ZDN model). It assumes a planar shock followed by a reaction zone after an induction delay period. This model permits the computation of the dynamic parameters when a model for the physical processes involved is given. However, results from such computations using the ZDN model for the detonation structure are quite far from those obtained from experiments, mainly because the ZDN structure is unstable and is never observed experimentally except under transient conditions. This is in accord with experimental observations that all self-sustained detonations have a three-dimensional cellular structure. A quantitative theory for the real cellular detonation structure has yet to be developed. Thus, it seems rather surprising that a century after the detonation phenomenon was first identified, and over eighty years since the successful Chapman-Jouguet theory for the prediction of the equilibrium properties was proposed, there is still no quantitative theory for estimating these dynamic detonation parameters. They remain as experimental parameters, and their measurements are not without some fundamental difficulties either.

Detonation research up until the early 1970s was mostly aimed at furthering our understanding of the physical and chemical processes involved in the various aspects of the detonation phenomena (i.e. transition from deflagration to detonation, onset of detonation, blast initiation, shock interaction processes leading to the so called multiheaded or cellular structure, etc.). Thus, sensitive mixtures of fuels (e.g. H_2, C_2H_2) with pure oxygen were used in general, and the experiments were usually carried out at subatmospheric initial pressures for convenience. Measurements of the dynamic parameters were also mostly made on fuel-oxygen mixtures at subatmospheric initial pressures, since the objective of these measurements was to study the dependence of these dynamic detonation parameters on the other experimental variables. As a result, there are very few experimental data on these dynamic parameters, particularly on practical systems, to permit even empirical correlations to be made.

In Volume 5 of this series, Oppenheim & Soloukhin (1973) gave a very comprehensive historical summary of detonation research up to that time, as well as an extensive bibliography of the early works. Of particular interest in their article are the beautiful laser-schlieren cinematographic records of the various physical processes associated with the phenomena of transition and initiation, as well as the detailed cellular structure of the detonation front itself. These exceptional experimental records provide a good qualitative understanding of the detonation phenomena, and readers are encouraged to refer to this article for a physical background on the subject. Other noteworthy review articles summarizing the detailed structure of cellular detonations are those of Edwards (1969) and Strehlow (1968). The recent monograph by Fickett & Davis (1979) also gives a good comprehensive review of the stability and structure of cellular detonations.

More recent reviews emphasizing the direct initiation and the transition from deflagration to detonation have been given by Lee (1977) and Lee & Moen (1980).

It may be concluded that by the early 1970s all aspects of the detonation phenomena were quite well understood, at least on a qualitative basis. There is, however, a strong deficiency in reliable experimental data on the dynamic detonation parameters, particularly in fuel-air mixtures at atmospheric initial pressures that are of practical interest. There are also very few results on fuel-air detonations due to the fact that very-high-energy sources such as solid explosive charges must be used for the initiation as a result of their relative insensitivity. Also, the size of the apparatus required necessitates that the experiments be carried out in the field. However, as a result of the current interest in accidental fuel-air explosions, a number of large-scale experiments have been carried out in the past decade on fuel-air detonations. We emphasize here the discussion of these new results and their correlations, and point out the progress made toward the prediction of the dynamic detonation parameters.

THE DETONATION CELL SIZE λ

Perhaps the most important dynamic parameter is the average size λ of a cell of the detonation front. Because of its fundamental significance, it is worthwhile to review briefly the essential features of a cellular detonation front. Figure 1 shows the pattern made by the normal reflection of a detonation on a glass plate coated lightly with carbon soot, which may be from either a wooden match or a kerosene lamp. The cellular structure of the detonation front is quite evident. If a similarly soot-coated polished metal (or mylar) foil is inserted into a detonation tube, the passage of the detonation wave will leave a characteristic "fish-scale" pattern on the smoked foil. Figure 2 is a sequence of laser-schlieren records of a detonation wave propagating in a rectangular tube. One of the side windows has been coated with smoke, and the fish scale pattern formed by the propagating detonation front is very well illustrated. The detailed shock configuration of the cellular detonation front itself is illustrated by the interferogram shown in Figure 3. The direction of propagation of the detonation is toward the right. As can be seen in the sketch at the top left corner, there are two triple points. At the first triple point A, AI and AM represent the incident shock and Mach stem of the leading front, while AB is the reflected shock. Point B is the second triple point of another three-shock Mach configuration on the reflected shock AB; the entire shock pattern represents what is generally referred to as a double Mach reflection. The hatched lines denote the reaction front, while the dash-dot lines represent the shear discontinuities or slip lines associated with the triple-shock Mach configurations. The entire front ABCDE is generally referred to as the transverse wave, and it propagates normal to the direction of the detonation motion (down in the present case) at about the sound speed C_1 of the hot product gases. It has been shown conclusively that it is the triple-point regions at A and B that "write" on the smoked foil. The exact mechanics of how the triple-point region does the writing is not clear. It has been postulated that the high shear at the slip discontinuity causes the soot particles to be erased. Figure 4 shows a schematic of the motion of the detonation front. The fish-scale pattern is a record of the trajectories of the triple points. It is important to note the cyclic motion of the detonation front. Starting at the apex of the cell at A, the detonation shock front is highly overdriven, propagating at about

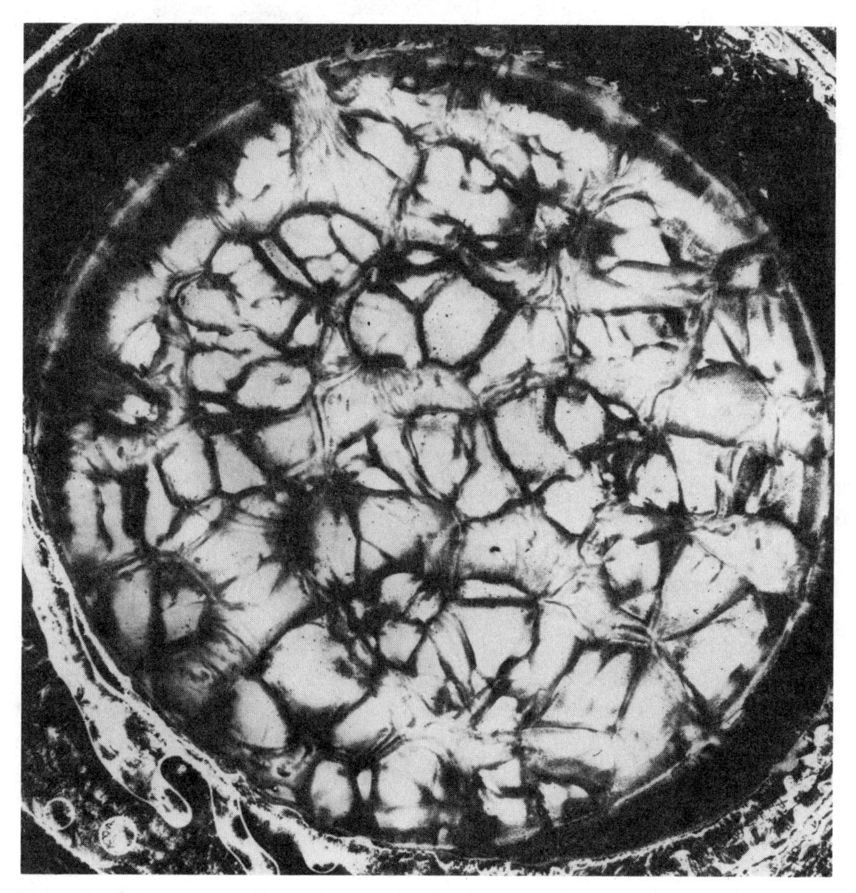

Figure 1 End-on pattern from the normal reflection of a cellular detonation on a smoked glass plate.

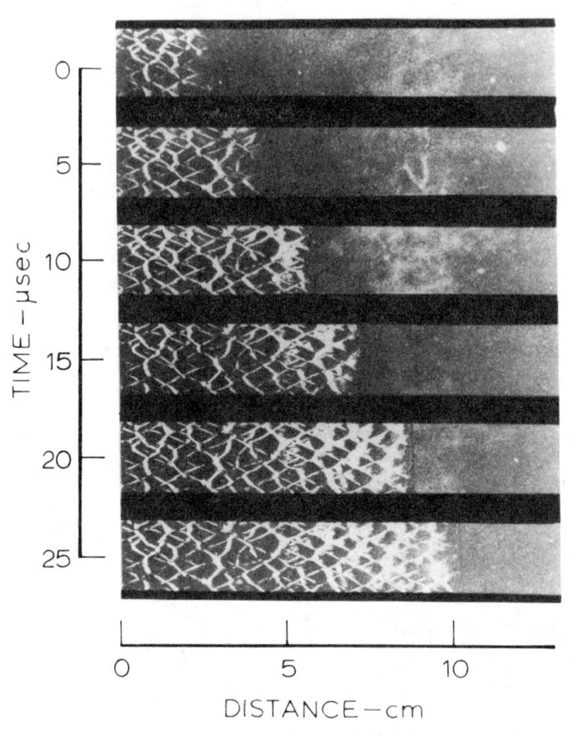

Figure 2 Laser-schlieren cinematography of a propagating detonation in low-pressure H_2-O_2 mixtures, with fish-scale pattern shown on a soot-coated window (courtesy of A. K. Oppenheim).

1.6 times the equilibrium Chapman-Jouguet detonation velocity. Toward the end of the cell at D, the shock has decayed to about 0.6 times the Chapman-Jouguet velocity before it is impulsively accelerated back to its highly overdriven state when the transverse waves collide to start the next cycle again. For the first half of the propagation from A to BC, the wave serves as the Mach stem to the incident shocks of the adjacent cells. During the second half from BC to D, the wave then becomes the incident shock to the Mach stems of the neighboring cells. Details of the variation of the shock strength and chemical reactions inside a cell can be found in a recent paper by Libouton et al. (1981). AD is usually defined as the length L_c of the cell, and BC denotes the cell diameter (also referred to as the cell width or the transverse-wave spacing). The average velocity of the wave is close to the equilibrium Chapman-Jouguet velocity.

We thus see that the motion of a real detonation front is far from the steady and one-dimensional motion given by the ZDN model. Instead, it proceeds in a cyclic manner in which the shock velocity fluctuates within a cell about the equilibrium Chapman-Jouguet value. Chemical reactions are essentially complete within a cycle or a cell length. However, the gasdynamic flow structure is highly three-dimensional, and full equilibration of the transverse shocks, so that the flow becomes essentially one-dimensional, will probably take an additional distance of the order of a few more cell lengths.

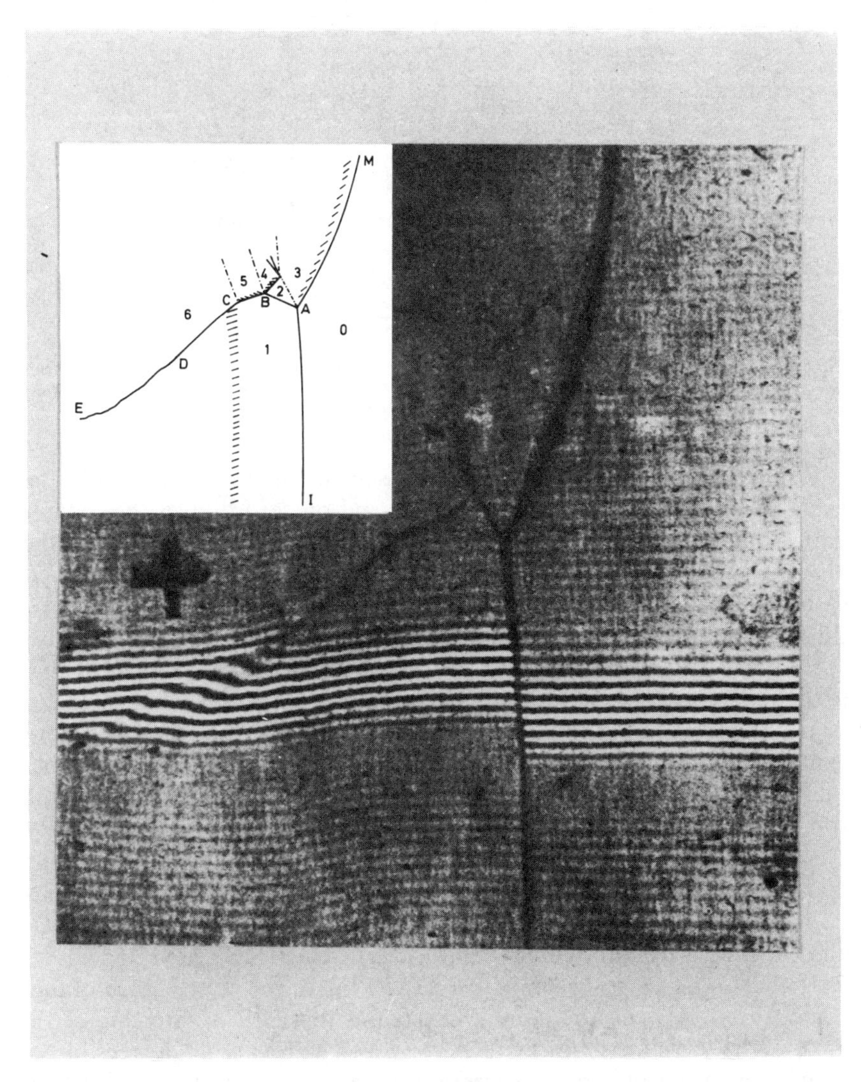

Figure 3 Interferogram of the detailed double Mach-reflection configuration of the structure of a cellular detonation front (courtesy of D. H. Edwards).

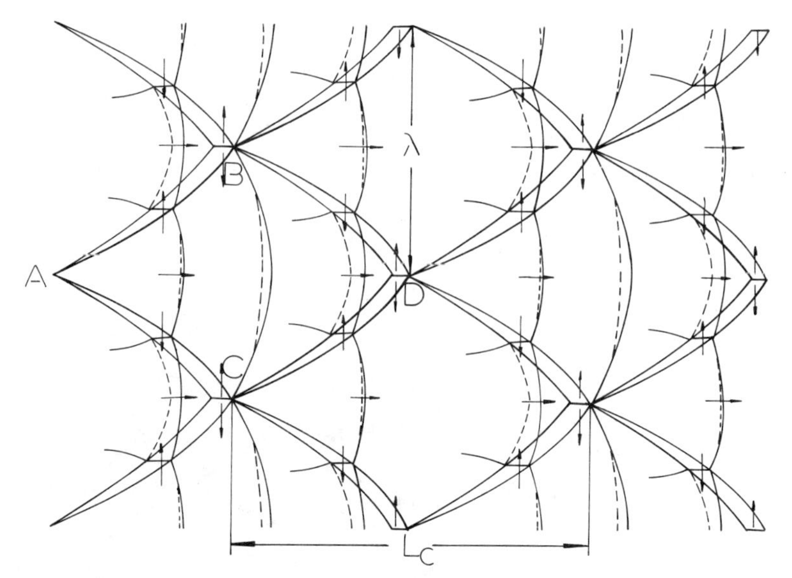

Figure 4 Schematic of the propagation of a cellular detonation front, showing the trajectories of the triple points.

From both the cellular end-on or the axial fish-scale smoke foil, the average cell size λ can be measured. The end-on record gives the cellular pattern at one precise instant. The axial record, however, permits the detonation to be observed as it travels along the length of the foil. It is much easier by far to pick out the characteristic cell size λ from the axial record; thus, the end-on pattern is not used, in general, for cell-size measurements.

Early measurements of the cell size have been carried out mostly in low-pressure fuel-oxygen mixtures diluted with inert gases such as He, Ar, and N_2 (Strehlow & Engel 1969). The purpose of these investigations is to explore the details of the detonation structure and to find out the factors that control it. It was not until very recently that Bull et al. (1982) made some cell-size measurements in stoichiometric fuel-air mixtures at atmospheric pressure. Due to the fundamental importance of the cell size in the correlation with the other dynamic parameters, a systematic program has been carried out by Knystautas to measure the cell size of atmospheric fuel-air detonations in all the common fuels (e.g. H_2, C_2H_2, C_2H_4, C_3H_6, C_2H_6, C_3H_8, C_4H_{10}, and the welding fuel MAPP) over the entire range of fuel composition between the limits (Knystautas et al. 1983). Stoichiometric mixtures of these fuels with pure oxygen, and with varying degrees of N_2 dilution at atmospheric pressures, were also studied (Knystautas et al. 1982). To investigate the pressure dependence, Knystautas et al. (1982) have also measured the cell size in a variety of stoichiometric fuel-oxygen mixtures at initial pressures $10 \lesssim p_0 \lesssim 200$ torr. Figure 5 gives the typical U-shaped curves of the variation of the cell size λ for some of the common hydrocarbon fuel-air mixtures with the equivalence ratio ϕ (i.e. $\phi = 1$ for stoichiometric composition) of the mixture. The minimum cell size usually occurs at about the most detonable composition ($\phi = 1$). The cell size λ is representative of the sensitivity of the mixture. Thus, in descending order of

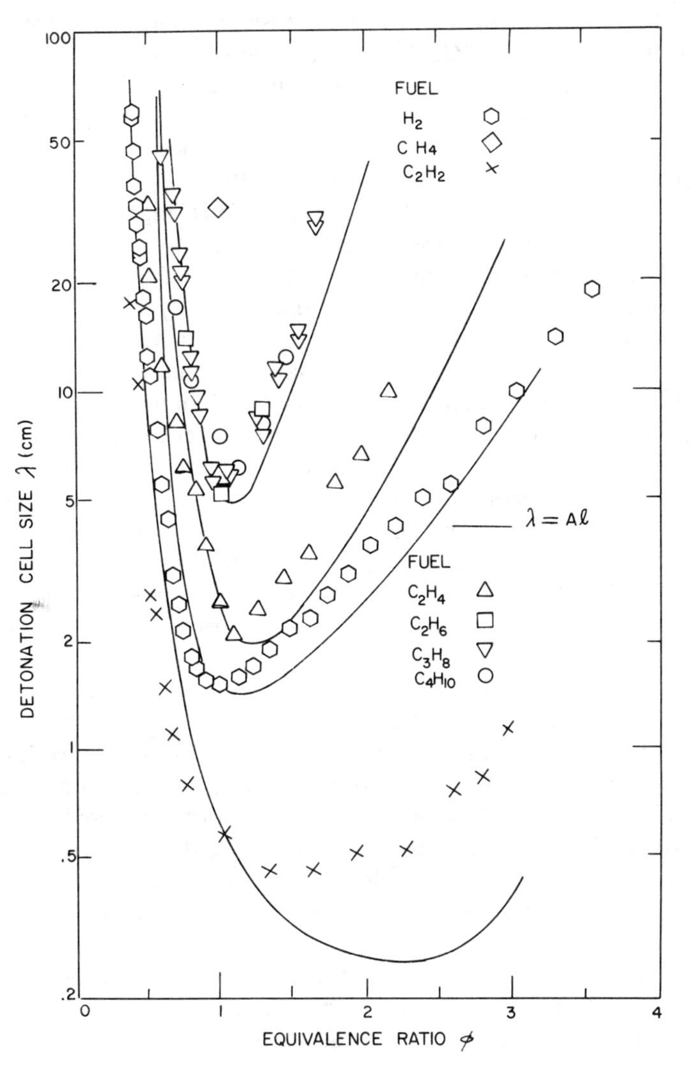

Figure 5 Cell size of fuel-air mixtures at atmospheric pressure.

sensitivity, we have C_2H_2, H_2, C_2H_4, and the alkanes C_3H_8, C_2H_6, and C_4H_{10}. Methane (CH_4), although belonging to the same alkane family, is exceptionally insensitive to detonation, with an estimated cell size $\lambda \simeq 33$ cm for stoichiometric composition as compared with the corresponding value of $\lambda \simeq 5.35$ cm for the other alkanes. That the cell size λ is proportional to the induction time of the mixture had been suggested by Shchelkin & Troshin (1965) long ago. However, to compute an induction time requires that the model for the detonation structure be known, and no theory exists as yet for the real three-dimensional structure. Nevertheless, one can use the classical ZDN model for the structure and compute an induction time or, equivalently, an induction-zone length l. While this is not expected to correspond to the cell size λ (or cell length L_c), it may elucidate the dependence of λ on l itself (e.g. a linear dependence $\lambda = Al$ as suggested by Shchelkin & Troshin 1965). Westbrook (1982, Westbrook & Urtiew 1982) has made computations of the induction-zone length l using the ZDN model, but his calculations are based on a constant-volume process after the shock, rather than integration along the Rayleigh line. Very detailed kinetics of the oxidation processes are employed. By matching with one experimental point, the proportionality constant A can be obtained. This is done in Figure 5 by using Westbrook's data for l and matching at $\phi = 1$. The solid lines in the figure represent the linear relationship $\lambda = Al$, where l is computed from detailed kinetics. The general dependence of λ on ϕ is reproduced qualitatively, but quantitatively the agreement is not as good as it appears because errors are hidden by the very steep nature of the U-shaped curve. The agreement, however, is within one order of magnitude. The constant A differs for different gas mixtures (e.g. $A = 10.14$ for C_2H_4, $A = 52.23$ for H_2); thus, the three-dimensional gasdynamic processes cannot be represented by a single constant alone over a range of fuel composition for all the mixtures. The chemical reactions in a detonation wave are strongly coupled to the details of the transient gasdynamic processes, with the end product of the coupling being manifested by a characteristic chemical length scale λ (or equivalently L_c) or time scale $t_c = \lambda/C_1$ (where C_1 denotes the sound speed in the product gases, which is approximately the velocity of the transverse waves) that describes the global rate of the chemical reactions. Since $\lambda \simeq 0.6L_c$ and $C_1 \simeq 0.5D$, where D is the Chapman-Jouguet detonation velocity, we have $t_c \simeq L_c/D$, which corresponds to the fact that the chemical reactions are essentially completed within one cell length (or one cycle).

It appears that a correct model for the cellular structure must be a time-dependent one in which the nonlinear coupling mechanism between gasdynamics and chemical kinetics can be properly modeled. Computer codes for transient reacting flows with shock waves are currently available.

The feasibility of modeling the real cellular structure of a two-dimensional detonation front has also been demonstrated recently by Taki & Fujiwara (1981) and Oran et al. (1981). However, an extension to three dimensions with the appropriate spatial resolutions to account for the details of the complex double Mach configurations, although possible in principle, would require computer time and storage capacity much in excess of what even the current fast machines are capable of. Thus, the use of numerical simulation as a tool for obtaining cell-size data would not be practical, and direct experimental measurement remains as the most convenient means of determining the cell size λ. Although other methods have been tried to measure the cell size directly, the simple smoke-foil technique has not been improved upon to date and still remains as the only successful method. For near-limit fuel-air detonations, where the cells are very large and foils on the order of meters in length have to be used, the actual deposition of a uniform coating of soot from numerous burners poses a rather difficult engineering problem. A more serious problem in the measurement of cell size is the actual interpretation of the fish-scale pattern. For most fuel-air mixtures, this pattern is highly irregular, and thus the selection of the "correct" cell size requires a certain amount of experience. This introduces a subjective element into the measurement of λ, which has to be resolved. The use of long foils so that the wave is recorded over a long travel length certainly facilitates the interpretation of the foil. Alternatively, if a large number of experiments are made in order to accumulate an ensemble of records under identical conditions, then this also improves the accuracy of the measurement. However, other techniques must be developed to facilitate the measurement of this important fundamental dynamic parameter, which characterizes the real structure of the detonation wave.

THE CRITICAL TUBE DIAMETER

Another important dynamic parameter that has received considerable attention in recent years is the so-called critical tube diameter. Experimentally, it is found that if a planar detonation wave propagating in a circular tube emerges suddenly into an unconfined volume containing the same mixture, the planar wave will transform into a spherical wave if the tube diameter d exceeds a certain critical value d_c (i.e. $d \geq d_c$). If $d < d_c$, the expansion waves will decouple the reaction zone from the shock, and a spherical deflagration wave results. Excellent schlieren records of these transmission phenomena for $d \geq d_c$ and $d < d_c$ can be found in the review by Oppenheim & Soloukhin (1973).

Perhaps the most important progress in recent years has been in the linking together of the various dynamic parameters. This work originates from the demonstration of a universal correlation between the critical tube diameter and the cell size, an observation first made some 20 years ago by Mitrofanov & Soloukhin (1965). In studying the diffraction of a planar detonation wave as it emerges from a circular tube into unconfined space containing the same mixture, they noticed that the critical tube diameter is about 13 times the cell size of the mixture (i.e. $d_c = 13\lambda$). For a square tube, the critical width W_c is found to be of the order of 10 times the cell size (i.e. $W_c = 10\lambda$). Since they based their observation on one particular mixture of stoichiometric acetylene and oxygen over a narrow range of initial pressures (i.e. $p_0 \lesssim 100$ torr), the significance of this observation was not recognized. Over a decade elapsed before Edwards et al. (1979) brought up this observation again. In repeating Mitrofanov & Soloukhin's experiments, they confirmed the earlier result that $d_c = 13\lambda$. Furthermore, they also pointed out that there is no reason why this correlation should not be applicable to all other detonative systems. Moen et al. (1981) followed up on this suggestion and demonstrated the validity of the 13λ correlation for stoichiometric ethylene-oxygen mixtures with nitrogen dilution. A systematic experimental program was then carried out by Knystautas et al., in which cell sizes and critical tube diameters were simultaneously measured for all the common hydrocarbons (i.e. H_2, C_2H_2, C_2H_4, C_3H_6, C_2H_6, C_3H_8, CH_4, and MAPP). Cell sizes were measured in stoichiometric mixtures of these fuels with pure oxygen at low initial pressures ($p_0 \lesssim 150$ torr), as well as in stoichiometric fuel-oxygen mixtures at atmospheric pressure, but with different degrees of nitrogen dilution. A comparison with the direct measurements of the critical tube diameter confirmed the universality of the 13λ correlation. A comprehensive report of the detailed results of this important study can be found in Knystautas et al. (1981). The cell-size measurement has recently been further extended to fuel-air mixtures over the range of fuel concentration between the limits. We have seen some of these cell-size data already in Figure 5. There have also been a number of large-scale field experiments on the direct measurements of the critical tube diameter in fuel-air mixtures (Rinnan 1982, Guirao et al. 1982, Moen et al. 1983). The results are summarized in Figure 6, which shows the dependence of the critical tube diameter d_c on the equivalence ratio ϕ (i.e. mixture composition) for some of these common fuel-air mixtures. The solid curves are the cell-size data from Figure 5 multiplied by 13. As can be observed, the agreement is sufficiently good to confirm the universal nature of the $d_c = 13\lambda$ correlation. Some doubt has been expressed by Moen et al. (1983), who simultaneously measured both cell size and critical tube diameter. They reported that, based on their own cell-size data, $14\lambda \lesssim d_c \lesssim 24\lambda$. However, their critical tube diameter data agree perfectly with the 13λ correlation when the cell-size data obtained independently by Knystautas et al. (1983) are used. Knystautas et al's cell-size data are judged more reliable in that very long smoke foils were used and usually a number of records for the same mixture were taken to ensure a correct interpretation of the true cell size. Furthermore, independent estimates of the cell size have also been obtained from the frequencies of the pressure oscillations from the pressure profile recorded by the transducers mounted on the detonation tube. An assessment of the data available thus far indicates that $d_c = 13\lambda$ is indeed a general relationship, at least for the range of fuels tested. It should be pointed out that the $d_c = 13\lambda$ correlation is extremely useful in practice. The cell size λ can be determined relatively easily from laboratory-scale

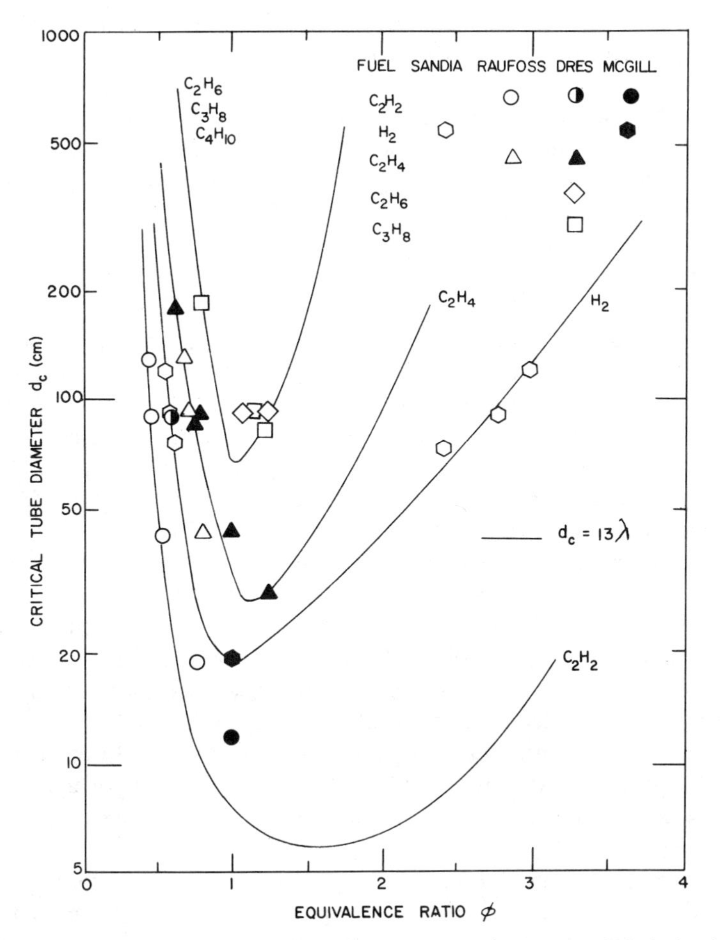

Figure 6 Comparison of experimental data for the critical tube diameter in fuel-air mixtures with the 13λ correlation.

detonation tubes under easily controlled conditions, thus enabling the critical tube diameter to be obtained immediately. Direct measurements of the critical tube diameter require rather large-scale field experiments, in general, for the relatively insensitive fuel-air mixtures.

There are no quantitative theories for the prediction of the critical tube diameter as yet. From physical considerations, a qualitative criterion for the transmission can be developed. When the detonation emerges from the tube, rarefaction waves are generated at the circumference that propagate toward the tube axis as the high-pressure detonation products expand radially outwards. These waves cool the shocked gases and thus increase the induction time, causing the reaction zone to decouple from the leading shock front. The characteristic time for this gasdynamic quenching process will be of the order of the tube radius divided by the sound speed of the product gases (i.e. $t_c = R_c/C_1$). It seems reasonable to assume that if the effective thickness of the detonation wave (i.e. from the leading shock to the equilibrium Chapman-Jouguet plane) is Δ_H, then the detonation must propagate a distance of at least $2\Delta_H$ prior to the rarefaction waves reaching the tube axis for it not to be quenched. In this way, at least a small detonation core near the tube axis is not influenced by the gasdynamic quenching, and this core serves as a kernel that subsequently develops into a spherical wave. The time t_D for the detonation to propagate an axial distance of $2\Delta_H$ will simply be $2\Delta_H/D$, where D is the Chapman-Jouguet velocity. Equating the two characteristic times (i.e. $t_D = t_c$) gives $R_c = 2C_1\Delta_H/D$. Since, for most detonable mixtures, the detonation velocity $D \simeq 2C_1$, we see that the above criterion gives $R_c \simeq \Delta_H$ or the critical tube diameter $d_c \simeq 2\Delta_H$. The effective thickness, or the so-called hydrodynamic thickness, of a detonation wave is reported by Edwards et al. (1976) to be between 2.5 to 4 cell lengths, or 5 to 8 cell diameters. They base this result on measurements of the decay of the transverse pressure vibrations (i.e. equilibration of the transverse shocks) in oxyhydrogen and oxyacetylene mixtures. Their results are in accord with earlier independent measurements by Vasiliev et al. (1972). Taking an averaged value of Δ_H from Edwards et al's observations (i.e. $\Delta_H \simeq 6.5\lambda$), we see that the proposed criterion ($t_c/t_D = 1$) leads to the result $d_c \simeq 2\Delta_H = 13\lambda$, which is in perfect agreement with experiment. From the above discussions, the critical-tube-diameter problem is thus reduced to the resolution of the hydrodynamic or effective thickness of the cellular detonation front. Knowledge of Δ_H permits d_c to be determined.

The recent experimental work of Liu et al. (1983) has also produced some interesting results that shed further light on the critical-tube-diameter problem. In studying the transmission through an orifice plate, instead of from a straight tube, it is found that the critical orifice diameter is identical to the critical tube diameter. Thus, the upstream pressure enhancement from the reflected detonation off the orifice plate does not influence the transition process from a planar to a spherical wave. The reflected detonation generates a radially converging shock wave, which, like the rarefaction waves, propagates toward the tube axis at about the sound speed of the product gases as a result of the high temperatures (and, hence, weak shock). The fact that there is no difference between the corresponding critical orifice and the critical tube diameters suggests that the unattenuated core near the tube axis, rather than the details of the exact gasdynamic flow structure behind the diffracted wave, dominates the development of the spherical detonation.

Experiments on the transmission through different-shaped orifice plates (i.e. square, rectangular, triangular, and elliptical) indicate that the 13λ correlation still holds if an effective diameter d_{eff}, which is the mean value of the largest and smallest dimensions that characterize the orifice opening, is used (i.e. $d_{eff} = 13\lambda$). For a square orifice of side length W, the smallest characteristic dimension would be W, while the largest would be the diagonal $\sqrt{2}W$. Thus, the effective diameter of the square orifice would be $d_{eff} = 1/2(W + \sqrt{2}W) \simeq 1.2W$. Hence, $d_{eff} = 13\lambda$ gives $W_c = (13/1.2)\lambda \simeq 10.8\lambda$, in close agreement with the earlier observations of $W_c = 10\lambda$ by Mitrofanov & Soloukhin (1965) and Edwards et al. (1979). The effective diameter is also the mean value of the diameter of the circumscribed and inscribed circles of the orifice opening. It would appear that the smallest dimension should be more important, since it controls the time for the rarefaction to penetrate the wave. However, when the two characteristic dimensions are not too different, the expansion is three-dimensional, and thus the mean value would be more appropriate to characterize the gasdynamic process.

In generalizing the results, one could also introduce the concept of wave curvature. The critical condition for transmission could then be stated in terms of a minimum radius of curvature of the diffracted wave. If the rarefaction waves give rise to a curvature of the diffracted detonation exceeding a certain critical value, then failure of the wave results. The maximum curvature (or minimum radius of curvature R) would be of the order of about the hydrodynamic thickness Δ_H (i.e. $R = \Delta_H = 6.5\lambda$) for the three-dimensional case of the transformation from arbitrarily shaped planar to spherical geometry.

For rectangular orifices (or slots) where the aspect ratio L/W becomes large (e.g. $L/W \gtrsim 7$) and the planar wave transforms to a cylindrical wave, Liu et al. (1983) found that the smaller length W now controls the transmission process, rather than the mean value of the two characteristic dimensions L and W. Based on the wave-curvature concept, the corresponding radius of curvature of a two-dimensional cylindrical wave would be one half that of the spherical wave for the same curvature. Thus, if $R = \Delta_H$ for the three-dimensional case, $R = \Delta_H/2 = 3.25\lambda$ for the two dimensional case. This is in agreement with the experimental result of Liu that $W_c \simeq 3.25\lambda$.

Since Liu's experiments are based on rectangular slot orifices of $L/W \lesssim 7$, some large-scale experiments on two-dimensional rectangular channels with $L/W \simeq 35$ have been carried out recently by Benedick et al. (1983) to verify the $W_c \simeq 3\lambda$ limit. Their experiments used H_2-air and C_2H_4-air mixtures at atmospheric pressures, instead of the H_2-O_2-N_2 and C_2H_4-O_2-N_2 mixtures used by Liu. The results, however, confirmed the two-dimensional limit of $W_c = 3\lambda$ found by Liu. Furthermore, a half-channel experiment was also performed that demonstrated symmetry about the centerline, in that a value of $W_c/2 = 1.5\lambda$ was obtained.

These very recent experimental findings are extremely interesting. It is clear that the critical-tube-diameter problem provides much insight into the coupling mechanisms between gasdynamics and chemical kinetics. The sudden emergence from confined to unconfined space, in essence, subjects the detonation to a finite perturbation. Whether it survives and readjusts to the new geometry or fails will define the minimum requirements for the self-sustained propagation of the wave under its new environment. The recent study of Murray & Lee (1983) on detonation failure when the detonation transmits from a rigid steel tube to a thin plastic tube is of particular interest in this respect. By using plastic tubes of different wall thickness, the severity of the gasdynamic expansion can be controlled, and the failure criterion in this case can then be related to the wave curvature. It should be noted that even in rigid tubes, the negative displacement thickness of the boundary layer causes flow divergence much like the yielding plastic wall experiment of Murray & Lee (1983). Thus, rarefaction waves are generated at the walls, which then propagate toward the tube axis and cause the wave to be curved. Direct measurements of the radius of curvature of a detonation wave have been carried out recently by Desbordes et al. (1981). They reported a radius of curvature of about 1 m for a planar detonation in a 10-cm-diameter tube. The corresponding curvature of the wave is much smaller than the critical value for the failure of a spherical wave (i.e. $1/R \simeq 1/6.5\lambda$). Thus, it appears that the minimum radius of curvature of a spherical wave would be about the hydrodynamic thickness $\Delta_H \simeq 3.5L_c \simeq 6.5\lambda$. For waves with radii of curvature much larger than the critical value, a velocity deficit results. Thus, failure in rigid and plastic tubes, as well as sudden expansion into unconfined space, are all similar processes and differ only by the degree of the transverse expansion itself. Since the lateral expansion results in a curved front, the propagation criterion under different boundary conditions can be unified under the wave-curvature concept.

CRITICAL ENERGY FOR DIRECT INITIATION

Experimentally, it is found that for a given mixture at given initial conditions, a definite quantity of energy must be used to initiate a detonation "instantaneously." By "instantaneously" is meant that the initial strong blast wave generated by the powerful igniter upon the rapid deposition of its energy decays asymptotically to a Chapman-Jouguet detonation. If the igniter energy is less than a certain critical value, the reaction zone progressively decouples from the blast as it decays and a deflagration results. Since the transition from deflagration to detonation under unconfined conditions is extremely difficult, if even possible, spherical detonations are almost always initiated in practice via the direct or blast-initiation mode. Since the subject of direct initiation has been thoroughly reviewed previously (Lee 1977), it suffices to consider only the recent developments and results.

The interest in unconfined fuel-air detonations in recent years has led to a number of experimental studies on the measurement of the critical initiation energy. The most extensive of these measurements for fuel-air mixtures are those by Bull et al. (1978). The dependence of the critical energy (or, equivalently, the weight of the solid explosive charge used) on the equivalence ratio ϕ is, like all dynamic parameters, in the form of U-shaped curves with the minimum around stoichiometric composition (i.e. $\phi = 1$). The minimum energy at $\phi = 1$ has been used as a relative measure of the sensitivity of the various fuels to detonation (Matsui & Lee 1978). Also, by specifying somewhat arbitrarily an upper limit of the critical energy above which the mixture is rendered nondetonable, the U-shaped curve can be used to define the detonability limits for unconfined detonations. Although there are a number of experimental parameters that may influence the initiation energy (e.g. the type of igniter, its geometry, energy-time characteristics, etc.), it can be concluded that as long as the time for the energy deposition is short compared with the characteristic time of the blast wave [i.e. R_0/C_0, where $R_0 = (E_0/p_0)^{1/3}$ is the explosion length and C_0 is the sound speed], the blast wave can be considered as an ideal point spherical blast characterized by the total energy E_0 only. Thus, experimental measurements of the critical energy in fuel-air mixtures using concentrated solid explosive charges satisfy this condition, and we need not consider other details of the source itself.

The various theories for the critical energy have already been reviewed (Lee 1977). They all show a cubic dependence of the critical energy E_0 on the induction time τ for spherical detonations (i.e. $E_0 \sim \tau^3$). However, the lack of an overall chemical length or time that characterizes the real three-dimensional cellular structure of the detonation wave prevents all these theoretical efforts from being completely quantitative. The use of an experimental data point is usually necessary to evaluate a certain constant of proportionality between E_0 and τ^3. As an example, consider the simplest theory, based on Zel'dovich's criterion (see Lee 1977) that when the initiating blast has decayed to the Chapman-Jouguet strength, the decay time must be of the order of the induction time. Using strong-blast theory, Lee (1977) derived the following expression for the critical energy E_0:

$$E_0 = k_j \rho_0 I D^{j+3}[(j+3)/2]^{j+1}\tau^{j+1}, \tag{1}$$

where $k_j = 1, 2\pi, 4\pi$ for $j = 0, 1, 2$, corresponding to the planar, cylindrical, and spherical geometry, respectively; I is a numerical constant that is a function of the specific heat ratio γ (for $\gamma = 1.4$, $I = 0.423, 0.626, 1.212$ for $j = 2, 1, 0$, respectively); ρ_0 is the initial density; D is the Chapman-Jouguet velocity, and τ is a characteristic chemical time. If the induction time evaluated at the shock temperature corresponding to a Chapman-Jouguet detonation is used for τ, Equation (1) gives values for E_0 about three orders of magnitude smaller than the experimental values. A more appropriate overall chemical time can be derived from the cell length, which is approximately twice the cell diameter λ, because chemical reactions are essentially complete within one cell length. Thus, we may define

$$\tau_c = 2\lambda/D,$$

and Equation 1 then becomes

$$E_0 = k_j \rho_0 I (j+3)^{j+1} D^2 \lambda^{j+1}, \tag{2a}$$

$$E_0 = 500\pi\rho_0 I D^2 \lambda^3 \quad \text{for } j = 2. \tag{2b}$$

A knowledge of the cell diameter λ from experiment permits the evaluation of the critical energy E_0. The Chapman-Jouguet velocity D and the value for the constant I can be determined easily when the initial state and the mixture composition are specified.

Another simple relationship linking the cell diameter to the critical energy can be obtained via the critical tube diameter. Based on the work-done concept, Lee & Matsui (1977) derived a simple expression for E_0 and d_c as

$$E_0 = \frac{\pi p_1 u_1}{24 c_1} d_c^3, \tag{3}$$

where p_1, u_1, and c_1 are the pressure, particle velocity, and sound speed at the Chapman-Jouguet plane. Using the Rankine-Hugoniot conditions across a Chapman-Jouguet detonation, Equation (3) may also be written as

$$E_0 = \frac{\pi \rho_0 D^2}{24\gamma(\gamma+1)} d_c^3 = \frac{2197\pi\rho_0 D^2 \lambda^3}{24\gamma(\gamma+1)} \tag{4}$$

when the 13λ correlation is used for d_c.

Recently, Lee et al. (1982) have proposed a more direct link between the initiation energy and the cell size via the critical tube diameter. The idea is based on the requirement of a minimum surface energy before a planar wave can evolve into a spherical wave without failure. This minimum surface energy should correspond to the area of the critical tube, $\pi d_c^2/4$. Thus, in blast initiation, the blast energy E_0 must be such that when the wave has decayed to the Chapman-Jouguet strength, the surface energy of the blast sphere must be at least proportional to $\pi d_c^2/4$. Equating directly the surface area of the blast sphere to the area of the critical tube, we write

$$4\pi R_c^2 = \pi d_c^2/4,$$

and hence,

$$R_c = d_c/4,$$

where R_c is the blast radius when its strength has decayed to the Chapman-Jouguet value (i.e. $R_s \rightarrow R_c$, $M_s \rightarrow M_{CJ}$, where M_s is the shock Mach number). Using strong-blast theory, one can easily determine the blast energy as a function of the blast radius, and the resultant expression is given by

$$\begin{aligned} E_0 &= 4\pi\gamma p_0 M_{CJ}^2 I(13\lambda/4)^3, \\ &= (2197/16)\pi\rho_0 I D^2\lambda^3, \end{aligned} \tag{5}$$

which is quite similar to Equations (2b) and (4). Using the cell-size data of Knystautas et al. (1982, 1983), the critical energies have been computed for the common fuel-air mixtures where experimental data are available (J. E. Elsworth, private communication). The results indicate that Equation (5), based on the surface-energy concept, gives the best correspondence. Figure 7 shows a comparison of the predicted results using Equation (5) with those obtained by Elsworth from direct experimental measurements of the critical weight. The critical energy in Figure 7 has been expressed as an equivalent charge weight of tetryl (1 g tetryl is equivalent to 4270 J) for easy comparison with Elsworth's experimental data. The agreement, in general, is reasonably good in view of the simplicity of the model. No experimental data of critical energies have been reported for C_2H_2-air mixtures to permit a comparison with the present prediction from Equation (5). For most practical situations, Equation (5) can be used to predict critical initiation energies (or charge weights) with quite acceptable accuracy.

DETONABILITY LIMITS

Composition limits refer to the minimum and maximum fuel concentration in which a self-sustained detonation can propagate. From the previous

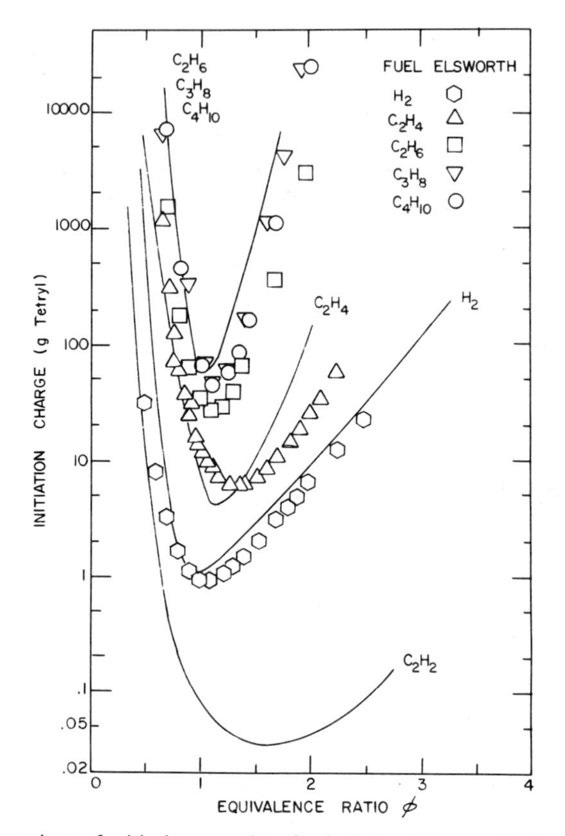

Figure 7 Comparison of critical-energy data for fuel-air mixtures with the surface-energy theory.

discussion, it is clearly meaningless to specify the composition limit without giving the details of the boundary conditions or the environment in which the wave propagates. Thus, for a planar detonation propagating in a rigid tube of circular cross section, the tube diameter must be specified simultaneously with the composition limits. Strictly speaking, the nature of the wall also influences the detonation propagation. Thus, the wall material and its surface roughness must also be specified. In general, different tubes of different diameters, or of different geometrical cross-sectional area, wall roughness, etc., will all have their respective lean and rich limits. Cylindrical and spherical detonations accordingly will also have their particular limiting concentrations. It appears that to be rid of the boundary problem completely, only cylindrical or spherical detonation limits can be considered as "true limits" characteristic of the mixture. We shall return to our discussion of the limits for unconfined detonations later.

Consider the case of "smooth" circular tubes where only the tube diameter need be specified in relation to the limiting composition. There exists the problem of an experimentally derived operational definition of the limits. Since very strong initiators must be used for the near-limit mixtures, the first question to be resolved is how long must the tube be before the influence of the initiator can be considered negligible and the wave is truly self-sustained. Experiments indicate that the effect of the initiator takes an extremely long time to decay (e.g. hundreds of tube diameters). There are strong indications that in confined tubes the single-head spinning mode can be initiated in mixtures outside the limits for a given tube size if a powerful igniter is used. Once initiated, the single-head spinning detonation does not appear to decay at all as it propagates down the tube. It is analogous to an unstable rocket motor resonating at the fundamental transverse mode of instability as it travels down the tube at the Chapman-Jouguet velocity. The experiments of Wolanski et al. (1981) on methane-air detonation are an example of this, since the 5-cm-diameter tube used is much smaller than that required for the propagation of methane-air detonations. Yet, no noticeable attenuation has been observed, and over a wide range of methane concentration, the detonation propagates at the fundamental spinning mode. The existence of such phenomena makes it extremely difficult to establish the limits experimentally. Perhaps there correspond certain limits where, irrespective of the strength of the igniter, a supersonic wave at the fundamental spinning mode propagating at about the Chapman-Jouguet velocity is not possible. However, there exists a range of composition between the first onset of the fundamental spinning mode in a given tube and the composition when it is impossible to initiate a supersonic combustion wave. The question then arises as to whether the supersonic combustion waves in this range are truly self-sustained detonations. That the single-head spinning detonation is the lowest possible stable mode in a given tube (thus the limit) was proposed long ago by Dove & Wagner (1960), among others. However, the recent study by Donato (1982) perhaps sheds some light on this question of what is considered as truly self-sustained single-headed spin. In studying the stability of the near-limit detonations, Donato found that by adding a finite perturbation (such as a few turns of a wire spiral inside the tube) to destroy the spinning wave, the detonation will recover and continue to propagate as a spinning wave only when the composition of the mixture corresponds to the first onset of single-headed spinning detonation in that particular tube. In other words, although the use of a stronger igniter can initiate a single-head spinning detonation wave below the composition when spinning detonation is first observed, the wave is not stable to the finite perturbation by the wire spirals. Once destroyed, the wave will not transit back to the detonation mode. The more recent studies by Lee et al. (1983) of turbulent-flame accelerations in tubes and transition to detonation confirm the results of Donato. It was found that transition to detonation is only possible in mixtures bounded by the composition limits that correspond to the first onset of single-head spin. In other words, transition to detonation is not possible in mixtures where the composition is poorer than that corresponding to the onset of single-headed structure. For the onset of a single-head spin, the tube circumference corresponds to one complete cell λ. Thus, we have $\pi D = \lambda$, which for a given tube diameter is the critical value of λ. Hence, the corresponding mixture composition can be determined. For example, for stoichiometric H_2-air mixtures, $\lambda = 1.5$ cm. Thus, the minimum tube diameter that could sustain a stable spinning detonation would be $D \simeq 0.5$ cm. For the alkanes, the minimum diameter would be of the order of $D \simeq 1.7$ cm for stoichiometric composition. It is of interest to note that according to the cell size reported by Moen et al. (1983) for methane (i.e. $\lambda \simeq 33$ cm), a minimum tube diameter for stable detonation propagation in stoichiometric methane-air mixtures would be of the order of at least 10 cm. Thus, the results reported by Wolanski et al. (1981) in a 5-cm tube should correspond to overdriven transient waves only. This is supported by the fact that they observed single-head spinning waves over the whole range of fuel concentrations studied. Based on stability considerations, the composition limit can now at least be defined experimentally.

With the cell size known as a function of composition, the limits for any given circular tube can now be specified by the criterion $\lambda = \pi D$. For two-dimensional planar channels, Vasiliev (1982) recently found that the critical condition for a stable wave corresponds to a channel width w equal to λ. Thus, $w = \lambda$ represents the criterion for two-dimensional rectangular channels of large aspect ratios. What remains to be done is an experimental study of the conditions for stable propagation in tubes of arbitrary geometries. It appears that a unified criterion for limits in tubes can be derived on the basis of the minimum wave-curvature concept discussed earlier.

The question of the limits for unconfined cylindrical and spherical detonations is much more difficult. Here there are no boundaries to be considered, and the answer must lie within the detailed processes of the cellular structure. In particular, the limits must be determined from the conditions whereby the cells can multiply in an unconfined wave. Stable propagation requires that the averaged cell size be constant for a given mixture. Hence, the detonation cells must multiply in a diverging wave front in accordance with the rate at which the surface area increases. The mechanism for the multiplication of the cells is in the transient processes of double Mach reflections (in particular, how the second Mach stem is formed and develops to form a new cell). Even in nonreacting media, the problem of describing various types of triple-shock interactions has not been fully resolved. For the cellular-detonation case, further complexities are introduced by the addition of flame fronts into the already complex shock pattern. The interaction process is also highly transient, and similar types of approximations may not apply. Apart from the fundamental limit mechanism of cell reproduction for unconfined detonations, it is also difficult to formulate the necessary operational definition for the experimental measurements of limits for diverging waves. It appears that the current method of arbitrarily specifying an upper initation energy above which the mixture is rendered nondetonable is reasonable, at least from a practical point of view. As such, the cell size also enters through the initiation energy. Specifying an upper value for the initiation energy is then equivalent to specifying a maximum value of the cell size beyond which the mixture is too insensitive to be detonable. Although there are other means of specifying reasonable limits for unconfined detonation from the practical point of view, such as the use of the size of the detonation kernel as the minimum detonable volume, the limit problem remains fundamentally linked to the nature of the cellular structure.

CONCLUSIONS

The cell size has been demonstrated to be the most fundamental parameter characterizing the dynamic detonation properties. It has also been demonstrated that a knowledge of the detonation cell size permits the dynamic parameters to be determined reasonably well. However, there is much room for refining the existing relationships that link λ to these dynamic parameters. The 13λ correlation between the cell size and the critical tube diameter d_c is presently a purely empirical relationship. Its validity is based on experiments on the common hydrocarbon fuels. Although the ranges of initial conditions and mixture compositions studied are fairly wide, all the fuels studied have similar kinetics of hydrocarbon oxidation. Thus it seems worthwhile to test the $d_c = 13\lambda$ correlation in other detonable systems, such as H_2-Cl_2 or CS_2-O_2 mixtures, where other fuel molecules are involved. Also, if the 13λ correlation is related to the concept of an effective or hydrodynamic thickness of the detonation wave, then the use of various inert diluents to modify the sound speed in the product gases should influence the equilibrations of the transverse wave and, hence, the hydrodynamic thickness. This may have an effect on the correlation $d_c = 13\lambda$. Strehlow (1968) has already demonstrated that the use of heavy argon dilution has significant influence on the "regularity" of the cellular structure of H_2-O_2 detonations. Further study of the transmission from different geometries, as well as the transmission from rigid to flexible wall tubes, could provide additional insight into the physics behind the $d_c = 13\lambda$ correlation. A correct theoretical model to explain this simple correlation will pave the way to the development of quantitative theories for the other dynamic parameters.

Regarding the prediction of λ from the given kinetics and physical properties of the mixture, it is the opinion of the author that "brute force" numerical computations will not lead to the solution of the problem. The feasibility of numerical simulations and the propagation of two-dimensional cellular detonations have already been demonstrated. The extension to the simulation of three-dimensional cellular detonations is, in principle, possible. It is, however, not clear what additional physics can be learned from these complex computations. Certainly, the use of these codes to compute the cell-size data from the basic kinetics of the system is not feasible from the economics standpoint. It would be much simpler to obtain cell-size data experimentally. There is, however, an urgent need to improve both the technique and accuracy for cell-size measurements and to eliminate the "experience factor" in the identification of the dominant cell size in an irregular pattern. If a relationship between the cell size and another more easily measured dynamic parameter can be firmly established, then perhaps the cell size can be determined indirectly from the measurement of this parameter.

Thus, it seems that if λ can be treated as a fundamental parameter characterizing the detonation mode of combustion, then we are well on the way to developing predictive theories for the important dynamic parameters of practical interest. The cell size λ, like the fundamental kinetic-rate constants of the elementary reactions, can be considered as a fundamental property of the mixture. Adopting such a point of view, the theory of gaseous detonation can be said to be very close to completion after a century since the initial discoveries.

Acknowledgments

Although numerous colleagues and students, past and present, have contributed directly or indirectly to this article, R. Knystautas, C. Guirao, I. Moen, Y. K. Liu, P. Thibault, M. Donato, and S. Murray should be singled out for their invaluable input. Edith Provost is responsible for turning my illegible scribbles into a manuscript. I also wish to thank N. Manson, D. H. Edwards, H. G. Wagner, and J. Elsworth for their comments on a draft of the manuscript. It is unfortunate that time does not permit their valuable suggestions to be incorporated into the final draft.

Literature Cited

Baker, W. E., Cox, P. A., Westine, P. S., Kulesz, J. J., Strehlow, R. A. 1983. *Explosion Hazards and Evaluation.* Amsterdam: Elsevier

Bartknecht, W. 1978. *Explosionen: Ablauf und Schutzmassnahmen.* Berlin/Heidelberg/New York: Springer

Benedick, W., Knystautas, R., Lee, J. H. 1983. *Large-scale experiments on the transmission of fuel-air detonations from two-dimensional channels.* Presented at Int. Colloq. Dyn. Explos. React. Syst., 9th, Poitiers

Bull, D. C., Elsworth, J. E., Hooper, G. 1978. Initiation of spherical detonation in hydrocarbon-air mixtures. *Astronaut. Acta* 5:997–1008

Bull, D. C., Elsworth, J. E., Shuff, P. J., Metcalfe, E. 1982. Detonation cell structures in fuel/air mixtures. *Combust. Flame* 45:7–22

Desbordes, D., Manson, N., Brossard, J. 1981. *Influence of walls on pressure behind self-sustained expanding cylindrical and planar detonation in gases.* Presented at Int. Colloq. Gasdyn. Explos. React. Syst., 8th, Minsk

Donato, M. 1982. *The influence of confinement on the propagation of near limit detonation waves.* PhD thesis. McGill Univ., Montreal

Dove, J. E., Wagner, H. G. 1960. A photographic investigation of the mechanism of spinning detonation. *Symp. (Int.) Combust., 8th,* pp. 589–600. Pittsburgh, Pa: Combust. Inst.

Edwards, D. H. 1969. A survey of recent work on the structure of detonation waves. *Symp. (Int.) Combust., 12th,* pp. 819–28. Pittsburgh, Pa: Combust. Inst.

Edwards, D. H., Jones, A. J., Phillips, D. E. 1976. The location of the Chapman-Jouguet surface in a multiheaded detonation wave. *J. Phys. D* 9:1331–42

Edwards, D. H., Thomas, G. O., Nettleton, M. A. 1979. The diffraction of a planar detonation wave at an abrupt area change. *J. Fluid Mech.* 95:79–96

Fickett, W., Davis, C. D. 1979. *Detonation.* Berkeley: Univ. Calif. Press

Gugan, K. 1978. *Unconfined Vapour Cloud Explosions.* Rugby, Engl: Inst. Chem. Eng.

Guirao, C. M., Knystautas, R., Lee, J. H. S., Benedick, W., Berman, M. 1982. Hydrogen-air detonations. *Symp. (Int.) Combust., 19th,* pp. 583–90. Pittsburgh, Pa: Combust. Inst.

Knystautas, R., Lee, J. H. S., Moen, I. 1981. Determination of critical tube diameter for C_2H_2-air and C_2H_4-air mixtures. *Chr. Michelsens Inst. Rep.*, Bergen, Norway

Knystautas, R., Lee, J. H. S., Guirao, C. 1982. The critical tube diameter for detonation failure in hydrocarbon-air mixtures. *Combust. Flame* 48:63–83

Knystautas, R., Guirao, C., Lee, J. H. S., Sulmistras, A. 1983. *Measurement of cell size in hydrocarbon-air mixtures and predictions of critical tube diameter, critical initiation energy and detonability limits.* Presented at Int. Colloq. Dyn. Explos. React. Syst., 9th, Poitiers

Lee, J. H. S. 1977. Initiation of gasesous detonation. *Ann. Rev. Phys. Chem.* 28:75–104

Lee, J. H. S., Guirao, C. M., eds. 1982. *Proc. Spec. Meet. Fuel-Air Explos.* Univ. Waterloo Press

Lee, J. H. S., Matsu, H. 1977. A comparison of the critical energies for direct initiation of spherical detonations in acetylene-oxygen mixtures. *Combust. Flame* 28:61–66

Lee, J. H. S., Moen, I. O. 1980. The mechanism of transition from deflagration to detonation in vapor cloud explosions. *Prog. Energy Combust. Sci.* 6:359–89

Lee, J. H. S., Knystautas, R., Guirao, C. M. 1982. The link between cell size, critical tube diameter, initiation energy and detonability limits. See Lee & Guirao 1982, pp. 157–87

Lee, J. H., Knystautas, R., Freiman, A. 1983. Flame acceleration in H_2-air mixtures. *Combust. Flame.* Submitted for publication

Libouton, J. C., Dormal, M., Van Tiggelen, P. J. 1981. Reinitiation process at the end of a detonation cell. *Proc. Gasdyn. Detonations Explos., Prog. Astronaut. Aeronaut.*, ed. J. R. Bowen, N. Manson, A. K. Oppenheim, R. I. Soloukhin, 75:358–69

Liu, Y. K., Lee, J. H. S., Knystautas, R. 1983. Effect of geometry on the transmission of detonation through an orifice. *Combust. Flame.* In press

Matsui, H., Lee, J. H. 1978. On the measure of the relative detonation hazards of gaseous fuel-oxygen and air mixtures. *Symp. (Int.) Combust., 17th,* pp. 1269–80. Pittsburgh, Pa: Combust. Inst.

Mitrofanov, V. V., Soloukhin, R. I. 1965. The diffraction of multi-front detonation waves. *Sov. Phys.-Dokl.* 9:1055

Moen, I. O., Donato, M., Knystautas, R., Lee, J. H. S. 1981. The influence of confinement on the propagation of detonations near the detonability limit. *Symp. (Int.) Combust., 18th,* pp. 1615–23. Pittsburgh, Pa: Combust. Inst.

Moen, I. O., Murray, S. B., Bjerketvedt, D., Rinnan, A., Knystautas, R., Lee, J. H. 1982. Diffraction of detonation from tubes into a large fuel-air explosive cloud. *Symp. (Int.) Combust., 19th,* pp. 635–45. Pittsburgh, Pa: Combust. Inst.

Moen, I. O., Thibault, P., Funk, J., Ward, S., Rude, G. M. 1983. *Detonation length scales for fuel-air explosives.* Presented at Int. Colloq. Dyn. Explos. React. Syst., 9th, Poitiers

Murray, S. B., Lee, J. H. 1983. *The influence of yielding confinement on large-scale ethylene-air detonations.* Presented at Int. Colloq. Dyn. Explos. React. Syst., 9th, Poitiers

Oppenheim, A. K., Soloukhin, R. I. 1973. Experiments in gasdynamics of explosions. *Ann. Rev. Fluid Mech.* 5:31–58

Oran, E. S., Boris, J. P., Young, T., Flanigan, M., Burks, T., Picone, M. 1981. Numerical simulations of detonations in hydrogen-air and methane-air mixtures. *Symp. (Int.) Combust., 18th,* pp. 1641–49. Pittsburgh, Pa: Combust. Inst.

Rinnan, A. 1982. Transmission of detonation through tubes and orifices. See Lee & Guirao 1982, pp. 553–64

Shchelkin, K. I., Troshin, Y. K. 1965. *Gasdynamics of Combustion.* Baltimore, Md: Mono Book. Corp.

Strehlow, R. A. 1968. Gas phase detonations: Recent developments. *Combust. Flame* 12:81–101

Strehlow, R. A., Engel, C. D. 1969. Transverse waves in detonation. II. Structure and spacing in H_2-O_2, C_2H_2-O_2, C_2H_4-O_2, and CH_4-O_2 systems. *AIAA J.* 7:3, 492

Taki, S., Fujiwara, T. 1981. Numerical simulation of triple shock behavior of gaseous detonation. *Symp. (Int) Combust., 18th,* pp. 1671–81. Pittsburgh, Pa: Combust. Inst.

Vasiliev, A. A. 1982. Geometric limits of gas detonation propagation. *Fiz. Goreniya Vzryva* 18:132–36

Vasiliev, A. A., Gavrilenko, T. P., Topchian, M. E. 1972. On the Chapman-Jouguet surface in multi-headed gaseous detonations. *Astronaut. Acta* 17:499–502

Westbrook, C. 1982. Chemical kinetics of hydrocarbon oxidation in gaseous detonations. *Combust. Flame* 46:191–210

Westbrook, C., Urtiew, P. 1982. Chemical kinetic prediction of critical parameters in gaseous detonations. *Symp. (Int.) Combust., 19th,* pp. 615–23. Pittsburgh, Pa: Combust. Inst.

Wolanski, P., Kauffman, C. W., Sichel, M., Nicholls, J. A. 1981. Detonation of methane-air mixtures. *Symp. (Int.) Combust., 18th,* pp. 1651–61. Pittsburgh, Pa: Combust. Inst.

Ann. Rev. Fluid Mech. 1984. 16: 337–63

SUPERCRITICAL AIRFOIL AND WING DESIGN

H. Sobieczky

Institut für Theoretische Strömungsmechanik, DFVLR-AVA, Göttingen, Federal Republic of Germany

A. R. Seebass

Aerospace Engineering Sciences, University of Colorado, Boulder, Colorado 80309

Introduction

The need for higher efficiency has forced the operating conditions of modern transport aircraft into the high subsonic speed regime. For turbojet-powered aircraft, which enjoy a propulsive efficiency proportional to the flight Mach number, the overall aircraft efficiency is also proportional to the Mach number. Increases in the Mach number, however, mean that eventually transonic effects are encountered, and some of these offset the propulsive efficiency gains that accrue from higher speeds. In particular, as the flight speed exceeds the Mach number at which the local flow speed on some portion of the aircraft just equals the sound speed there (i.e. the so-called *critical Mach number*), it is nearly universal that the recompression of the flow from supersonic to subsonic speeds occurs through a shock wave. While the shock wave has its own inherent losses, its predominant adverse effect derives from the boundary-layer separation that inevitably occurs as the shock wave becomes stronger and moves farther aft on the wing surface. Airfoils and wings that either mitigate these adverse effects, or avoid them altogether well into the transonic regime, have come to be termed *supercritical airfoils* or *supercritical wings*. Thus, when we speak of supercritical airfoils and wings we do not simply mean those that operate at supercritical Mach numbers, but in particular that class of airfoils and wings that retain their efficiency at supercritical Mach numbers. This then raises the question of whether shock-free supersonic-to-subsonic compres-

0066-4189/84/0115-0337$02.00

sions can be achieved, and if so, can practical airfoils and wings be designed that avoid the recompression shock wave?

By combining the hodograph solutions for a compressible source and vortex, Ringleb (1940) was able to construct a compressible corner flow and show that acceleration from subsonic to supersonic speeds and back again could be accomplished without a shock wave. But this early result did not yield theoretically designed shock-free airfoils or wings. Before any experimental verification could confirm this theoretical concept, mathematical results indicated that such shock-free flows were isolated from one another. This led to questions about the stability of such flows, and there were, consequently, serious doubts about their practical value. Also, the transonic regime came to be considered a mere transition phase to supersonic operating conditions, and one that would eventually become of only transient interest. We now know, however, that flight at supersonic speeds does not yet provide sufficient savings in travel time to compensate for the considerably increased cost in fuel that comes with it and its attendant wave drag. Despite their technical success, supersonic transports are far from a commercial success. As turbojet fuel has become more expensive we have experienced a transition from a time when fuel costs were only a minor contributor to the direct operating cost of aircraft, to the present where they are a major contributor to these costs. A number of technologies that promise improved fuel efficiency for transport aircraft are advancing rapidly, and it seems likely that the transport aircraft of the 1990s will achieve double the seat miles per gallon of the current transport-aircraft fleet. One of the technologies contributing to this advance is addressed here, namely, the design of supercritical airfoils and wings.

The development of numerical algorithms and their implementation on faster computers with larger memories has provided rapid progress in the solution of the partial differential equations that govern or approximate aerodynamic flows. These computational tools have become an important element in the design of modern aircraft, even though at the moment they are limited to approximate calculations of the whole flow field, or to Reynolds-averaged simulations of local flow fields.

The most important steps toward the development of supercritical wings were the pioneering experiments by Pearcey (1962) and Whitcomb & Clark (1965), who demonstrated the practicality of supercritical airfoils. The first analytically designed shock-free airfoil was achieved by Nieuwland (1967), using hodograph techniques. Computational tools to design such airfoils were soon provided by Garabedian & Korn (1971), using the complex extension of real-space, and by Boerstoel & Huizing (1974), using hodograph techniques.

Airfoils and wings that operate supercritically and that do so without shock waves obviously avoid the adverse effects usually associated with supercritical flight Mach numbers. Generally, the best efficiency will be obtained when such airfoils are operated at a Mach number or a lift coefficient slightly higher than that for which shock-free flow exists. This slight increase in the flight Mach number provides an increase in the aircraft's efficiency proportional to the increment Mach number. The losses associated with the shock wave that immediately appears have been determined experimentally to increase with a rather high power of the increment in Mach number; thus, small increases in lift coefficient or a free-stream Mach number above those for which shock-free flow occurs lead to an improved efficiency. Eventually, however, these losses overtake the linear gains, and this trade-off determines the most efficient flight Mach number for a given lift coefficient. Because the operating conditions for maximum efficiency involve a weak shock wave, one can argue that it is not essential to find shock-free flows to obtain efficient designs. This is no doubt true. But the relative ease with which shock-free flows can be found in both two and three dimensions provides a direct path to efficient airfoil and wing designs.

Because of the past success of airfoils that were designed to be shock-free at specific operating conditions and the ease with which such airfoils, and indeed wings, can be found, we adopt the view that at supercritical Mach numbers shock-free configurations are of paramount importance in achieving the maximum gains possible through careful aerodynamic design. Thus, we restrict this review principally to a delineation of the methods of finding shock-free airfoils and wings.

We begin by alerting the reader to the possibility of actually seeing shock waves on aircraft wings, and then briefly discuss the main phenomena that occur in transonic flow about airfoils. For ease of explanation, we restrict our modeling of the flow to that given by potential theory. Within the framework of this theory, we briefly discuss hodograph techniques for shock-free airfoil design and then delineate more fully a method that has become known as the fictitious-gas technique. We then follow with some remarks about the practical use of shock-free wings and aerodynamic efficiency, and discuss their implementation in novel design concepts.

The rather narrow subject of shock-free flows treated here complements earlier reviews of theoretical, numerical, and experimental studies in transonic flow. The reader should be aware of the review by Nieuwland & Spee (1973), who recount the first theoretical developments related to finding supercritical airfoils. A comprehensive review of the hodograph method for shock-free design was provided by Boerstoel (1976). Concepts outlined in the present review have been implemented in a number of the computer programs reviewed by Caughey (1982). The use of these potential-flow algorithms for numerical optimization, which is another route to finding efficient airfoil and wing designs, is reviewed by Hicks (1981) and by Lores & Hinson (1982).

Flight at High Subsonic Mach Numbers

Airplane passengers may have the opportunity to observe transonic flow phenomena on the wing of their aircraft. If the Sun's rays reach the wing surface from the proper direction, then the density difference across the shock wave can produce a shadow line on the wing surface that holds a steady position under many flight conditions. Figure 1 shows such a shadow line on the surface of an aircraft wing. The shock—observed here near the wing root—may extend over a large portion of the aircraft wing. That shown in the figure has a bifurcation that stems from the coalescing of two shock waves coming from the wing-fuselage junction.

A cambered wing or wing section in a uniform free stream will cause circulatory flow. If the subsonic flow speed is sufficiently high, i.e. supercritical, a local supersonic flow field will form on the upper surface of the wing. As the flow leaves the supersonic region it is decelerated to subsonic velocity in all but exceptional cases through a shock wave. The strength, extent, and position of the shock wave on the wing or airfoil depend on the flight Mach number and the flow circulation, as well as on the wing or airfoil geometry. At the shock foot the boundary layer is thickened by the abrupt pressure rise, and it may even be separated. Even a boundary layer that remains attached will evidence the adverse effect of the shock wave upon it and influence the quality of the flow separation that forms the airfoil's or wing's wake. This interaction at the trailing edge is actually the driving mechanism for the global coupling of all the effects involved, in that the effective wing shape seen by the outer flow is decambered by the viscous flow and less circulation is produced; this so-called viscous circle converges toward an airfoil flow with reduced lift (Figure 2).

As a consequence of their entropy production, there is an intrinsic drag associated with the shock waves that we refer to as *wave drag*. This effect is small for weak shock waves and does not adversely affect the lift of airfoils or wings. But the shock wave has an adverse effect on the boundary layer and may even cause it to separate, which consequently affects the wing's lift. It is the goal of supercritical airfoil and wing design to produce both airfoils and wings that avoid all but weak shock waves, thus eliminating any noticeable adverse effect of increased drag and any serious compromise of the airfoil's or wing's lift. Such designs must also include the best elements of subsonic design aerodynamics, such as tailored pressure distributions, to avoid separation near the airfoil's or wing's trailing edge.

Figure 1 Shadow line of a shock wave on an aircraft wing. This mirror-image view has the flow from left to right.

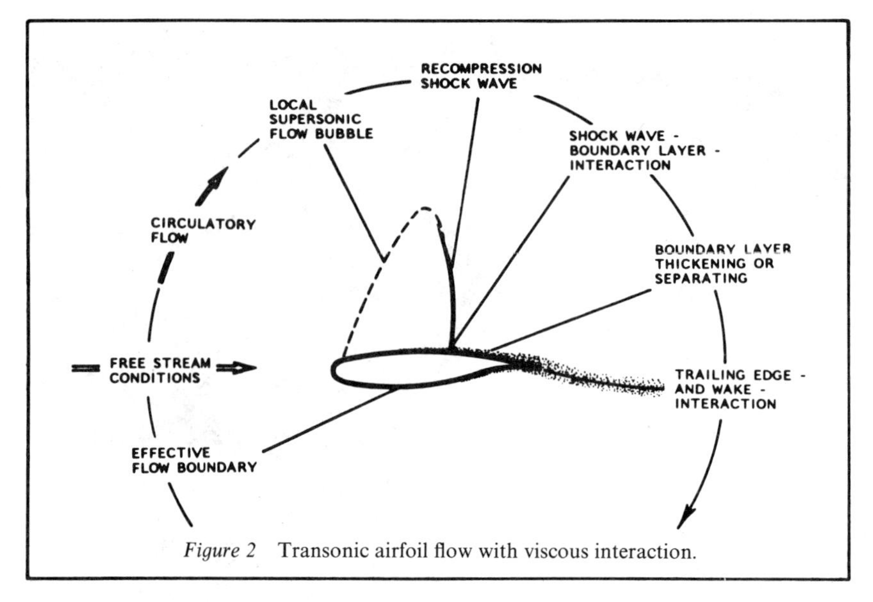

Figure 2 Transonic airfoil flow with viscous interaction.

Compressible Flow Models

The flows of interest occur at high Reynolds numbers and, as a consequence, viscous effects are confined to boundary layers, shock waves, and wakes. For flows in which the boundary layer remains attached over most of the airfoil or wing, the inviscid flow is the correct first approximation to the flow field. The governing equations for inviscid flows are simply the conservation of mass, momentum, and energy. These equations are referred to as the *Euler equations.* Because the flows of interest here are initially irrotational, they will remain irrotational outside the boundary layer provided they are shock-free. Weak shock waves introduce vorticity downstream of them that is proportional to the cube of the pressure jump across the shock wave, and, hence, for shock-free flows or flows with only weak shocks embedded in them, the irrotational approximation for the inviscid flow remains a good one. Because the flow is irrotational, we may derive the velocity field from the gradient of a scalar potential ϕ. Thus we write

$$\mathbf{q} = \nabla\phi,$$

where $\mathbf{q}$ is the velocity vector. With the momentum and energy equations essentially satisfied by the introduction of a scalar potential, we may write the single equation governing the flow field as the conservation of mass:

$$\nabla\cdot(\rho\nabla\phi) = 0, \tag{1}$$

where the density ρ is related to the square of the flow speed by the usual Bernoulli relation

$$\frac{\rho}{\rho_*} = \left[\frac{\gamma+1}{2} - \frac{(\gamma-1)q^2}{2a_*^2}\right]^{1/(\gamma-1)}, \tag{2}$$

and where the subscript $(\ )_*$ refers to sonic conditions, a is the sound speed, and γ is the ratio of specific heats.

For two-dimensional flows these nonlinear equations can be transformed into linear equations through either the hodograph or Legendre transformations. If the usual hodograph variables of the flow speed q and the flow deflection angle θ are replaced by the Prandtl-Meyer turning angle ν and the flow deflection angle θ, where

$$\nu = \int_{a_*} |1 - M^2|^{1/2} \frac{dq}{q},$$

the equations take their canonical form in the velocity potential ϕ and the usual stream function ψ, defined by

$$\psi_x = -\rho\phi_y \quad \text{and} \quad \psi_y = \rho\phi_x.$$

Then we find

$$\phi_\nu = \pm K(\nu)\psi_\theta \tag{3}$$

and

$$\phi_\theta = K(\nu)\psi_\nu, \tag{4}$$

with the negative (positive) sign in Equation (3) applying for subsonic (supersonic) flow. Here, the coefficient $K(\nu)$ is related to the flow properties by

$$K(\nu) \equiv |1 - M^2|^{1/2}/\rho.$$

This system of two equations and two unknowns is equivalent to the characteristic relation

$$d\phi \mp K d\psi = 0$$

along the real characteristics ($\nu \mp \theta = \text{constant}$) that exist in locally supersonic flow, as well as along their complex extension into the subsonic domain.

The first advances in the design of shock-free airfoils came from this hodograph formulation because the linearity of Equations (3) and (4) allows the superposition of exact solutions to this system. Another advantage of the hodograph formulation is that it provides the framework within which one can determine the analytical structure of local elements of the flow field.

Its disadvantages are that the airfoil shape cannot be prescribed, and prescribing the pressure, say, results in an overdetermined system.

The Nature of Shock-Free Flows

We can illustrate some of the specific requirements for a flow to be shock-free by considering the hodograph image, or more precisely the ν, θ-image, of the supersonic region of the flow over an airfoil when this flow is terminated by a shock wave and when it is terminated by a smooth compression. Figure 3 shows on the left an airfoil with a supersonic region whose recompression is obtained mainly through a weak shock wave. The right-hand side shows the flow past a similar airfoil that has been carefully designed to obtain a shock-free recompression. The supersonic flow field of Figure 3 is delineated in both the physical and hodograph planes by the expansion and recompression characteristics or Mach waves that emanate from the sonic line. In the physical plane these characteristics meet the sonic line in a cusp where the flow crosses the sonic line; here, the flow is perpendicular to the axis of this cusp. We may, if we wish, model viscous effects at the foot of the shock wave (point F) by allowing for a discontinuous change in the flow direction at the shock foot, as indicated. This so-called viscous-ramp model has been found to be a simple but reasonably effective device for describing the effect of the shock on the boundary-layer flow and vice versa. In the absence of this viscous-flow modeling, the shock wave must meet the airfoil surface normally and curve upstream from that point. The curvature of the shock wave, in this case at its foot, is infinite; this corresponds to a logarithmic singularity in the pressure gradient with distance along the airfoil surface behind the shock wave.

The present example is a theoretical reconstruction of an airfoil tested in a wind tunnel. If we first examine the nature of the flow field with the shock-wave recompression, we can see that the shock tip P is embedded in the supersonic domain, and that the sonic line meets the shock wave below its tip downstream of the shock at point Q with the two images representing pre- and postshock conditions. As we move down the shock wave toward the shock foot F, we pass through the point G where the shock becomes normal. With no flow deflection across the shock the two images have the same ordinate θ. Finally, we arrive at the shock foot F with its corresponding images both ahead of and behind the shock wave. The airfoil boundary is denoted by the shaded line. The isotachs (constant speed lines) have a saddle point in the flow field behind the shock corresponding to the point R. While the hodograph image of the flow with the recompression shock is a complex one, most of the local details may be described analytically. These details are discussed in Sobieczky (1975) and in

Sobieczky & Stanewsky (1976). We see that if a shock-free flow is to be obtained, the recompression characteristics that emanate from the sonic line must not coalesce in the physical plane, and this can only happen if the sonic-line shape and the flow deflection there are very carefully tailored

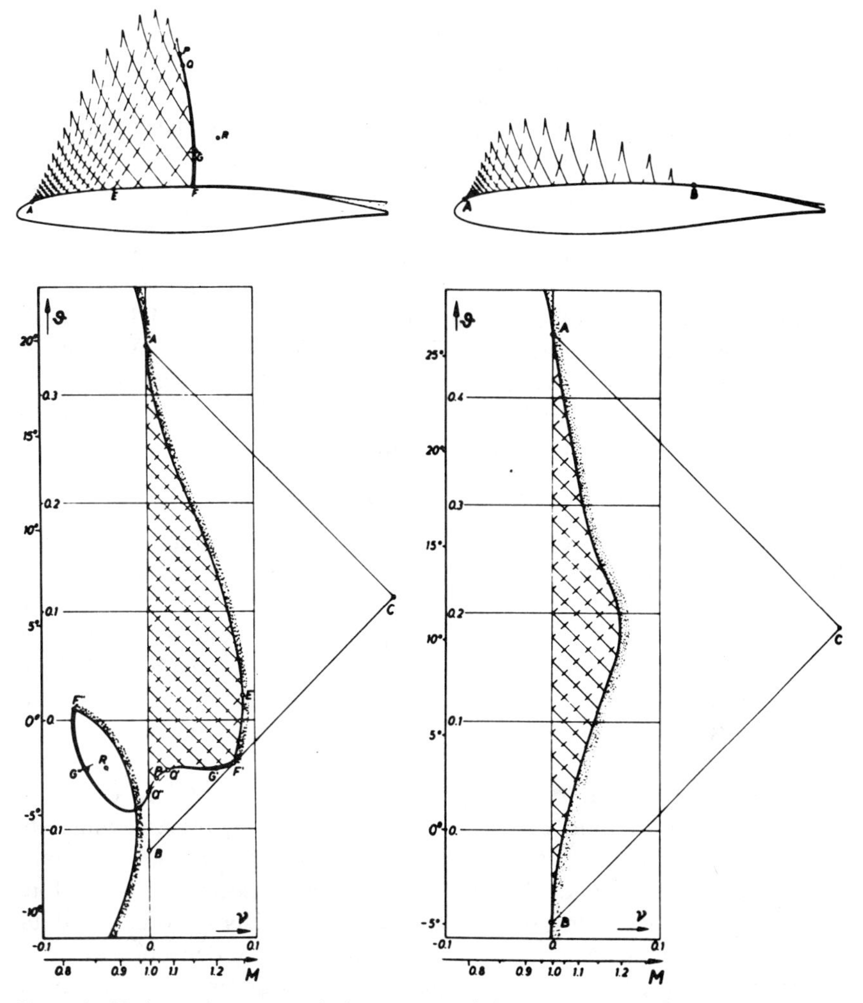

Figure 3 Hodograph structure of a local supersonic flow domain with and without a shock: sonic flow condition near airfoil leading edge (A); shock wave (*left*) forms in flow field (P) and on viscous-flow ramp (F); normal shock point (G); sonic flow postshock conditions (Q). Postshock subsonic domain shows saddle point (R) in the physical plane; this causes branch-point mapping and a second Riemann sheet in the hodograph plane. Shock-free flow (*right*) with smooth recompression.

to one another. Any small perturbation to the outer flow field or to the airfoil surface will affect this balance, and the disturbances traveling downstream along the reflected Mach waves will result in their coalescence, which requires a discontinuous connection with downstream subsonic flow. This connection is physically possible through the Rankine-Hugoniot shock relations or, in the framework of potential theory, its isentropic approximation.

This simplified illustration of how carefully the airfoil geometry must be arranged in order to obtain shock-free flow was demonstrated in the classical theorems of Morawetz (1956, 1957, 1958) that showed that such shock-free solutions were mathematically isolated from one another. This result led many to the incorrect conclusion that shock-free flows would be of no practical interest, since any small disturbance would immediately result in a flow with a shock wave. However, careful experimental investigations showed that certain types of shock-free flows are quite insensitive to perturbations, and that the shock waves that result from small perturbations are themselves weak. From this early experimental work it was recognized that the systematic development of theoretical shock-free design tools could lead to both airfoils and wings with increased aerodynamic efficiency over a range of operating conditions.

Integration of Local Supersonic Flow Fields

For a shock-free flow to be achieved, the flow deflection angle at the sonic line for the flow entering the supersonic region, as well as that for the flow exiting the sonic region, must be such that, when one calculates the supersonic flow field from it, the body streamline AB is found before a singularity in the mapping from the hodograph plane to the physical plane is reached. Such singularities are termed *limit lines* and their occurrence corresponds to the coalescence of characteristics or Mach waves in the physical plane. Suppose the flow deflection angle is prescribed on the sonic line between points A and B. The analytical continuation of these data into the subsonic or elliptic domain determines the flow there, while its extension by characteristics into the supersonic or hyperbolic domain determines the nature of the flow field between the sonic line and the airfoil. One way to construct a transonic flow would be to prescribe sonic-line data in the hodograph plane, or the sonic-line shape and flow deflection angle in the physical plane. To obtain the solution for the local supersonic region of the flow, one must march away from the sonic line, using the flow deflection as initial data in two dimensions and the other two velocity components in three dimensions. In two-dimensional supersonic flow both directions are timelike and, therefore, suited for developing a marching algorithm. In other words, we may model our problem by the simple two-dimensional

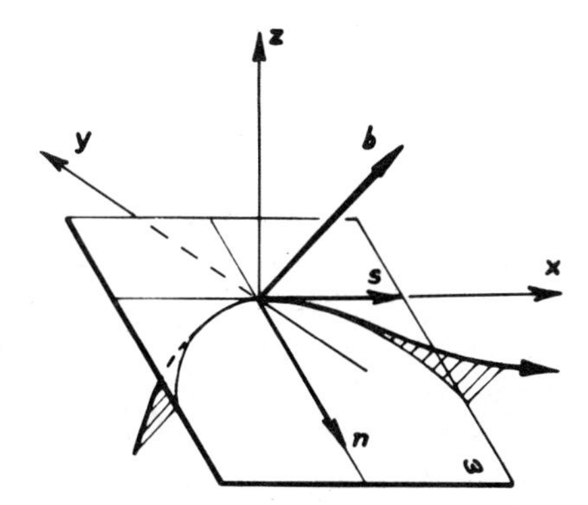

Figure 4 Orthonormal triad for a three-dimensional streamline. The binormal vector **b** is normal to the osculating plane ω, defined by the tangent vector **s** and the normal vector **n**.

wave equation

$$\phi_{xx} - \phi_{yy} = 0;$$

here it is clear that both x and y may be considered timelike. In three dimensions this is no longer the case, for the wave equation takes the form

$$\phi_{xx} - \phi_{yy} - \phi_{zz} = 0; \tag{5}$$

in this representation only the x-direction is timelike. We may rotate the coordinates into the local normal and binormal vectors to the flow streamline (Figure 4); in these coordinates the wave equation, with s associated with x locally, becomes

$$\phi_{ss} - \phi_{nn} = R,$$

where the term on the right-hand side does not contain second derivatives. Once again the flow direction and its normal in the osculating plane to the streamline are timelike directions. But a timelike marching from the sonic surface in three-dimensional flows requires the definition of the solution-dependent vector at every point. In two-dimensional flow the hodograph characteristics grid sketched in Figure 3 guarantees marching along the local normal vector. The integration of Equation (5) in the y-direction, given data in the x,z-plane, is an ill-posed problem, and any numerical calculations must be carried out with this in mind. For a further discussion, see Fung et al. (1980) and Seebass (1982).

Design Methods for Shock-Free Airfoils and Wings

Since supercritical wing technology has become an important concept in the quest for greater aerodynamic efficiency, there has been an increasing demand for design methods that will produce shock-free airfoils. These airfoils are useful for the section definition in the design of swept supercritical wings. While these wings will not be shock-free because the three-dimensional character of the flow destroys the desired shock-free balance, the shock waves on such wings will be much weaker than those on wings with conventional airfoil sections. Aircraft-industry designers have therefore gained practical experience in how to use supercritical airfoil results in the design of supercritical wings. Despite the value of this airfoil technology in wing design, there is also a need for three-dimensional methods for shock-free design.

We begin our review of design methods for shock-free airfoils and wings chronologically. Nieuwland & Spee (1973) have reviewed the earlier work in this field in Volume 5 of this series.

Hodograph and Inverse Design Methods for Airfoils

In 1976 Boerstoel provided a detailed review of the mathematical background of the few methods for shock-free airfoil design available at that time. Thus, these methods are only briefly reviewed here. The first methods employed particular solutions to the hodograph equations. Nieuwland's (1967) early results came from the superposition of a suitable set of such solutions. The nature of the branch points connecting different Riemann sheets, the far-field singularity, and other mathematically well-defined but complicated details of the hodograph mapping are carefully delineated in this review. But the complicated nature of the hodograph image of the real flow has kept hodograph methods from becoming popular tools for designers. On the other hand, they were the only systematic methods available for about a decade, and the experimental verification of their validity proved their practical value. Nieuwland's superposition of particular solutions was further advanced by Boerstoel & Huizing (1974). They used a large set of particular solutions and determined the coefficients required for a desirable linear combination by a numerical method. A variety of airfoils were supplied by this research, which was particularly focused on the design of very thick shock-free airfoils. Boerstoel (1976) provided an estimate of the maximum airfoil thickness that one could achieve as a function of the design Mach number and lift coefficient; his values have been well confirmed by the results of subsequent design methods, and are shown in Figure 5, along with a typical airfoil design.

Garabedian & Korn (1971) developed another, and conceptually quite different, method for the hodograph design of shock-free airfoils. They extended the hodograph equations in their characteristic form [equivalent to Equations (3) and (4)] into the complex domain. This allows them to solve the elliptic part of the flow field by calculating the solution to the hyperbolic equations that derive from this extension into a complex four-

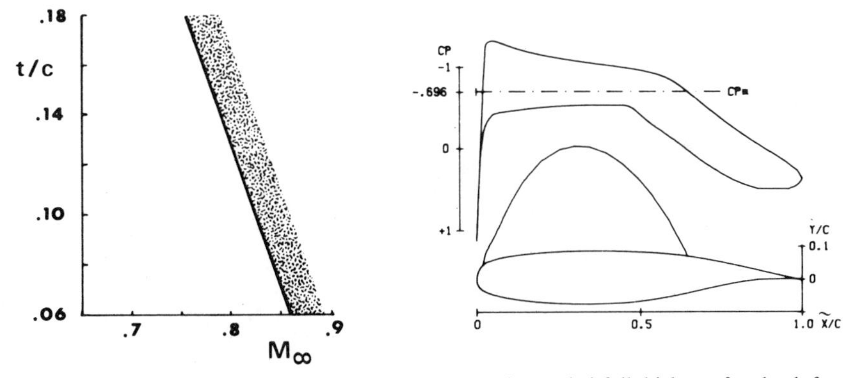

Figure 5 Estimated maximum free-stream Mach number and airfoil thickness for shock-free flows with $C_L = 0$ using Boerstoel's hodograph method; design example of a thick supercritical airfoil ($M_\infty = 0.721$, $t/c = 0.163$, $C_L = 0.595$).

dimensional space. The computational algorithm is therefore relatively fast; the extraction of the real solution is performed by the use of conjugate coordinates. A crucial element in this endeavor was the continuation of the solution past the sonic surface provided by Swenson (1968). The real hodograph image of the solution is obtained by Schwartz' reflection principle. Detailed steps can be found in Bauer et al. (1972, 1975, 1977). The first version of this method was too complicated to find wide acceptance in the aircraft industry, and the choice of the complex initial conditions required mathematical, not physical, intuition. The first results obtained were for airfoils with cusped trailing edges in inviscid flow, but later open trailing-edge designs that allowed the subtraction of the boundary-layer displacement thickness were provided. Improved trailing-edge shapes to avoid viscous separation were introduced by Garabedian (1975). A later version of the method (Bauer et al. 1977) incorporated the transformation of the hodograph image of the flow onto a circle; this avoided a complicated procedure for searching for the airfoil streamline. More importantly, the input for the far-field initial conditions is automatically determined by the designer supplying a desired pressure distribution for the airfoil. Thus, inverse airfoil design is possible with this method, requiring experience only in choosing the pressure distributions and in modifying them to achieve better results. A recent result of this method is provided in Figure 6. The pressure distribution is close enough to that prescribed for this to be considered an effective design tool.

The hodograph structure of various transonic flow phenomena was studied by Sobieczky (1971a). These studies provided simple analytical expressions for local and asymptotic flow details by applying the small perturbation approximation to the hodograph equations. Sobieczky also

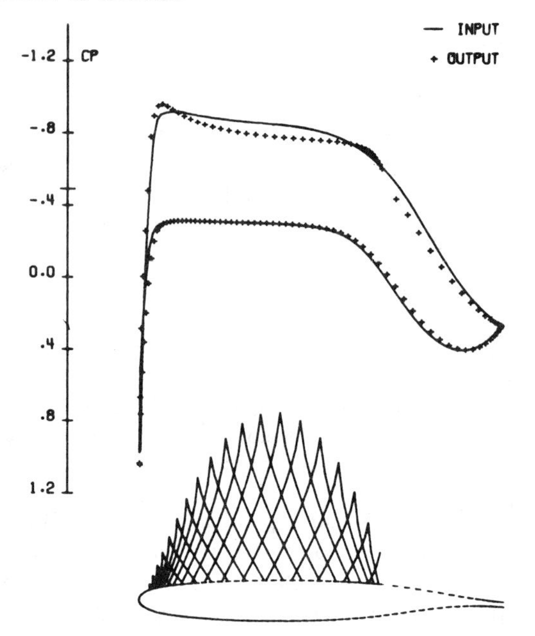

Figure 6 Shock-free airfoil resulting from a specified input pressure distribution using the hodograph method of Garabedian & Korn.

developed a rheoelectric analog computer to solve the compressible flow problem posed by these small perturbation equations. This analog technique, coupled with the calculation of the supersonic flow by the method of characteristics, also provided some of the early shock-free airfoil designs (Sobieczky 1971b, 1975). The rheoelectric hodograph ("rheograph") was especially useful because it supplied a simple electrostatic analog to which one's physical intuition could easily be applied. This method has a long history in France and was employed by Rigaut (1968) in computing transonic flows. Such an approach was reviewed by Sobieczky (1979), where it was noted that a fast Poisson solver could be used to replace the analog devised.

Eberle (1976) replaced the analog flow computation by a panel method, and Hassan et al. (1981) subsequently implemented a fast Poisson solver as suggested earlier by Sobieczky. In this method one inputs a desired Mach-number distribution on the subsonic portion of the airfoil and then finds the airfoil that achieves this pressure distribution. As in Garabedian's method, this airfoil might not be satisfactory, and the Mach-number distribution must be modified and another airfoil found.

Other inverse design methods have also been developed. Carlson (1976)

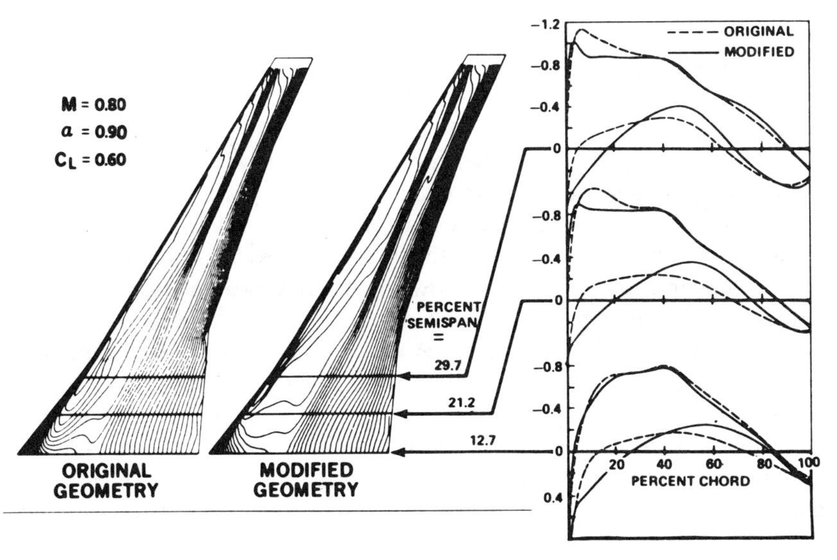

Figure 7 Comparison of original and modified isobars and section pressure distributions of wings designed with Henne's inverse method.

prescribes the leading-edge geometry of the airfoil and the pressure distribution on the remaining portion of the chord length. The input pressure distribution may or may not have a shock wave in it. Some inverse methods that provide shock-free flow at the airfoil may result in a concave surface geometry with a shock wave absent on the airfoil surface, but embedded within the flow field. Because the shock wave is above the airfoil, its adverse effects on the boundary layer are mitigated. Both Volpe & Melnik (1981) and Garabedian & McFadden (1982) have developed successful inverse methods. Garabedian & McFadden's method can also be applied to wing designs with shock waves. An attractive wing design method was developed by Henne (1981) through an extension of Tranen's (1974) inverse method to three dimensions. An example is provided in Figure 7. Despite the changes in the pressure distribution near the root, a shock wave persists on the outboard portion of the wings.

A Direct Method for Shock-Free Design

Sobieczky's two-step procedure for the rheograph plane, which consists of solving the elliptic flow-field problem first to find the flow deflection angle on the sonic line, and then using the initial data on the sonic line to integrate into the supersonic region to see if the body streamline can be continued without meeting a limit line, motivates a both simpler and more general procedure for shock-free airfoil and wing design.

Suppose one takes the data on the sonic line and continues them analytically into the supersonic flow field. This analytic continuation, of course, requires that the equations remain elliptic there, but this would be true if the gas law were changed to some other form at the sonic line. This interpretation of Sobieczky's hodograph method suggests a very attractive alternative: Given a reliable transonic analysis algorithm for solving the potential equation (1), the density law (2) may be altered in the supersonic domain to provide a fictitious gas that is identical to the gas of Equation (2) in the subsonic portion of the flow, but that has a relationship to the velocity in the supersonic flow field that results in elliptic behavior. This device, then, can be used to generate sonic-line or sonic-surface data directly in the physical plane. For example, we might suppose that by some mysterious process the gas density freezes at its sonic value upon crossing the sonic line. With the density constant in the supersonic domain, the equation that governs the flow there is Laplace's equation and is obviously elliptic. Of course there are a wealth of possibilities for this fictitious-gas law. Among them is the simple relation

$$\frac{\rho}{\rho_*} = \left[1 + \frac{1}{P}\left(\frac{q}{a_*} - 1\right)\right]^{-P} \quad \text{for} \quad q > a_*.$$

Here a value of $P = 1$ leads to parabolic equations, while $P = 0$ provides the incompressible limit for the fictitious subsonic flow, viz. $\rho = \rho_*$. The effect of varying P is depicted in Figure 8. If we were working with the quasi-linear form of the governing equations rather than their conservative form, we would have to choose a fictitious relationship between the sound speed and the flow speed rather than a fictitious density–flow speed relationship. Thus, a relatively simple method of generating smooth sonic-line or sonic-surface data directly in the physical plane is readily available by replacing the correct-gas law with a fictitious one in the computation of the flow field.

Of course these data must be used to calculate the real supersonic flow field beneath this sonic line or sonic surface with the correct density law. There is no guarantee that such a calculation will provide a smooth airfoil or wing surface. A limit line or limit surface may intervene before the streamline or stream surface that defines the continuation of the airfoil or wing is found. In that event, different sonic-line or sonic-surface data must be generated. Provided we can find a suitable marching procedure for three dimensions to replace the method of characteristics used in two dimensions, we then have the physical-plane analog of Sobieczky's rheograph for finding both shock-free airfoils and wings. This extension of the original hodograph procedure to the physical plane was suggested by Sobieczky (1978) and subsequently implemented by Sobieczky et al. (1979). The relationship between the elliptic continuation in the hodograph plane and

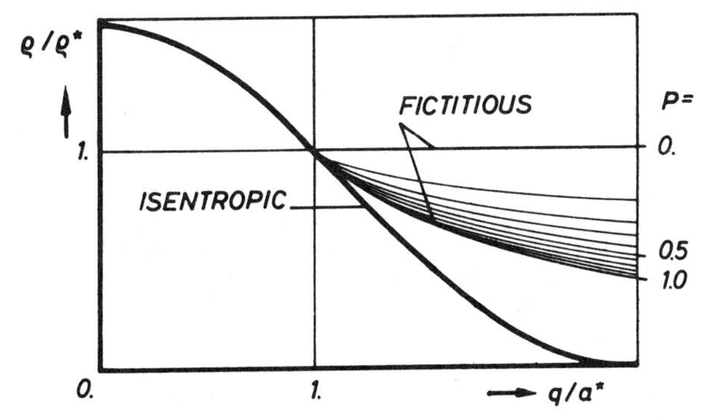

Figure 8 Isentropic and ficitious density-velocity relationship.

the fictitious gas in the physical space was reviewed by Sobieczky (1979). Various finite-difference and finite-volume analysis codes were adopted to include the fictitious gas as a shock-free design tool. These are discussed in Sobieczky et al. (1979), Yu (1980), Fung et al. (1980), Nakamura (1981), and Raj et al. (1982), among others. The method was applied to airfoil designs including viscous interactions by Nebeck et al. (1980).

A typical result of the fictitious-gas method including viscous effects, as calculated by the Grumfoil algorithm of Melnik et al. (1977), is shown in Figure 9. A general-aviation airfoil was redesigned to be shock-free by calculating the flow using one of the fictitious-gas laws, and this produces the sonic line 2. The data on the sonic line are then used to calculate the correct supersonic flow field using the method of characteristics. These calculations produce a new airfoil shape between the upstream and

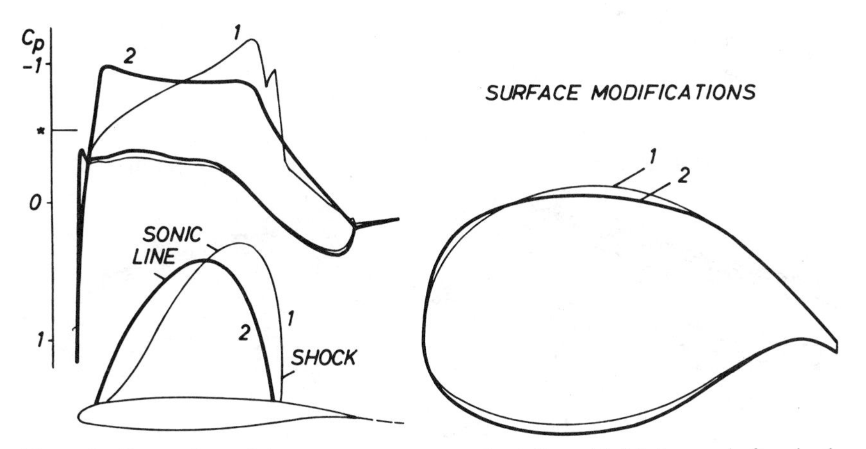

Figure 9 Comparison of the pressure on a general-aviation airfoil before and after shock-free redesign. $M_\infty = 0.77$, $C_L = 0.5$, $\mathrm{Re} = 6 \times 10^6$.

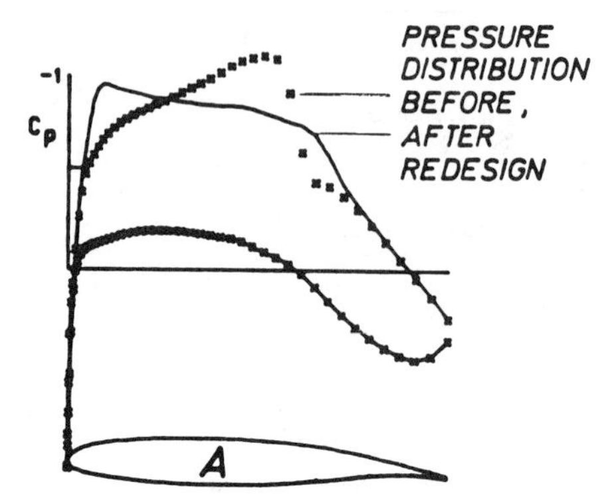

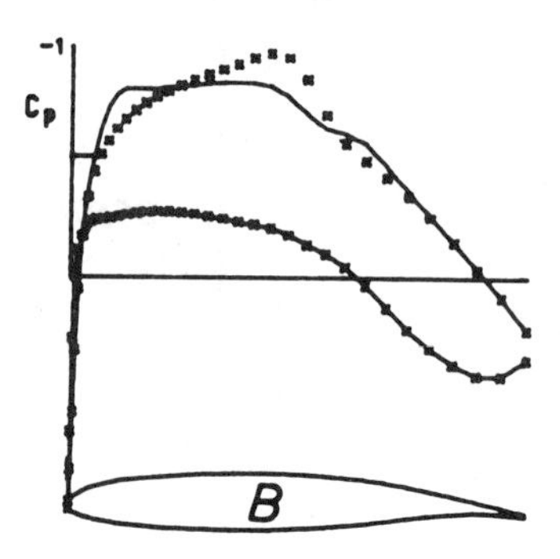

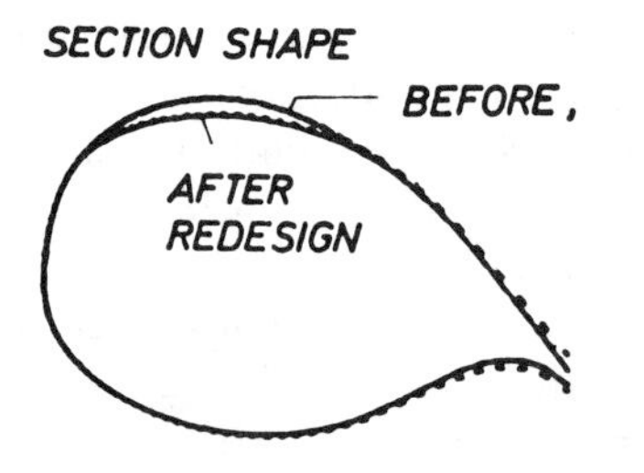

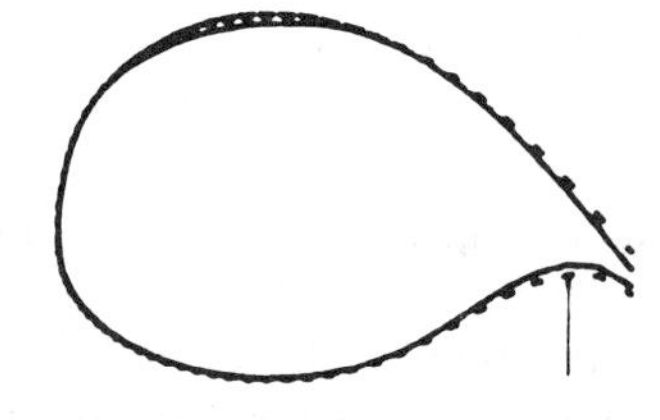

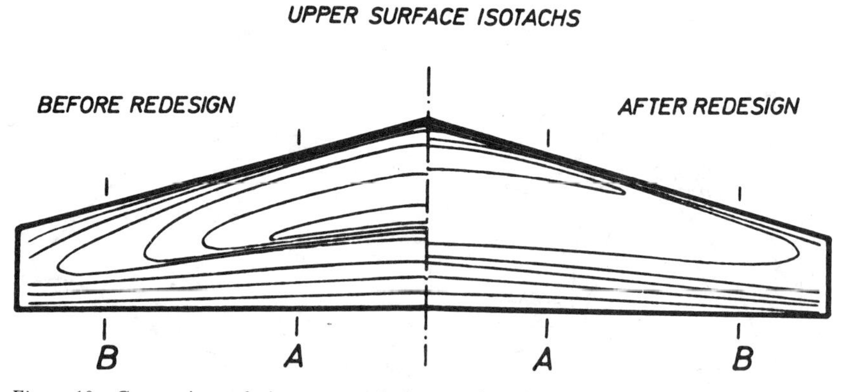

Figure 10 Comparison of wing pressure isobars and section pressure coefficients before and after shock-free redesign. $M_\infty = 0.77$, $C_L = 0.6$, $Re = 6 \times 10^6$.

downstream junctions of the sonic line. With this new airfoil shape 2 determined, the pressure is also known there. Or one may use it and recalculate the flow past it by using the correct-gas law. Both procedures result in the pressure distribution 2, verifying the method. For comparison purposes the pressure on an equally thick airfoil 1 that results in the same lift coefficient is also shown. Here the vertical section coordinates have been magnified five times. Further details may be found in Cosentino (1983). One of the principal virtues of this method is that one can take an airfoil section with otherwise very desirable qualities and modify it slightly to be shock free, and this can usually be done for relatively high Mach numbers and lift coefficients. The constraint on Mach number and lift coefficient depends on the airfoil geometry. If one tries to increase the Mach number to too large a value for a fixed lift coefficient or, conversely, tries to increase the lift coefficient to too high a value for a fixed Mach number, then the characteristic calculation will result in a limit line intervening between the sonic line and the airfoil. If the characteristic calculation is carried out in the hodograph variables, this will not affect the characteristic calculation, but the mapping of the results back to the physical plane will result in a multivalued stream function.

The concept of a fictitious gas not only allows the extension of Sobieczky's hodograph method to the physical plane, but also to three dimensions. There is a special difficulty in three dimensions, as we have already mentioned. In three dimensions the marching algorithm is implemented by approximating the osculating plane. Provided that the wing's aspect ratio is sufficiently large, then a constant spanwise station has been found to be a satisfactory approximation to the osculating plane. Numerical experiments by Fung et al. (1980) have shown that as long as the spanwise gradients are small when compared with those in the chordwise direction, then the approximation of the osculating plane by a spanwise station is satisfactory. A recent result of this method is shown in Figure 10. The pressure distribution is shown at two spanwise stations before and after redesign to be shock-free, as are the changes needed in the two sections there (in conjunction with the other changes along the wing span) to produce these pressure distributions. The section scale heights have been magnified five times in order to easily depict these changes. Also shown are the changes that occur in the isobar pattern on the top surface of the wing.

Aerodynamic Efficiency of Supercritical Airfoils and Wings

As we noted earlier, the aerodynamic efficiency of turbojet-powered aircraft is proportional to the Mach number. More specifically, the number of pounds of fuel burned per aircraft mile is directly proportional to the aerodynamic drag. At shock-free flow operating conditions the aerodynamic drag consists only of viscous and induced drag, as is the case in subcritical flow. The aerodynamic efficiency at these conditions is dominated by the linear dependence of the efficiency on the lift coefficient and Mach number. A shock-free airfoil, therefore, is no doubt operating below its true optimum efficiency at its shock-free design point. As discussed earlier, the actual optimum is dictated by the growth of the drag when the shock waves become dominant in the "viscous circle." A practical airfoil-design case is illustrated in Figure 11. This example and the wing discussed in the next section were designed by Redeker & Schmidt (1980). Shock-free design conditions were at $M_\infty = 0.73$ and $C_L = 0.6$. These values are found to be in good agreement both in numerical computation and in experiment, as indicated in the figure. Figure 11 also depicts curves of constant aerodynamic efficiency that were obtained from the wind-tunnel test. It can be seen that the optimal performance of the airfoil is obtained at a higher lift

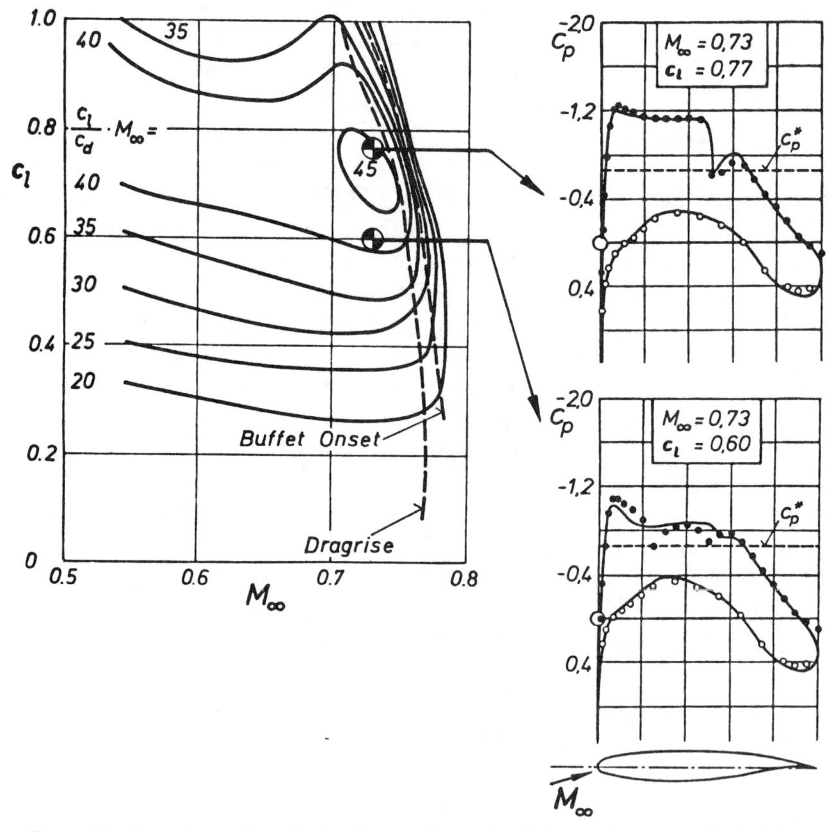

Figure 11 Experimental results for the aerodynamic efficiency of a supercritical airfoil.

coefficient, namely, $C_L \simeq 0.7$. At this Mach number and lift coefficient there is a weak shock wave that sits on the airfoil, but the increases in C_L that have accrued have not been offset by the very slight increases in drag that have occurred in obtaining this lift coefficient. This example shows that shock-free design is an effective method to find nearly optimum performance designs; these designs will be optimum at Mach numbers or lift coefficients slightly above those of the shock-free design point.

Practical swept-wing design technology uses airfoil sections for the definition of the wing. Although shock-free flow is usually not produced by this method, good efficiencies are obtained if a suitable airfoil is selected for the wing design. This is illustrated here with a supercritical wing design based on the previous airfoil example. This wing, shown in Figure 12, is a supercritical design for a transport aircraft. It evidences the trend toward higher aspect ratio and thicker wings with reduced sweep. All these are

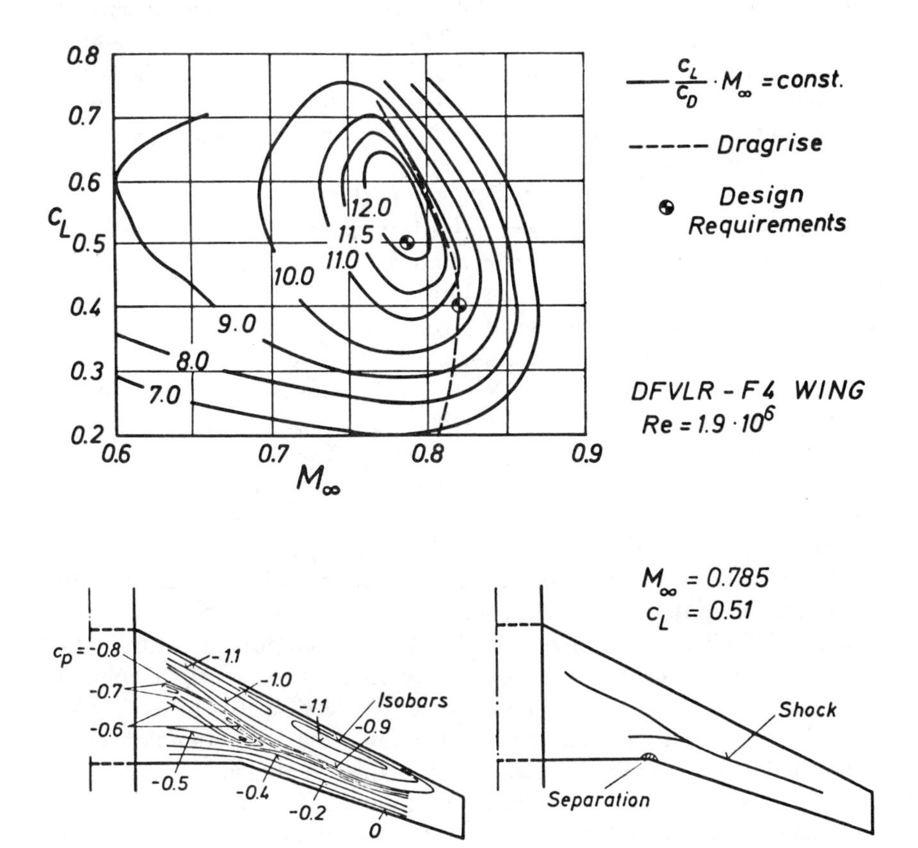

Figure 12 Experimental results for the aerodynamic efficiency of a supercritical wing.

possible through the use of the supercritical flow with reduced shock strengths. A detailed review of the role of supercritical wings in the aerodynamic design of commercial transport aircraft has been given by Lynch (1982).

Verification by Computation and Experiment

The verification of the shock-free design procedure discussed here awaits the analysis of some of the results by more advanced computational codes and experimental investigation. Computational analysis should vindicate these results if they have been obtained with a satisfactory analysis algorithm. Probably the principal shortcoming in present shock-free wing design is that the algorithms coupling viscous and inviscid effects are not as advanced as those for airfoils. One can, of course, also turn to experiment, but experiments suffer from the usual difficulties of transonic testing, namely, wind-tunnel wall effects. Of course all wings reside on aircraft fuselages and, while the application of the ideas reviewed here to wing-body combinations is in principle straightforward, it has yet to be effected in practice. Since very small changes in the wing shape have fairly dramatic effects on the flow, the inclusion of the fuselage in the computations seems essential in the design of advanced wings using a fictitious-gas procedure. The calculation of the elliptic flow field with the fictitious gas uses central differences, and this provides second-order accuracy. The addition of the method of characteristics to the supersonic part of the flow also retains this second-order accuracy, and thus the flow field is computed rather accurately. In three dimensions a leapfrog marching method provides local second-order accuracy; but the ill-posed nature of this problem makes the global accuracy indeterminate. Very small changes in airfoil or wing geometry, as noted earlier, have rather dramatic effects on the pressure distribution. The supersonic expansion near the nose of the airfoil is an especially delicate one. Figure 13 shows a typical computational grid used in the design and analysis of an airfoil. We can see that the method of characteristics has defined rather intensely the upper surface of the airfoil near the supersonic expansion; if these results are to be tested through numerical computation, then the conforming grid must resolve this detail in order to duplicate accurately the local upper-surface curvature. And analysis algorithms are only first-order accurate in the supersonic domain. Another difficulty that will always be present in the case of coupled viscous-inviscid interactions is that one must assume that the boundary layer computed with the fictitious gas models well the actual boundary layer that results. This is not precisely the case, of course, but the airfoil designs obtained by this procedure have proved quite satisfactory under computational tests.

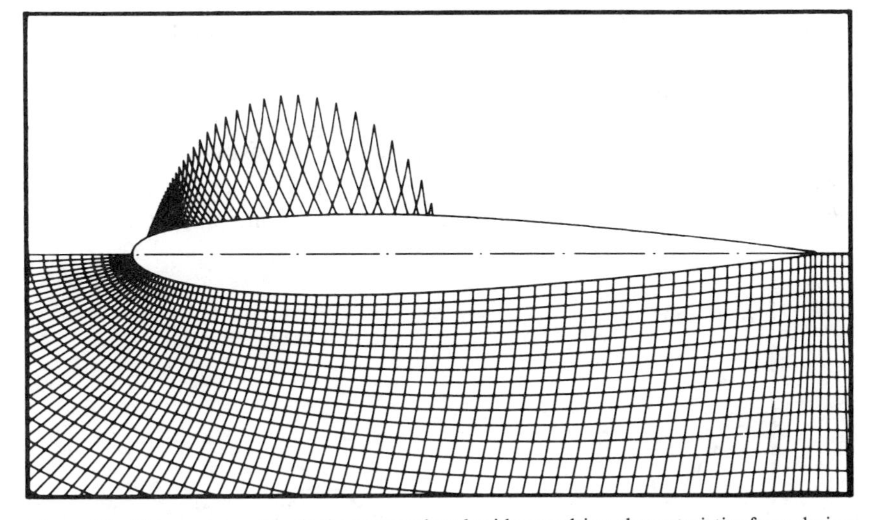

Figure 13 Comparison of typical computational grids: resulting characteristics from design and the surface-conforming grid from analysis.

A more serious and difficult problem is the experimental verification of these shock-free designs. Two practical requirements give rise to this problem. They are well known, and to a large part define the main problem of transonic testing. First, the proper representation of the viscous effects requires that the Reynolds number of the model and that of the real configuration be the same. This requires either a large model, a high wind-tunnel pressure, a low flow temperature, or some combination of all three. Fortunately, new facilities such as the National Transonic Facility at NASA Langley should make Reynolds-number simulation less difficult. Second, the wind-tunnel walls must properly model the far-field flow conditions, or else wind-tunnel wall effects will disturb the flow and produce erroneous results. This requires either large wind-tunnel cross sections compared with the model size in ventilated walls, or wind-tunnel walls that are adaptable so that they adjust to represent the stream surface bounding the correct far field. There is much progress in adaptive-wall technology, and this advance will also help provide reliable wind-tunnel verification of supercritical wing designs.

Advanced Technology Concepts

The need for increased aerodynamic efficiencies has led to some new ideas for future aircraft development. Drag reduction is the key to such improvements, and the transonic drag rise is delayed through the shock-free design procedure reviewed here. But the predominant drag is the viscous drag, and this can best be reduced by maintaining laminar flow over the wing surface. New ideas for boundary-layer control and removal offer the prospect of improving aircraft efficiency by as much as 40% on long-range flights. A combination of this technology with shock-free supercritical flows has been studied by Pfenninger et al. (1980).

The use of active controls, with their influence on aerodynamics, also offers potential improvements in efficiency. Active controls can be used to

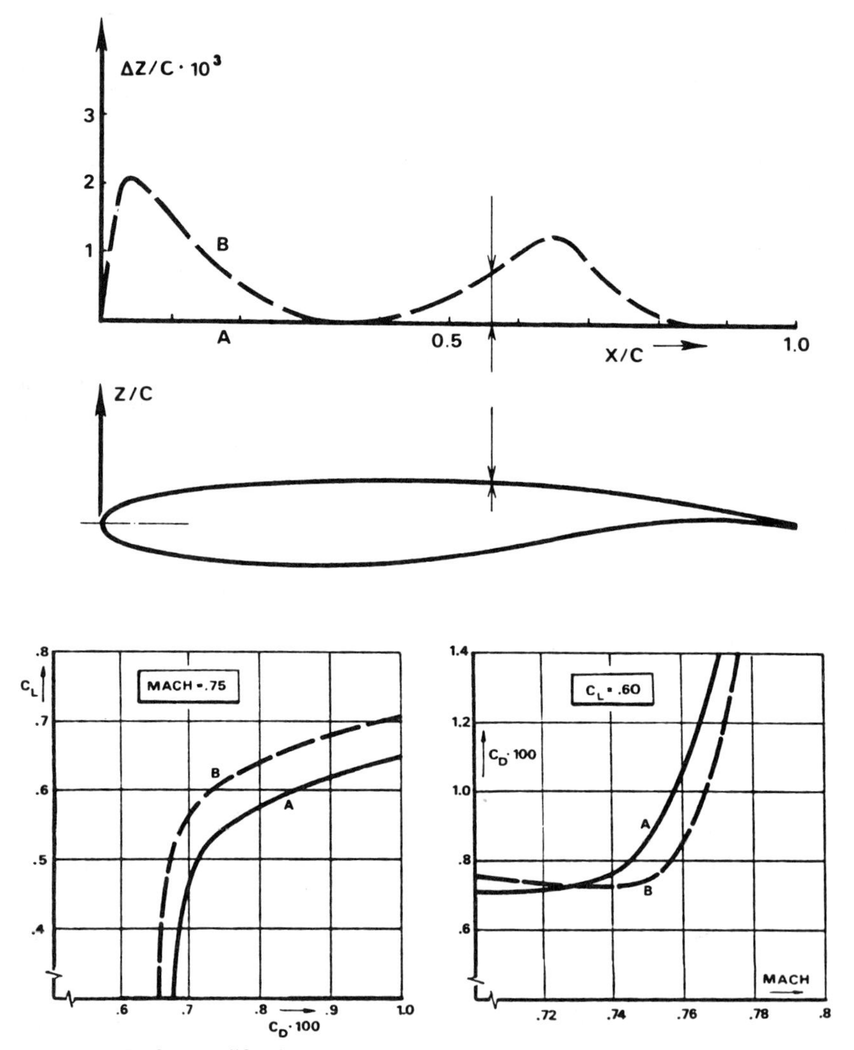

Figure 14 Surface modifications to an airfoil that shifts the drag rise toward a higher Mach number and lift coefficient.

reduce stability margins and gust loadings, resulting in reductions in both trim drag and structural weight. The technology for active control can also be employed to make small modifications to airfoil shapes should it be desirable to maintain their supercritical efficiency over a range of Mach numbers and lift coefficients. Figure 14 shows an airfoil that has been designed for $M_\infty = 0.73$ and $C_L = 0.55$. The two small bumps added on the upper surface shift the shock-free design conditions to $M_\infty = 0.75$ and $C_L = 0.6$. The surface modifications required are comparable to the size of the boundary-layer displacement thickness. Thus the technology that can be used to remove and control the boundary layer can also be applied to maintaining nearly shock-free flow over a range of operating conditions, as proposed by Sobieczky & Seebass (1980).

We should also remark that the method has been extended by S. S. Sritharan (1982) to the transonic cross-flow of supersonic conical wings, providing a method that can be used to eliminate embedded cross-flow shocks in the design of supersonic conical wings.

Conclusions

Methods of finding shock-free transonic flows have been briefly reviewed in this article. These methods are theoretically justified and numerically exact for airfoils. Within the confines of the accuracy of numerical algorithms, shock-free supercritical wings can also be found, provided their spanwise gradients are small compared with their chordwise ones. Both inverse methods and the nearly direct fictitious-gas design procedure provide effective tools for the design of supercritical aircraft wings. The latter has received recent attention because of its simplicity and ease of implementation. It provides a method of designing supercritical wings with the accuracy with which you can compute the aerodynamic flow over these wings. As the capability of numerical algorithms improves, the design of fully integrated fuselages and wings that have relatively high aerodynamic efficiency will be possible. This technology, when coupled with a number of other technologies such as active control, laminar-flow control, and composite primary structures, should make it possible to double the fuel efficiency of present transport aircraft before the end of the century.

Literature Cited

Bauer, F., Garabedian, P., Korn, D. 1972. *Supercritical Wing Sections, Lect. Notes Econ. Math. Syst. No. 66*. Berlin: Springer

Bauer, F., Garabedian, P., Korn, D., Jameson, A. 1975. *Supercritical Wing Sections II, Lect. Notes Econ. Math. Syst. No. 108*. Berlin: Springer

Bauer, F., Garabedian, P., Korn, D., Jameson, A. 1977. *Supercritical Wing Sections III, Lect. Notes Econ. Math. Syst. No. 150*. Berlin: Springer

Boerstoel, J. W. 1976. Review of the application of hodograph theory to transonic airfoil design and theoretical and experimental analysis of shock-free aerofoils. *Symp. Transsonicum II*, ed. K. Oswatitsch, D. Rues, pp. 109–33. Berlin: Springer

Boerstoel, J. W., Huizing, G. H. 1974. Transonic airfoil design by an analytic hodograph method. *AIAA Pap. 74-539*

Carlson, L. A. 1976. Transonic airfoil design using Cartesian coordinates. *NASA CR-2578*

Caughey, D. A. 1982. The computation of transonic potential flows. *Ann. Rev. Fluid Mech.* 14:261–83

Cosentino, G. B. 1983. Modifying a general aviation airfoil for supercritical flight. *J. Aircr.* 20:377–79

Eberle, A. 1976. An exact hodograph method for the design of supercritical wing sections. *Symp. Transsonicum II*, ed. K. Oswatitsch, D. Rues, pp. 314–21. Berlin: Springer

Fung, K.-Y., Sobieczky, H., Seebass, A. R. 1980. Shock-free wing design. *AIAA J.* 18:1153–58

Garabedian, P. R. 1975. On the design of airfoils having no boundary layer separation. *Adv. Math.* 15:164–68

Garabedian, P. R., Korn, D. G. 1971. Numerical design of transonic airfoils. In *Numerical Solution of Partial Differential Equations*, ed. B. Hubbard, 2:253–71. New York: Academic

Garabedian, P. R., McFadden, G. 1982. Design of supercritical swept wings. *AIAA J.* 20:289–91

Hassan, A., Seebass, A. R., Sobieczky, H. 1981. Transonic airfoils with a given pressure distribution. *AIAA Pap. 81-1235*

Henne, P. A. 1981. Inverse transonic wing design method. *J. Aircr.* 18:121–27

Hicks, R. M. 1981. Transonic wing design using potential-flow codes—successes and failures. *SAE Pap. 810565*

Lores, M. E., Hinson, B. L. 1982. Transonic design using computational aerodynamics. *Prog. Astronaut. Aeronaut.* 81:377–402

Lynch, F. T. 1982. Commercial transports—aerodynamic design for cruise performance efficiency. *Prog. Astronaut. Aeronaut.* 81:81–147

Melnik, R. W., Chow, R., Mead, H. R. 1977. Theory of viscous transonic flow over airfoils at high Reynolds numbers. *AIAA Pap. 77-680*

Morawetz, C. S. 1956. On the non-existence of continuous transonic flows past airfoils. I. *Commun. Pure Appl. Math.* 9:45–68

Morawetz, C. S. 1957. On the non-existence of continuous transonic flows past airfoils. II. *Commun. Pure Appl. Math.* 10:107–31

Morawetz, C. S. 1958. On the non-existence of continuous transonic flows past airfoils. III. *Commun. Pure Appl. Math.* 11:129–44

Nakamura, M. 1981. A method for obtaining shockless transonic flows past two-dimensional airfoils whose profiles are modified from a given arbitrary profile. *Trans. Jpn. Soc. Aeronaut. Space Sci.*, pp. 195–213

Nebeck, H. E., Seebass, A. R., Sobieczky, H. 1980. Inviscid-viscous interactions in the nearly direct design of shock-free supercritical airfoils. *Pap. 3, AGARD CP-291, Conf. Comput. Viscous-Inviscid Interact., Colorado Springs*

Nieuwland, G. Y. 1967. Transonic potential flow around a family of quasi-elliptical sections. *NLR TR T 172*

Nieuwland, G. Y., Spee, B. M. 1973. Transonic airfoils: Recent developments in theory, experiment, and design. *Ann. Rev. Fluid Mech.* 5:119–50

Pearcey, H. H. 1962. The aerodynamic design of section shapes for swept wings. *Advances Aeronaut. Sci.* 3:277–322

Pfenninger, W., Reed, H. L., Dagenhart, J. R. 1980. Design considerations of advanced supercritical low drag suction airfoils. *Prog. Astronaut. Aeronaut.* 72:249–71

Raj, P., Miranda, L. R., Seebass, A. R. 1982. A cost effective method for shock-free supercritical wing design. *J. Aircr.* 19:283–89

Redeker, G., Schmidt, N. 1980. Design and experimental verification of a transonic wing for a transport aircraft. *Pap. 13, AGARD CP-285, Conf. Subsonic/Transonic Configuration Aerodyn., Munich*

Rigaut, F. 1968. Détermination analogique de profils d'aile en régime transsonique. *Pap. 7, AGARD CP-35, Conf. Transonic Aerodyn., Paris*

Ringleb, F. 1940. Exakte Lösungen der Differentialgleichungen einer adiabatischen Gasströmung. *Z. Angew. Math. Mech.* 20:185–98. See also *J. R. Aeronaut. Soc.* 46:403–4 (1942)

Seebass, A. R. 1982. Shock-free configurations in two- and three-dimensional transonic flow. In *Transonic, Shock, and Multidimensional Flows: Adv. Sci. Comput.*, pp. 17–36. New York: Academic

Sobieczky, H. 1971a. Exakte Lösungen der ebenen gasdynamischen Gleichungen in Schallnaehe. *Z. Flugwiss.* 19:197–214

Sobieczky, H. 1971b. Rheoelektrische Analogie zur Darstellung transsonischer Stroemungen. *DLR FB 71-26*

Sobieczky, H. 1975. Entwurf überkritischer Profile mit Hilfe der rheoelektrischen Analogie. *DLR FB 75-43*

Sobieczky, H. 1978. Die Berechnung lokaler räumlicher Überschallfelder. *Z. Angew. Math. Mech.* 58T:331–33

Sobieczky, H. 1979. Related analytical, analog and numerical methods in transonic airfoil design. *AIAA Pap. 79-1556*

Sobieczky, H., Seebass, A. R. 1980. Adaptive airfoils and wings for efficient transonic flight. *ICAS Pap. 80-11.2, ICAS Congr., 12th, Munich*

Sobieczky, H., Stanewsky, E. 1976. The design of transonic airfoils under consideration of shock wave boundary layer interaction. *ICAS Pap. 76-14, ICAS Congr., 10th, Ottawa*

Sobieczky, H., Yu, N. J., Fung, K.-Y., Seebass, A. R. 1979. New method for designing shock-free transonic configurations. *AIAA J.* 17:722–29

Sritharan, S. S. 1982. *Nonlinear aerodynamics of conical wings.* PhD thesis. Univ. Ariz., Tucson

Swenson, E. V. 1968. Geometry of the complex characteristics in transonic flow. *Commun. Pure Appl. Math.* 27:175–85

Tranen, T. L. 1974. A rapid computer-aided transonic airfoil design method. *AIAA Pap. 74-501*

Volpe, G., Melnik, R. E. 1981. The role of constraints in the inverse design problem for transonic airfoils. *AIAA Pap. 81-1233*

Whitcomb, R. T., Clark, L. 1965. An airfoil shape for efficient flight at supercritical Mach numbers. *NASA TM X-1109*

Yu, N. J. 1980. Efficient transonic shock-free wing redesign procedure using a fictitious gas method. *AIAA J.* 18:143–48

Ann. Rev. Fluid Mech. 1984. 16: 365–424

PERTURBED FREE SHEAR LAYERS

Chih-Ming Ho and Patrick Huerre

Department of Aerospace Engineering, University of Southern California, Los Angeles, California 90089-1454

言天下之至動而不可亂也 周易繫辭[1]

Denique si semper motus connectitur omnis,
et vetere exoritur semper novus ordine certo,
nec declinando faciunt primordia motus
principium quoddam quod fati foedera rumpat,
ex infinito ne causam causa sequatur,
libera per terras unde haec animantibus exstat...?

Lucretius, *De Rerum Natura*[2]

1. INTRODUCTION

This review is about free shear layers of the kind that are formed by the merging of two streams initially separated by a thin surface: the flow is sketched in Figure 1. Intensive mixing occurs in the velocity-gradient region between the two free streams, and such layers are often referred to as

[1] *Speaking of the most restless phenomenon in the universe,*
one should not be distracted by its randomness

I Chin

[2] *Once more, if movement always is to other movement linked*
And if the new comes ever from the old,
As in determinist argument;
If atoms in their swerve do not fresh start
To break the bonds of Fate;
If cause may follow cause from infinite time,
Whence comes free will for living things on earth?

Lucretius, *Of the Nature of Things*

0066-4189/84/0115-0365$02.00

mixing layers. This simple flow configuration is a generic model arising in numerous natural phenomena and in artificial devices, such as combustors and gas lasers. From a practical point of view, one often seeks to manipulate the downstream evolution of shear flows to enhance the efficiency of certain industrial processes. In the present paper, the mixing layer is viewed more fundamentally as the prototype flow of a wider class of inviscidly unstable free shear flows that also includes jets and wakes. For the most part, we focus our attention on homogeneous, incompressible, two-dimensional mixing layers. However, some of the peculiar processes exhibited by two-dimensional or axisymmetric jets are briefly alluded to. Other related topics, such as aerodynamic-noise generation or the interaction between shear layers and downstream solid boundaries, are discussed in recent review articles by Goldstein (1984) and Rockwell (1983).

Two approaches are currently used to study turbulent shear flows. In the more classical theories, turbulence is viewed as an essentially random process, which can only be adequately described *statistically*. The flow field is then decomposed into a mean and a fluctuating part. The main difficulty resides in the derivation of a closed system of governing equations for the mean and fluctuating flows. For recent discussions of statistical theories of turbulence, the reader is referred to Monin & Yaglom (1971, 1975) and Lumley (1978, 1981), among others.

Another approach consists in viewing *quasi-deterministic* vortex structures as the main building blocks of turbulent mixing layers. Corrsin (1943) and Townsend (1947) were the first to notice that the interface separating turbulent and nonturbulent fluid is sharp and corrugated, and three decades later Kovasznay (1970) related this interface to the presence of large-scale vortex structures in boundary layers. Crow & Champagne (1971) discovered that the shear layer of a jet can support orderly vortical structures and operates as a finely tuned amplifier of upstream disturbances. Brown & Roshko (1974) subsequently confirmed that large-scale

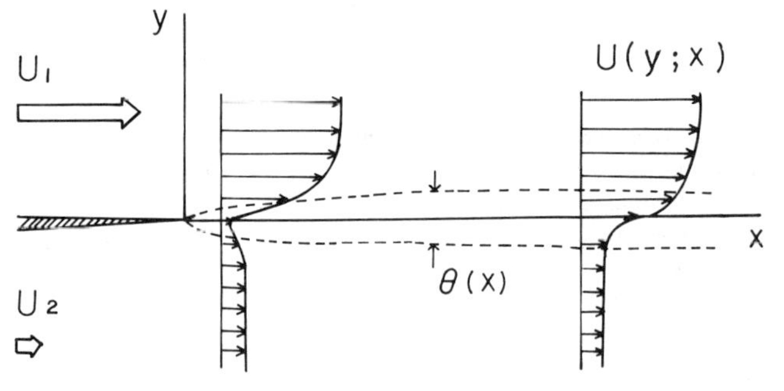

Figure 1 Sketch of spatially developing mixing layer.

coherent structures are indeed intrinsic features of turbulent mixing layers at high Reynolds numbers. Furthermore, sequential mergings of vortices provide the primary mechanism for the spreading of the layer in the downstream direction, as underscored by the experiments of Winant & Browand (1974). These observations have had a major impact on our understanding of turbulence in free shear flows, as discussed, for instance, in the reviews of Roshko (1976), Cantwell (1981), and Laufer (1975, 1983). In the present survey, we unequivocally adopt the coherent-structure point of view. Strong reservations on the part of some researchers [in particular P. Bradshaw and his co-workers (Chandrsuda et al. 1978)] do exist, however, regarding the indefinite persistence of two-dimensional orderly structures at very high Reynolds numbers.

Among the many experimental and theoretical studies of coherent structures, one may distinguish two main modes of approach. A first school would lean toward a description in terms of the evolution of the vorticity field in physical space. The structures are then considered as vortices of a certain characteristic size that undergo different types of nonlinear interactions. General accounts of vorticity dynamics can be found in Saffman & Baker (1979), Zabusky (1981), and Aref (1983). One of the main advantages of this viewpoint is that the results can be compared, using some caution, with flow-visualization studies conducted in the laboratory. The application of vortex methods to direct two-dimensional numerical simulations does yield a realistic fully nonlinear modeling of the roll-up of the shear layer and of vortex mergings. The extraction of detailed information regarding the intrinsic scales of the motion tends, however, to be somewhat more involved. On the experimental side, it should be mentioned that the vorticity field cannot presently be measured satisfactorily with available instruments, and various approximation schemes have to be used.

Other workers have chosen instead to rely on classical hydrodynamic stability theory: the unsteady mixing layer is then conceptualized as a superposition of interacting instability waves that propagate and amplify in the downstream direction. The flow is therefore examined in Fourier space rather than in physical space, and large-scale vortices are themselves composed of several instability waves of different frequency. As a result, measured data in frequency space cannot easily be related to the evolution of the orderly structures, unless a technique such as flow visualization is used to make this crucial connection. As we shall see, linear stability analysis has been shown to describe very satisfactorily the initial development of the mixing layer preceding the establishment of fully rolled-up vortices. There are also strong indications that its use in conjunction with an appropriate phenomenological model can also help to determine the main length scales of the flow much farther downstream. In contrast with the vortex approach, it is impossible to use stability analyses to model strongly nonlinear phenomena. Some of the observed wave interactions may only be weakly nonlinear, and analytical techniques can be developed to handle such effects. Truly nonlinear processes, however, have to be simulated numerically. General surveys on the stability of mixing layers have been written by Michalke (1972), Liu (1981), Huerre (1984), and Morkovin (1984).

Each individual vortical structure occurring in naturally excited shear layers is known to be subjected to spatial jitter as it travels along the stream. As a result, a large number of probes have to be used at any instant to obtain a detailed knowledge of the flow field. The application of low-level forcing to free shear layers was initially meant to overcome this problem. An excitation provided a clear phase reference and thereby alleviated the need for many simultaneous measurements. The low-level forcing disturbance was thought to have no particular effects on the dynamics of the flow, other than a decrease in broadband noise. It has only recently been realized that the evolution of mixing layers is highly susceptible to very low-amplitude disturbances. Thus, artificially generated perturbations can be used advantageously not only to deepen our basic understanding of shear-layer dynamics, but also to alter significantly the downstream development of the flow.

Naturally or artificially excited shear layers are necessarily perturbed flows. Without prejudice to other interpretations, our tendency has been therefore to emphasize hydrodynamic instability-wave concepts. This article is subdivided into four main sections. The basic mechanisms governing the development of mixing layers are reviewed in Section 2. In Sections 3 and 4, we assess our current knowledge of the effects of operational parameters and artificial forcing on the characteristics of the flow. Finally, Section 5 is devoted to speculations regarding global effects and the feedback mechanism.

2. DYNAMICAL PROCESSES IN FREE SHEAR LAYERS

Within the conceptual framework of stability theory, two laminar streams of velocity U_1 and U_2 ($U_1 > U_2$) are assumed to give rise, by viscous diffusion, to a weakly diverging *steady* basic flow $U(y;x)$, as shown in Figure 1. The variables x, y, z, and t respectively denote the streamwise, cross-stream, and spanwise coordinate, and time. The momentum thickness $\theta(x)$ initially increases as the square root of x. The flow can conveniently be characterized by two nondimensional parameters R and

Re_0. The velocity ratio $R = \Delta U/(2\bar{U})$ measures the relative magnitude of the total shear $\Delta U = U_1 - U_2$, as compared with the average velocity $\bar{U} = (U_1 + U_2)/2$. When $R = 0$, the flow reduces to a wake. When $R = 1$, only one stream is present, as in the initial mixing region of a jet. The Reynolds number $\mathrm{Re}_0 = \bar{U}\theta_0/\nu$, based on the initial momentum thickness θ_0 and kinematic viscosity ν, accounts for viscous effects.

2.1 *Linear Instability Regime*

The development of mixing layers downstream of a splitter plate is initially dominated by a linear instability mechanism. For general reviews of the hydrodynamic stability of parallel flows, the reader is referred to Lin (1955), Drazin & Howard (1966), Betchov & Criminale (1967) and Drazin & Reid (1981).

The basic vorticity distribution, which possesses a maximum, is inviscidly unstable to small perturbations via the Kelvin-Helmholtz instability mechanism. Thus, two-dimensional waves grow exponentially with downstream distance and are observed to roll up into vortices. If the basic flow $U(y)$ is taken to be strictly parallel and disturbances are infinitesimal, the perturbation stream function can be cast into the form

$$\psi(x, y, t) = \phi(y) \exp\{i(\alpha x - \omega t)\} + \text{c.c.}, \tag{1}$$

where $\omega = 2\pi f$ is the angular frequency, α is the wave number, and c.c. denotes the complex conjugate.

In the inviscid limit $\mathrm{Re} \equiv \infty$, the eigenfunction $\phi(y)$ is obtained by solving the linear Rayleigh equation. Most calculations have been conducted for the hyperbolic-tangent velocity profile $U(y;R) = \bar{U}[1 + R \tanh(y/2\theta)]$. Michalke (1964, 1965a,b) has determined numerically the stability characteristics of both temporal (ω complex, α real) and spatial (ω real, α complex) waves in the limit $R = 1$. More recently, Monkewitz & Huerre (1982) have studied the theoretical dependence of spatially growing waves on R for both the hyperbolic-tangent profile and the Blasius shear layer calculated by Lock (1951). The equivalent problem for axisymmetric jets has been investigated by Michalke & Hermann (1982). The nondimensional spatial growth rate $(-\alpha_i\theta)$ and phase velocity $c_r/\bar{U} = \omega/(\alpha_r\bar{U})$ are found to vary with Strouhal number $\mathrm{St} = f\theta/\bar{U}$ in the fashion displayed in Figure 2. The Strouhal number $\mathrm{St}_n = 0.032$ of the most amplified wave corresponds to the natural frequency f_n of the mixing layer: it changes by only 5% between $R = 0$ and $R = 1$. The maximum amplification rate $(-\alpha_i\theta)_{max}$ increases approximately linearly with R, and the associated phase velocity is equal to the average velocity $\bar{U}$ of the two streams. In the limit of small shear $R \ll 1$, the spatial growth rate $(-\alpha_i)$ and its temporal counterpart ω_i are related via the transformation $(-\alpha_i)/R \sim \omega_i/\bar{U}$, sug-

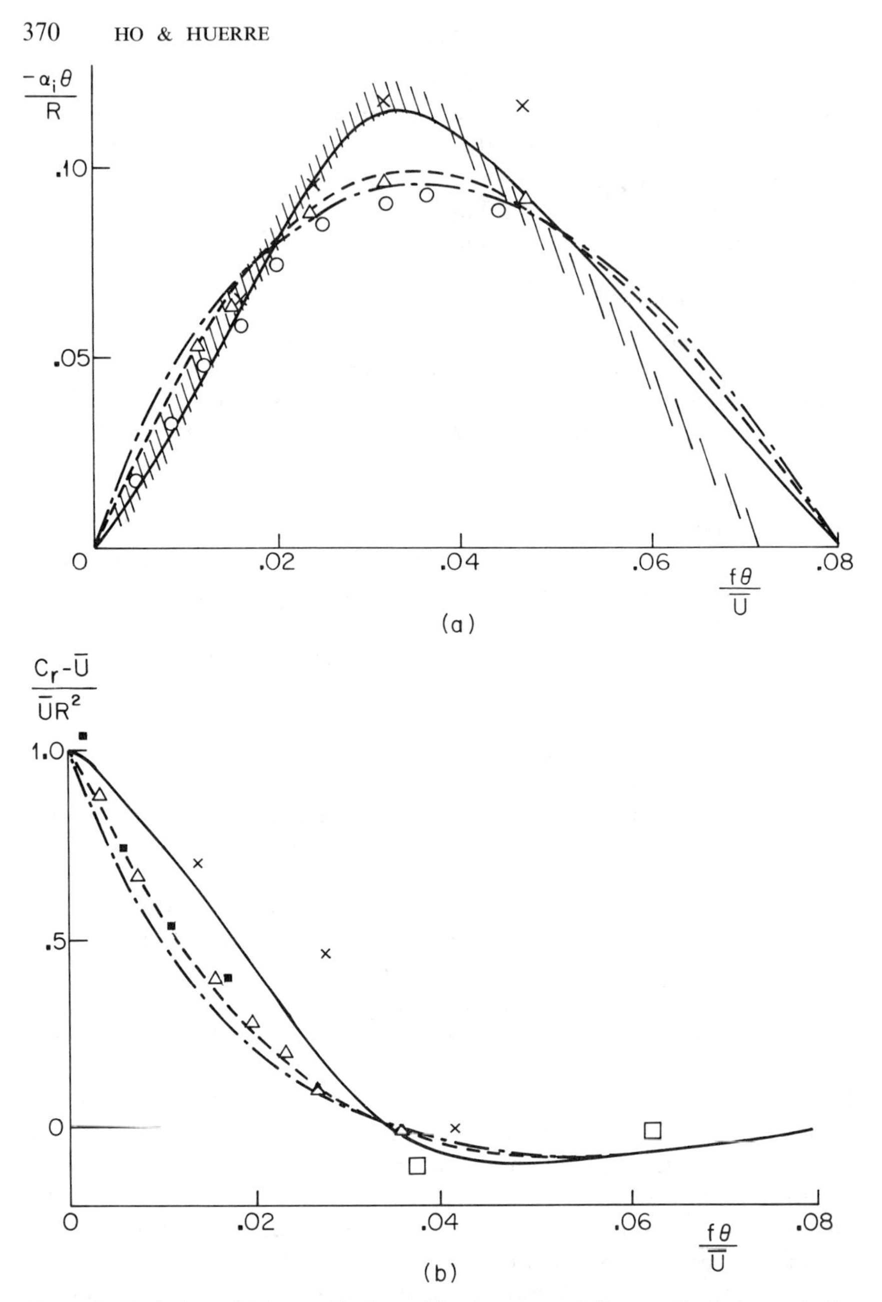

Figure 2 Variations of (*a*) normalized amplification rate and (*b*) normalized phase velocity with Strouhal number $f\theta/\overline{U}$. Linear stability theory (from Monkewitz & Huerre 1982): — $R = 1$; —·— $R = 0.5$; ——— $R \ll 1$. Experiments: □ $R = 1$ (Sato 1960); ○ $R = 1$ (Freymuth 1966); × $R = 0.72$ (Miksad 1972); \\\ $R = 1$ (Fiedler et al. 1981); △ $R = 0.31$ (Ho & Huang 1982); ■ $R = 1$ (Drubka 1981).

gested by Gaster (1962). As a consequence, the $R \ll 1$ curve in Figure 2*a* simply represents the variations of the temporal amplification rate ω_i with wave number. We also note (Figure 2*b*) that, in contrast with temporal waves, spatial waves are *dispersive* below St_n and nondispersive above.

Experimental investigations of the initial instability region (Sato 1956, 1959, Browand 1966, Freymuth 1966, Miksad 1972, Mattingly & Chang 1974, Ho & Huang 1982) have been aimed at testing these predictions. In unforced mixing layers, the peak of the power spectrum immediately downstream of the trailing edge is found to occur at the calculated natural frequency f_n given by $St_0 = f_n\theta_0/\bar{U} = 0.032$. In most cases, however, the flow is forced (see Section 4.1) at a specific frequency f_f within the linearly unstable range. As an example, Figure 3 summarizes the response of a

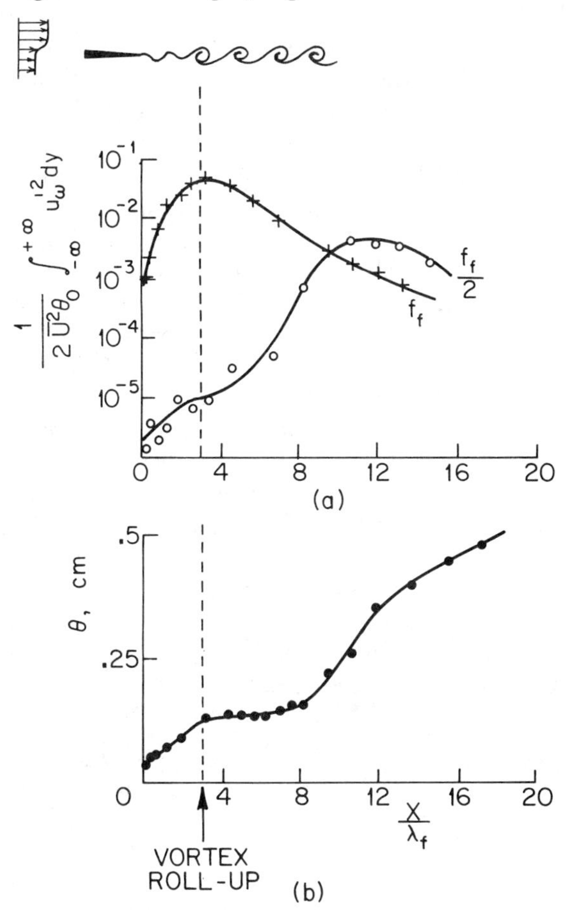

Figure 3 Evolution of (*a*) spectral components and (*b*) momentum thickness θ with downstream distance (from Ho & Huang 1982). $\frac{1}{2}f_n < f_f \leq f_n$; $R = 0.31$. u'_ω is the narrow-band velocity fluctuation centered at $\omega = 2\pi f$. $\lambda_f = \lambda_r$.

mixing layer to an excitation f_f close to f_n: the variations of the momentum thickness $\theta(x)$ and of the main frequency components are represented along the streamwise direction. Throughout Section 2, our discussion is restricted to the case $f_f = f_n$, which serves as an illustration. Measurements made at a given downstream station compare very favorably with the spatial stability properties of the local parallel velocity profile at the forcing Strouhal number $\mathrm{St}_f = f_f\theta(x)/\bar{U}$. The studies of Michalke (1965a,b) and Freymuth (1966), in particular, have clearly demonstrated that only a spatial analysis could account for the observed cross-stream distributions of fluctuating velocity. The use of temporal theory to describe forced instability waves in mixing layers with $0 < R < 1$ is therefore not valid, as confirmed theoretically by Huerre & Monkewitz (1983). To conclude this brief account of the linear stability of parallel shear layers, we note that three-dimensional spatial instability waves have smaller spatial growth rates than their two-dimensional counterparts (Michalke 1969), a feature in agreement with the predominantly two-dimensional character of the instability-wave evolution (see Section 2.4).

It should be recognized, however, that the basic flow is really not parallel, its thickness θ increasing with downstream distance. Furthermore, the velocity profile $U(y;x)$ is wakelike immediately downstream of the plate trailing edge and gradually relaxes to a shear-layer distribution. Thus, to comply with the parallel-flow and small-disturbance assumptions, comparisons between measurements and calculations should be made one instability wavelength downstream of the trailing edge (see Section 4.2) *and* before any significant nonlinear interactions have occurred. The choice of a station satisfying both these criteria is, in practice, a subjective matter, but it directly influences the predicted value of f_n, as given by $f_n\theta(x)/\bar{U} = 0.032$.

To overcome this ambiguity, spatial linear instability theory has been extended to slowly diverging flows of small spreading rate $d\theta/dx = O(\mu)$, with $\mu \ll 1$ (Crighton & Gaster 1976, Plaschko 1979, Gaster et al. 1983). For instance, the parallel hyperbolic-tangent profile can be generalized to the self-similar family $U(y;X) = \bar{U}[1+R\tanh\{y/2\theta(X)\}]$, where $X \equiv \mu x$ is a slow space scale characterizing the basic flow variations along x. The perturbation stream function then takes the form

$$\psi(x,y,t) \sim A(X)\phi(y;X)\exp\left\{i\left(\int_0^X \alpha(\mu\xi)\,d\xi - \omega t\right)\right\} + \text{c.c.} \tag{2}$$

The complex amplitude function $A(X)$, which accounts for nonparallel effects, satisfies a linear-evolution equation of the form $dA/dX + p(X)A = 0$, where $p(X)$ is obtained from the detailed calculations. The local stability properties $\alpha(X)\theta(X)$ and $\phi(y;X)/\theta(X)$ are still given by the parallel-flow results of Figure 2 applied at the Strouhal number $\mathrm{St}(X) = f\theta(X)/\bar{U}$. According to this theory, the total amplification rate becomes

$$\frac{1}{|\psi|}\frac{\partial|\psi|}{\partial x} = -\alpha_i(X) + \mu\left\{\frac{1}{A(X)}\frac{d|A(X)|}{dX} + \frac{1}{|\phi(y;X)|}\frac{\partial|\phi(y;X)|}{\partial X}\right\}. \tag{3}$$

It depends not only on the downstream location X but also on the cross-stream coordinate y. One can also show that different flow quantities (pressure, velocity) are characterized by different amplification rates. A qualitative understanding of this approach can be achieved by neglecting temporarily the $O(\mu)$ correction terms of (3) and by reasoning from Figure 2*a*. For a given value of the frequency f, the local Strouhal number $\mathrm{St}(X)$ increases with increasing thickness $\theta(X)$ downstream. As a result, the local amplification rate $-\alpha_i(X)$ ultimately decreases to zero at the location $\mathrm{St}(X) = 0.079$, where the wave is locally neutral. Thus, unlike parallel flows, slowly diverging flows yield a maximum wave amplitude that occurs at the locally neutral downstream station. This is an approximate reasoning, which ignores the contributions of $O(\mu)$ in Equation (3). Surprisingly enough, this procedure has not yet been applied to the amplitude development in the initial instability region. It has, however, been used to model the evolution of large-scale structures in turbulent mixing layers (Section 2.4) and jets (Section 2.8).

2.2 *Nonlinear Instability Regime*

Beyond the region of exponential growth, Kelvin-Helmholtz instability waves evolve into a periodic array of compact spanwise vortices moving at the average velocity $\bar{U}$ with a wavelength $\lambda_n = \bar{U}/f_n$. A striking illustration of this phenomenon is provided by the flow-visualization pictures of Roberts et al. (1982) shown in Figure 4. The recirculating regions, which contain most of the vorticity, are usually referred to as *Kelvin cat's eyes*. They are connected to each other via thin vorticity layers or *braids*. The main features of the roll-up process and the ensuing "climax state" have been confirmed by many two-dimensional numerical simulations of temporally evolving shear layers (Amsden & Harlow 1964, Christiansen 1973, Acton 1976, Patnaik et al. 1976, Aref & Siggia 1980, Riley & Metcalfe 1980, Corcos & Sherman 1983). Experimental observations indicate that the roll-up process is predominantly two-dimensional and is completed at the downstream station where the fundamental component at frequency f_n reaches its maximum amplitude (Figure 3*a*). It is accompanied by the generation of a subharmonic component $f_n/2$, which in the case of Figure 3 is three orders of magnitude below the fundamental. Thus, the *linear* stability theory of slowly diverging flows outlined in Section 2.1 may only approximately describe the equilibration of the fundamental f_n.

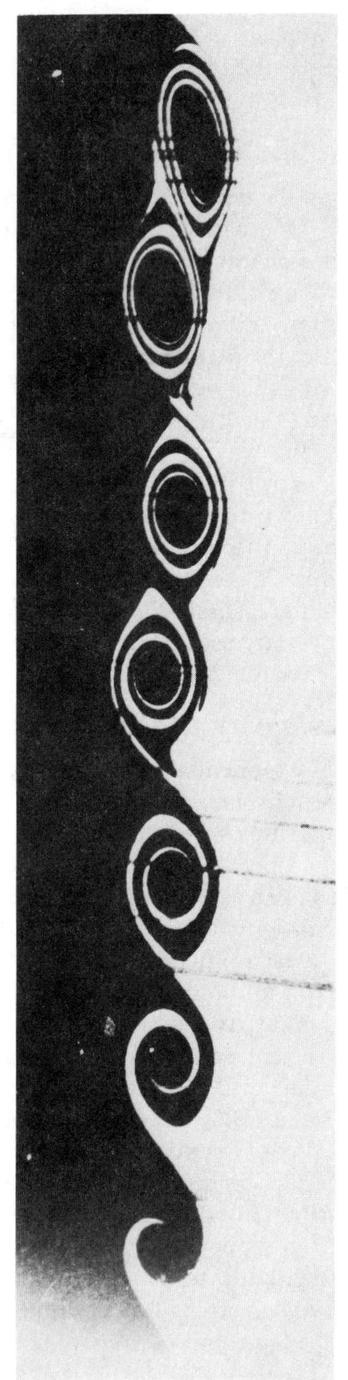

Figure 4 Roll-up of spanwise vortices (from Roberts et al. 1982). $f_f = f_n$; $R < 1$.

These experimental results do suggest, however, that a *weakly nonlinear* self-interaction approach neglecting the subharmonic $f_n/2$ might be appropriate to model the saturation mechanism of f_n. For general reviews of nonlinear stability theory in fluids, the reader may consult, among others, Stuart (1971), Stewartson (1974), and Maslowe (1981). The specific case of free shear layers is discussed in detail by Huerre (1984). The method takes into account the distortion of the basic flow as well as the generation of higher harmonics, but it relies on two strong assumptions: wave amplitudes are small, and amplification rates remain low during the entire wave development. Additionally, most investigations are restricted to an effectively parallel basic flow. Applications of this formulation to homogeneous free shear layers at large Reynolds number have been limited to nearly monochromatic waves in the vicinity of the neutral mode in Figure 2. The perturbation stream function is again written in terms of a complex amplitude $A(X, T)$ as

$$\psi(x, y, t) \sim \varepsilon A(X, T)\phi_1(y)\exp\left\{\frac{i}{2\theta}(x-\bar{U}t)\right\}+\text{c.c.}, \tag{4}$$

where the subscript 1 denotes the neutral mode. The slow space and time scales X and T are suitably related to the small-amplitude parameter ε, and the nonlinear partial differential equation satisfied by $A(X, T)$ is obtained by making use of singular perturbation methods. The nature of the finite-amplitude motion is very sensitive to the structure of the critical layer, where the phase speed of the wave nearly matches the basic flow velocity. The role of the critical layer in shear flows has been reviewed by Stuart (1971), Stewartson (1981), and Huerre & Redekopp (1983). From a physical standpoint, this region contains the Kelvin cat's eyes of Figure 4. The balance between nonlinear effects and viscous dissipation in the layer can be characterized by a critical-layer Reynolds number $\text{Re}_{\text{CL}} \equiv \text{Re}\ \varepsilon^{3/2}$, based on its thickness $\varepsilon^{1/2}$ and a typical perturbation velocity ε. Within the classical theory (Lin 1955), $\text{Re}_{\text{CL}} \ll 1$, and the critical layer is viscous dominated. Benney & Bergeron (1969) and Davis (1969) were the first to investigate the structure of nonlinear critical layers for singular neutral modes in the limit $\text{Re}_{\text{CL}} = \infty$. These ideas have subsequently been extended by Haberman (1972) to arbitrary values of Re_{CL}, for which viscous and nonlinear phenomena coexist in the layer. The application of these concepts to marginally unstable waves of the form (4) leads to the following results: when $\text{Re}_{\text{CL}} \ll 1$, i.e. at low-amplitude levels, $A(X, T)$ satisfies a Stuart-Landau equation

$$\frac{dA}{dT} = \sigma A + b|A|^2 A, \tag{5}$$

where σ and b are complex constants. The cubic nonlinear term is destabilizing with Real $b > 0$ (Huerre 1980), a result that corrects the earlier study of Schade (1964). When $\text{Re}_{\text{CL}} \gg 1$, i.e. for large fluctuations levels, $A(X, T)$ satisfies an equation admitting finite-amplitude oscillatory solutions (Benney & Maslowe 1975). Finally, at finite values of Re_{CL}, spatial waves experience decreasing growth rates as they amplify, ultimately giving rise to algebraically growing waves at large distances (Huerre & Scott 1980). When viscous effects are important throughout the flow, namely, $\text{Re} = O(1)$, critical-layer considerations do not apply, and the amplitude A satisfies a Stuart-Landau equation (5) with a stabilizing (Real $b < 0$) cubic nonlinear term (Maslowe 1977a). The experimentally determined values of Re_{CL} associated with the evolution of the fundamental do encompass these different regimes, as noted by Miksad (1972). But there is presently no theory available that would link them in a specific manner during the course of the amplitude development. Furthermore, all the above analytical studies unfortunately fail to describe the behavior of the Strouhal number St_{n} of the most amplified wave, which is well below the neutral Strouhal number. This conceptual difficulty can be partially remedied if one adds a stabilizing mechanism, such as a density gradient [see Stewartson (1981) for detailed references] or nearby solid boundaries (Huerre 1983). All linearly unstable waves then fall within reach of the weakly nonlinear formalism. The introduction of such extraneous agents might well reproduce the main qualitative features of finite-amplitude waves in constant-density free shear layers. It is, however, doubtful that they will lead to quantitative predictions of the saturation level reached by the natural frequency f_{n}.

The emergence of a subharmonic component at half the frequency of the fundamental constitutes one of the most striking features in mixing-layer dynamics, as underscored by the experiments of Sato (1959), Wille (1963), Browand (1966), Freymuth (1966), Miksad (1972), and Ho & Huang (1982). When the forcing frequency f_{f} is sufficiently close to f_{n}, as in Figure 3, the subharmonic first reaches a plateau at the station where the fundamental equilibrates. This is followed further downstream by a sharp increase in the subharmonic amplification rate. The excitation of a large subharmonic disturbance can be accounted for by appealing to a *secondary subharmonic resonance* mechanism that arises when the fundamental f_{n} has reached its equilibrium level. In the simplest formulation (Lamb 1932), the row of vortex structures with wavelength λ_{n} is modeled by an array of point vortices. A straightforward linear analysis conducted in the comoving frame indicates that this configuration is most unstable to a disturbance of wavelength $2\lambda_{\text{n}}$. For a more satisfactory representation of the fundamental, we exploit the results of the qualitative reasoning outlined in Section 2.1: the natural component in its equilibrated state can be reasonably well approximated by a neutral mode. Following Kelly (1967) and the recent discussion of Monkewitz (1982), one then examines the linear temporal instability of the *periodic* flow

$$U(y)+\varepsilon\left\{A\phi_1'(y)\exp\left[\frac{i}{2\theta}(x-\bar{U}t)-i\beta\right]+\text{c.c.}\right\} \tag{5a}$$

to a subharmonic wave of the form

$$\psi(x,y,t)\sim\varepsilon B(T)\phi_{1/2}(y)\exp\left[\frac{i}{4\theta}(x-\bar{U}t)+\omega_{i1/2}t\right]+\text{c.c.} \tag{5b}$$

In the above relations, $\omega_{i1/2}$ and $\phi_{1/2}(y)$ are, respectively, the subharmonic temporal amplification rate and eigenfunction, β is the phase difference between the two waves, and $T=\varepsilon t$ is a slow time scale. Note that the subharmonic component (5*b*), close to the most amplified wave at $\alpha\theta=0.22$, is the only wave that can reproduce itself via quadratic interactions with the periodic component: this subharmonic resonance mechanism requires that both waves have approximately the same phase speed $\bar{U}$. The amplitude function $B(T)$ is found to satisfy the equation

$$\frac{dB}{dT}=c\exp(-i\beta)AB^*, \tag{6a}$$

where c is a constant and B^* denotes the complex conjugate of B. The initial total subharmonic growth rate $\omega_{i1/2}+\{d(\ln|B|)/dT\}$ is proportional to the periodic amplitude A and varies continuously from a maximum when both waves are in phase ($\beta=0$) to a minimum when they are out of phase ($\beta=\pi$). These results are in qualitative agreement with numerical simulations (Patnaik et al. 1976, Riley & Metcalfe 1980) and with experimental observations (Zhang et al. 1983). To allow more detailed comparisons with experiments, future theoretical studies could be aimed at extending Kelly's analysis to a spatially developing subharmonic. Spatial waves are dispersive (Figure 2*b*) and, as emphasized by Petersen (1978), subharmonic resonance can only occur downstream of the station given by $f_n\theta(x)/2\bar{U}=\text{St}_n$, where both waves become approximately nondispersive. According to experiments (Figure 3*a*) the subharmonic growth rate does increase significantly around that same downstream location, as long as the flow is not forced at the subharmonic frequency itself. In more elaborate analyses of subharmonic resonance arising, for example, in stratified shear layers (Kelly 1968, Maslowe 1977b, Weissman 1979), the subharmonic B, governed by the amplitude equation (6a), is allowed to react back nonlinearly on the fundamental A through a second coupled equation

$$\frac{dA}{dT}=gB^2\exp(i\beta). \tag{6b}$$

Applications of this approach to homogeneous free shear layers have been hampered by the fact that the linear growth rate $\omega_{i1/2}$ cannot be regarded as small.

The periodic row of Kelvin cat's eyes (Figure 4) can be more satisfactorily modeled by the exact solution of the two-dimensional Euler equations discovered by Stuart (1967). In the present notation, the associated stream function is

$$\psi(x, y, t) = \frac{\bar{U}a}{4\pi}\left[y + 2R \ln\left\{\cosh\left(\frac{2\pi y}{a}\right) - \rho \cos\left(\frac{2\pi}{a}(x - \bar{U}t)\right)\right\}\right], \quad (7)$$

where a is the distance between consecutive vortices and ρ is a vorticity concentration parameter: in the limit $\rho \to 1$, the row of point vortices is recovered, while an expansion of (7) for small ρ reduces to the periodically perturbed parallel tanh shear layer (5a). Browand & Weidman (1976) have compared their observations of fully rolled-up vortices with Stuart's solution (7) and found that the value $\rho = 0.25$ did provide a good overall fit of the measured vorticity distribution. A linear stability analysis recently conducted by Pierrehumbert & Widnall (1982) reveals that the periodic flow (7) is also most unstable to a disturbance of wavelength $2a$: the subharmonic temporal growth rate increases by approximately 20% as the basic vorticity field (7) changes from a parallel tanh shear layer ($\rho = 0$) to point vortices ($\rho = 1$).

2.3 *Vortex-Interaction Mechanisms*

It is now well established that the selective growth of the subharmonic component leads farther downstream to the *pairing* of neighboring vortices, as shown in the flow-visualization pictures of Figure 5. This remarkable interaction was first identified experimentally by Wille (1963) and subsequently documented by Freymuth (1966). But Winant & Browand (1974) were the first to demonstrate that successive mergings of vortices were indeed the primary process governing the streamwise growth of the mixing layer. Thus, the vorticity initially contained in the basic velocity profile is being constantly redistributed into larger and larger vortices, their wavelength and strength being doubled after each interaction. Equivalently, the passage frequency is halved after each coalescence (Kibens 1980), and the spectrum exhibits a shift toward lower frequencies with increasing distance from the trailing edge. Furthermore, the pairing location, defined by the cross-stream alignment of the two vortices, occurs at the downstream station where the subharmonic reaches a maximum (Ho & Huang 1982). In a jet with $R = 1$, amplification rates are high and this location is close to the station where the subharmonic first becomes nondispersive (Petersen 1978). Under controlled conditions (see Section 4.4

for a more detailed discussion), the first few pairings are accompanied by an approximate doubling in momentum thickness θ, the variations of θ between mergings being negligible. This feature is more readily observed in Figure 19*b* than in Figure 3*b*. But in both instances merging interactions eventually occur randomly along the downstream direction and give rise to a linear spreading rate. Beyond the initial roll-up phase, vortex interactions therefore govern in large part the rate of entrainment of surrounding fluid. However, this point of view remains the subject of debate: for instance, the recent digital image analysis of a flow-visualization film performed by Hernan & Jimenez (1982) indicates that most of the entrainment takes place during the growth stage of the large eddies and not during pairing. Finally, we emphasize that interactions between vortices are not necessarily limited to pairings, but may also involve three or more vortices as is further discussed in Section 4.4.

A quasi-analytical description of the vortex-merging mechanism is presently unavailable, and some of the future research effort could be directed at analyzing vortex amalgamations in terms of the spatial weakly nonlinear theories discussed in Section 2.2. However, as pointed out by Huerre (1980), there exist conceptual difficulties in retaining the small growth-rate assumption when the underlying medium is slowly diverging. Alternatively, one may abandon the weakly nonlinear formalism altogether and choose instead to use the *energy integral approach* of Stuart (1958). This has been the path taken by Liu and his coworkers. For instance, in a recent application of the method, Nikitopoulos (1982) and Nikitopoulos & Liu (1982) have considered the spatial nonlinear interaction of a fundamental

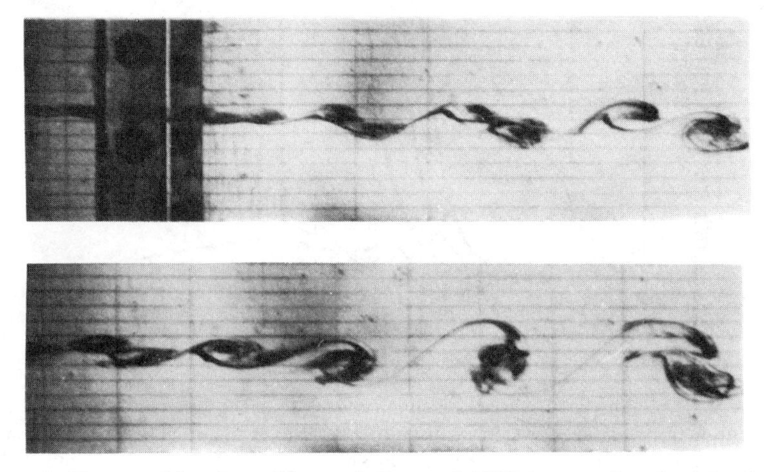

Figure 5 Vortex pairing (from Winant & Browand 1974) in an unforced mixing layer. $R = 0.48$.

wave of frequency f_n

$$\psi_A(x, y, t) = A(x)\phi_A\left(\frac{y}{\theta(x)}\right) \exp(-i\omega_n t - i\beta) + \text{c.c.}, \tag{8a}$$

and its subharmonic

$$\psi_B(x, y, t) = B(x)\phi_B\left(\frac{y}{\theta(x)}\right) \exp\left\{-i\frac{\omega_n}{2}t\right\} + \text{c.c.}, \tag{8b}$$

with a phase difference β between the two waves. The energy method usually requires "shape assumptions" to be made. In the present case the

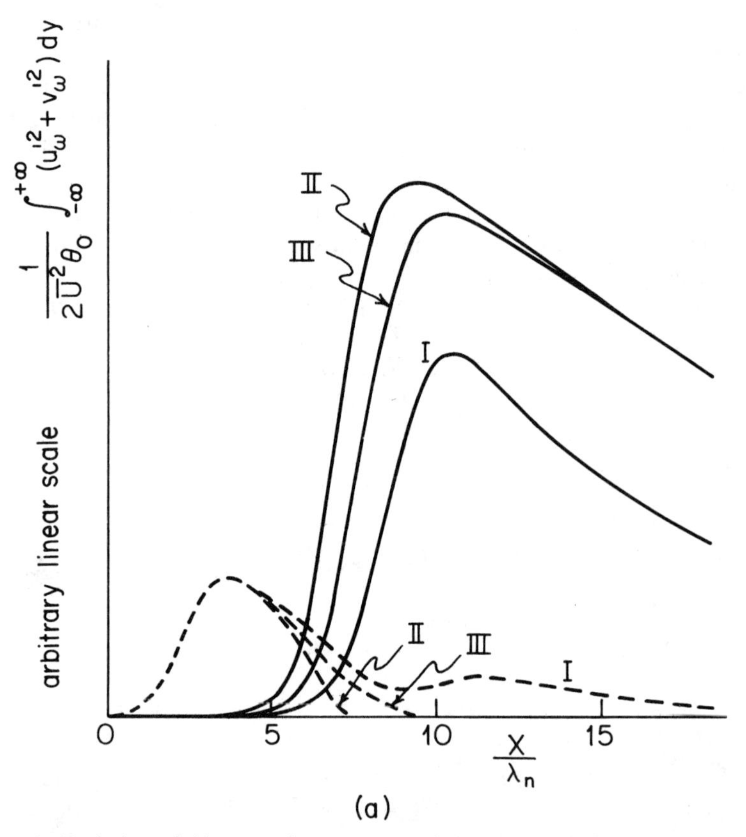

Figure 6 Evolution of (*a*) spectral components f_n (– – –) and $f_n/2$ (———) and (*b*) momentum thickness and downstream distance. Energy integral formulation of Nikitopoulos (1982). At $x = 0$, the ratio of the initial amplitudes of $f_n/2$ and f_n is 2.6×10^{-2}. The initial phase difference between the two waves is (I) $\beta = \pi$, (II) $\beta = 0$. In case (III), the two waves are decoupled. $R = 0.31$.

mean flow is taken to be self-similar of the form $U(y,x) = \bar{U}[1 + R \tanh\{y/2\theta(x)\}]$, and the cross-stream distribution functions $\phi_A(y/\theta)$ and $\phi_B(y/\theta)$ are assumed to be given by linear parallel stability theory applied to the mean-flow profile at the station x. The nonlinear evolution equations describing the development of the local thickness $\theta(x)$ and of the amplitudes $|A(x)|$ and $|B(x)|$ are then determined from the time-averaged integral kinetic-energy equations for the mean flow and the two wave components. Typical results are shown in Figure 6 for $\beta = 0$, $\beta = \pi$, and the decoupled case. As expected from Kelly's analysis, the larger the phase difference β, the smaller the subharmonic peak. It is also worth noting that artificially decoupling both waves by neglecting the interaction terms does not considerably alter the development of each wave. This feature may be viewed as indirect evidence that the interaction is indeed only weakly nonlinear. Furthermore, the evolution of the momentum thickness $\theta(x)$, as well as the streamwise location of the wave-amplitude maxima, is in qualitative agreement with the measurements of Figure 19. We note, however, that the energy-integral formulation cannot presently account for nonlinear variations in the phase of the two waves.

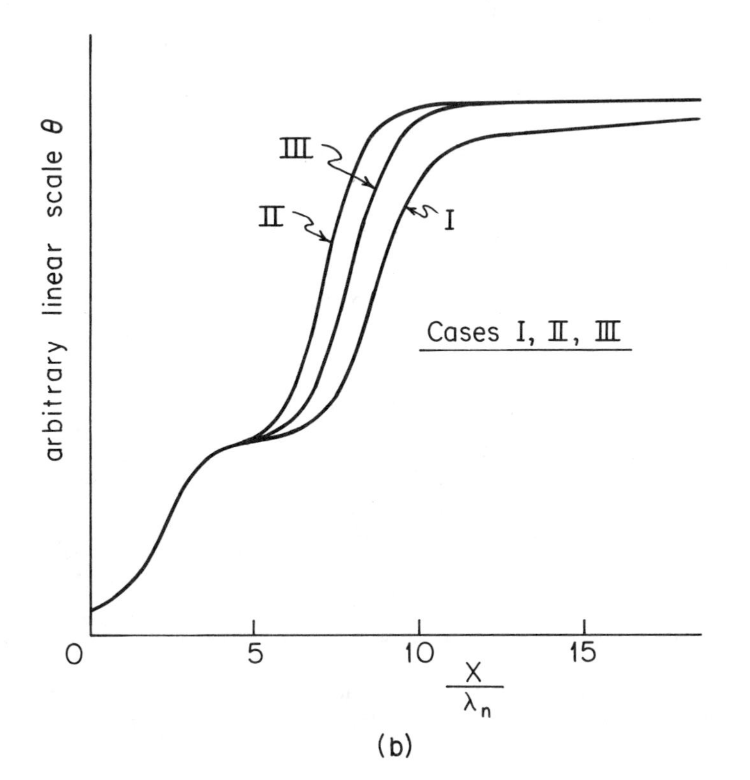

(b)

Direct computations of the vortex-pairing interaction in mixing layers have been conducted by Patnaik et al. (1976), Acton (1976), Ashurst (1979), Delcourt & Brown (1979), Riley & Metcalfe (1980), and Corcos & Sherman (1983). In all cases except Ashurst (1979), the flow is assumed to be spatially periodic and evolves in time: in qualitative agreement with the subharmonic resonance mechanism, the smaller the initial phase difference between the fundamental and subharmonic, the faster the coalescence of the two vortices. A pairing interaction is illustrated in Figure 7 for three values of the phase difference. In the limiting case where the two components are in antiphase ($\beta = \pi$), vortex pairing is actually suppressed and replaced by a *shredding interaction* (Patnaik et al. 1976), in which the cores of the fundamental are destroyed in the straining field of the subharmonic. A large strain rate is generated during the pairing interaction ($\beta = 0$), but the two vortices still maintain their identity for relatively large times and are unaffected by viscous diffusion even when $\Delta U\theta/\nu$ is as low as 50 (Corcos & Sherman 1983). Note that flow-visualization techniques based on the detection of streak lines may well obscure the persistence of two distinct structures after pairing, as stressed by Williams & Hama (1980). The onset of small scales (Section 2.6) may also smear the identity of each vortex. The amalgamation mechanism can also be displayed numerically by appli-

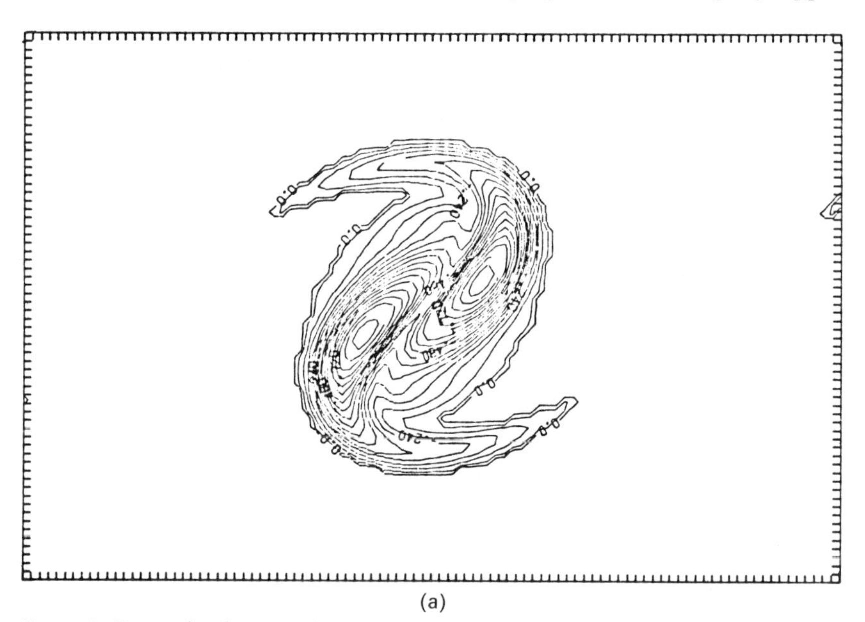

(a)

Figure 7 Interaction between f_n and $f_n/2$ in a temporally developing mixing layer, using numerical simulation of Riley & Metcalfe (1980). The initial phase difference is (*a*) $\beta = 0$, (*b*) $\beta = \pi/2$, and (*c*) $\beta = \pi$.

cation of suitable perturbations to two isolated corotating vortices of uniform vorticity (Overman & Zabusky 1982). The structures, which are solutions of the two-dimensional Euler equations, are observed to merge into a single rotating elliptical vortex with filamentary arms.

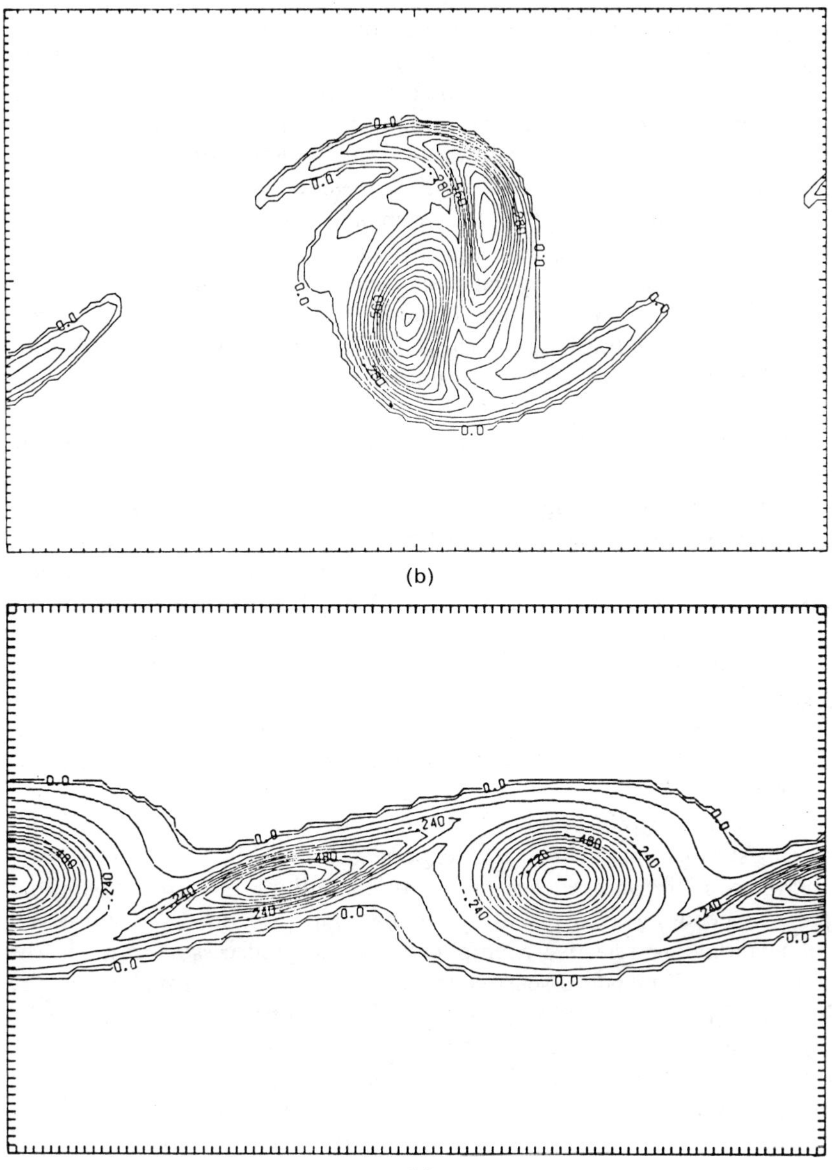

(b)

(c)

A different dynamical process, referred to as *tearing*, has been suggested by Moore & Saffman (1975) to explain the growth of mixing layers. The tearing mechanism has recently been linked by Pierrehumbert & Widnall (1981) to the energetic properties of a steady arrangement of streamwise periodic vortex structures with uniform vorticity, simultaneously discovered by Saffman & Szeto (1981). The basis for the tearing process relies on the fact that the ratio between the average vortex radius and the streamwise spacing is found to possess a maximum. Thus, as the vortex radius increases by "turbulent entrainment" of irrotational fluid, this ratio soon reaches its maximum allowable value. Any additional entrainment must then be accommodated by a corresponding increase in spacing through breakup of a vortex by the strain of its neighbors. Tearing, as noted by Corcos & Sherman (1983), implicitly assumes that the dynamical states of the flow are restricted to a particular set of steady configurations. It also presumes that the subharmonic instability has been somehow inhibited. None of these conditions are borne out by numerical simulations. Tearing is in fact rarely observed in experimental situations (Hernan & Jimenez 1982).

In the absence of satisfactory nonlinear theories, one presently relies on approximate analyses to describe the evolution of mixing layers and to establish the main scaling relationships. For instance, in the phenomenological model suggested by Ho (1981), the time-averaged momentum thickness $\theta(x)$ is assumed to increase by vortex pairings only, these events taking place at fixed downstream stations. By invoking experimental results such as those of Figure 19*b*, one considers $\theta(x)$ to be constant between pairings and to double instantaneously at pairing locations. Except for these obviously *nonlinear* effects, the unsteady response of the shear layer is supposed to be completely governed by linear stability theory. This means that between two consecutive pairings, say $j-1$ and j, the subharmonic of frequency $f_n/2^j$ is assumed to grow by a constant factor G at the maximum amplification rate pertaining to the thickness θ_{j-1}. At the jth pairing station, the thickness θ_{j-1} experiences a step increase to $\theta_j = 2\theta_{j-1}$ and the subharmonic reaches a maximum amplitude, its Strouhal number $f_j\theta_j/U$ becoming locally neutral. At that same location, a new $(j+1)$th subharmonic is selectively amplified by its predecessor through the resonance mechanism discussed in Section 2.2. A new cycle is thereby initiated, in which the sudden change in momentum thickness has made the original subharmonic neutrally stable while the new subharmonic becomes the locally most amplified wave. From this drastically simplified model of mixing-layer dynamics, one can easily deduce that the streamwise spacing between pairings doubles along the flow. Furthermore, the spreading rate is found to be

$$\frac{d\theta}{dx} = \frac{(-\alpha_i\theta)_{\max}}{\ln G} \approx 0.12\frac{R}{\ln G}, \tag{9}$$

where use has been made of the approximate linear dependence of the maximum amplification rate on the velocity ratio R (see Section 2.1). It is worth emphasizing that changes in the mean flow induced by Reynolds stresses have to be *postulated* or *determined from experiments.* Important *nonlinear* effects, such as changes in subharmonic growth rates caused by resonance with the fundamental [Equations (6a) and (6b)] have also been deliberately ignored. It is nonetheless comforting to notice that linear stability theory does provide approximate scaling laws for what is essentially a nonlinear dynamical process: measured spreading rates as well as merging distances are found to be in good agreement with the model. For an alternate evolution model in terms of a row of point vortices, one may consult Jimenez (1980).

Instead of relying on a purely *deterministic* approach, as in the preceding analyses, one can seek to develop *statistical* models in which the turbulent mixing layer is assumed to be in statistical equilibrium. Such formulations, which might very well reproduce the average characteristics of the mixing layer after several pairing generations, have been suggested by Takaki & Kovasznay (1978), Tam & Chen (1979), Aref & Siggia (1980), and Bernal (1981).

2.4 *Large-Scale Structures*

It should be made clear that interacting spanwise vortices are observed not only in the early laminar stages of the flow evolution but also farther downstream in the turbulent region, where they coexist with a fine-scale motion. For instance, large-scale structures can clearly be identified in the plan- and side-view shadowgraph pictures of Konrad (1976) shown in Figure 8. These features persist up to values of Re at least as high as 10^7 (Dimotakis & Brown 1976), well into the self-preserving region of classical theory. By making use of conditional-sampling techniques, it is possible, as shown by Browand & Weidman (1976) at lower Reynolds numbers, to extract from the turbulent flow ensemble-averaged states characterizing different phases of the coherent-structure evolution. From the results, one may infer that the large-scale vortices undergo sequential mergings in the same manner as their low-Reynolds-number counterparts. But pairing locations (in unforced shear layers) are now randomly distributed along the stream and give rise to a linear spreading rate.

The persistence of an organized quasi-two-dimensional motion in high-

Figure 8 Coherent structures at high Reynolds number. Plan- and side-view shadowgraph pictures from Konrad (1976). $R = 0.45$.

Reynolds-number mixing layers has been the subject of some debate. Chandrsuda et al. (1978), in particular, have argued that a significant level of free-stream turbulence or the tripping of the boundary layers on the splitter plate will lead to the disappearance of the two-dimensional structures. The resulting flow field will then be three-dimensional, as in the classical description of turbulence. Two separate issues need to be examined: the persistence of large-scale structures on the one hand, and their quasi-two-dimensional character on the other. Regarding the first point, Wygnanski et al. (1979) have demonstrated experimentally that, even in the presence of strong random perturbations, primary large-scale vortices re-emerge from the three-dimensional background sufficiently far downstream (see also Browand & Ho 1983). These observations have been confirmed by the three-dimensional numerical simulations of Riley & Metcalfe (1980). Regarding the second point, we note that, according to Browand & Troutt (1980), the spanwise correlations asymptotically scale with the local thickness $\theta(x)$ in the limit of large x. By taking 20% *average* velocity correlation contours as a measure of eddy size, these authors find that the structures extend approximately $2\theta(x)$ along the stream and only $6\theta(x)$ along the span at $R = 0.8$. Nonetheless, instantaneously most structures span the entire apparatus, the low value of the averaged lateral scale resulting from the presence of spanwise deformations in the rolls.

Finally, we note that the simulations of Riley & Metcalfe (1980) do not reveal a strong lateral coherence: the large-scale structures are only approximately two-dimensional. The two- or three-dimensional character of the structures seems to subjectively depend on the eyes of the beholder!

From the above experimental observations, it is increasingly clear that the evolution of vortical structures in both laminar and turbulent shear layers is governed by essentially the same dynamical processes. It is therefore appealing to apply to turbulent shear layers the same hydrodynamic stability concepts that had originally been developed to describe the roll-up of laminar shear layers. Within this framework, the large-scale structures are then viewed as instability waves propagating on the pseudolaminar flow defined by the time-averaged velocity field. We stress that such a formalism is meaningful only if there exists a large disparity of length scales between the coherent structures and the fine-grained motion (see Figure 8), a requirement that is almost always satisfied in practical high-Reynolds-number flows. As shown by Fiedler et al. (1981), the quasi-parallel inviscid linear approximation (see Figure 2*a*) satisfactorily predicts the local amplification rate of large-scale structures generated by low-frequency forcing. The more refined slowly diverging flow analysis conducted by Gaster et al. (1983) has been very successful in reproducing the entire development of the wave amplitude. Even more remarkably, the measured cross-stream distribution of fluctuating velocity close to the maximum-amplitude location is adequately described by quasi-parallel linear theory. An example of comparison between theory and experiment is displayed in Figure 9. Natural extensions of these analyses to the description of pairing events have already been discussed. There remains to be considered, however, an aspect peculiar to turbulent mixing layers, namely the interaction between the large-scale coherent motion and the fine-scale turbulence. A recent review by Liu (1981) discusses this topic in the context of the energy-integral method outlined in Section 2.3. The large-scale structure is then modeled by the monochromatic instability wave (8a), and a suitable shape assumption is made regarding the nature of the fine-grained turbulence. To isolate the three components of the flow, namely, the mean flow, the large-scale flow, and the fine scales, the usual time average is supplemented by a conditional average as suggested by Hussain & Reynolds (1970). Amplitude equations for the three components are then derived from the respective integrated energy equations. Temporally evolving and spatially evolving mixing layers have been examined respectively by Liu & Merkine (1976) and Alper & Liu (1978). In more recent developments of the method (Gatski & Liu 1980), the combined motion of the mean and large-scale structure field is computed exactly from the time-

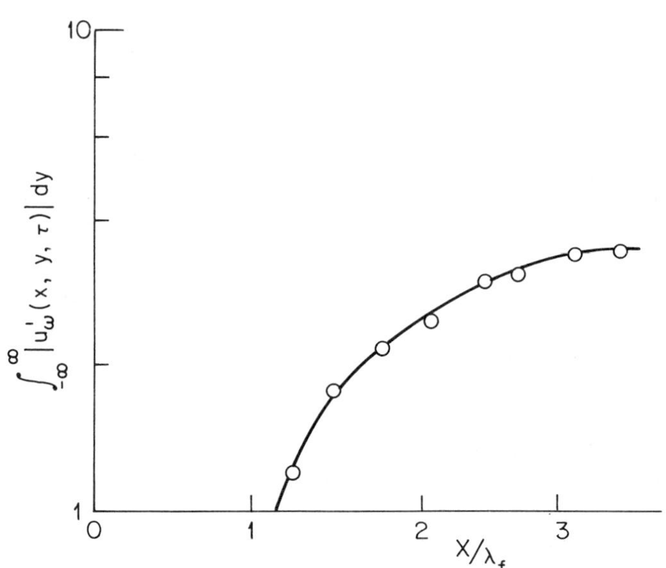

Figure 9 Evolution of forced component with downstream distance in a turbulent mixing layer (from Gaster et al. 1983). $f_f \ll f_n$; $R = 0.43$. ○ experimental data; —— slowly diverging theory.

averaged Navier-Stokes equations. In all instances, a high level of fine-scale motion is found to gradually suppress the large-scale structure and to slightly enhance the spreading rate, as intuitively expected from eddy-viscosity concepts. We emphasize that such a result does not signify the total demise of large scales at high turbulence levels: the thicker mixing layer can still support a most unstable Kelvin-Helmholtz wave of lower frequency, so that a new structure will necessarily be generated.

2.5 *Three-Dimensional Structures*

The occurrence of significant three-dimensional effects was first discovered by Miksad (1972). For a general overview of these phenomena, the reader is referred to Roshko (1981). The plan-view shadowgraph pictures of Figure 8 taken by Konrad (1976) clearly reveal the existence of periodically distributed streamwise streaks. The dye streaks are generated by secondary counterrotating streamwise vortices superposed on the primary rolls. The average spanwise spacing of the secondary structures increases with downstream distance and remains of the same order of magnitude as the local thickness. More importantly, the spanwise wavelength appears to be rather insensitive to irregularities in the boundary layers on the splitter plate: it is the result of a genuine internal instability. The frame-by-frame movie observations of Bernal (1981) have provided us with a detailed knowledge of the morphology of the secondary motion, as shown in Figure 10. But it is not yet clear whether this system of mature sinuous vortices represents the finite-amplitude stage of the "wiggle" detected around the first pairing location by Breidenthal (1981).

Ironically, the earliest theoretical analysis of streamwise structures was conducted by Benney & Lin (1960) and Benney (1961) in a mixing layer at a time when experimental evidence of the existence of such structures was available only in boundary layers. The approach, which is by now familiar to the reader (Section 2.2), consists in studying the weakly nonlinear interactions between a two-dimensional wave of wave number α and a three-dimensional wave of *identical* streamwise wave number and of spanwise wave number γ. A mean (x-independent) spanwise distortion of the basic tanh y profile is then nonlinearly generated, which takes the form of two horizontal layers of counterrotating streamwise vortices. Each layer is located on either side of the inflection point of the basic flow. If A and B denote the respective amplitudes of the two-dimensional and three-dimensional components, the spanwise wave number associated with the counterrotating eddies varies from γ to 2γ as the ratio $|B|/|A|$ increases from zero to infinity. It is unclear whether such a configuration can be detected experimentally. We note, however, that according to Stuart (1962) A and B are coupled via evolution equations of the form

$$\frac{dA}{dT} = \sigma_A A + b_A|A|^2 A + c_A|B|^2 A + d_A A^* B^2, \tag{10a}$$

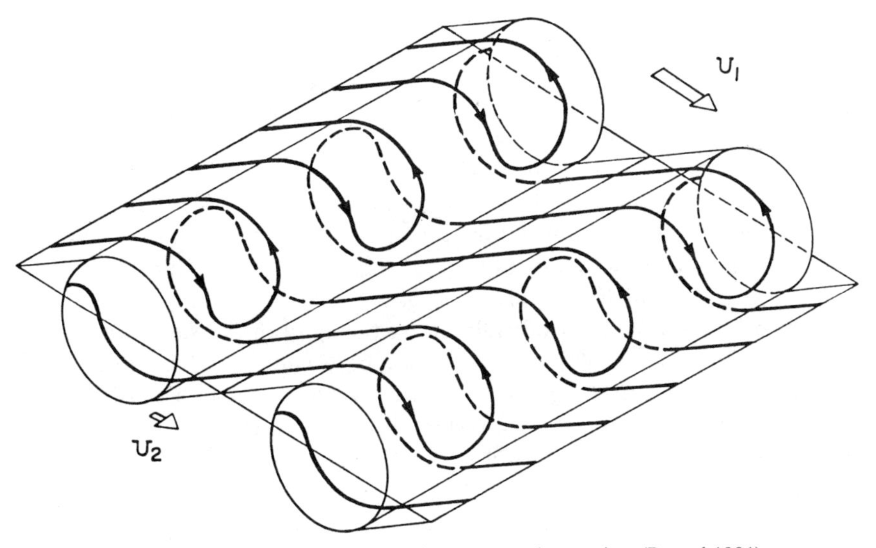

Figure 10 Morphology of secondary streamwise vortices (Bernal 1981).

$$\frac{dB}{dT} = \sigma_B B + b_B|A|^2 B + c_B|B|^2 B + d_B B^* A^2, \tag{10b}$$

when T is a suitably defined slow time scale. The complete streamwise structure is therefore determined by the superposition of the steady mean spanwise distortion *and* the finite-amplitude three-dimensional wave of amplitude $B(T)$. No detailed calculations of the constants arising in (10a) and (10b) have been performed for mixing-layer profiles, and it is presently not possible to say whether $B(T)$ reaches a steady equilibrium value or remains periodic for different phase relations between the two waves.

Another interaction model initially proposed by Craik (1971, 1980) in the context of boundary-layer transition is likely to be relevant to mixing layers. The formulation is essentially a three-dimensional generalization of the subharmonic evolution model described in Section 2.2. In contrast with the Benney-Lin model, the spanwise wave number is selected so that the three wave number pairs $(\alpha, 0)$, $(\alpha/2, \gamma)$ and $(\alpha/2, -\gamma)$ form a *resonant triad*, and the amplitudes $A(T)$ and $B(T)$ satisfy equations of the form (6a) and (6b). Note that the quadratic interaction terms present in those evolution equations are potentially more influential than the cubic interactions of the Benney-Lin model. The formalism of Craik has recently been shown to describe the development of Tollmien-Schlichting waves in a Blasius boundary-layer (Thomas & Saric 1981). Although the amplification rates of Kelvin-Helmholtz waves are much higher than those of Tollmien-Schlichting waves, these weakly nonlinear models could profitably be adapted to mixing layers.

Alternatively, following Pierrehumbert & Widnall (1982), the two-dimensional flow can be assumed to have equilibrated to a state described by an array of Stuart vortices (7) of wavelength a. The streamwise structures then arise through a secondary three-dimensional instability of this primary finite-amplitude flow. Two principal modes of instability can be distinguished. In the *translative mode*, the periodic base flow is linearly perturbed by a three-dimensional wave of streamwise wavelength a and spanwise wave number γ. The temporal growth rate σ varies with γ, as shown in Figure 11*a*. The primary spanwise vortex tubes are found to be deformed periodically along the span in the same way as in Figure 10. For a value of the concentration parameter $\rho = 0.25$, the growth rate reaches a maximum when the spanwise spacing is two thirds of the Kelvin-Helmholtz streamwise wavelength, a value in good agreement with the experimental observations of Bernal (1981). Furthermore, the growth-rate curve is relatively flat around its maximum. In the *helical-pairing mode*, the streamwise wavelength of the three-dimensional perturbations is increased to $2a$, and the growth rate is given by Figure 11*b*. When $\gamma = 0$, the

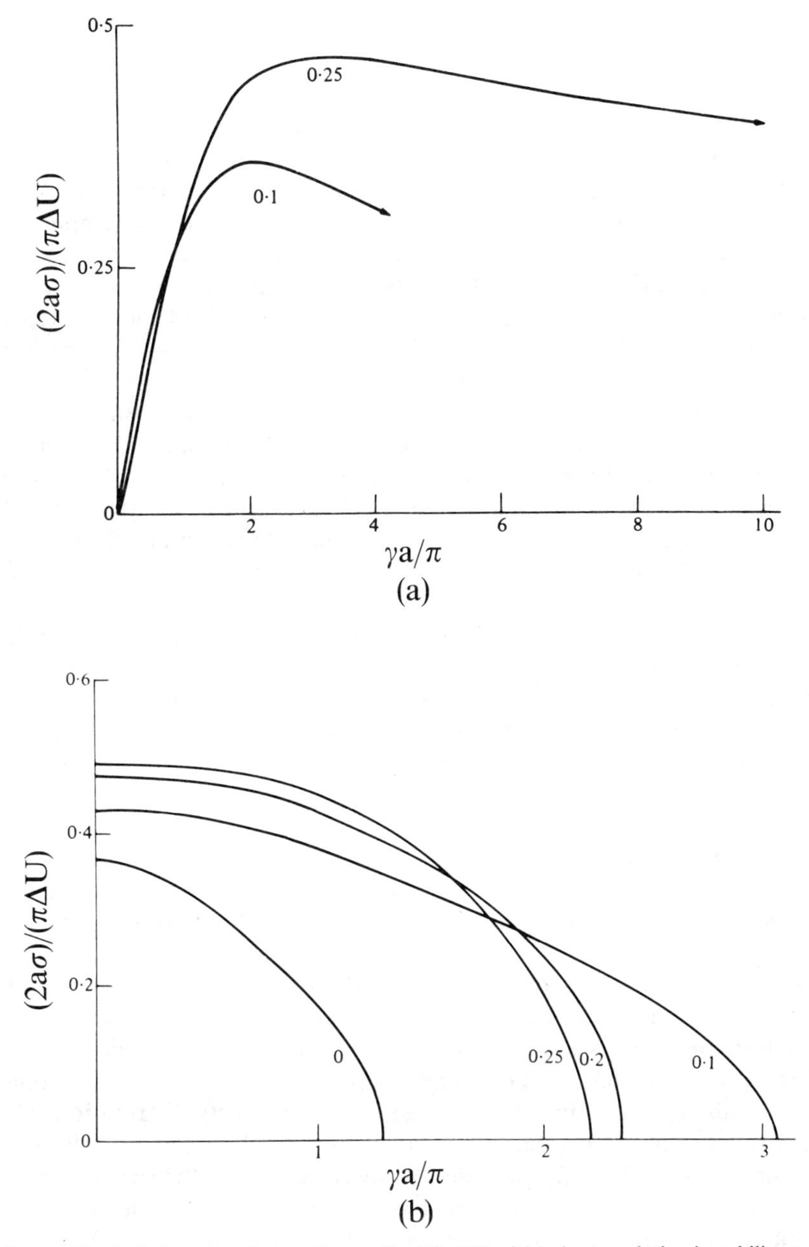

Figure 11 Scaled temporal growth rate $(2a\sigma)/(\pi\Delta U)$ of (*a*) the translative instability and (*b*) the helical-pairing instability against nondimensional spanwise wave number $(\gamma a)/\pi$ (from Pierrehumbert & Widnall 1982). Each curve refers to a specific value of the concentration parameter ρ.

characteristics of the two-dimensional subharmonic instability are recovered (Section 2.2). When $\gamma \neq 0$, pairings occur locally along the span, thereby resulting in the phase dislocations observed by Browand & Troutt (1980). As noted by Pierrehumbert & Widnall (1982), the existence of a cutoff at a finite value of $\gamma\theta$ implies that helical pairing cannot completely destroy the two-dimensional structures: a given γ is necessarily stabilized further downstream as $\gamma\theta$ ultimately crosses the cutoff value. We emphasize that these two instabilities are not intrinsically different from the preceding nonlinear interaction models. In the limit $\rho \to 0$, Stuart vortices take the form of a periodically perturbed tanh y mixing layer, and the translative and helical-pairing instabilities reduce respectively to the Benney-Lin instability and the Craik resonant-triad instability.

The generation of three-dimensionality, as exemplified by Breidenthal's wiggle, can also be understood in terms of the fully nonlinear dynamics of vortex filaments. In a private communication, Browand has recently conjectured that *solitary waves* might be generated by the sidewall boundary layers and subsequently propagate in the form of helical twists along the axis of the spanwise structures. In the same spirit, Aref & Flinchem (1983) have examined numerically the evolution of solitary waves on a single vortex filament embedded in a tanh y mixing layer. Under the combined effects of induction and of interaction with the continuous shear, solitary waves propagate along the vortex tube. Simultaneously, these initially localized entities are found to radiate families of waves. The resulting spanwise deformation of the tube is almost periodic and bears a striking qualitative similarity with the wiggle observed by Breidenthal (1981).

Three-dimensional numerical simulations have been performed by Couet (1979), Riley & Metcalfe (1980), Cain et al. (1981), and Brachet & Orszag (1983). The calculations are, by necessity, still relatively coarse but display the same qualitative features as laboratory experiments. Since current schemes are restricted to temporal mixing layers that are streamwise and spanwise periodic, the effects of the velocity ratio cannot be properly handled. In an interesting extension of classical linear stability theory, Corcos & Lin (1983) have examined in particular the three-dimensional *linear* stability of the time-developing spanwise vorticity distribution. The growth rate of the translative instability is found to remain relatively constant during the roll-up phase. However, pairing of the spanwise rolls creates within the cores strong centers of energy transfer from the three-dimensional flow back to the two-dimensional motion. Thus, vortex mergings considerably delay the linear development of streamwise vortices. Furthermore, the Reynolds stresses induced by the three-dimensional disturbances have a negligible effect on the development of the two-dimensional dynamics. This result confirms the suggestion of Corcos (1979) regarding the secondary nature of the streamwise structures when compared with the primary two-dimensional motion. The *fully nonlinear* calculations of Brachet & Orszag (1983) display essentially the same features, but the three-dimensional motion keeps increasing in magnitude. No steady or time-periodic regime is reached, a feature requiring further study.

Other aspects of the three-dimensional secondary flow can also be investigated by relying on approximate dynamical models. For instance, in the analysis of Lin (1981) and Lin & Corcos (1983), the streamwise vorticity is assumed to be preferentially generated in the braids by the strain field of two neighboring spanwise structures. All streamwise gradients are neglected. Thus, the finite-amplitude instability of vortex layers in the presence of a simple strain is obtained numerically from the two-dimensional equations of motion in the cross-stream y-z plane. It is found that, for a large class of spanwise-periodic initial conditions, such layers are pathologically unstable and collapse into discrete round vortices with a radius of the order of the Taylor microscale $\theta/\mathrm{Re}^{1/2}$. A typical stage in the evolution of the counterrotating structures is shown in Figure 12. Calculated "dye patterns" in the y-z plane display a characteristic mushroom shape that is very similar to the cross-stream cuts visualized experimentally by Bernal (1981).

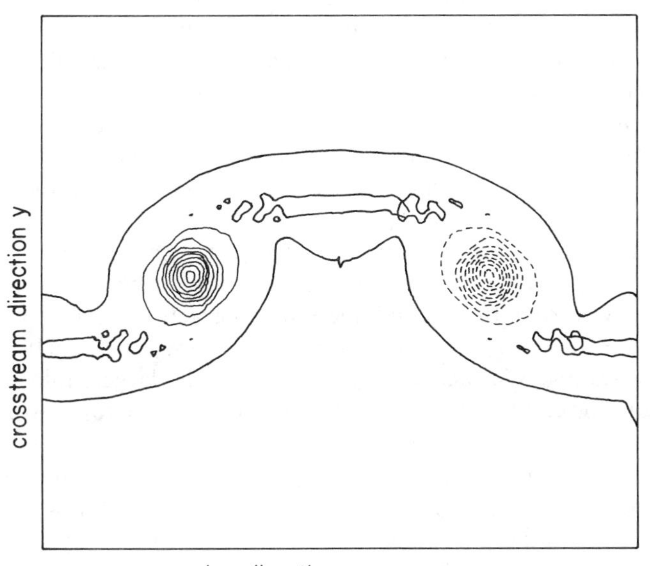

Figure 12 Calculated isovorticity contours of streamwise structures in a y-z plane cut through the braids (from Lin 1981). —— positive vorticity; – – – negative vorticity.

2.6 *Small-Scale Transition*

Initially laminar mixing layers eventually become turbulent in a process known as the small-scale transition. The shadowgraph pictures of Figure 8 suggest that the onset of small-scale eddies takes place within the cores of the large structures and is fairly localized along the stream.

Various criteria have been introduced to characterize quantitatively the location of the small-scale transition. Sato (1956, 1960), who initially investigated this question, chose as a measure the departure of the amplitude evolution from its initial exponential growth. Bradshaw (1966) selected instead the location where the peak turbulence level or peak Reynolds stress reaches a maximum. In chemically reacting flows, one can alternatively consider the "mixing transition" taking place at the streamwise station where the "mixedness" P/δ sharply increases (Konrad 1976, Breidenthal 1981). In this context, P/δ measures the amount of product formed in a shear layer initially carrying two distinct chemicals. Finally, the magnitude of the roll-off exponent in the velocity spectrum provides a convenient indicator of the small-scale onset in inert mixing layers (Jimenez et al. 1979, Huang & Ho 1983). The flow is considered to be turbulent when the exponent reaches the value $-5/3$. A clearly identifiable inertial subrange is then established, which extends over approximately one decade of frequency. These different transition criteria are compared in Figure 13. The discrete points all pertain to data taken in air, whereas the dashed curve represents a suitable average of mixedness measurements in liquids. Note that in terms of the normalized coordinate Rx/λ_n the first three pairings take place on the average at 4, 8, and 16 units, respectively. The transition process is found to span a few λ_n's. As observed by Bradshaw (1966), the peak turbulence level reaches a maximum at the first merging location and relaxes to its asymptotic value at the second merging location, precisely where the roll-off exponent becomes equal to $-5/3$. By changing the free-stream velocities U_1 and U_2, Huang & Ho (1983) have deliberately varied the locations of pairing events and checked that transition in air occurs consistently during the second merging interaction at the normalized distance $Rx/\lambda_n = 8$. Furthermore, the data of Jimenez et al. (1979) follow the same scaling. We therefore speculate that the large strain rates that result from the coalescence of the orderly structures are responsible for the generation of small-scale eddies.

As expected on intuitive grounds, the local Reynolds number $\mathrm{Re} = \bar{U}\theta/\nu$ should be sufficiently large for the flow to sustain a fine-scale motion. Konrad (1976) initially suggested a value of the transition Reynolds number of the order of 5×10^3. On the basis of a wider set of data taken in *air*, it appears, however, that the onset of small scales does not occur at a precise value of Re, but rather in a range $3 \times 10^3 < Re < 5 \times 10^3$. For unknown reasons, typical transition Reynolds numbers in *liquids* are found to be much lower and fall in the range $750 < Re < 1700$ (Jimenez et al. 1979, Breidenthal 1981). Note also from Figure 13 that in liquids the fine-scale motion is established farther downstream around $Rx/\lambda_n = 12$. It should be emphasized that such a discrepancy between the case of liquids and gases cannot be attributed to differences in chemistry or to the choice of transition criterion: the normalized transition locations of liquid flow are found to be the same when determined by the mixedness method or the roll-off exponent method.

Theoretical analyses of the actual transition process are scarce. Energy-integral methods (Liu & Merkine 1976, Alper & Liu 1978, Gatski & Liu 1980) indicate a large energy transfer from the mean flow to the fine-scale motion as the large-scale disturbance reaches a maximum. However, these analyses are restricted to a single monochromatic large-scale instability wave. Lin & Corcos (1983) have proposed, instead, to relate the "mixing" transition to the dynamics of the three-dimensional structures in the braid region. Large increases in the mixed volume of a passive scalar are obtained

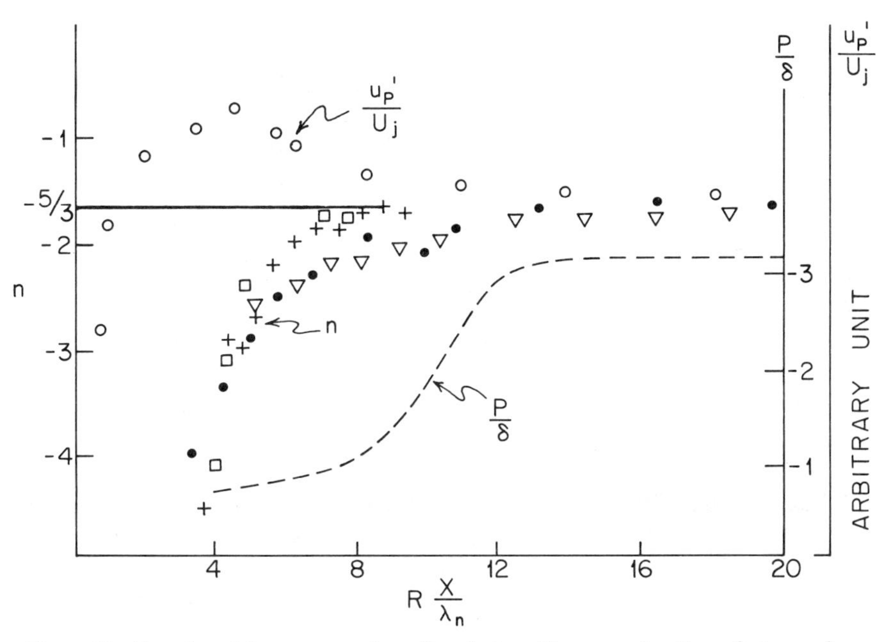

Figure 13 Experimental measures of small-scale transition as a function of streamwise distance. u_p'/U_j: Peak turbulence level; ○ $R = 1$ (Bradshaw 1966). n: slope of the inertial subrange; ● $R = 1$ (Jimenez et al. 1979); □ $R = 0.45$; + $R = 0.69$; ▽ $R = 1$ (Huang & Ho 1983). P/δ: "Mixedness" ratio; ——— $0.14 < R < 0.45$ (Breidenthal 1981).

from the calculations, as streamwise vorticity layers collapse into round vortices, but the resulting scaling predictions are, according to the authors, inconclusive. We stress that no theoretical information is currently available regarding the expected contribution of pairing interactions to the small-scale transition. For a phenomenological model of the mixing process beyond the onset of small-scales, the reader is referred to Broadwell & Breidenthal (1982).

2.7 *Preferred Mode in Jets*

A fundamental difference between a jet and a mixing layer is the existence of a second length scale that is the diameter D of the axisymmetric nozzle or the height H of the two-dimensional nozzle. In both kinds of jets, a potential flow of velocity U_j issuing from the exit plane comes to an end about $5D$ (or $3H$) downstream, and axisymmetric (or two-dimensional) orderly structures can clearly be identified in the shear layer surrounding the potential core. As in the simple mixing layer, the large-scale vortices experience successive amalgamations that eventually lead to small-scale transition. Beyond the tip of the potential core, these vortical features are more difficult to identify, and their morphology is still unknown (Tso et al. 1981).

The forcing experiments of Crow & Champagne (1971) have indicated that the response of a circular jet to a monochromatic excitation, measured as the gain or ratio of the peak amplitude to initial amplitude of the wave along the jet axis, reaches a maximum for a Strouhal number $f_p D/U_j \approx 0.3$. The passage frequency of the vortical structures at the end of the potential core is given by f_p, and the corresponding fluctuating flow field is commonly referred to as the *preferred mode* or jet-column mode. It should be distinguished from the *shear-layer mode* of higher frequency f_n associated with the most amplified wave of the initial velocity profile. In qualitative agreement with observations, the slowly diverging linear stability analysis of Crighton & Gaster (1976) does single out a mode with a maximum gain in pressure amplitude at a Strouhal number of $f_p D/U_j \approx 0.4$. Note that this theory is implicitly nonlinear, since the measured mean-velocity profiles used in the calculations already include the effects of Reynolds stresses.

The preferred Strouhal number in circular jets is found to vary between 0.25 and 0.5 in different experiments. This scatter may be due to data-processing techniques (Petersen 1978) or to the nature of the residual noise present in individual facilities (Gutmark & Ho 1983). Recent experimental data, taken under very clean conditions and presumably less affected by background noise, have shed some light on this question. According to the results of Kibens (1981) displayed in Figure 14, the preferred Strouhal number $f_p D/U_j$ is approximately proportional at low jet velocities, i.e. low

$D/(2\theta_0)$, to the Strouhal number $f_n\theta/\bar{U}$ of the shear-layer mode. For higher flow rates, however, f_pD/U_j remains constant and equal to 0.44, the change of scaling behavior occurring at $D/2\theta_0 = 120$. Similar conclusions have been reached by Drubka (1981) in the range $115 < D/(2\theta_0) < 175$. Ho & Hsiao (1983) have further confirmed the validity of Kibens' result in the case of plane jets, provided that the nozzle radius $D/2$ is replaced by the height of the slit H. Thus, below a critical value of the ratio $D/(2\theta_0)$, or H/θ_0, the vortex passage frequency f_p at the end of the potential core scales with the initial shear-layer momentum thickness θ_0. It is simply a fraction of f_n, as dictated by the sequence of amalgamations taking place in the mixing layer (see Figure 23). Above the critical value of the length scale ratio, f_p scales with the characteristic size of the nozzle, although the initial passage frequency f_n still decreases to f_p via the same number of vortex mergings. There is presently no satisfactory explanation of this Strouhal-number locking. Ho & Hsiao (1983) have conjectured that the critical value is linked to the appearance of the small-scale transition in the mixing layer. The

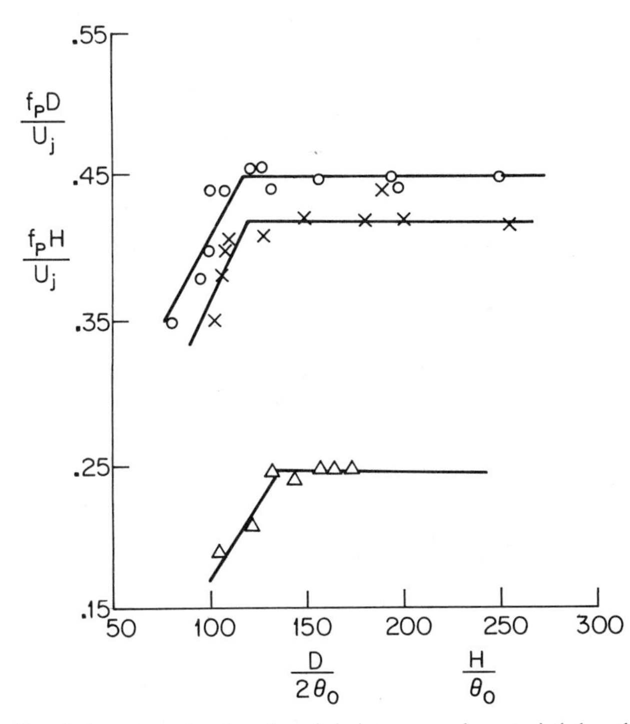

Figure 14 Strouhal number of preferred mode in jets versus characteristic length scale ratio. Axisymmetric jets: ○ Kibens (1981); × Drubka (1981). Two-dimensional jets: △ Ho & Hsiao (1983).

corresponding increase in turbulent diffusion would then lead to a less concentrated vorticity distribution in the primary structures. According to the results of Pierrehumbert & Widnall (1982) shown in Figure 11*b*, this could in turn reduce the growth rate of the pairing instability and keep the number of mergings constant.

2.8 *Symmetry-Breaking Modes in Jets*

Experiments in circular geometries with low background noise indicate that the coherent vortex rings generated close to the nozzle exit are initially axisymmetric, but that the energy content of helical disturbances increases significantly farther downstream (Browand & Laufer 1975, Drubka 1981). As a support for these observations, the spatial parallel-flow calculations of Michalke (1971) have shown that the axisymmetric mode is indeed slightly more unstable than any of the spinning modes, as long as the basic velocity distribution possesses a potential core. In contrast, the "fully-developed" bell-shape profiles prevailing downstream of the potential core can only support helical instability waves (Batchelor & Gill 1962). Furthermore, according to linear slowly diverging flow theory extended to spinning modes (Plaschko 1979), the axisymmetric mode suffers a larger total gain than helical modes at the higher Strouhal numbers that dominate the flow near the nozzle lip. At lower Strouhal numbers, i.e. near the end of the potential core, the situation is reversed, and the first helical mode achieves the largest maximum amplitude (see also Mattingly & Chang 1974). We also note that the disintegration of the originally axisymmetric structures can also be attractively explained by the azimuthal instabilities taking place on mature vortex rings (Widnall et al. 1974).

It is generally observed (Fuchs & Michel 1977, Drubka 1981, Cohen et al. 1983) that helical components appear closer to the nozzle exit at higher jet velocities. We can only speculate that changes in the characteristics of the background noise or possibly a stronger feedback from the downstream flow are responsible for this peculiar phenomenon.

Finally, the development of the sinuous mode and varicose mode in two-dimensional jets is found to exhibit features that are similar to the evolution of axisymmetric and helical modes in circular jets, as observed by Sato (1960) and Cervantes de Gortari & Goldschmidt (1981).

3. INFLUENCE OF OPERATIONAL PARAMETERS

In this section, we specifically examine how the operational parameters, namely, the velocity ratio, the Reynolds number, and the initial turbulent or laminar state of the boundary layer, may influence the dynamics of free shear layers.

3.1 *Velocity Ratio*

A mixing layer is characterized by two distinct velocity scales: the velocity difference ΔU provides a measure of the growth rate of the orderly structures, and the average velocity $\bar{U}$ is approximately equal to their convection velocity. For mixing layers that evolve *in time* from some initial state, such as those usually studied in direct numerical simulations, the convection velocity can be rigorously scaled out by a Galilean transformation. A similar relationship is not available for forced shear layers that develop *spatially* in response to excitations applied at the trailing edge, i.e. at a *fixed* streamwise location. Thus, the velocity ratio enters the dynamics in a nontrivial manner. Only in the limit of small R can the temporal and spatial versions of the flow be related by a Galilean transformation. As noted in Section 2.1, the effect of R is particularly apparent in the linear stability characteristics of mixing layers (Monkewitz & Huerre 1982) and jets (Michalke & Hermann 1982).

An approximate scaling law governing the variations of mixing-layer spreading rate $d\theta/dx$ with velocity ratio R is easily obtained from the following heuristic argument (Brown & Roshko 1974). In a time Δt, a spanwise vortical structure travels a distance $x = \bar{U}\Delta t$, and its cross-stream length scale, proportional to ΔU, increases by an amount $\Delta U \Delta t$. The spreading rate $d\theta/dx$ is therefore linearly related to the velocity ratio $R = \Delta U/(2\bar{U})$, as already obtained from the phenomenological model of Ho (1981) outlined in Section 2.3. The experimental data plotted in Figure 15 do follow this approximate scaling law. To leading order, the velocity ratio R is seen to act on the dynamics of the spanwise rolls as a simple stretching parameter. The lower R, the more extended the flow development. Accordingly, whenever it is meaningful, the streamwise coordinate in the figures has been rescaled in the form Rx/λ. Finally, we note that very little is known regarding the dependence of the geometry of the streamwise vortices on R. However, as discussed in Section 2.6, the downstream location for the onset of small scales is proportional to $1/R$.

3.2 *Reynolds Number*

Although the evolution of the orderly structures is essentially inviscid, viscous effects play an important role both in determining the laminar or turbulent state of the splitter-plate boundary layers and in the transition to small scales in the mixing layer.

For sufficiently low free-stream velocities, the boundary layers at the trailing edge are laminar, their momentum thickness varying as $U_1^{-1/2}$ and $U_2^{-1/2}$, respectively. Thus, on account of viscous diffusion, the initial thickness θ_0 is roughly proportional to $\bar{U}^{-1/2}$ for reasonably large values of

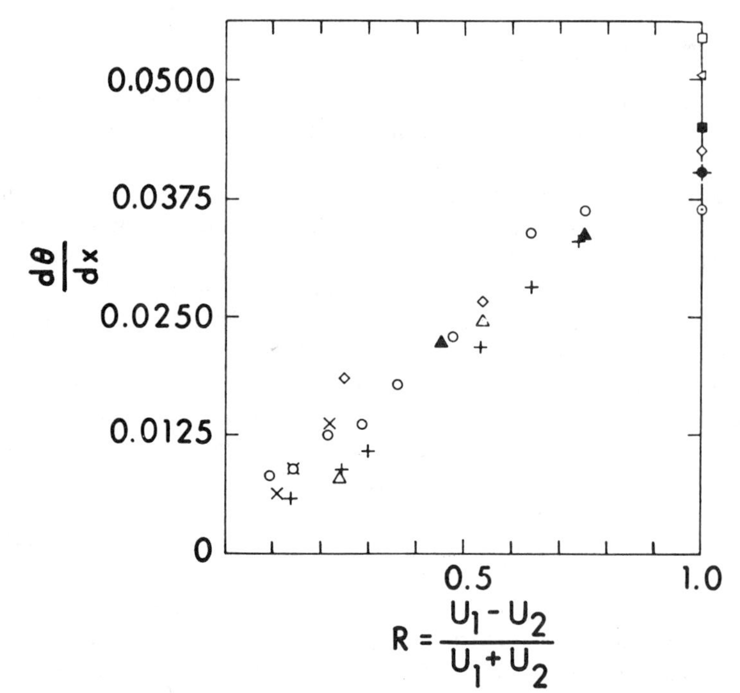

Figure 15 Dependence of mixing-layer spreading rate $d\theta/dx$ on velocity ratio R (from Brown & Roshko 1974). ● Liepmann & Laufer (1947); ⊙ Reichardt (1942); ○ Miles & Shih (1968); ◇ Mills (1968); × Pui (1969); □ Wygnanski & Fiedler (1970); + Spencer & Jones (1971); ◁ Sunyach (1971); △ Yule (1972); ■ Patel (1973); ▲ Brown & Roshko (1974).

R, and the natural frequency f_n, given by $f_n\theta_0/U = 0.032$, scales with $\bar{U}^{3/2}$. Correspondingly, the initial Reynolds number $\mathrm{Re}_0 = \bar{U}\theta_0/\nu$ varies as $\bar{U}^{1/2}$. As the average velocity $\bar{U}$ exceeds a critical value, the boundary layers become turbulent and lead to a turbulent mixing layer immediately downstream of the trailing edge (Section 3.3). It is known that flat-plate boundary layers in a zero pressure gradient undergo transition at a Reynolds number of the order of 10^3 (Tani 1969), but the precise value depends on many factors, such as the level of free-stream turbulence, the design of the contraction leading to the test section, etc. Furthermore, relatively few experimental data are available for the case of natural (untripped) transition in the presence of a pressure gradient. According to Browand (1983, private communication), the boundary layer at the trailing edge of the splitter plate also reaches transition around $\mathrm{Re}_0 \approx 10^3$. Above this value, the initial shear layer becomes turbulent and, for presently unexplained reasons, the most amplified Strouhal number St_n changes from a value of 0.032 for a laminar flow to 0.044–0.048 (Hussain & Zaman 1981, Drubka 1981).

Linear stability calculations based on the Orr-Sommerfeld equation (Betchov & Szewczyk 1963, Morris 1976) have indicated that shear layers are not completely stabilized by viscous dissipation, even at low values of Re. Indeed, calculated amplification rates are very much insensitive to variations in Reynolds number, provided Re > 50. In most experiments, the initial Reynolds number Re_0 is equal to 100 or more, and vortex roll-up as well as merging interactions are found not to be directly affected by viscosity, provided the boundary layers are laminar. By the same token, increasing the Reynolds number by two orders of magnitude does not significantly alter the downstream variation of the momentum thickness (Browand & Ho 1983).

In contrast, the organized three-dimensional secondary motion and the small-scale transition do involve significant Reynolds-number effects (Sections 2.5, 2.6). The average spacing between streamwise vortices scales with the local thickness $\theta(x)$, as dictated by inviscid stability considerations but, according to the model proposed by Lin & Corcos (1983), the vortex radius should be of the order of the Taylor microscale $\theta/Re^{1/2}$. No detailed measurements have yet been undertaken, which would confirm the latter prediction. Small-scale transition is also dependent on local Reynolds number: $Re(x)$ needs to be of the order of 3×10^3 (Huang & Ho 1983) or more to sustain a well-established inertial subrange; otherwise, viscous dissipation will not permit the establishment of fine-scale motion.

3.3 *Laminar-Turbulent Boundary Layer*

The downstream development of shear layers is known to be very sensitive to the laminar or turbulent state of the boundary layers on the splitter plate (Bradshaw 1966, Wygnanski & Fiedler 1970). A simple measure of this sensitivity is provided by the variations of mixing-layer thickness $\theta(x)$ (Figure 16) and peak turbulence level (Figure 17) with streamwise distance.

In almost all experiments, the flow is made turbulent by tripping the boundary layer, and the initial Reynolds number Re_0 is usually of the order of 10^3 or less ($Re_0 = 240$–1000 in Oster et al. 1977, $Re_0 = 110$–450 in Hussain & Zedan 1978, $Re_0 = 210$–450 in Drubka 1981). As seen in Figure 16, the resulting turbulent shear layer is characterized by two distinct spreading rates, the shift from the initial value to the final higher value taking place at a normalized streamwise distance $Rx/\lambda_n \approx 10$. Nonetheless, in the same facility the asymptotic spreading rate of turbulent mixing layers ($R \neq 1$) is usually found to be much lower than the corresponding laminar value (Oster et al. 1977, Browand & Latigo 1979). The situation is more confusing in jets ($R = 1$): spreading rates are either enhanced (Batt et al. 1970, Oster et al. 1977) in a half-jet or reduced (Drubka 1981) in a circular jet by tripping the boundary layer.

In a mixing layer with laminar upstream conditions, the peak turbulence

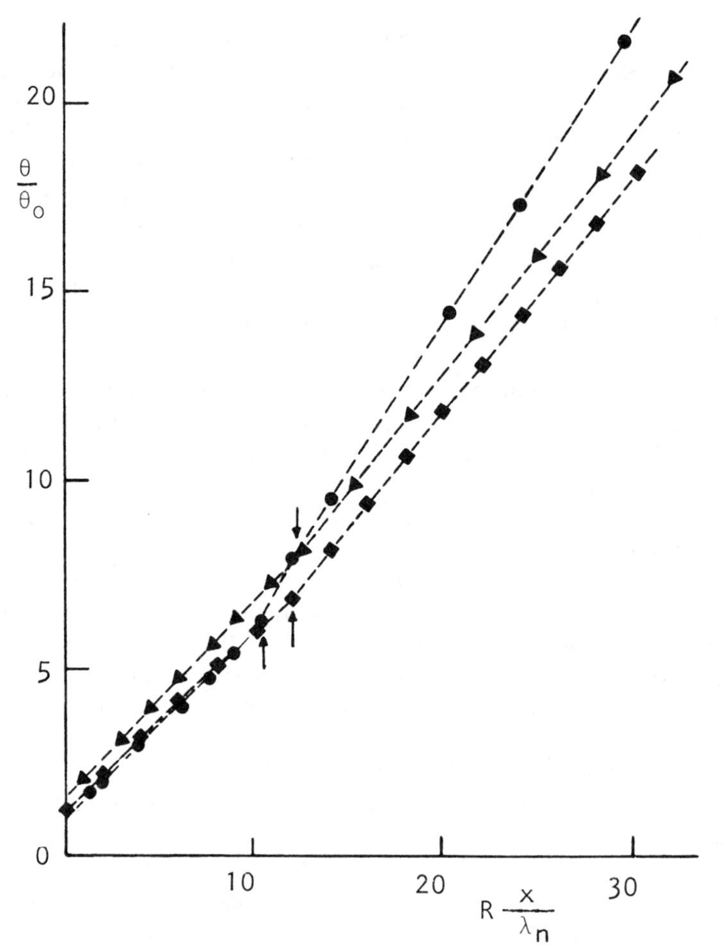

Figure 16 Variation of momentum thickness with downstream distance in the case of a turbulent boundary layer. ● $R = 1$, Foss (1977); ▲ $R = 1$, Hussain & Zedan (1978); ■ $R = 0.7$, Browand & Latigo (1979).

intensity $u'_p/\Delta U$ or u'_p/U_j increases along x (see Figure 17) and reaches a maximum around the first merging station. As a result of the small-scale transition, it eventually settles farther downstream to an asymptotic value characteristic of "fully developed" turbulent mixing layers. The peak fluctuation intensity of a *turbulent* boundary layer at $x = 0$ relaxes monotonically to approximately the same constant value as in the laminar boundary-layer case, but there is no overshoot (Bradshaw 1966, Hussain & Zedan 1978, Browand & Latigo 1979, Drubka 1981). It is interesting to note that for both laminar and turbulent initial conditions the asymptotic turbulence level is reached at the same streamwise station $Rx/\lambda_n \approx 10$,

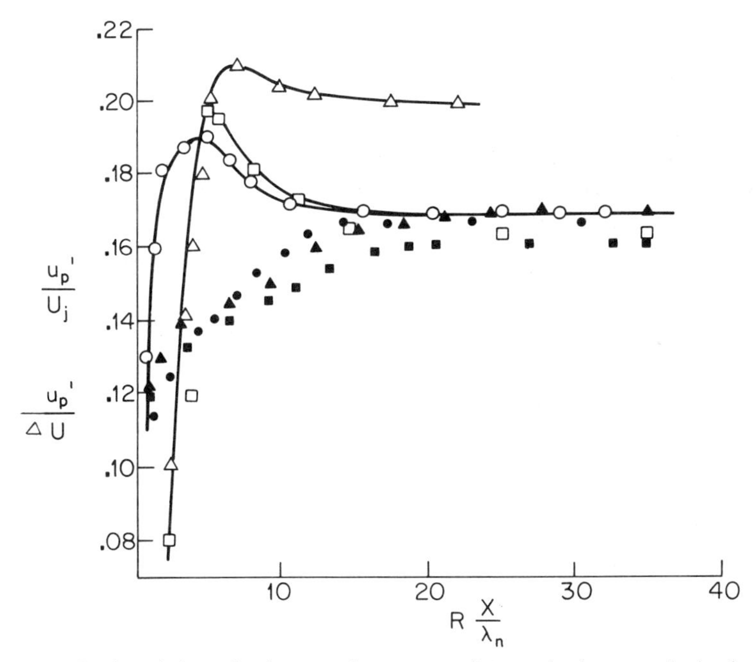

Figure 17 Peak turbulence level versus downstream distance in the case of a laminar or turbulent boundary layer. ○ $R = 1$, laminar boundary layer, ● turbulent boundary layer, arbitrary units (Bradshaw 1966); ▲ $R = 1$, turbulent boundary layer (Hussain & Zedan 1978); □ $R = 0.7$, laminar boundary layer, ■ $R = 0.7$, turbulent boundary layer (Browand & Latigo 1979); △ $R = 1$, laminar boundary layer (Jimenez et al. 1979).

which also coincides approximately with the breaking point in the spreading rates of Figure 16. Note also that in both cases the local Reynolds number at that same station is about 2–7 × 10^3 (Re = 3.7 × 10^3 in Foss 1977, Re = 1.6–3.1 × 10^3 in Hussain & Zedan 1978, Re = 6.8 × 10^3 in Browand & Latigo 1979) and of the same order of magnitude as the small-scale transition Reynolds number for an initially laminar shear layer (Section 2.6).

Thus, it appears that laminar and turbulent boundary layers take the same distance downstream of the trailing edge to reach a self-similar state characteristic of turbulent shear layers (compare Figures 13 and 17). However, since the spreading rates differ, the concept of a single universal asymptotic state does not seem to apply. As an additional cautionary statement, we add that the measurements and even trends of several experiments do not agree with each other. The root cause of such discrepancies might very well lie in the long-lasting effect of the particular initial conditions prevailing in different facilities, as is discussed in Section 4.7.

4. SENSITIVITY TO ARTIFICIAL EXCITATION

4.1 *Methods of Excitation*

According to Sections 2.3 and 2.5, sequential mergings of coherent spanwise rolls are believed to be responsible for most of the entrainment into the shear layer and for the occurrence of small-scale transition. Hence, in order to manipulate most effectively the development of the flow, imposed excitations should as a first requirement be *spatially coherent* along the span.

Forcing is usually applied in two ways: mechanically with a ribbon or an oscillating flap, or acoustically by means of loudspeakers. In experiments conducted at low Mach numbers, the mismatch between the speed of sound and the phase speed of instability waves is large, and the only manner in which acoustic waves can effectively feed energy into the fluctuating vorticity field is at the trailing edge (Morkovin & Paranjape 1971). In high-speed flows, however, the shear layer can directly be excited acoustically without the mediation of a splitter plate (Tam 1978). Mechanical forcing acts in a more straightforward way: a large portion of the input energy is converted into instability waves.

The acoustic excitation frequencies commonly used in experiments are such that the acoustic wavelength is much larger than the typical spanwise extent of the test section. In other words, the criterion of spatial coherence is easily met in the acoustic forcing technique. Furthermore, loudspeakers can cover a wide range of frequencies from 20 Hz to 20 kHz. In contrast, the use of vibrating ribbons is limited to frequencies of the order of 100 Hz. In many experiments, sinusoidal acoustic excitations are therefore often preferred as a means of forcing.

4.2 *Trailing-Edge Receptivity*

Under this name, we include the class of mechanisms that at low Mach numbers govern the conversion of acoustic excitations into vortical instability waves at the sharp trailing edge of the splitter plate (Morkovin & Paranjape 1971). Since these issues have recently been reviewed by Crighton (1981) in the context of aeroacoustics (see also the forthcoming survey by the same author in Volume 17 of this series), the present discussion need not be exhaustive.

The rate at which vorticity is shed from the splitter plate, i.e. the magnitude of the initial instability wave in the shear layer, is primarily determined mathematically by *postulating* that an *unsteady Kutta condition* applies at the trailing edge ($x = 0$). According to this condition, a vortical instability wave should be added to the forced acoustic field of such magnitude as to keep pressure and velocity bounded at the trailing edge.

Theoretical investigations of this question have been limited to low Strouhal numbers $St_0 = f\theta_0/\bar{U} \ll 1$, in which case the instability wavelength is much larger than the shear-layer thickness and the mean flow can be approximated by a vortex sheet. The first linear inviscid study was conducted by Orszag & Crow (1970) for the case of the eigenfunction only, i.e. without forcing. Daniels (1978) has performed a detailed asymptotic analysis of the same eigenfunction problem, which indicates that the Kutta condition is indeed applicable to unsteady trailing-edge flows. Under very specific scaling assumptions, the multilayered inner viscous flow derived by Daniels in the immediate vicinity of the trailing edge is found to match with the inviscid solution of Orszag & Crow that satisfies the full Kutta condition. As emphasized by Crighton (1981), the experiments of Bechert &

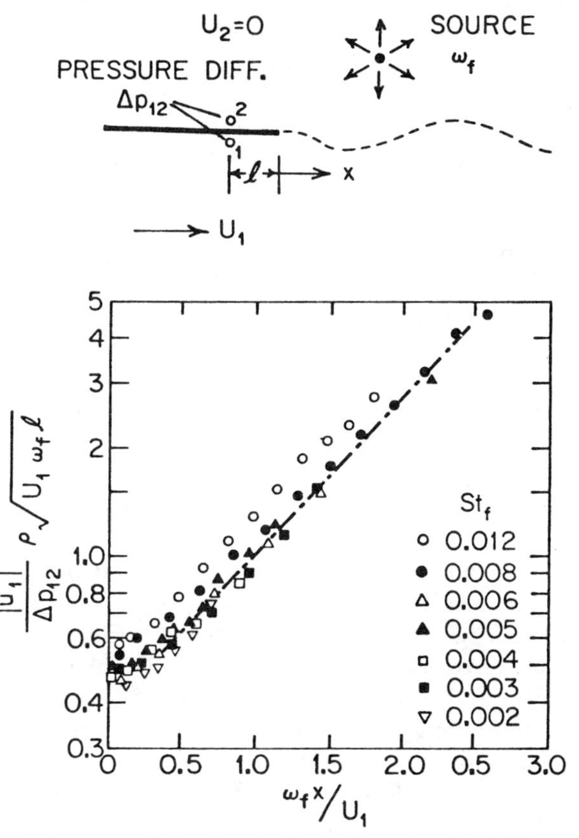

Figure 18 Mixing-layer receptivity close to the trailing edge: instability-wave evolutions with downstream distance (from Bechert 1983). The mixing layer is forced acoustically at a frequency $\omega_f = 2\pi f_f$. $R = 1$. —·—·— Semi-infinite vortex sheet analysis. Points refer to experiments.

Pfizenmaier (1975) have confirmed that the measured shape of the dividing streamline issuing from the trailing edge is consistent both with the Kutta condition and with Daniel's analysis. The unsteady flow leaves the splitter plate tangentially, so that the pressure gradient on the deflected streamline approaches zero at the trailing edge. More specifically, the dividing streamline has a shape of the form $y \sim x^{3/2}$ in the inviscid region and $y \sim x^{1/3}$ in the inner viscous region when only one stream is present ($R = 1$).

The efficiency of the trailing-edge conversion mechanism has been studied extensively in connection with the jet-noise problem. Bechert (1980), among others, has investigated the coupling effects arising between an incident acoustic wave traveling down a circular nozzle and instability waves on the jet shear layers. At low Helmholtz numbers, $f_f D/a_0 \ll 1$ (a_0 is the speed of sound), the acoustic power radiated into the far-field was measured to be as much as 20 db below the net acoustic power transmitted down the nozzle. Since no significant reflected waves were detectable, most of the incident energy at low Helmholtz numbers was converted into vortical waves. Furthermore, a low-Mach-number cylindrical vortex-sheet model predicted very satisfactorily the total attenuation experienced by acoustic waves as a result of the conversion process.

Bechert (1983) has recently examined theoretically and experimentally the receptivity of a semi-infinite vortex sheet to a sound excitation of frequency f_f (see also Bechert & Michel 1975). An elegant analytical model, valid for small amplitudes, small Strouhal number, and small Mach number, is applied to the flow configuration shown in Figure 18. The spatially growing instability waves propagating along the vortex sheet are found to be influenced by the trailing edge in a near-field region of the order of one hydrodynamic wavelength $\lambda_f \equiv \bar{U}/f_f$. The parameter Δp_{12}, which denotes the pressure difference across the plate a distance l from the edge, provides a convenient reference level to characterize the acoustic excitation. Thus, in order to relate the flow response to the imposed sound field, Bechert suggests scaling the magnitude $|u_1|$ of the streamwise fluctuating velocity on the flow side of the layer as $|u_1|\rho\sqrt{U_1\omega_f l}/\Delta p_{12}$. A comparison between the theory and carefully conducted experiments is displayed in Figure 18. The agreement is excellent at low Strouhal numbers St_f. At higher St_f, however, the stability characteristics of the true finite-thickness mixing layer are no longer approximated by the vortex-sheet model, and the theory breaks down.

Most of our present understanding of trailing-edge receptivity is limited to small amplitudes, small Mach numbers, and small Strouhal numbers, where the Kutta condition is applicable. The assumption of small Strouhal numbers is particularly restrictive, since it excludes the natural frequency of free shear layers. Crighton (1981) suggests that in such a case local separation might occur and the Kutta condition would break down.

4.3 *Mode Competition*

One of the recurring themes in this review has been the interpretation of the dynamics of orderly structures in terms of a collection of single-frequency instability waves. The order of magnitude of the amplification rates seems to be well predicted by linear stability analysis (Section 2.1) and, in certain cases (Gaster et al. 1983), the total gain in amplitude is satisfactorily obtained from linear slowly diverging flow theories. In general situations, however, different frequency components not only extract energy from the mean flow, but they also exchange energy with one another through a variety of nonlinear interaction mechanisms (Sections 2.2–2.4). The downstream development of coherent structures is therefore "dictated" by a continuing nonlinear mode competition taking place in Fourier space between a finite number of waves. The important stages of the competition can, for instance, be visualized as in Figures 3 and 19. We now briefly mention some consequences of this phenomenon before giving a more complete discussion in the next few sections.

Mode competition most simply arises between an artificially generated single-frequency wave and natural broadband background noise, as shown by Freymuth (1966) and Miksad (1972). The presence of the monochromatic excitation suppresses the broadband component and leads to a very organized train of vortical structures. Single-frequency forcing can also considerably delay the appearance of another discrete frequency component (for instance, the naturally excited subharmonic $f_f/2$, as shown in Figure 3). Similarly, when a shear layer is forced at two frequencies (Miksad 1973), the respective equilibration levels of each wave differ markedly from those for single-frequency excitation. They are also greatly dependent on the initial wave amplitudes.

The spatial evolution of the flow therefore depends on the frequency, amplitude, and phase of each forced spectral component. Available experimental data do not presently cover the entire range of forcing parameter space, and in the next sections we separately consider the effects of frequency, amplitude, and phase, although such a separation is clearly not justified.

4.4 *Forcing Frequency*

The sensitivity of shear layers to initial conditions is most readily illustrated by the effects of a varying excitation frequency f_f (Ho & Huang 1982). The initial *vortex-formation frequency* f_r is seen to correspond to that

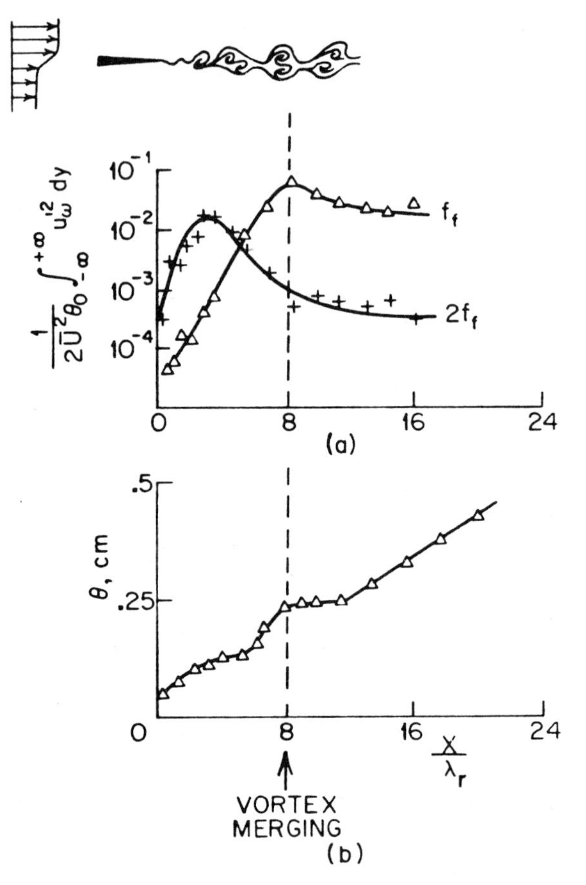

Figure 19 Evolution of (*a*) spectral components and (*b*) momentum thickness versus downstream distance (from Ho & Huang 1982). $f_n/4 < f_f \leq f_n/2$; $R = 0.31$; $\lambda_f = 2\lambda_r$.

particular harmonic of the forcing frequency f_f that is nearest to but smaller than f_n. In the range $(1/2)f_n < f_f \ll 2f_n$, f_r is found to be equal to f_f, and vortices form at the forcing frequency. The downstream evolution of the corresponding spectral components is shown in Figure 3. The higher initial level of the fundamental f_f temporarily suppresses the growth of the subharmonic $f_f/2$, and vortex pairing is delayed. When $(1/3)f_n < f_f < (1/2)f_n$, the response frequency f_r jumps to the first harmonic $2f_f$ closest to the natural frequency f_n (Figure 19). In other words, the excitation becomes the first subharmonic of the initial vortex passage frequency f_r; neighboring vortices are laterally displaced and subsequently wrap around each other to form a single structure. In contrast with the previous case, vortex pairing is promoted. Similarly, further reductions in the forcing frequency f_f lead to successive *frequency-locking* stages in which f_f becomes the second and third subharmonic of f_r. *Subharmonic forcing* can thereby result in the simultaneous coalescence of as many as three or four vortices. Consequently, the shear-layer spreading rate can be either enhanced by promoting multiple-vortex amalgamations, or reduced by delaying them. *Mixing layers can be manipulated effectively with very low forcing levels,* 0.01–0.1% of the average velocity $\bar{U}$, *provided that the excitation is applied at the proper frequency.*

The dynamics of the spanwise vortices is intimately related to the amplitude evolution of the instability waves, as illustrated in Figures 3 and 19. For instance, in the range $(1/3)f_n < f_f < (1/2)f_n$, the roll-up process is completed at the station where the component $f_r = 2f_f$ reaches its peak amplitude. As the wave of frequency f_f amplifies, vortices are displaced until they become aligned perpendicular to the stream. This location coincides with the maximum level of the subharmonic f_f, where the local Strouhal number $f_f\theta/\bar{U}$ becomes neutrally stable. The pairing of vortices is thereby directly related to the development of the subharmonic. Similar arguments can be made in the other modes of excitation.

At forcing frequencies much lower than f_n, say, less than one tenth of f_n, and at high forcing amplitudes, many small vortices of streamwise spacing $\bar{U}/f_n$ are observed to roll into large-scale structures on the downstream side of the long-wave crests. The small vortices lying on the upstream side of the crest are stretched by the resulting strain and constitute the braids of the large structures. In other cases, short instability waves of frequency f_n propagate on the long wave and directly form large vortices without going through the small-vortex stage. Variations in the flow pattern depend on the relative frequencies, amplitudes, and phases of the forced long wave and the naturally excited short wave. At any rate, a train of large vortices is formed at the low frequency f_f. This qualitatively new type of dynamical process was identified by Ho & Nosseir (1981) as a *collective interaction.* The original forced-jet experiments of Crow & Champagne (1971) belong to this regime, the excitation frequency f_f being of the order of the preferred jet frequency $f_p \ll f_n$. It is also interesting to note that the vortex simulations of Aref & Siggia (1980) give qualitatively the same picture: clusters of small vortices agglomerate to produce large-scale structures.

A similar phenomenon occurs in the experimental study of turbulent mixing layers conducted by Oster & Wygnanski (1982). Under a low-frequency excitation, the momentum thickness is initially found to increase sharply as a result of the collective interaction process. Farther downstream, θ reaches a plateau and finally resumes its linear growth. In the region of constant thickness, the *Reynolds stresses* $(-\overline{uv})$ become *negative* across the entire shear layer, as illustrated in Figure 20. Thus, there

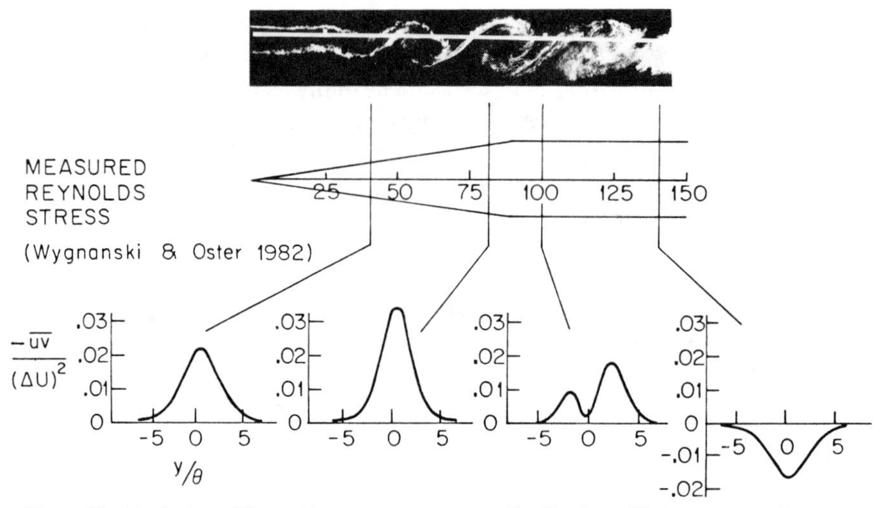

Figure 20 Evolution of Reynolds-stress cross-stream distribution with downstream distance in a forced turbulent mixing layer (from Oster & Wygnanski 1982 and Browand & Ho 1983). $f_f \ll f_n$. The measurements were performed at $R = 0.25$, but the visualization refers to $R = 0.43$. Comparison is only qualitative.

is in this part of the flow a transfer of energy from the fluctuations back to the mean flow, a characteristic behavior that has been confirmed numerically by Riley & Metcalfe (1980). Browand & Ho (1983) have given a simple kinematic explanation of the process by examining the orientation changes of elliptically shaped vortices, or *vortex nutation* in the terminology of Zabusky & Deem (1971). When the vortical structures are tilted as in Figure 21*a*, the corresponding streamwise and cross-stream velocity components generate a positive Reynolds stress and energy is extracted from the mean flow by the fluctuating field. The situation is reversed when the angle of inclination is as illustrated in Figure 21*b*. The reader may check from Figure 20 that this interpretation is indeed consistent. In the absence of vortex pairing, periodic exchanges of energy between the mean flow field and the fluctuating vorticity field can arise, as demonstrated by the numerical investigation of Brachet & Orszag (1983). From the preceding argument, this energy transfer is associated with the periodic nutation of the vortices. Furthermore, nonlinear stability analyses indicate that finite-amplitude periodic oscillations are possible close to the neutral Strouhal number (Benney & Maslowe 1975, Huerre 1977) and for the most amplified wave (Miura & Sato 1978). Note that the kinematic argument is also applicable to vortex-pairing events: according to Figure 19, the major axis of the elliptically shaped vorticity distribution rotates through the vertical during pairing and changes from the top configuration to the bottom configur-

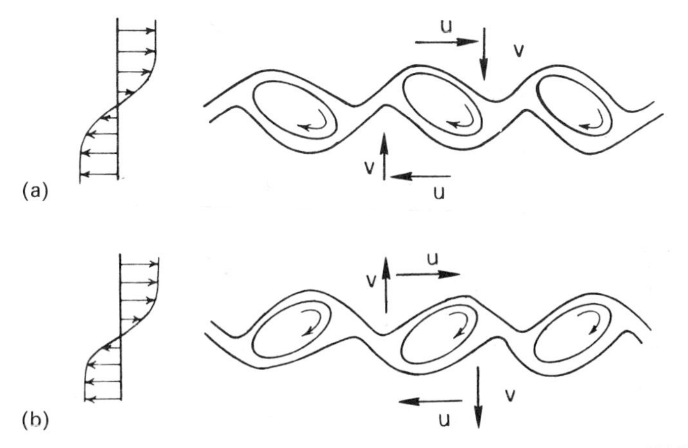

Figure 21 Mechanics of vortex nutation (from Browand & Ho 1983).

ation in Figure 21. Consequently, the energy transfer is reversed at that streamwise station, and the subharmonic energy goes through a maximum. This observation is in full agreement with the change of sign in the Reynolds-stress distribution calculated in the course of an amalgamation by Aref & Siggia (1980).

4.5 *Forcing Level*

The formation and mergings of vortices are known to occur at the peaks of the fundamental component and its subharmonics. Thus, by varying the excitation intensity at a given frequency f_f, one may expect to alter the evolution of the orderly structures. Unfortunately, too few measurements have been made to allow us to make any definite statements regarding the effects of forcing level. In the experiments of Laufer & Zhang (1983), a jet is forced close to the natural frequency f_n at very low levels, with u'/U_j estimated to be of order 10^{-7}. The peak of the fundamental is found to remain relatively constant with increasing excitation level, the resulting decrease in gain indicating the presence of nonlinear effects. In contrast, the subharmonic $f_f/2$ develops in a linear manner with a constant gain. The application of *low-level forcing* leads to a more organized flow evolution, but the locations of successive mergings appear to be insensitive to the excitation intensity. The measurements of Freymuth (1966), also conducted in a jet and at the same forcing frequency $f_f \sim f_n$, give some indication of the response at *higher forcing* levels, of the order of 10^{-4}. As seen from Figure 22, the location of the peak of f_n moves upstream, and its maximum amplitude saturates to a constant level with increasing forcing level. It can be inferred from the flow visualizations of Ho & Huang (1982) that under subharmonic forcing the location of vortex pairing, and hence the peak of

f_f, also tends to be displaced upstream. In contrast with the high sensitivity of the flow to excitation frequency, adjustments in forcing level do *not* provide an efficient way to shift merging locations. As an example, note that in jets the initial amplitude at $f_f = f_n/2$ needs to be increased by a factor of 37 in order to move the vortex-pairing location one wavelength λ_n upstream. *Very high excitation levels* of the order of 17% have been applied to circular jets by Reynolds & Bouchard (1981): vortex rings then keep their identity and roll around each other three or four times before merging into a single structure. At very low frequencies, $f_f \ll f_n$, comparatively higher initial amplitude levels are necessary to force the mixing layer into a collective interaction regime. Such large perturbations ($u'/U_j = 10^{-2}$ in Crow & Champagne 1971, 10^{-3} in Moore 1977, 10^{-2} in Oster & Wygnanski 1982) are required to offset the small amplification rates prevailing in the low-Strouhal-number range. The forced wave may thereby dominate the evolution of the flow, offering yet another example of mode competition between the two frequencies f_f and f_n. Furthermore, the formation of the large vortices can be shifted upstream by increasing the forcing level, just as in the higher-frequency case of Figure 22.

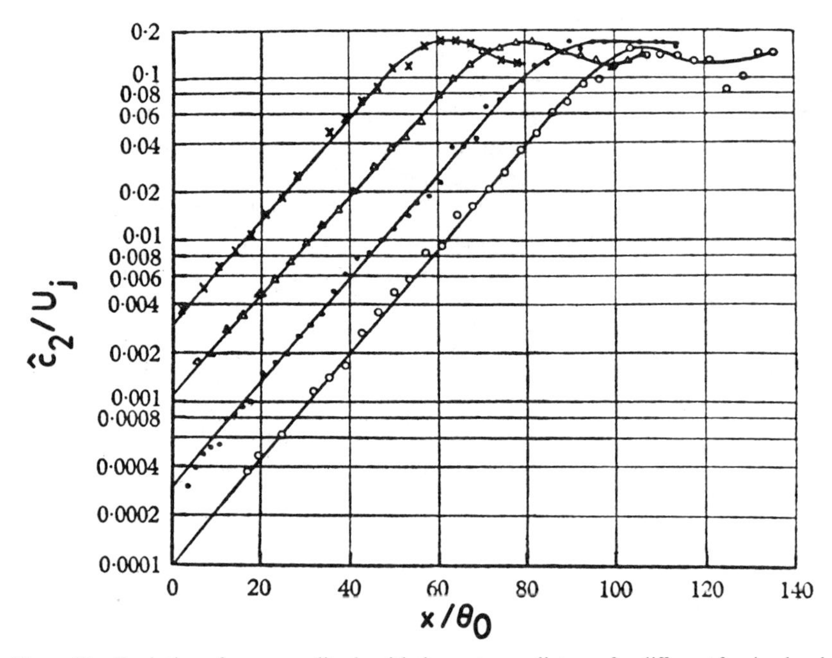

Figure 22 Evolution of wave amplitude with downstream distance for different forcing levels (from Freymuth 1966). $\hat{C}_2$ is the peak of the eigenfunction. $f_f = 0.74 f_n$; $R = 1$. ○ 70 dB; ● 80 dB; △ 90 dB; × 100 dB.

4.6 *Relative Phase of Frequency Components*

The phase difference β between the fundamental f_n and its subharmonic $f_n/2$ affects not only the amplification rate of the subharmonic (Figure 6*a*), but also the nature of the vortex interaction (Figure 7). In mixing layers forced at both frequencies (Zhang et al. 1983), measured subharmonic growth rates are found to decrease by as much as 30% when β is varied between 0 and π. This observation is in qualitative agreement with the models of Kelly (1967) and Nikitopoulos & Liu (1982) mentioned in Sections 2.2 and 2.3. Furthermore, according to the simulations of Patnaik et al. (1976) and Riley & Metcalfe (1980), a vortex-pairing interaction occurs for all values of β except those close to π. When both waves are in phase ($\beta = 0$), the interaction involves two vortices of equal size; but as β is increased, one structure becomes thinner than the other during the pairing process (see Figure 7*b*). Finally, in the antiphase case ($\beta = \pi$), pairing is replaced by a shredding interaction (Figure 7*c*). We note that shredding seldom takes place in mixing layers excited in antiphase at f_n and $f_n/2$ (Zhang et al. 1983). It is observed instead that merging events are no longer localized as for other values of β. Ambient background noise possibly detunes the effective phase difference away from π, which precludes shredding and results instead in randomly distributed merging locations.

4.7 *Facility Forcing*

The measurements of characteristic flow properties that have been repeatedly taken in the past few decades in various facilities are found to differ by factors greatly exceeding experimental uncertainty. A scatter as large as 100% can exist among different estimates of the spreading rate (Brown & Roshko 1974, Oster & Wygnanski 1982), the jet-mode Strouhal number (Petersen 1978), or the initial shear-layer Strouhal number (Hussain & Zedan 1978). Facility-related parameters that might contribute to these discrepancies are the intensity and the scale of turbulence in the initial boundary layer (Batt et al. 1970, Oster et al. 1977, Hussain & Zedan 1978, Browand & Latigo 1979, Drubka 1981), the perturbations generated by the fan blades (Fiedler & Thies 1978), free-stream turbulence (Pui & Gartshore 1979), and organ-pipe resonances (Gaster 1971, Fiedler & Thies 1978).

Wind tunnels contain potential acoustic resonators, such as turbulence manipulators (honeycombs, straws, etc.; Way et al. 1973) or the system formed by the stagnation chamber, the contraction, and the test section. Infinitesimal acoustic perturbations present in any facility may then trigger a cavity resonance with a harmonic content in the unstable frequency range of the initial shear layer. According to Gutmark & Ho (1983), disturbance levels as low as 10^{-5} can generate, within a certain range of velocities, 100% changes in the spreading rate, shear-layer Strouhal number, and preferred Strouhal number. The sensitivity of mixing layers to initial conditions is such that experimentalists will have to live with the many extraneous perturbations existing in individual flow facilities. *"Natural" mixing layers are by nature excited flows.*

4.8 *Sources of Randomness*

The application of a well-defined excitation is known to provide as one of its main advantages an unambiguous *phase reference*. In other words, the evolution of the flow downstream of the trailing edge is locked in phase with the forcing signal. Vortex interactions take place at fixed streamwise locations, and measurements taken at a fixed point in space can easily be related to a specific stage in the dynamics of the vorticity field. In this manner, a well-designed forcing experiment may unfold the main mechanisms governing the development of mixing layers. However, since a small level of forcing can significantly alter the flow, care should be exercised in making quantitative comparisons between different naturally or artificially excited mixing layers.

It should be emphasized that, even in the presence of forcing, the temporal evolution of the flow does not remain phase-locked indefinitely far downstream. It is usually observed that absolute control of the flow by forcing is limited to the first few vortex amalgamations. In certain jet experiments (Kibens 1980, Drubka 1981), the shear-layer natural frequency f_n is chosen as a multiple of the preferred frequency f_p of the form $f_n = 2^3 f_p$, so that exactly three pairings take place between the nozzle lip and the tip of the potential core. In such a case, initial phase information is preserved until the third vortex amalgamation, which is probably due to the presence of a feedback loop (see Section 5.1). At any rate, forcing is much more effective in permanently altering the asymptotic spreading rate of free shear layers than it is in phase-locking the flow very far downstream.

It is tempting to conjecture that the loss of phase reference experienced by the vortical structures during their evolution is related in some manner to the intrinsic chaotic behavior displayed by certain deterministic dynamical systems (see, for instance, Lanford 1982). In this context, Aref (1983) has recently emphasized that the motion of an initially periodic row of finite vortices is expected to display a gradual loss of phase coherence. The amplification of infinitesimal disturbances destroys the periodicity, in the sense that pairing events are no longer in phase along the row. Alternatively, from the point of view of wave theory, one can speculate that the phenomenon is similar to the onset of *phase turbulence* encountered in particular nonlinear amplitude evolution equations (Kuramoto 1978). This term denotes a class of motions in which the loss of periodicity can primarily be ascribed to chaotic behavior in the phase of the waves but not in their amplitude. In any case, the self-generated loss of phase reference inherent in the dynamics of large-scale structures will result in a broadband spectrum in the low-frequency range.

Two additional factors contribute to the observed noise. The small-scale transition discussed in Section 2.6 is another intrinsic process linked to pairing events; it is responsible for the emergence of a continuous spectrum at high frequencies. Finally, the ambient background noise constitutes an obvious external source of random fluctuations. We note that these three processes are likely to be coupled; for instance, small scales may induce high-frequency phase jitter in the dynamics of the orderly structures.

5. THE GLOBAL FEEDBACK EFFECT

In most of the dynamical processes discussed so far in this review, the shear layer has been assumed to develop in the streamwise direction according to the *local* conditions prevailing at every downstream station. In other words, we have considered the x-coordinate to be a timelike variable and have deliberately ignored any influence of the downstream flow on the upstream flow. Dimotakis & Brown (1976) have suggested, however, that the upstream influence from all downstream large-scale vortices can equally be felt at the trailing edge. According to their argument, the $1/x$ decrease of induced velocity expected from the Biot-Savart law is exactly compensated by a corresponding linear increase of the circulation of the vortices. Thus, the development of free shear layers should also be viewed globally by including long-range coupling effects between the trailing edge and the dominant flow events.

5.1 *Feedback Mechanism*

Laufer & Monkewitz (1980) have chosen in particular to consider the global *feedback mechanism* associated with the sudden change of circulation during merging interactions. Each vortex-pairing event along the stream is assumed to be linked to the trailing edge via a *feedback loop*, consisting of a downstream-propagating subharmonic instability wave and an upstream-propagating acoustic wave. Furthermore, according to Ho & Nosseir (1981), the phase difference between the two waves is necessarily of the form $2N\pi$, N an integer. If x_j denotes the location of the jth vortex merging, λ_j the corresponding subharmonic wavelength, and λ_a the acoustic wavelength, the following constraint must then be satisfied:

$$\frac{x_j}{\lambda_j} + \frac{x_j}{\lambda_a} = N. \tag{11}$$

In a low subsonic shear layer ($\lambda_j \ll \lambda_a$), the acoustic branch of the loop can be neglected, and vortex-merging stations are simply given by $x_j = \lambda_j N$. If P vortices are involved in each merging interaction, the wavelength λ_j is related to the initial instability wavelength $\lambda_r \equiv \bar{U}/f_r$ by the relation $\lambda_j = P^j \lambda_r$, and vortex-merging locations are simply given by

$$x_j = NP^j \lambda_r. \tag{12}$$

As seen from Figure 23, this relation leads to excellent predictions in a jet forced in the $P = 2$ pairing mode, provided that one assumes $N = 2$.

It can be recalled from Section 2.3 that *local* linear stability arguments have been incorporated by Ho (1981) into a phenomenological evolution model of the mixing layer. An alternate *global* model can be constructed by exploiting instead the consequences of the feedback equation (12). As before, the thickness θ is assumed to increase by a factor 2 in the roll-up phase and by a factor P during each merging interaction, so that $\theta = 2P^j \theta_0$. The spreading rate is then deduced from (12) in the form

$$\frac{d\theta}{dx} = \frac{2}{N} \frac{f_r \theta_0}{\bar{U}} = 2 \frac{\mathrm{St}_r}{N}, \tag{13}$$

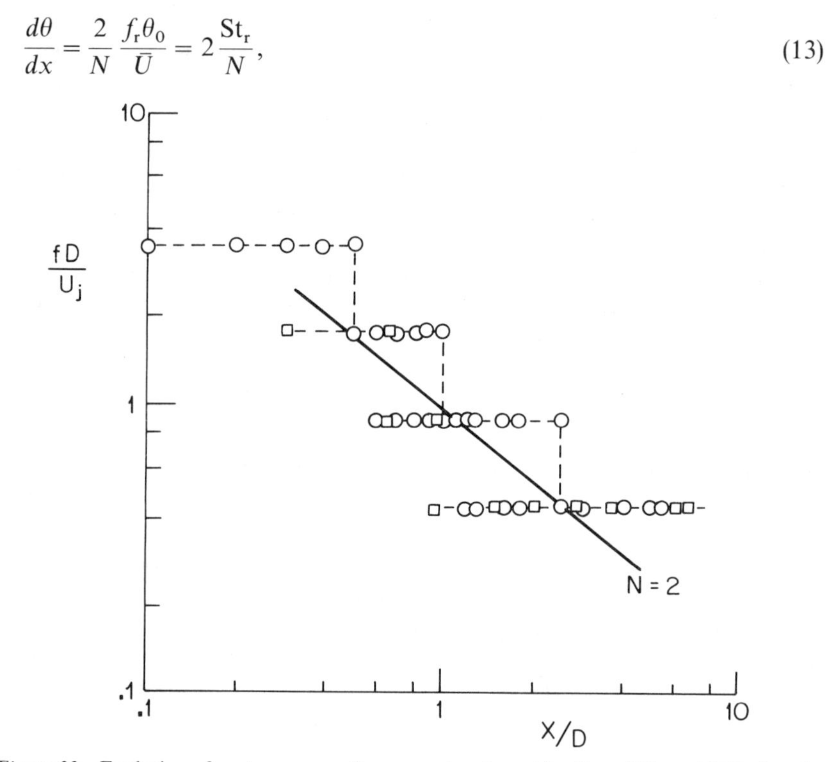

Figure 23 Evolution of vortex passage frequency in a forced jet (from Kibens 1980). $f_f \sim f_n$; $R = 1$. ○ measured in shear layers; □ measured on jet centerline. —— Feedback equation (Laufer & Monkewitz 1980) with $N = 2$.

which the reader may compare with the scaling relation (9) obtained solely from local stability considerations. The dependence of N on the velocity ratio R must be inferred from experiments: according to Huang & Ho (1983), N falls in discrete steps from 5 to 2 as R increases between 0 and 1, the length of each step approximately covering an interval of 0.25 in R. Both relations (9) and (13) then lead to spreading-rate variations with R that are consistent with the measured data in Figure 15.

The importance of the feedback mechanism remains to be assessed, both experimentally and theoretically. Nonetheless, a few clues are already available. The feedback model (12) is the only one that correctly states that the distance to the first pairing event is the same as the distances between the first and second pairings, merging distances being doubled thereafter. In contrast with the local evolution approach, this particular prediction requires no information about the variations in the transverse-flow scale. Furthermore, experiments by Drubka (1981) indicate that pressure fluctuations at the outer edge of the initial shear layer are dominated by a strong first-subharmonic component. The first two pairings are indeed likely to be the most effective in exciting the shear layer, since they are spatially coherent. Finally, additional time scales are often found to be present in the unsteady flow field at the trailing edge. Such phenomena might be due to feedback from the end of the experimental apparatus (Dimotakis & Brown 1976) or, in the case of jets, to feedback from the preferred mode at the end of the potential core (Laufer & Monkewitz 1980, Monkewitz 1983).

5.2 *Rescaled Shear Layer*

The scaling relationships implied by the previous models may to some degree be tested by suitably renormalizing the downstream variations of momentum thickness under very different flow conditions. Guided by earlier linear stability arguments, we first rescale θ as a local Strouhal number $f_f\theta/\bar{U}$ at the forcing frequency f_f. A proper choice of streamwise coordinate is more ambiguous. Recall that the early evolution of mixing layers is dominated by vortex amalgamations that adjust the wavelength from λ_r to λ_f. According to the feedback equation (12), this process is completed at a downstream distance $N\lambda_f$. Thus, in order to obtain a unified picture of the region $0 < x < N\lambda_f$, the streamwise coordinate should be normalized as $x/N\lambda_f$. Such a scaling is illustrated in Figure 24 at different values of f_f and R. The increase of local Strouhal number in the range $0 < x/(N\lambda_f) < 1$ is followed by a plateau extending to approximately $x/(N\lambda_f) = 1.5$. Note that, as expected from stability considerations, the level of the plateau always coincides with the neutral Strouhal number 0.08 in Figure 2*a*. Father downstream, linear spreading resumes at varying rates.

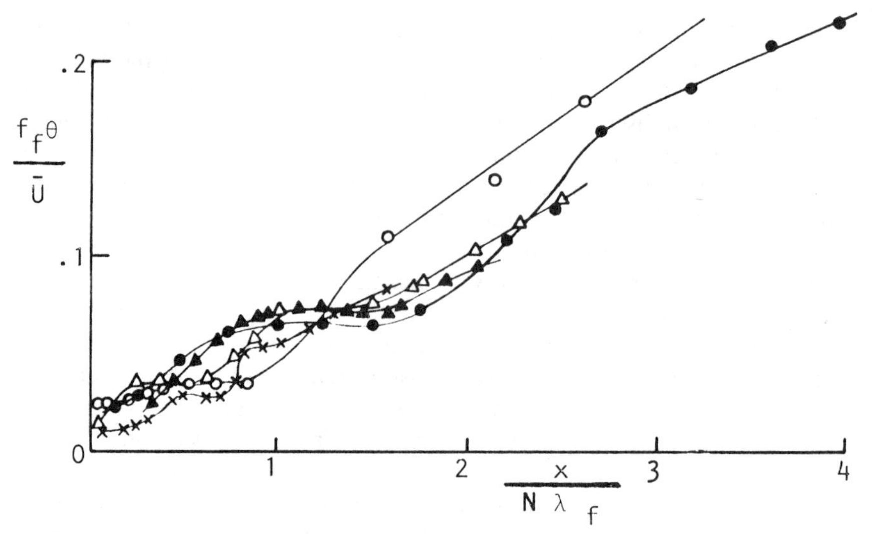

Figure 24 Rescaled free shear layer: variations of momentum thickness with rescaled downstream distance. ● $f_f = 0.65 f_n$, △ $f_f = 0.43 f_n$, × $f_f = 0.24 f_n$; $R = 0.31$, $N = 4$ (Ho & Huang 1982). ▲ $f_f \ll f_n$; $R = 0.25$, $N = 4$ (Oster & Wygnanski 1982). ○ $f_f = f_n$; $R = 1$, $N = 2$, natural forcing (Drubka 1981).

This normalization is not meant to imply the existence of a unique spreading profile; it simply illustrates the extent of the region controlled by artificial excitations. For comparison, a naturally forced mixing layer is also plotted in Figure 24. A similar collapse of experimental data would be obtained with the alternate streamwise coordinate Rx/λ_f resulting from the local evolution model.

6. PARTING REMARKS

In this review, we have deliberately avoided a detailed discussion of the random component of the motion, whether in the form of ambient noise or generated intrinsically in the dynamics of the large-scale structures. A new approach that would encompass both the random and non-random aspects of mixing layers but that would still be based on deterministic dynamics remains an ambitious but desirable goal.

ACKNOWLEDGMENTS

This work would not have been possible without the lively and stimulating research environment created by our friend John Laufer at USC. We are very grateful for his critical review of the manuscript, completed a few days before his untimely passing.

We wish to thank F. K. Browand and L. G. Redekopp for their helpful comments and their kind encouragement. We are also very much indebted to F. B. Hsiao, F. Munoz, and A. Nassiri for their help in preparing the draft.

The work was supported by the Office of Naval Research Contract N00014-77-C-0314 (CMH), the Air Force Office of Scientific Research Contract F49620-82-K-0019 (CMH), and the National Science Foundation Grant MEA-8120904 (PH).

Literature Cited

Acton, E. 1976. The modelling of large eddies in a two-dimensional shear layer. *J. Fluid Mech.* 76:561–92

Alper, A., Liu, J. T. C. 1978. On the interactions between large-scale structure and fine grained turbulence in a free shear flow. II. The development of spatial interactions in the mean. *Proc. R. Soc. London Ser. A* 359:497–523

Amsden, A. A., Harlow, F. H. 1964. Slip instability. *Phys. Fluids* 7:327–34

Aref, H. 1983. Integrable, chaotic, and turbulent vortex motion in two-dimensional flows. *Ann. Rev. Fluid Mech.* 15:345–89

Aref, H., Flinchem, E. P. 1983. Dynamics of a vortex filament in a shear flow. *J. Fluid Mech.* Submitted for publication

Aref, H., Siggia, E. D. 1980. Vortex dynamics of the two-dimensional turbulent shear layer. *J. Fluid Mech.* 100:705–37

Ashurst, W. T. 1979. Numerical simulation of turbulent mixing layers via vortex dynamics. In *Turbulent Shear Flows I*, ed. F. Durst et al., pp. 402–13. Berlin/Heidelberg/New York: Springer

Batchelor, G. K., Gill, A. E. 1962. Analysis of the stability of axisymmetric jets. *J. Fluid Mech.* 14:529–51

Batt, R. G., Kubota, T., Laufer, J. 1970. Experimental investigation of the effect of shear flow turbulence on a chemical reaction. *AIAA Pap. No. 70-721*

Bechert, D. W. 1980. Sound absorption caused by vorticity shedding demonstrated with a jet flow. *J. Sound Vib.* 70: 389–405

Bechert, D. W. 1983. A model of the excitation of large scale fluctuation in a shear layer. *AIAA Pap. No. 83-0724*

Bechert, D. W., Michel, U. 1975. The control of a thin free shear layer with and without a semi-infinite plate by a pulsating flow field. *Acustica* 33(5):287–307

Bechert, D. W., Pfizenmaier, E. 1975. Optical compensation measurements on the unsteady exit condition at a nozzle discharge edge. *J. Fluid Mech.* 71:123–44

Benney, D. J. 1961. A non-linear theory for oscillations in a parallel flow. *J. Fluid Mech.* 10:209–36

Benney, D. J., Bergeron, R. F. 1969. A new class of nonlinear waves in parallel flows. *Stud. Appl. Math.* 48:181–204

Benney, D. J., Lin, C. C. 1960. On the secondary motion induced by oscillations in a shear flow. *Phys. Fluids* 3:656–57

Benney, D. J., Maslowe, S. A. 1975. The evolution in space and time of nonlinear waves in parallel shear flows. *Stud. Appl. Math.* 54:181–205

Bernal, L. P. 1981. *The coherent structure of turbulent mixing layers. I. Similarity of the primary vortex structure. II. Secondary streamwise vortex structure.* PhD thesis. Calif. Inst. Technol., Pasadena

Betchov, R., Criminale, W. O. 1967. *Stability of Parallel Flows.* New York/London: Academic

Betchov, R., Szewczyk, A. 1963. Stability of a shear layer between parallel streams. *Phys. Fluids* 6(10):1391–96

Brachet, M. E., Orszag, S. A. 1983. Secondary instability of free shear flows. *J. Fluid Mech.* Submitted for publication

Bradshaw, P. 1966. The effect of initial conditions on the development of a free shear layer. *J. Fluid Mech.* 26:225–36

Breidenthal, R. 1981. Structure in turbulent mixing layers and wakes using a chemical reaction. *J. Fluid Mech.* 109:1–24

Broadwell, J. E., Breidenthal, R. E. 1982. A simple model of mixing and chemical reaction in a turbulent shear layer. *J. Fluid Mech.* 125:397–410

Browand, F. K. 1966. An experimental investigation of the instability of an incompressible separated shear layer. *J. Fluid Mech.* 26:281–307

Browand, F. K., Ho, C. M. 1983. The mixing layer: an example of quasi two-dimensional turbulence. *J. Méc.* In press

Browand, F. K., Latigo, B. O. 1979. Growth of the two-dimensional mixing layer from

a turbulent and non-turbulent boundary layer. *Phys. Fluids* 22(6): 1011–19
Browand, F. K., Laufer, J. 1975. The role of large scale structures in the initial development of circular jets. In *Turbulence in Liquids*, ed. J. L. Zakin, G. K. Patterson, pp. 33–44. Princeton, N.J.: Science
Browand, F. K., Troutt, T. R. 1980. A note on spanwise structure in the two-dimensional mixing layer. *J. Fluid Mech.* 97: 771–81
Browand, F. K., Weidman, P. D. 1976. Large scales in the developing mixing layer. *J. Fluid Mech.* 76: 127–44
Brown, G. L., Roshko, A. 1974. On density effects and large structure in turbulent mixing layers. *J. Fluid Mech.* 64: 775–816
Cain, A. B., Reynolds, W. C., Ferziger, J. H. 1981. A three-dimensional simulation of transition and early turbulence in a time-developing mixing layer. *Intern. Rep. No. TF-14*, Stanford Univ., Calif.
Cantwell, B. J. 1981. Organized motion in turbulent flow. *Ann. Rev. Fluid Mech.* 13: 457–515
Cervantes de Gortari, J., Goldschmidt, V. W. 1981. The apparent flapping motion of a turbulent plane jet—further experimental results. *J. Fluids Eng.* 103: 119–26
Chandrsuda, C., Mehta, R. D., Weir, A. D., Bradshaw, P. 1978. Effect of free-stream turbulence on large structure in turbulent mixing layers. *J. Fluid Mech.* 85: 693–704
Christiansen, J. P. 1973. Numerical simulation of hydrodynamics by the method of point vortices. *J. Comput. Phys.* 13: 363–79
Cohen, J., Gutmark, E., Wygnanski, I. 1983. On modal distribution of coherent structures in a jet. *AIAA J.* Submitted for publication
Corcos, G. M. 1979. The mixing layer: deterministic models of a turbulent flow. *Intern. Rep. No. FM-79-2*, Univ. Calif., Berkeley
Corcos, G. M., Lin, S. J. 1983. Deterministic models of the shear layer. Part II: the origin of the three-dimensional motion. *J. Fluid Mech.* Submitted for publication
Corcos, G. M., Sherman, F. S. 1976. Vorticity concentration and the dynamics of unstable free shear layers. *J. Fluid Mech.* 73: 241–64
Corcos, G. M., Sherman, F. S. 1983. The mixing layer: deterministic models of a turbulent flow. Introduction and Part 1. The two-dimensional flow. *J. Fluid Mech.* Submitted for publication
Corrsin, S. 1943. Investigations of flow in an axially symmetric heated jet of air. *NACA Advis. Conf. Rep. 3123*
Couet, B. 1979. Evolution of turbulence by three-dimensional numerical particle-vortex tracing. *Intern. Rep. SU-IPR No. 793*, Stanford Univ., Calif.
Craik, A. D. D. 1971. Non-linear resonant instability in boundary layers. *J. Fluid Mech.* 50: 393–413
Craik, A. D. D. 1980. Nonlinear evolution and breakdown in unstable boundary layers. *J. Fluid Mech.* 99: 247–65
Crighton, D. G. 1981. Acoustics as a branch of fluid mechanics. *J. Fluid Mech.* 106: 261–98
Crighton, D. G., Gaster, M. 1976. Stability of slowly diverging jet flow. *J. Fluid Mech.* 77: 397–413
Crow, S. C., Champagne, F. H. 1971. Orderly structure in jet turbulence. *J. Fluid Mech.* 48: 547–91
Daniels, P. G. 1978. On the unsteady Kutta condition. *Q. J. Mech. Appl. Math.* 31: 49–75
Davis, R. E. 1969. On the high Reynolds number flow over a wavy boundary. *J. Fluid Mech.* 36: 337–46
Delcourt, B. A. G., Brown, G. L. 1979. The evolution and merging structure of a vortex sheet in an inviscid and viscous fluid modelled by a point vortex method. *Proc. Symp. Turbul. Shear Flows, 2nd*, pp. 14.35–40
Dimotakis, P. E., Brown, G. L. 1976. The mixing layer at high Reynolds number: large-structure dynamics and entrainment. *J. Fluid Mech.* 78: 535–60
Drazin, P. G., Howard, L. N. 1966. Hydrodynamic stability of parallel flow of inviscid fluid. *Adv. Appl. Mech.* 9: 1–89
Drazin, P. G., Reid, W. H. 1981. *Hydrodynamic Stability*. Cambridge Univ. Press
Drubka, R. E. 1981. *Instabilities in near field of turbulent jets and their dependence on initial conditions and Reynolds number.* PhD thesis. Ill. Inst. Technol., Chicago
Fiedler, H. E., Thies, H. J. 1978. Some observations in a large two-dimensional shear layer. In *Structure and Mechanisms in Turbulence, Lecture Notes in Physics*, ed. H. E. Fiedler, 75: 108–17. Berlin/Heidelberg/New York: Springer
Fiedler, H. E., Dziomba, B., Mensing, P., Rösgen, T. 1981. Initiation, evolution and global consequences of coherent structures in turbulent shear flows. In *The Role of Coherent Structures in Modelling Turbulence and Mixing, Lecture Notes in Physics*, ed. J. Jimenez, 136: 219–51. Berlin/Heidelberg/New York: Springer
Foss, J. 1977. The effects of the laminar/turbulent boundary layer states on the development of a plane mixing layer. *Proc. Symp. Turbul. Shear Flows, 1st*, pp. 11.33–42
Freymuth, P. 1966. On transition in a separated laminar boundary layer. *J. Fluid Mech.* 25: 683–704
Fuchs, H. V., Michel, U. 1977. Experimental

evidence of turbulent source coherence affecting jet noise. *AIAA Pap. No. 77-1348*

Gaster, M. 1962. A note on the relation between temporally-increasing and spatially-increasing disturbances in hydrodynamic stability. *J. Fluid Mech.* 14:222–24

Gaster, M. 1971. Some observations on vortex shedding and acoustic resonances. *CP No. 1141*, Natl. Phys. Lab. Middlesex, Engl.

Gaster, M., Kit, E., Wygnanski, I. 1983. Large scale structures in a forced turbulent mixing layer. *J. Fluid Mech.* Submitted for publication

Gatski, T. B., Liu, J. T. C. 1980. On the interactions between large-scale structures and fine-grained turbulence in a free shear layer. III. A numerical solution. *Philos. Trans. R. Soc. London* 293(1403):473–509

Goldstein, M. E. 1984. Aeroacoustics of turbulent shear flows. *Ann. Rev. Fluid Mech.* 16:263–85

Gutmark, E., Ho, C. M. 1983. On the preferred modes and the spreading rates of jets. *Phys. Fluids.* In press

Haberman, R. 1972. Critical layers in parallel flows. *Stud. Appl. Math.* 51:139–61

Hernan, M. A., Jimenez, J. 1982. Computer analysis of a high-speed film of the plane turbulent mixing layer. *J. Fluid Mech.* 119:323–45

Ho, C. M. 1981. Local and global dynamics of free shear layers. In *Numerical and Physical Aspects of Aerodynamic Flows*, ed. T. Cebeci, pp. 521–33. Berlin/Heidelberg/New York: Springer

Ho, C. M., Hsiao, F. B. 1983. Evolution of coherent structures in a lip jet. In *Structure of Complex Turbulent Shear Layers*, ed. R. Dumas, L. Fulachier, pp. 121–36. Berlin/Heidelberg/New York: Springer

Ho, C. M., Huang, L. S. 1982. Subharmonics and vortex merging in mixing layers. *J. Fluid Mech.* 119:443–73

Ho, C. M., Nosseir, N. S. 1981. Dynamics of an impinging jet. Part 1. The feedback phenomenon. *J. Fluid Mech.* 105:119–42

Huang, L. S., Ho, C. M. 1983. Small scale transition in two-dimensional mixing layer. In preparation

Huerre, P. 1977. Nonlinear instability of free shear layers. In *Laminar-Turbulent Transition, AGARD CP No. 224*, pp. 5-1–12

Huerre, P. 1980. The nonlinear stability of a free shear layer in the viscous critical layer regime. *Philos. Trans. R. Soc. London* 293(1408):643–75

Huerre, P. 1983. Finite-amplitude evolution of mixing layers in the presence of solid boundaries. *J. Méc.* In press

Huerre, P. 1984. Basic instability mechanisms in mixing layers. In preparation

Huerre, P., Monkewitz, P. 1983. Absolute and convective instabilities in free shear layers. *J. Fluid Mech.* Submitted for publication

Huerre, P., Redekopp, L. 1983. Nonlinear evolution equations and critical layers. *Lect. Appl. Math.* 20:79–96

Huerre, P., Scott, J. F. 1980. Effects of critical layer structure on the nonlinear evolution of waves in free shear layers. *Proc. R. Soc. London Ser. A* 371:509–24

Hussain, A. K. M. F., Reynolds, W. C. 1970. The mechanics of an organized wave in turbulent shear flow. *J. Fluid Mech.* 41:241–58

Hussain, A. K. M. F., Zaman, K. B. M. Q. 1981. The "preferred mode" of the axisymmetric jet. *J. Fluid Mech.* 110:39–71

Hussain, A. K. M. F., Zedan, M. F. 1978. Effects of the initial condition on the axisymmetric free shear layer: effects of the initial momentum thickness. *Phys. Fluids* 21(7):1100–12

Jimenez, J. 1980. On the visual growth of a turbulent mixing layer. *J. Fluid Mech.* 96:447–60

Jimenez, J., Martinez-Val, R., Rebollo, M. 1979. On the origin and evolution of three-dimensional effects in the mixing layer. *Intern. Rep. DA-ERO 79-G-079*, Univ. Politec. Madrid

Kelly, R. E. 1967. On the stability of an inviscid shear layer which is periodic in space and time. *J. Fluid Mech.* 27:657–89

Kelly, R. E. 1968. On the resonant interaction of neutral disturbances in two inviscid shear flows. *J. Fluid Mech.* 31:789–99

Kibens, V. 1980. Discrete noise spectrum generated by an acoustically excited jet. *AIAA J.* 18:434–41

Kibens, V. 1981. The limit of initial shear layer influence on jet development. *AIAA Pap. No. 81-1960*

Konrad, J. H. 1976. An experimental investigation of mixing in two-dimensional turbulent shear flows with applications to diffusion-limited chemical reactions. *Intern. Rep. CIT-8-PU*, Calif. Inst. Technol., Pasadena

Kovasznay, L. S. G. 1970. The turbulent boundary layer. *Ann. Rev. Fluid Mech.* 2:95–112

Kuramoto, Y. 1978. Diffusion-induced chaos in reaction systems. *Suppl. Prog. Theor. Phys.* 64:346–67

Lamb, L. 1932. *Hydrodynamics.* New York: Dover

Lanford, O. E. III. 1982. The strange attractor theory of turbulence. *Ann. Rev. Fluid Mech.* 14:347–64

Laufer, J. 1975. New trends in experimental turbulence research. *Ann. Rev. Fluid Mech.* 7:307–26

Laufer, J. 1983. Deterministic and stochastic aspects of turbulence. *J. Appl. Mech.* In press

Laufer, J., Monkewitz, P. A. 1980. On turbulent jet flow in a new perspective. *AIAA Pap. No. 80-0962*

Laufer, J., Zhang, J. Q. 1983. Unsteady aspects of a low Mach number jet. *Phys. Fluids* 26:1740–50

Liepmann, H. W., Laufer, J. 1947. Investigation of free turbulent mixing. *NACA Tech. Note No. 1257*

Lin, C. C. 1955. *The Theory of Hydrodynamic Stability*. Cambridge Univ. Press

Lin, S. J. 1981. *The evolution of streamwise vorticity in the free shear layer*. PhD thesis. Univ. Calif., Berkeley

Lin, S. J., Corcos, G. M. 1983. The mixing layer: deterministic models of a turbulent flow. Part III: The effect of plane strain on the dynamics of streamwise vortices. *J. Fluid Mech.* Submitted for publication

Liu, J. T. C. 1981. Interaction between large-scale coherent structures and fine-grained turbulence in free shear flows. In *Transition and Turbulence*, pp. 167–214. New York/London: Academic

Liu, J. T. C., Merkine, L. 1976. On the interactions between large-scale structure and fine-grained turbulence on a free shear flow. I: The development of temporal interactions in the mean. *Proc. R. Soc. London Ser. A* 352:213–47

Lock, R. C. 1951. The velocity distribution in the laminar boundary layer between parallel streams. *Q. J. Mech.* 4:42–63

Lumley, J. L. 1978. Computational modeling of turbulent flows. *Adv. Appl. Mech.* 18:123–76

Lumley, J. L. 1981. Coherent structures in turbulence. In *Transition and Turbulence*, pp. 215–41. New York/London: Academic

Maslowe, S. A. 1977a. Weakly nonlinear stability of a viscous free shear layer. *J. Fluid Mech.* 79:689–702

Maslowe, S. A. 1977b. Weakly nonlinear stability theory of stratified shear flows. *Q. J. R. Meteorol. Soc.* 103:769–83

Maslowe, S. A. 1981. Shear flow instabilities and transition. In *Hydrodynamic Instabilities and the Transition to Turbulence*, ed. H. L. Swinney, J. P. Gollub, pp. 181–228. Berlin/Heidelberg/New York: Springer

Mattingly, G. E., Chang, C. C. 1974. Unstable waves on an axisymmetric jet column. *J. Fluid Mech.* 65:541–60

Michalke, A. 1964. On the inviscid instability of the hyperbolic tangent velocity profile. *J. Fluid Mech.* 19:543–56

Michalke, A. 1965a. Vortex formation in a free boundary layer according to stability theory. *J. Fluid Mech.* 22:371–83

Michalke, A. 1965b. On spatially growing disturbance in an inviscid shear layer. *J. Fluid Mech.* 23:521–44

Michalke, A. 1969. A note on spatially growing three-dimensional disturbances in a free shear layer. *J. Fluid Mech.* 38:765–67

Michalke, A. 1971. Instabilität eines Kompressiblen runden Freistrahls unter Berücksichtigung des Einflusses der Strahlgrenzschichtdicke. *Z. Flugwiss.* 19:319–28

Michalke, A. 1972. The instability of free shear layers. *Prog. Aerosp. Sci.* 12:213–39

Michalke, A., Hermann, G. 1982. On the inviscid instability of a circular jet with external flow. *J. Fluid Mech.* 114:343–59

Miksad, R. W. 1972. Experiments on the nonlinear stages of free shear layer transition. *J. Fluid Mech.* 56:695–719

Miksad, R. W. 1973. Experiments on nonlinear interactions in the transition of a free shear layer. *J. Fluid Mech.* 59:1–21

Miles, J. B., Shih, J. S. 1968. Similarity parameter for two-stream turbulent jet-mixing region. *AIAA J.* 6(7):1429–30

Mills, R. D. 1968. Numerical and experimental investigations of the shear layer between two parallel streams. *J. Fluid Mech.* 33:591–616

Miura, A., Sato, T. 1978. Theory of vortex nutation and amplitude oscillation in an inviscid shear instability. *J. Fluid Mech.* 86:33–47

Monin, A. S., Yaglom, A. M. 1971. *Statistical Fluid Mechanics*, ed. J. L. Lumley, Vol. 1. Cambridge, Mass: MIT Press

Monin, A. S., Yaglom, A. M. 1975. *Statistical Fluid Mechanics*, ed. J. L. Lumley, Vol. 2. Cambridge, Mass: MIT Press

Monkewitz, P. A. 1982. On the effect of the phase difference between fundamental and subharmonic instability in a mixing layer. *Intern. Rep.*, Univ. Calif., Los Angeles

Monkewitz, P. A. 1983. On the nature of the amplitude modulation of jet shear layer instability waves. *Phys. Fluids*. In press

Monkewitz, P. A., Huerre, P. 1982. The influence of the velocity ratio on the spatial instability of mixing layers. *Phys. Fluids* 25:1137–43

Moore, C. J. 1977. The role of shear-layer instability waves in jet exhaust noise. *J. Fluid Mech.* 80:321–67

Moore, D. W., Saffman, P. G. 1975. The density of organized vortices in a turbulent mixing layer. *J. Fluid Mech.* 69:465–73

Morkovin, M. 1984. *Guide to Experiments on Instabilities and Laminar Turbulent Transition in Shear Layers.* New York: AIAA

Morkovin, M., Paranjape, S. V. 1971. Acoustic excitation of shear layers. *Z. Flugwiss.* 9:328–35

Morris, P. J. 1976. The spatial viscous instability of axisymmetric jets. *J. Fluid Mech.* 77:511–29

Nikitopoulos, D. E. 1982. *Nonlinear interaction between two instability waves in a developing shear layer.* MS thesis. Brown Univ., Providence, R.I.

Nikitopoulos, D.. Liu, J. T. C. 1982. Mode interactions in developing shear flows. *Bull. Am. Phys. Soc.* 27(9):1192

Orszag, S. A., Crow, S. C. 1970, Instability of a vortex sheet leaving a semi-infinite plate. *Stud. Appl. Math.* 49:167–81

Oster, D., Wygnanski, I. 1982. The forced mixing layer between parallel streams. *J. Fluid Mech.* 123:91–130

Oster, D., Wygnanski, I., Fiedler, H. E. 1977. Some preliminary observations on the effect of initial conditions on the structure of the two-dimensional turbulent mixing layer. In *Turbulence in Internal Flows*, ed. S. N. B. Murthy, pp. 67–87. Washington, DC: Hemispheres

Overman, E. A. II, Zabusky, N. J. 1982. Evolution and merger of isolated vortex structures. *Phys. Fluids* 25:1297–1305

Patel, R. P. 1973. An experimental study of a plane mixing layer. *AIAA J.* 11(1):67–71

Patnaik, P. C., Sherman, F. S., Corcos, G. M. 1976. A numerical simulation of Kelvin-Helmholtz waves of finite amplitude. *J. Fluid Mech.* 73:215–40

Petersen, R. A. 1978. Influence of wave dispersion on vortex pairing in a jet. *J. Fluid Mech.* 89:469–95

Pierrehumbert, R. T., Widnall, S. E. 1981. The structure of organized vortices in a free shear layer. *J. Fluid Mech.* 102:301–13

Pierrehumbert, R. T., Widnall, S. E. 1982. The two- and three-dimensional instabilities of a spatially periodic shear layer. *J. Fluid Mech.* 114:59–82

Plaschko, P. 1979. Helical instabilities of slowly divergent jets. *J. Fluid Mech.* 92:209–15

Pui, N. K. 1969. *The plane mixing layer between parallel streams.* MSc thesis. Univ. Br. Columbia, Vancouver

Pui, N. K., Gartshore, I. S. 1979. Measurements of the growth rate and structure in plane turbulent mixing layers. *J. Fluid Mech.* 91:111–30

Reichardt, H. 1942. Gesetzmässigkeiten der freien Turbulenz. In Schlichting, H. 1960. *Boundary Layer Theory*, pp. 599. New York: McGraw-Hill

Reynolds, W. C., Bouchard, E. E. 1981. The effect of forcing on the mixing-layer region of a round jet. In *Unsteady Turbulent Shear Flows*, ed. R. Michel, J. Cousteix, R. Houdeville, pp. 402–11. Berlin/Heidelberg/New York: Springer

Riley, J. J., Metcalfe, R. W. 1980. Direct numerical simulation of a perturbed turbulent mixing layer. *AIAA Pap. No. 80-0274*

Roberts, F. A., Dimotakis, P. E., Roshko, A. 1982. Kelvin-Helmholtz instability of superposed streams. In *Album of Fluid Motion*, ed. M. Van Dyke, p. 85. Stanford, Calif: Parabolic

Rockwell, D. 1983. Oscillations of impinging shear layers. *AIAA J.* 21(5):645–64

Roshko, A. 1976. Structure of turbulent shear flows: a new look. *AIAA J.* 14(10):1349–57

Roshko, A. 1981. The plane mixing layer:flow visualization results and three dimensional effects. In *The Role of Coherent Structures in Modelling Turbulence and Mixing, Lecture Notes in Physics*, ed. J. Jimenez, 136:208–17. Berlin/Heidelberg/New York: Springer

Saffman, P. G., Baker, G. R. 1979. Vortex interactions. *Ann. Rev. Fluid Mech.* 11:95–122

Saffman, P. G., Szeto, R. 1981. Structure of a linear array of uniform vortices. *Stud. Appl. Math.* 65:223–48

Sato, H. 1956. Experimental investigation in the transition of laminar separated layer. *J. Phys. Soc. Jpn.* 11:702–9

Sato, H. 1959. Further investigation on the transition of two-dimensional separated layers at subsonic speed. *J. Phys. Soc. Jpn.* 14:1797–1810

Sato, H. 1960. The stability and transition of a two-dimensional jet. *J. Fluid Mech.* 7:53–80

Schade, H. 1964. Contribution to the nonlinear stability theory of inviscid shear layers. *Phys. Fluids.* 7:623–28

Spencer, B. W., Jones, B. G. 1971. Statistical investigation of pressure and velocity fields in the turbulent two-stream mixing layer. *AIAA Pap. No. 71-613*

Stewartson, K. 1974. Some aspects of nonlinear stability theory. *Fluid Dyn. Trans.* 7:101–28

Stewartson, K. 1981. Marginally stable inviscid flows with critical layers. *IMA J. Appl. Math.* 27:133–75

Stuart, J. T. 1958. On the non-linear mechanics of hydrodynamic stability. *J. Fluid Mech.* 4:1–21

Stuart, J. T. 1962. Nonlinear effects in hydrodynamic stability. *Proc. Int. Congr. Appl. Mech., 10th*, ed. F. Rolla, W. T. Koiter, pp. 63–97.

Stuart, J. T. 1967. On finite amplitude oscillations in laminar mixing layers. *J. Fluid Mech.* 29:417–40

Stuart, J. T. 1971. Nonlinear stability theory. *Ann. Rev. Fluid Mech.* 3:347–70

Sunyach, M. 1971. *Contribution à l'étude des frontières d'écoulements turbulents libres.* DSc thesis. Univ. Claude-Bernard, Lyon

Takaki, R., Kovasznay, L. S. G. 1978. Statistical theory of vortex merger in the two-dimensional mixing layer. *Phys. Fluids* 21(2): 153–56
Tam, C. K. W. 1978. Excitation of instability waves in a two-dimensional shear layer by sound. *J. Fluid Mech.* 89: 357–71
Tam, C. K. W., Chen, K. C. 1979. A statistical model of turbulence in two-dimensional mixing layers. *J. Fluid Mech.* 92: 303–26
Tani, I. 1969. Boundary-layer transition. *Ann. Rev. Fluid Mech.* 1: 169–96
Thomas, A. S. W., Saric, W. S. 1981. Harmonic and subharmonic waves during boundary layer transition. *Bull. Am. Phys. Soc.* 26(9): 1252
Townsend, A. A. 1947. Measurements in the turbulent wake of a cylinder. *Proc. R. Soc. London Ser. A* 190: 551–61
Tso, J., Kovasznay, L. S. G., Hussain, A. K. M. F. 1981. Search for large-scale coherent structures in the nearly self-preserving region of a turbulent axisymmetric jet. *J. Fluids Eng.* 103: 503–8
Way, J. L., Nagib, H. M., Tan-atichat, J. 1973. On aeroacoustic coupling in free-stream turbulence manipulators. *AIAA Pap. No. 73-1015*
Weissman, M. A. 1979. Nonlinear wave packets in the Kelvin-Helmholtz instability. *Philos. Trans. R. Soc. London Ser. A* 290: 639–85
Widnall, S. E., Bliss, D. B., Tsai, C. Y. 1974. The instability of short waves on a vortex ring. *J. Fluid Mech.* 66: 35–47
Wille, R. 1963. Beiträge zur Phänomenologie der Freistrahlen. *Z. Flugwiss.* 11: 222–33
Williams, D. R., Hama, F. R. 1980. Streaklines in a shear layer perturbed by two waves. *Phys. Fluids* 23: 442–47
Winant, C. D., Browand, F. K. 1974. Vortex pairing, the mechanism of turbulent mixing-layer growth at moderate Reynolds number. *J. Fluid Mech.* 63: 237–55
Wygnanski, I., Fiedler, H. E. 1970. The two-dimensional mixing region. *J. Fluid Mech.* 41: 327–61
Wygnanski, I., Oster, D., Fiedler, H., Dziomba, B. 1979. On the perserverance of a quasi-two-dimensional eddy-structure in a turbulent mixing layer. *J. Fluid Mech.* 93: 325–35
Yule, A. J. 1972. Two-dimensional self-preserving turbulent mixing layers at different free stream velocity ratios. *Aeronaut. Res. Counc. R & M 3683*
Zabusky, N. J. 1981. Computational synergetics and mathematical innovation. *J. Comput. Phys.* 43(2): 195–249
Zabusky, N. J., Deem, G. S. 1971. Dynamical evolution of two-dimensional unstable shear flows. *J. Fluid Mech.* 47: 353–79
Zhang, Y. Q., Ho, C. M., Monkewitz, P. A. 1983. The two-dimensional mixing layer with bi-modal excitation. In preparation

SUBJECT INDEX

D

E

F

G

H

I

CUMULATIVE INDEXES

CONTRIBUTING AUTHORS, VOLUMES 1–16

CHAPTER TITLES, VOLUMES 1–16